W0049943

The Physics and Chemistry of Aqueous Ionic Solutions

NATO ASI Series

Advanced Science Institutes Series

A series presenting the results of activities sponsored by the NATO Science Committee, which aims at the dissemination of advanced scientific and technological knowledge, with a view to strengthening links between scientific communities.

The series is published by an international board of publishers in conjunction with the NATO Scientific Affairs Division

A Life Sciences B Physics	Plenum Publishing Corporation London and New York
C Mathematical and Physical Sciences	D. Reidel Publishing Company Dordrecht, Boston, Lancaster and Tokyo
D Behavioural and Social Sciences E Engineering and Materials Sciences	Martinus Nijhoff Publishers Dordrecht, Boston and Lancaster
F Computer and Systems Sciences G Ecological Sciences H Cell Biology	Springer-Verlag Berlin, Heidelberg, New York, London, Paris, and Tokyo

Series C: Mathematical and Physical Sciences Vol. 205

The Physics and Chemistry of Aqueous Ionic Solutions

edited by

M.-C. Bellissent-Funel

Laboratoire Léon Brillouin (CEA-CNRS),
Centre d'Etudes Nucléaires de Saclay, France

and

G. W. Neilson

H. H. Witts Physics Laboratory,
University of Bristol, U.K.

D. Reidel Publishing Company

Dordrecht / Boston / Lancaster / Tokyo

Published in cooperation with NATO Scientific Affairs Division

Proceedings of the NATO Advanced Study Institute on
The Physics and Chemistry of Aqueous Ionic Solutions
Cargèse, Corsica (France)
June 22-July 5, 1986

Library of Congress Cataloging in Publication Data

NATO Advanced Study Institute on the Physics and Chemistry of Aqueous Ionic Solutions (1986: Cargèse, Corsica)
The physics and chemistry of aqueous ionic solutions.

(NATO ASI series. Series C, Mathematical and physical sciences; vol. 205)
"Proceedings of the NATO Advanced Study Institute on the Physics and Chemistry of Aqueous Ionic Solutions, Cargèse, Corsica (France), June 22–July 5, 1986"— CIP t.p. verso.
1. Ionic solutions—Congresses. I. Bellissent-Funel, M.-C. (Marie-Claire), 1943– II. Neilson, G. W. III. North Atlantic Treaty Organization. Scientific Affairs Division. IV. Title. V. Series: NATO ASI series. Series C, Mathematical and physical sciences; vol. 205.
QD543.N38 1986 541.3′422 87–12722
ISBN-13: 978-94-010-8236-5 e-ISBN-13: 978-94-009-3911-0

DOI: 10.1007/978-94-009-3911-0

Published by D. Reidel Publishing Company
P.O. Box 17, 3300 AA Dordrecht, Holland

Sold and distributed in the U.S.A. and Canada
by Kluwer Academic Publishers,
101 Philip Drive, Assinippi Park, Norwell, MA 02061, U.S.A.

In all other countries, sold and distributed
by Kluwer Academic Publishers Group,
P.O. Box 322, 3300 AH Dordrecht, Holland

D. Reidel Publishing Company is a member of the Kluwer Academic Publishers Group

Softcover reprint of the hardcover 1st edition 1987

TABLE OF CONTENTS

List of participants..........vii

Photograph..........xii

Preface and Acknowledgements..........xv

J.P. Hansen
Basic Statistical Theory of Liquids..........1

H.L. Friedman
Theory of Ionic Solutions at Equilibrium..........61

J.B. Hubbard
Non-Equilibrium Theories of Electrolyte Solutions..........95

J.E. Enderby
Diffraction Studies of Aqueous Ionic Solutions..........129

A.J. Dianoux
Neutron and NMR Spectroscopy of Aqueous Ionic Solutions..........147

P.A. Madden
Light Scattering and Related Topics..........181

P. Bopp
Molecular Dynamics Simulations of Aqueous Ionic Solutions..........217

M.P. Tosi, P. Ballone and G. Pastore
Structural Models of the Electrode/Electrolyte Interface..........245

R. Parsons
Electrode/Electrolyte Interfaces : Experimental Results..........255

J.-M. Victor
Basic Theory of Polyelectrolytes..........291

G. Jannink
Experiments on Polyelectrolyte Solutions..........311

E.U. Franck
Supercritical Studies..........337

P. Chieux
The Study of Aqueous Ionic Glasses..359

G. Michard
Natural Aqueous Solutions in the Earth................................379

D. Levesque
Theoretical Calculations of Ionic Solutions...........................399

P. Turq
Computer Simulation of Electrolyte Solutions..........................409

D. Andelman, F. Brochard and J.F. Joanny
Structured Monolayers of Charged and Polar Molecules at the Liquid/Air Interface..417

C. Monnin and J. Schott
Chemical Equilibrium Between Minerals and Natural Waters..........429

J.Klinger
Solid Aqueous Solutions..441

J. Dupuy, A. Elarby-Aouizerat, P. Claudy, J.-F. Jal, J.M. Letoffe and P. Chieux
Vitrification and Crystallization of Water 447

Posters..**453**
Postscript and Summary...**457**
Subject Index..**461**
Formula and Compound Index.......................................**471**

LIST OF PARTICIPANTS

BELLISSENT-FUNEL Marie-Claire
Laboratoire Léon Brillouin, CEN-Saclay
91191 Gif-sur-Yvette Cédex, France

NEILSON George W.
H.H. Wills Physics Lab. Univ. of Bristol
Tyndall Avenue, BS8 1TL Bristol, U.K.

AKDENIZ Zehra
I.T.U. Fen-Edebiyat Fakültesi
Maslak-Istanbul, Turkey

AOUIZERAT-ELARBY Anna
Dept. Physique des Matériaux, Univ.
Claude Bernard,
43 bd du 11 Novembre 1918
69622 Villeurbanne Cédex, France

ANDELMAN David
Exxon Research and Eng. Rte 22 East
Annandale NJ 08801, U.S.A.

ASHDOWN Steven
Dept. of Physical Chemistry, Univ. of
Bristol, Cantock's Close, Bristol, U.K.

AUTHIER Isabelle
Lab. Physique Thermique, ESPCI
10 rue Vauquelin, 75231 Paris Cédex 05
France

BARANYAI Andras
Theoretical Chemistry Lab.
Lorand Eotvos, Univ. Muzeum Krt. 6-8,
H 1088 Budapest, Hungary

BELLONI Luc
Dept. de Physico-Chimie, CEN-Saclay
91191 Gif-sur-Yvette Cédex, France

BENMOUNA Mustapha
INES Sciences Exactes, Dept. de
Physique
Tlemcen BP. 119, Algérie

BOPP Philippe
Institut für Physikalische Chemie,
Techn. Hoch., Petersenstrasse 20,
D-6100 Darmstadt, F.R.G.

BORGIS Daniel
Lab. de Physique Théorique des
Liquides, Univ. P. & M. Curie,
4 place Jussieu
75230 Paris Cédex 05, France

BROSETA Daniel
Lab. Physico-Chimie Macromoléculaire
(CNRS), ESPCI, 10 rue Vauquelin
75231 Paris Cédex 05, France

BULONE Donatella
Istituto per le Applicazioni Interdisci-
plinari della Fisica (CNR),
Via Archirafi 36, 90123 Palermo,
Italy

CABACO Maria Isabel
Centro Fisica Materia Condensada,
2 av. Prof. Gama Pinto, 1699 Lisboa
Codex, Portugal

CAILLOL Jean-Michel
Lab. Physique Théorique & Hautes
Energies, Bat. 210, Univ. Paris Sud,
91405 Orsay Cédex, France

CAPONETTI Eugenio
Istituto Chimica Fisica, Via Archirafi
26, 90144 Palermo, Italy

CAUCHETEUX Isabelle
Collège de France, Physique de la Matière Condensée, 11 place Marcelin-Berthelot, 75231 Paris Cédex 05, France

CHIEUX Pierre
Institut Laue-Langevin, 156 X
38042 Grenoble Cédex, France

COLLINS John
College of Natural and Agricultural Sci., Univ. of California, Riverside
CA 92521-0424, U.S.A.

CUNNINGHAM Daniel
Dept. of Chemistry, Univ. of Southampton, Southampton S09 5NH, U.K.

DANG Xuan
Dept. of Chemistry, Univ. of Houston, Central Campus, Houston TX 77004, U.S.A.

DELAPLANE Robert
University of Uppsala, Box 531,
S-751 21 Uppsala, Sweden

DEL VALLE GONZALES Alejandro
Dept. Cristalografia e Mineralogia
Univ. de Valladolid, 47011 Valladolid, Spain

DIANOUX Albert
Institut Laue-Langevin, 156 X
38042 Grenoble Cédex, France

DUBESSY Jean
CREGU, B.P. 23
54501 Vandoeuvre les Nancy Cédex, France

DUPUY Josette
Dept. Physique des Matériaux, Univ. Claude Bernard, 43 bd du 11 Nov. 1918
69622 Villeurbanne Cédex , France

ENDERBY John
Institut Laue-Langevin, 156 X
38042 Grenoble Cédex, France

ERIKSSON Anders
Uppsala University, Inst. of Chemistry
Box 531, S-751 21 Uppsala, Sweden

FINNEY John
Birkbeck College, Univ. of London
Malet Street, London WC1E 7HX, U.K.

FRANCK Ernst
Institute of Physical Chemistry,
Univ. of Karlsruhe, 12 Kaiser St.
D-7500 Karlsruhe, F.R.G.

FRIEDMAN Harold
Chemistry Dept. State Univ. of New York, Stony Brook,
New York 11794, U.S.A.

GAILLARD Jean-François
Lab. de Géochimie des Eaux, Univ. Paris VII, 2 place Jussieu,
75251 Paris Cédex 05, France

GOLDMAN Saul
Dept. of Chemistry and Biochemistry, Univ. of Guelph, Guelph, Ontario
N1G 2W1, Canada

GOMEZ-ESTEVEZ Juan
Dept. de Termologia, Fac. de Fisica, Univ. de Barcelona, Diagonal 645,
08028 Barcelona, Spain

GORDON Heather
Dept. of Chemistry and Biochemistry
Univ. of Guelph, Guelph,
Ontario N1G 2W1, Canada

GULLIDGE Philip
H.H. Wills Physics Lab., Univ. of Bristol
Tyndall Avenue, Bristol BS8 1TL, U.K.

HANSEN Jean-Pierre
Lab. de Physique Théorique des Liquides
Univ. Paris VI, 4 place Jussieu
75230 Paris Cédex 05, France

HERDMAN John
H.H. Wills Physics Lab., Univ. of Bristol
Tyndall Avenue, Bristol BS8 1TL, U.K.

HERNING Thierry
Société Rousselot, 8 rue Christophe Colomb, 75008 Paris, France

HOLMBERG Sven
Dept. of Physics, Uppsala Univ., Box 530 S-75121 Uppsala, Sweden

HUBBARD Joseph
Thermophysics Division, Bldg 221, Rm A311, US Dept. of Commerce, NBS Gaithersburg MD 20899, U.S.A.

JANNINK Gérard
Laboratoire Léon Brillouin, CEN-Saclay 91191 Gif-sur-Yvette Cédex, France

KJELLANDER Roland
Dept. of Applied Mathematics, Research School of Physical Sciences, Australian National Univ., Canberra ACT 260, Australia

KLINGER Jürgen
Lab. de Glaciologie & de Géophysique de l'Environnement du CNRS, B.P. 96 38402 St-Martin-d'Hères Cédex, France

KORKMAZ Mustafa
Hacetteppe Univ., Fac. of Engineering, Dept. of Physics Engineering, Beytepe, Ankara, Turkey

KRISTIANSSON Olof
Institut de Chimie, Univ. d'Uppsala Box 531, S-75121 Uppsala, Sweden

LEVESQUE Dominique
Lab. de Physique Théorique & Hautes Energies, Bat. 210, Univ. Paris Sud 91405 Orsay Cédex, France

MADDEN Paul
Dept. of Physical Chemistry, South Park Road, Oxford OX1 3Q2, U.K.

MAO Bingwei
Dept. of Chemistry, Univ. of Southampton Southampton, SO9 5NH, U.K.

MATHIEU Cécile
Dept. Physique des Matériaux, Univ. Claude Bernard, 43 bd du 11 Nov. 1918 69622 Villeurbanne Cédex, France

MICHARD Gil
Lab. de Géochimie des Eaux, Univ. Paris VII, 2 place Jussieu, 75251 Paris Cédex, 05, France

MIDDENDORF Dieter
The Clarendon Lab., Univ. of Oxford, Parks Road, Oxford OX1 3PU, U.K.

MOBERG Robert
Dept. of Physics, Uppsala Univ., Box 530 S-75121 Uppsala, Sweden

MONNIN Christophe
Lab. de Minéralogie & Cristallographie 38 rue des 36 Ponts, 31062 Toulouse, France

MONTAGUE Daniel
Dept. of Physics, Willamette Univ. Salem, Oregon 97301, U.S.A.

OBERTHUR Radulf
Institut Laue-Langevin, 156 X 38042 Grenoble Cédex, France

PARSONS Roger
Dept. of Chemistry, Univ. of Bristol, Bristol BS8 1TS, U.K.

PENDER Celia
Dept. of Chemistry, Univ. of Southampton, Southampton SO9 5NH, U.K.

PHILIPSE Albert
Van't Hoff Lab., Padualaan 8, Utrecht, The Netherlands

PODGORNIK Rudi
Institute Jozef Stefan, Jamova 39, 61000 Ljubljana, Yugoslavia

REMLER Dahlia
Physical Chemistry Lab., South Parks Road, Oxford OX1 3Q2, U.K.

RICHARDSON Charles
Physics Dept. Univ. of Arkansas,
Fayetteville AR 72701, U.S.A.

ROSI Barbara
Dipartimento di Fisica, Univ. di Parma
Via d'Azeglio 85, Italy

ROSINBERG Martin
Groupe Froment, ESPCI,
10 rue Vauquelin, 75005 Paris, France

RUSSIER Vincent
Physique des Liquides & Electrochimie,
LP 15 CNRS, 4 place Jussieu
75230 Paris Cédex 05, France

SALMON Philip
School of Mathematics & Physics,
Univ. of East Anglia, Norwich,
NR4 7TJ, U.K.

SANDSTROM Magnus
Dept. of Inorganic Chemistry, Royal
Inst. of Technology, S-10044 Stockholm
Sweden

SANTOS LOPES Maria
Centro de Electroquimica e cinetica da
Univ. de Lisboa, Fac. de Ciencias, rua
da Escola Politecnica, 1200 Lisboa,
Portugal

SCHAAF Pierre
Institut Charles Sadron (CRM-EAHP)
6 rue Boussingault, 67083 Strasbourg
France

SCHOSSELER François
Institut Charles Sadron (CRM-EAHP)
6 rue Boussingault, 67083 Strasbourg
France

SIMONIN Jean-Pierre
Lab. d'Electrochimie, Bat. F, Univ.
P. & M. Curie, 8 rue Cuvier,
75005 Paris, France

SKIPPER Neal
H.H. Wills Physics Lab., Univ. of Bristol
Tyndall Avenue, Bristol BS8 1TL, U.K.

SPOHR Eckhard
Max Planck Institut für Chemie,
Postfach 3060, D-6500 Mainz, F.R.G.

SVENSSON Bo
Physical Chemistry II, Chemical Center
POB 124, S-22100 Lund, Sweden

TANI Alessandro
Dipartimento di Chimica, Univ. di Pisa,
Via Risorgimento 85, 56100 Pisa, Italy

TIAN Zhong
Dept. of Chemistry, Southampton Univ.
Southampton SO9 5NH, U.K.

TOSI Mario
International Center for Theoretical
Phys., Dipartimento di Fisica Teorica,
Strada Costiera 11, 34100 Trieste, Italy

TUMEO Mark
Dept. of Civil Engineering, Univ. of
California, 206 Walker Hall,
Davis CA 95616, U.S.A.

TURKMAN Aysen
Dokuz Eylül Univ. Muhendislik Mimarlic
Fakultesi, Bornova Izmir, Turkey

TURNER Jacky
Dept. of Cristallography, Birkbeck Coll.
Univ. of London, Malet Street,
London WC1E 7HX, U.K.

TURQ Pierre
Propriétés Physico-Chimiques des
Electrolytes
Univ. Paris VI, 4 place Jussieu
75230 Paris Cédex 05, France

VAN AKEN George
Van't Hoff Lab., Padualaan 8, POB 80051
35008 TB, Utrecht, The Netherlands

VAN DER MAAREL Johan
Dept. of Physical and Macromolecular
Chemistry, POB 9502, 2300 RA Leiden
The Netherlands

VICTOR Jean-Marc
Lab. Physique Théorique des Liquides,
Univ. P. & M. Curie, 4 place Jussieu
75230 Paris Cédex 05, France

WALKER Patricia
H.H. Wills Physics Lab., Univ. of Bristol
Tyndall Avenue, Bristol BS8 1TL, U.K.

WEIS Jean-Jacques
Lab. de Physique Théorique & Hautes
Energies, Bat. 210, Univ. Paris Sud,
91405 Orsay Cédex, France

WILGOCKI Michal
Institute of Chemistry, Univ. of Wroclaw,
F. Joliot Curie 14, 50-383 Wroclaw,Poland

WOODWARD Clifford
Physical Chemistry II, Chemical Centre
POB 124, S-22100 Lund, Sweden

XIA Tai-he
Institute of Biophysics,
Academia Sinica, Beijing, China

XU Hong
Service Chimie Physique II, CP 231
ULB Campus Plaine, 1050 Bruxelles
Belgique

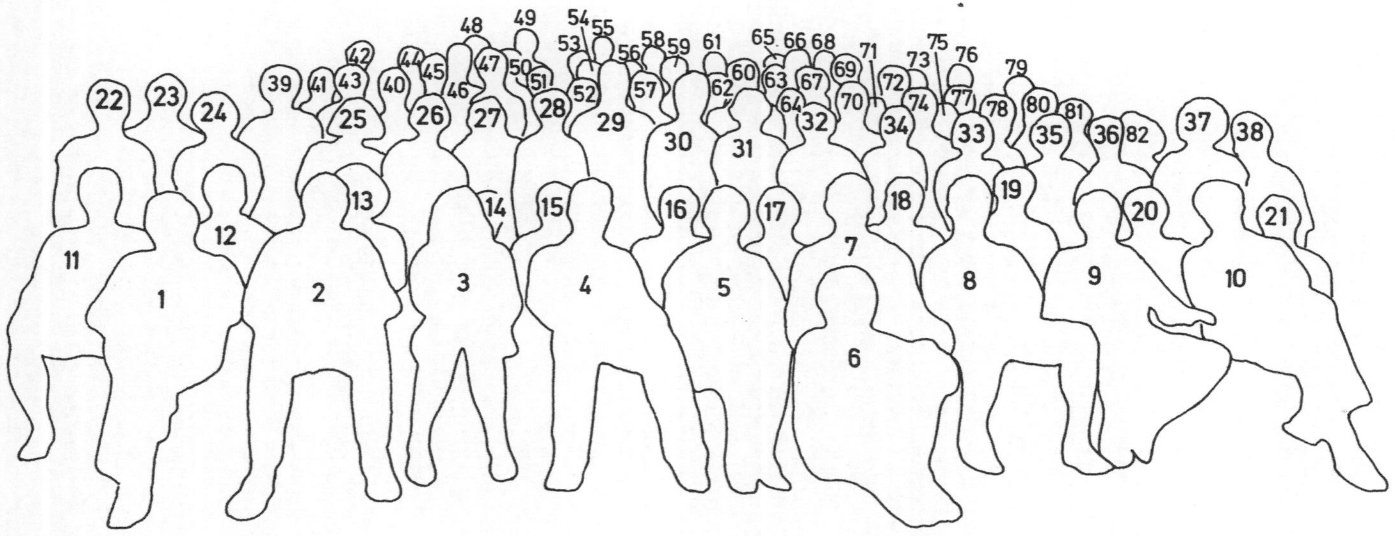

1. M. Tosi, 2. R. Parsons , 3. J. Dupuy, 4. P. Bopp, 5. M.-C. Bellissent-Funel, 6. H. Friedman,7. G. Neilson, 8. A.J. Dianoux,
9. J. Hubbard, 10. J.-P. Hansen, 11. M. Tumeo,12. P. Salmon, 13. J.M. Caillol, 14. M. Rosinberg, 15. D. Levesque,
16. P .Madden, 17. J. Finney,18. E.U. Franck, 19. G. Jannink, 20. C. Monnin, 21. G. Michard, 22. Z. Tian, 23. P. Gullidge,
24. B.Mao, 25. Z. Akdeniz, 26. M. Santo Lopes, 27. D. Bulone, 28. A. Turkman, 29. R. Moberg, 30. T. Herning, 31. S. Goldman,
32.D.Remler, 33. I. Authier, 34. I. Caucheteux, 35. H. Xu,36. H. Gordon, 37. A. Aouizerat-Elarby, 38. C. Mathieu,
39. L. Dang,40.A. Tani, 41. A. Del Valle Gonzales, 42. A. Baranyai, 43. J. Gomez-Estevez,44. R. Podgornik, 45. G. Van Aken
46. D. Montague, 47. S. Holmberg, 48. M.I. Cabaco, 49. E. Caponetti,50. T.H.Xia, 51. F. Chériot, 52. F. Schosseler,
53. M. Wilgocki,54.J.Herdman, 55. J.J. Weiss, 56. S. Ashdown, 57. E. Spohr, 58. C. Richardson, 59. A. Eriksson, 60. N. Skipper,
61. C. Pender,62.C. Woodward,63. J. Collins, 64. D. Borgis,65. P. Turq, 66. R. Kjellander, 67. D. Broseta, 68. J.-F. Gaillard,
69.J.-M.Victor,70. L.Belloni,71. J. Dubessy, 72. J.-P. Simonin, 73. P. Schaaf, 74.M. Korkmaz, 75. B. Rosi,
76. J. Van der Maarel, 77. R. Delaplane,78. D. Cunningham, 79. R. Oberthür, 80. A. Philipse, 81. B. Svensson, 82. V. Russier

D. Andelman, M. Benmouna, P. Chieux, J.E. Enderby, J. Klinger, O. Kristiansson, D. Middendorf,M. Sandström,
J. Turner, P. Walker are not on the photograph.

PREFACE AND ACKNOWLEDGEMENTS

As stated in 1977 by Professor de Gennes in his closing address to the participants of a previous NATO ASI on the Microscopic Structure and Dynamics of Liquids : "my own dream for future ASI's is to see future summer schools occurring as joint projects : liquid experts and solution chemists, liquid experts and electron chemists, etc." Taking up this challenge, we organised at Gargèse (Corsica) an ASI in June 1986 which was composed of scientists whose interests lay within the broad area of aqueous ionic solutions. The aims of the school were to provide both new and established researchers from physics, chemistry, biochemistry, electrochemistry and geochemistry with (i) a comprehensive background to the theory of ionic solutions and (ii) a substantial account of theoretical and experimental developments which have taken place in the last decade.

This book contains a written record of the proceedings of our ASI on the Physics and Chemistry of Aqueous Ionic Solutions. The Institute began with a series of lectures on the basic statistical mechanical theory of liquids and ionic solutions. These lectures were interspersed with presentations on the new and powerful methods of neutron scattering, X-ray diffraction, light scattering and n.m.r. The role which computer simulation plays in forging a link between the formal theory and experimental results was discussed in detail.

During the second half of the school special consideration was given to how the subject might develop in the next decade. To achieve as broad a perspective as possible lectures were given in a variety of areas including ionic solutions under critical conditions, ionic glasses, polyelectrolytes, electrochemistry, biochemistry and geochemistry. Lecturers were particularly careful to point out where their work might be of technological significance.

The summer school also benefitted from several seminars on topics of current interest, and these are presented towards the end of the book. Of particular interest were the significant advances being made in the theoretical development of solutions at interfaces, and dynamics of ions in the bulk.

A poster session was held with the main objective of encouraging younger participants to present their results. A summary of the contents of this session is given on page 453.

The ASI was brought to a close by Professors Friedman and Enderby who, drawing on their wealth of experience, pointed out the many achievements which have taken place in the past decade and how our view of Ionic Solutions has been transformed by the combined efforts of theorists and experimentalists. They invited us to contemplate, how much has still to be done, especially in understanding complex ionic systems where our knowlege of basic properties is still sparse.

Given the scope of this school it is clear that it could not have been undertaken without the help of many colleagues. Foremost among these is Dr. Philip Bopp who greatly assisted us before, during and after the ASI. His organisation of the poster session was much appreciated by students and lecturers alike. The assistance of Drs. Josette Dupuy and José Dianoux in helping us set up the lecturing programme, and in organising an exciting and entertaining social programme is gratefully acknowledged.

We thank Mrs M-F. Hanseler and Mrs F. Chériot for secretarial services which ensured the smooth running of the school. We also are extremely grateful to Mrs C. Pomeau for typing and arranging the final manuscript, and to Mrs Pols-v.d. Heijden of Reidel , who has been instrumental in assembling this final record of the ASI.

It is a pleasure to thank all lecturers and seminar speakers for presenting a vivid and enthusiastic account of their work and thereby stimulating many of us in our own endeavours in future.

Finally and most importantly we thank NATO and the other organisations including CEA, CNRS, ILL, ICI, NSF (USA), IBM-Deutschland, WELLA-AG for financial support which enabled us to provide funds for almost all participants.

M.-C. Bellissent-Funel - Paris

G.W. Neilson - Bristol

BASIC STATISTICAL THEORY OF LIQUIDS

J.P. HANSEN
Laboratoire de Physique Théorique des Liquides ;
(Unité associée au CNRS)
Université Pierre et Marie Curie ; 75 Paris Cedex05

1. FLUCTUATIONS AND CORRELATIONS IN THE BULK AND AT INTERFACES.

1.1. Simple and less simple liquids.

The title of this lecture course is over-ambitious, since there is no way in which a coherent and self-contained presentation of the statistical theory of liquids could fit into just four lectures. A more modest goal which we hope to achieve is to familiarize the audience with some of the fundamental concepts and tools of the theory of liquids and ionic solutions which will be used by other lecturers throughout this School, and to provide a guide to the literature. The main emphasis of the present lectures will be on the static and dynamic correlations in liquids, as can be probed by spectroscopic techniques like neutron or light scattering.

For the sake of clarity, the fundamental concepts will generally be introduced first for <u>simple</u>, one-component liquids made up of atoms with translational degrees of freedom only. The generalization to <u>complex</u>, molecular and multi-component liquids introduces complications which are often (but not always !) of a technical, rather than conceptual nature. Of course the additional, rotational and vibrational degrees of freedom lead to much richer physics, of which the dielectric behaviour, which is intimately connected to orientational correlations, is an obvious example.

Although a minimum amount of formalism is unavoidable in the statistical theory of liquids, diagrams, propagators and other headaches will be banned from these lectures. Much more complete discussions of the topics which we shall touch upon can be found in number of recent monographs (1 - 7) and review articles (8 - 11) which should be consulted for detailed developments.

M.-C. Bellissent-Funel and G. W. Neilson (eds.), The Physics and Chemistry of Aqueous Ionic Solutions, 1–59.

1.2. Static correlation functions and density profiles.

Although long-range order, characteristic of crystalline solids, is absent in liquids, the latter exhibit a considerable amount of short-range order on the scale of several atomic diameters. This short-range order, which plays a crucial role in our under-standing of the collective behaviour of liquids, can be characterized by pair and higher-order distribution functions, or, equivalently, by static (time-independent) correlation functions, which will be introduced below. We shall first restrict the discussion to simple, one-component liquids made up of N identical spherical particles (atoms) of diameter σ and mass m in a volume V. The corresponding phase space Γ_N of the positions $\vec{r}_i$ and momenta $\vec{p}_i$ $(1 \leq i \leq N)$ has 6N dimensions. Since the de Broglie thermal wavelength associated with translational motion of the atoms :

$$\Lambda = \frac{h}{\sqrt{2\pi m k_B T}} \tag{1.1}$$

is much smaller than their mean spacing $d \simeq (V/N)^{1/3}$ for temperatures T above the triple point (this excludes liquid Helium), quantum effects are negligible, and classical Statistical Mechanics applies in the liquid state. Note that the latter is a condensed state of matter where atoms "touch", so that $d \simeq \sigma$, or, in other words, the reduced number density $n = N\sigma^3/V$ is of the order of 1, comparable to its value in the solid state.

Although simplicity would seem to recommend consideration of <u>homogeneous</u> (i.e. translationally invariant) liquids first, it proves in fact more convenient to consider inhomogeneous liquids from the outset (12,9). Physically, liquids are homogeneous in the bulk, while they are inhomogeneous near interfaces where translational invariance is broken by some physical boundary (like the container walls or an electrode) or by phase coexistence (e.g. at the liquid-gas interface). Spatial inhomogeneity can also be induced by a spatially varying external field acting on the atoms of the liquid. Consequently we consider a hamiltonian of the form :

$$H_N = K_N(\{\vec{p}_i\}) + V_N(\{\vec{r}_i\}) + \Phi_N(\{\vec{r}_i\}) \tag{1.2}$$

where K_N is the sum of kinetic energies $p_i^2/2m$ of the N atoms, V_N is the total inter-atomic potential energy in a configuration $\{\vec{r}_i\}$, and Φ_N is the potential energy of interaction with the external field, which couples to the microscopic density $\rho(\vec{r})$:

$$\Phi_N(\{\vec{r}_i\}) = \sum_{i=1}^{N} \varphi(\vec{r}_i)$$

$$= \int \rho(\vec{r})\varphi(\vec{r})\, d\vec{r} \tag{1.3}$$

where :

$$\rho(\vec{r}) = \sum_{i=1}^{N} \delta(\vec{r} - \vec{r}_i) \tag{1.4}$$

We shall now consider statistical averages taken over a grand canonical ensemble characterized by fixed values of the temperature T, of the volume V and of the chemical potential μ. The equilibrium average of any phase space variable $A(\Gamma_N) = A(\{\vec{r}_i, \vec{p}_i\})$ is then given by :

$$\langle A \rangle = \frac{1}{\Xi} \mathrm{Tr}\, [f_o(\Gamma_N)\, A(\Gamma_N)] \tag{1.5}$$

where Tr denotes the classical trace, i.e. integration over phase space and summation over all values of N :

$$\mathrm{Tr} = \sum_{N=0}^{\infty} \frac{1}{h^{3N}\, N!} \int d\Gamma_N \tag{1.6}$$

Ξ is the grand canonical partition function :

$$\Xi = \mathrm{Tr}\, [\exp\{-\beta(H_N - \mu N)\}] \tag{1.7}$$

and $f_o(\Gamma_N)$ is the normalized equilibrium probability density :

$$f_o(\Gamma_N) = \frac{1}{\Xi} \exp\{-\beta(H_N - \mu N)\} \tag{1.8}$$

The integration over momenta in equ. (1.7) is trivial, so that :

$$\Xi = \sum_{N=0}^{\infty} \frac{z^N}{N!} Z_N \tag{1.9}$$

where $z = \exp(\beta\mu)/\Lambda^3$ is the fugacity and Z_N the N-particle configuration integral :

$$\begin{aligned} Z_N &= \int \exp\{-\beta[V_N + \Phi_N]\}\, d\vec{r}^N \\ &= \int \prod_{i=1}^{N} \exp\{-\beta\varphi(\vec{r}_i)\} \exp\{-\beta V_N\}\, d\vec{r}^N \end{aligned} \tag{1.10}$$

The equilibrium <u>one-particle density</u> is the statistical average of the microscopic density (1.4) :

$$\rho^{(1)}(\vec{r}) = \langle \rho(\vec{r}) \rangle \tag{1.11}$$

For a homogeneous liquid, translational invariance implies that $\rho^{(1)}(\vec{r})$ reduces to the number density $n = \langle N \rangle / V$. At the liquid -gas phase equilibrium, translational invariance is, most commonly, broken by the gravitational field

$\varphi(Z) = -mgZ$ (where Z is the vertical coordinate) which induces a horizontal, planar interface characterized by a <u>density profile</u> $\rho^{(1)}(Z)$. Below the interface $\rho^{(1)}$ tends rapidly towards the bulk density n_ℓ of the liquid, while above it goes over to the bulk density n_g of the coexisting gas phase. Conventionally the interface is located at the Gibbs dividing surface $Z = Z_G$, defined such that :

$$\int_{-\infty}^{Z_G} [\rho^{(1)}(Z) - n_\ell]\, dZ + \int_{Z_G}^{\infty} [\rho^{(1)}(Z) - n_g]\, dZ = 0 \qquad (1.12)$$

Near the triple point the interface is quite sharp, since its width is typically of the order of two atomic diameters. Similar considerations apply to the electric double layer near a planar electrode where the inhomogeneity is induced by the potential difference between the electrode and the bulk of the ionic solution.

Now, for a given interatomic potential energy V_N and a given chemical potential μ, the grand partition function (1.9), and hence the grand potential :

$$\Omega = -k_B T \ln \Xi \qquad (1.13)$$

are clearly <u>functionals</u> of the external potential $\varphi(\vec{r})$ or, equivalently of the "intrinsic" chemical potential :

$$\psi(\vec{r}) = \mu - \varphi(\vec{r}) \qquad (1.14)$$

According to the elementary rules of functional differentiation, it follows then directly from equs. (1.8), (1.9), (1.10) and (1.13) that :

$$\rho^{(1)}(\vec{r}) = -\frac{\delta\Omega[\psi]}{\delta\psi(\vec{r})} \qquad (1.15)$$

Next we define the two-body density :

$$\rho^{(2)}(\vec{r},\vec{r}') = \langle\rho(\vec{r})\rho(\vec{r}')\rangle - \rho^{(1)}(\vec{r})\,\delta(\vec{r}-\vec{r}') \qquad (1.16)$$

and the related density-density correlation function :

$$\begin{aligned} H(\vec{r},\vec{r}') &= \langle[\rho(\vec{r}) - \langle\rho(\vec{r})\rangle][\rho(\vec{r}') - \langle\rho(\vec{r}')\rangle]\rangle \\ &= \rho^{(2)}(\vec{r},\vec{r}') + \rho^{(1)}(\vec{r})\,\delta(\vec{r}-\vec{r}') - \rho^{(1)}(\vec{r})\,\rho^{(1)}(\vec{r}') \\ &= \rho^{(1)}(\vec{r})\,\rho^{(1)}(\vec{r}')\,h^{(2)}(\vec{r},\vec{r}') + \rho^{(1)}(\vec{r})\,\delta(\vec{r}-\vec{r}') \end{aligned}$$

In the absence of long-range correlations (like those due to long-range order in a crystal), it is clear that :

$$\lim_{|\vec{r}-\vec{r}'|\to\infty} \rho^{(2)}(\vec{r},\vec{r}') = \rho^{(1)}(\vec{r})\,\rho^{(1)}(\vec{r}') \qquad (1.18a)$$

$$\lim_{|\vec{r}-\vec{r}'|\to\infty} H(\vec{r},\vec{r}') = 0 \qquad (1.18b)$$

It is customary to define the dimensionless pair distribution function (p. d. f.) :

$$g^{(2)}(\vec{r},\vec{r}') = \frac{\rho^{(2)}(\vec{r},\vec{r}')}{\rho^{(1)}(\vec{r})\rho^{(1)}(\vec{r}')} = 1 + h^{(2)}(\vec{r},\vec{r}') \qquad (1.19)$$

whose asymptotic value is 1 according to equ. (1.18a). Again, in a homogeneous liquid, translational invariance implies that the 3 functions $\rho^{(2)}$, H and g depend only on the relative positions $\vec{r} - \vec{r}'$. If the liquid is moreover isotropic, the functions depend only on the relative distance $|\vec{r} - \vec{r}'|$.
All three functions $\rho^{(2)}$, $g^{(2)}$ and H are a measure of spatial (positional) pair correlations. In an ideal gas of non-interacting atoms $\rho^{(2)}$, $g^{(2)}$ and H reduce to n^2, 1 and 0 respectively.
Taking the functional derivative of equ. (1.15) it is easily verified that :

$$H(\vec{r},\vec{r}') = k_B T \frac{\delta\rho^{(1)}(\vec{r})}{\delta\psi(\vec{r}')} = -k_B T \frac{\delta^2\Omega[\psi]}{\delta\psi(\vec{r})\delta\psi(\vec{r}')} \qquad (1.20)$$

Similarily, higher order functional derivatives of the grand potential with respect to the external potential lead to higher order correlation functions, or n-particle densities. Ω appears thus as the generating functional for n-point correlation functions or densities.

There is, however, a second family of correlation functions which play a fundamental role in the theory of liquids, namely the direct correlation functions (d.c.f.), first introduced by Ornstein and Zernike in their study of spatial correlations near the liquid-gas critical point. The d.c.f. can be derived from a free energy functional, which is related to the grand potential by a Legendre transformation (13, 14) :

$$\begin{aligned} \mathcal{F}[\rho^{(1)}] &= \Omega[\psi] + \int \psi(\vec{r})\,\rho^{(1)}(\vec{r})\,d\vec{r} \\ &= \Omega[\psi] + \mu\int\rho^{(1)}(\vec{r})d\vec{r} - \int\rho^{(1)}(\vec{r})\varphi(\vec{r})d\vec{r} \end{aligned} \qquad (1.21)$$

In view of equ. (1.15) this Legendre transformation transforms from the "variable" $\psi(\vec{r})$ to the "variable" $\rho^{(1)}(\vec{r})$. Indeed, it can be proved that $\mathcal{F}$ is a unique functional of the one-particle density (14, 9) : for fixed V_N and μ there is only one external potential $\varphi(\vec{r})$ which can be associated with a given $\rho^{(1)}(\vec{r})$; in other words φ is uniquely determined by $\rho^{(1)}$. According to equ. (1.21), $\mathcal{F}$ is related to the usual Helmholtz free energy F by :

$$F = \mathcal{F}[\rho^{(1)}] + \int \rho^{(1)}(\vec{r})\ \varphi(\vec{r})d\vec{r} \tag{1.22}$$

i.e. $\mathcal{F}$ is the "intrinsic" part of the free energy ; it is conveniently separated into "ideal" and "excess" parts :

$$\mathcal{F}[\rho^{(1)}] = \mathcal{F}_{id}[\rho^{(1)}] + \mathcal{F}_{ex}[\rho^{(1)}] \tag{1.23a}$$

$$\beta\mathcal{F}_{id}[\rho^{(1)}] = \int \rho^{(1)}(\vec{r})\ \{\ln[\Lambda^3\rho^{(1)}(\vec{r})] - 1\}d\vec{r} \tag{1.23b}$$

The excess part is the generating functional of the n-particle d.c.f. of which the first two are defined by :

$$c^{(1)}(\vec{r}) = -\frac{\delta\beta\ \mathcal{F}ex[\rho^{(1)}]}{\delta\rho^{(1)}(\vec{r})} \tag{1.24a}$$

$$c^{(2)}(\vec{r},\vec{r}\,') = \frac{\delta c^{(1)}(\vec{r})}{\delta\rho^{(1)}(\vec{r}\,')} = \frac{\delta^2\beta\ \mathcal{F}_{ex}[\rho^{(1)}]}{\delta\rho^{(1)}(\vec{r})\ \delta\rho^{(1)}(\vec{r}\,')} \tag{1.24b}$$

In view of the above-mentioned uniqueness property, these d.c.f. must be intimately related to the n-particle densities or correlation functions introduced earlier. First note that :

$$\frac{\delta\beta\mathcal{F}_{id}[\rho^{(1)}]}{\delta\rho^{(1)}(\vec{r})} = \ln[\Lambda^3\rho^{(1)}(\vec{r})] \tag{1.25a}$$

$$\frac{\delta^2\beta\mathcal{F}_{id}[\rho^{(1)}]}{\delta\rho^{(1)}(\vec{r})\delta\rho^{(1)}(\vec{r}\,')} = \frac{1}{\rho^{(1)}(\vec{r})}\ \delta(\vec{r} - \vec{r}\,') \tag{1.25b}$$

Combination of equs. (1.23a), (1.24a) and (1.25a) yields :

$$\frac{\delta\beta\,\mathcal{F}[\rho^{(1)}]}{\delta\rho^{(1)}(\vec{r})} = \ln[\Lambda^3\rho^{(1)}(\vec{r})] - c^{(1)}(\vec{r}) \tag{1.26}$$

This functional derivative can also be calculated from equs. (1.21) and (1.15), with the result :

$$\frac{\delta\beta\ \mathcal{F}[\rho^{(1)}]}{\delta\rho^{(1)}(\vec{r})} = \int\frac{\delta\beta\Omega[\psi]}{\delta\psi(\vec{r}\,')}\ \frac{\delta\psi(\vec{r}\,')}{\delta\rho^{(1)}(\vec{r})}d\vec{r}\,' + \beta\psi(\vec{r}) + \int\frac{\delta\beta\psi(\vec{r}\,')}{\delta\rho^{(1)}(\vec{r})}\ \rho^{(1)}(\vec{r}\,')\ d\vec{r}\,'$$

$$= \beta\psi(\vec{r}) \tag{1.27}$$

Combining equs. (1.26) and (1.27) we arrive at :

$$\rho^{(1)}(\vec{r}) = z\ \exp\{-\beta\varphi(\vec{r}) + c^{(1)}(\vec{r})\} \tag{1.28}$$

Since for an ideal gas $\rho^{(1)}(\vec{r}) = z\{\exp -\beta\varphi(r)\ + c^{(1)}(\vec{r})\}$ describes the effect of correlations on the one-particle density. Note that for a homogeneous fluid, where $\varphi(\vec{r}) = 0$ and $\rho^{(1)}(\vec{r}) = n$, $c^{(1)}(\vec{r})$ reduces to $-\beta\mu_{ex} = -[\beta\mu - \ln(n\Lambda^3)]$. Similarily, if we differentiate equs. (1.26) and (1.27) with respect to $\rho^{(1)}(\vec{r}')$, and use equs. (1.24b), (1.25b) and (1.20), we obtain :

$$\frac{\delta\beta\psi(\vec{r})}{\delta\rho^{(1)}(\vec{r}')} = H^{-1}(\vec{r},\vec{r}') = \frac{1}{\rho^{(1)}(\vec{r})}\delta(\vec{r}-\vec{r}') - c^{(2)}(\vec{r},\vec{r}') \quad (1.29)$$

The two-body d.c.f. $c^{(2)}$ appears thus as the inverse of the density-density correlation function (1.20), the functional inverse being defined by an obvious generalization of a matrix inverse, i.e. :

$$\int H(\vec{r},\vec{r}'')\,H^{-1}(\vec{r}'',\vec{r}')d\vec{r}'' = \int H^{-1}(\vec{r},\vec{r}'')\,H(\vec{r}'',\vec{r}')d\vec{r}'' = \delta(\vec{r}-\vec{r}') \quad (1.30)$$

Substitution of (1.29) into (1.30) leads to the celebrated Ornstein-Zernike (OZ) relation (7) :

$$h^{(2)}(\vec{r},\vec{r}') = c^{(2)}(\vec{r},\vec{r}') + \int c^{(2)}(\vec{r},\vec{r}'')\rho^{(1)}(\vec{r}'')h^{(2)}(\vec{r}'',\vec{r}')d\vec{r}'' \quad (1.31)$$

For a homogeneous and isotropic fluid, all correlation functions depend only on the relative distance between two space points, and the OZ relation reduces to the familiar form:

$$h^{(2)}(r) = c^{(2)}(r) + n\,c^{(2)} * h^{(2)} \quad (1.32)$$

where * denotes a convolution product.

Returning for a moment to the grand potential, we introduce, following Mermin (14), for a fixed external potential φ, a functional of an arbitrary density $\tilde{\rho}^{(1)}(r)$ defined by :

$$\Omega_\varphi[\rho^{(1)}] = F[\rho^{(1)}] - \int \psi(\vec{r})\tilde{\rho}^{(1)}(\vec{r})d\vec{r} \quad (1.33)$$

where F is the free energy functional of equ. (1.21) for arbitrary $\rho^{(1)}$. It is easy to prove (14,9) that the equilibrium one-particle density $\rho^{(1)}$ minimizes the functional (1.33), which reduces to the grand potential Ω when $\tilde{\rho}^{(1)} = \rho^{(1)}$

$$\left.\frac{\delta\Omega_\varphi[\tilde{\rho}^{(1)}]}{\delta\tilde{\rho}^{(1)}(r)}\right|_{\tilde{\rho}^{(1)}=\rho^{(1)}} = 0 \quad (1.34)$$

$$\Omega_\varphi[^{(1)}] = \Omega \quad (1.35)$$

Equ. (1.34) formulates a variational principle which can be put to use in approximate calculations of density profiles (15). Alternative procedures for their calculation start from exact integro-differential equations which relate $\rho^{(1)}$ to the two-particle density $\rho^{(2)}$ or to the d.c.f. $c^{(2)}$.

Thus, taking the gradient of the logarithm of equ.(1.28) and using equ. (1.24b) we obtain :

$$\frac{\partial}{\partial r} \ln \rho^{(1)}(\vec{r}) + \frac{\partial}{\partial \vec{r}} \beta\varphi(r) = \frac{\partial}{\partial \vec{r}} c^{(1)} [\rho^{(1)};\vec{r}]$$
$$= \int d\vec{r}' \frac{\delta c^{(1)}[\rho^{(1)};\vec{r}]}{\delta \rho^{(1)}(\vec{r}')} \frac{\partial}{\partial \vec{r}'} \rho^{(1)}(\vec{r}')$$
$$= \int d\vec{r}' c^{(2)}(\vec{r},\vec{r}') \frac{\partial}{\partial \vec{r}'} \rho^{(1)}(\vec{r}') \qquad (1.36)$$

Note that equ. (1.36) does not depend explicitly on the interaction between particles (i.e. on V_N). Given a reasonable approximation for $c^{(2)}$ or $h^{(2)}$ (e.g. choosing $c^{(2)}(\vec{r},\vec{r}')$ to be equal to the d.c.f. of the homogeneous liquid at the density $n = 1/2[\rho^{(1)}(\vec{r}) + \rho^{(1)}(\vec{r}')]$) equ. (1.36) can, in principle, be solved by iteration to yield the density profile $\rho^{(1)}(\vec{r})$.

1.3. Link with thermodynamics.

The difference between the thermodynamic functions of a state of interest and some "reference" state can be obtained by thermodynamic integration. To that purpose we consider a transformation, at constant temperature, between an initial state characterized by the density $\rho_o^{(1)}(\vec{r})$ and the final state of density $\rho^{(1)}(\vec{r})$ Since the desired difference must be independent of the path chosen in density space, it is convenient to choose a linear path depending on a single parameter λ (16) :

$$\rho^{(1)}(\vec{r};\lambda) = \rho_o^{(1)}(\vec{r}) + \lambda\, [\rho^{(1)}(\vec{r}) - \rho_o^{(1)}(\vec{r})] = \rho_o^{(1)}(\vec{r}) + \lambda\Delta\rho^{(1)}(\vec{r}) \qquad (1.37)$$

where $0 \leq \lambda \leq 1$. Under these conditions, integration of equ. (1.24a) leads immediately to an expression for the difference of the excess part of the intrinsic free energies in the initial and final states :

$$F_{ex}[\rho^{(1)}] = F_{ex}[\rho_o^{(1)}] - k_B T \int_o^1 d\lambda \int d\vec{r}\, \Delta\rho^{(1)}(\vec{r})\, c^{(1)}\, [\rho^{(1)}(\vec{r};\lambda)] \qquad (1.38)$$

Contact with a well-known thermodynamic relation is easily made by considering the special case of a homogeneous fluid of density $\rho^{(1)}(\vec{r}) = n$; if the initial state is taken to be an ideal gas at the same temperature ($\rho_o^{(1)}(\vec{r}) = n \to 0$), equ. (1.38) leads immediately to the standard relation ;

$$F_{ex}(n) = \int_o^1 d\lambda\, V\, \mu_{ex}(\lambda n)$$
$$= N \int_o^n \mu_{ex}(n')\, dn' \qquad (1.39)$$

Proceeding as for equ. (1.38) we obtain similarily for the

one-particle d.c.f.

$$c^{(1)}[\rho^{(1)};\vec{r}] = c^{(1)}[\rho_o^{(1)};\vec{r}] + \int_o^1 d\lambda \int d\vec{r}'c^{(2)}[\rho^{(1)}(\lambda);\vec{r},\vec{r}']\Delta\rho^{(1)}(\vec{r}')$$

In the homogeneous case, where $c^{(2)}[\rho^{(1)}(\lambda) ; \vec{r},\vec{r}'] = c^{(2)}(\lambda n;\vec{r}-\vec{r}')$ and $c^{(1)}$ reduces to $-\beta\mu_{ex}$, equ. (1.40) yields :

$$\beta\mu_{ex} = \int_o^n dn' \int d\vec{r}'\; c^{(2)}(n',\vec{r}') \qquad (1.41)$$

Differentiating eq. (1.41) with respect to density and adding the ideal part, we arrive at the fundamental compressibility equation for homogeneous fluids (7) :

$$\frac{\chi_T^{(o)}}{\chi_T} = \left(\frac{\partial\beta P}{\partial n}\right)_T = n\left(\frac{\partial\beta\mu}{\partial n}\right)_T = 1 - n\int c^{(2)}(n;\vec{r})d\vec{r} \qquad (1.42)$$

where χ_T denotes the isothermal compressibility and $\chi_T^{(o)} = \beta/n$ its ideal gas value. Equ. (1.38) can be rewritten in the more generally valid form :

$$\begin{aligned}
F_{ex}[\rho^{(1)}] &= F_{ex}[\rho_o^{(1)}] - k_BT\int_o^1 d\lambda\int d\vec{r}\; \frac{\partial\rho^{(1)}(\vec{r};\lambda)}{\partial\lambda}\, c^{(1)}[\rho^{(1)}(\vec{r};\lambda)] \\
&= F_{ex}[\rho_o^{(1)}] - k_BT\int d\vec{r}\;\rho^{(1)}(\vec{r})\; c^{(1)}[\rho_o^{(1)};\vec{r}] \\
&\quad + k_BT\int_o^1 d\lambda\int d\vec{r}\int d\vec{r}'[\rho^{(1)}(\vec{r};\lambda) - \rho^{(1)}(\vec{r})]\; c^{(2)}[\rho^{(1)}(\lambda);\vec{r}']\;\frac{\partial\rho^{(1)}(\vec{r}',\lambda)}{\partial\lambda} \\
&= F_{ex}[\rho_o^{(1)}] - k_BT\int d\vec{r}\;\Delta\rho^{(1)}(\vec{r})\; c^{(1)}[\rho_o^{(1)};\vec{r}] \\
&\quad - k_BT\int d\lambda(\lambda-1)\int d\vec{r}\int d\vec{r}'\Delta\rho^{(1)}(\vec{r})\; c^{(2)}[\rho^{(1)}(\lambda);\vec{r},\vec{r}']\Delta\rho^{(1)}(\vec{r}')
\end{aligned} \qquad (1.43)$$

Integration by parts and equ. (1.40) were used in going from the 1st to the 2d equality in equ. (1.43), while the 3d equality follows from equ. (1.37) ; the first and second terms in that line vanish if the reference state is chosen to be the ideal gas. If moreover the final state is homogeneous, the total free energy per particle deduced from equ. (1.43) reads :

$$\frac{\beta F}{N} = \ln(n\Lambda^3) - 1 + n\int_o^1 d\lambda(\lambda-1)\; c^{(2)}(\lambda n;\vec{r})d\vec{r} \qquad (1.44)$$

Equ. (1.43) is the starting point for the modern density functional theory of freezing (17) : the coexisting fluid and solid phases are then characterized by a uniform and a periodic one-particle density respectively.

All formulae given so far do not depend explicitly on the interactions between atoms (i.e. V_N). These are generally assumed to be pair-wise additive, i.e. :

$$V_N(\{\vec{r}_i\}) = \sum_{i<j}\sum v(\vec{r}_i,\vec{r}_j) = \int d\vec{r} \int d\vec{r}\,' \; v(\vec{r},\vec{r}\,')\; \rho(\vec{r})[\rho(\vec{r}\,') - \delta(\vec{r}-\vec{r}\,')] \tag{1.45}$$

where $v(\vec{r}_i, \vec{r}_j)$ is the pair-potential acting between particles i and j ; if the latter are spherically symmetric (atoms), v depends only on the relative distance $r = |\vec{r}_i - \vec{r}_j|$. For a fixed external potential $\psi(\vec{r})$, the grand potential is, according to equ. (1.9), (1.10), (1.13) and (1.45), a functional of v ; the same is true of the functionals $\Omega\varphi$ and F of $\rho^{(1)}$, introduced in equ. (1.21) and (1.33). From (1.45) and the elementary rules of functional differentiation, it is then easy to verify that the two-particle density (1.16) is given by (7, 9) :

$$\rho^{(2)}(\vec{r},\vec{r}\,') = 2\,\frac{\delta\Omega}{\delta v(\vec{r},\vec{r}\,')} = 2\,\frac{\delta\Omega\varphi[\rho^{(1)}]}{\delta v(\vec{r},\vec{r}\,')} = 2\,\frac{\delta F[\rho^{(1)}]}{\delta v(\vec{r},\vec{r}\,')} \tag{1.46}$$

We do not reproduce here the standard relations which express the internal energy, the virial pressure and the surface tension of a fluid in terms of $\rho^{(2)}$.

The pair-potential v is generally made up of a harshly repulsive short-range part, v (reflecting the impenetrability of molecules), and a more smoothly varying long-range part, w, which can often be looked upon as a perturbation. The effect on the free energy of gradually "switching on" the perturbation, by considering the set of intermediate potentials :

$$v_\lambda(\vec{r},\vec{r}\,') = v_o(\vec{r},\vec{r}\,') + \lambda w(\vec{r},\vec{r}\,') \;;\; 0 \le \lambda \le 1 \tag{1.47}$$

is then easily derived from equ. (1.46) by integration :

$$F[\rho^{(1)},v] = F[\rho^{(1)},v_o] + \frac{1}{2}\int_o^1 d\lambda \int d\vec{r} \int d\vec{r}\,' \;\rho^{(2)}[v_\lambda;\vec{r},\vec{r}\,']\; w(\vec{r},\vec{r}\,') \tag{1.48}$$

This equation is the starting point of thermodynamic perturbation theory (7,8) ; to first order in w, $\rho^{(2)}[v_\lambda]$ is simply replaced by its value in the "reference system", $\rho^{(2)}[v_o]$, and this leads often to satisfactory results. An alternative perturbation scheme can be formulated in terms of the d.c.f. Denoting the one associated with the reference system of particles interacting via v_o by $c_o^{(2)}$, we obtain from equ. (1.24b) and (1.48) :

$$c^{(2)}(\vec{r},\vec{r}\,') = c_o^{(2)}(\vec{r},\vec{r}\,') - \frac{1}{2}\,\frac{\delta^2}{\delta\rho^{(1)}(\vec{r})\,\delta\rho^{(1)}(\vec{r}\,')} \int d\lambda \int d\vec{r} \int d\vec{r}\,' \;\rho^{(2)}[v_\lambda;\vec{r},\vec{r}\,']\; \beta w(\vec{r},\vec{r}\,') \tag{1.49}$$

The simplest approximation is to ignore correlations altogether in the second term on the r. h. s. of (1.49) through factorization :

$$\rho^{(2)}\,[v_\lambda;\,\vec{r},\vec{r}\,'] \simeq \rho^{(1)}(\vec{r})\,\rho^{(1)}(\vec{r}\,')$$

then :

$$c^{(2)}(\vec{r},\vec{r}\,') \simeq c_o^{(2)}(\vec{r},\vec{r}\,') - \beta w(\vec{r},\vec{r}\,') \tag{1.50}$$

This is a compact expression of the (generalized) "random phase approximation" (RPA) (7). Systematic improvements over (1.50) have been worked out by a number of authors (9). The RPA is expected to become exact when $|\vec{r} - \vec{r}\,'| \gg \sigma$ (the particle diameter), since it is generally accepted that $c^{(2)}$ behaves asymptotically as $-\beta v$ (7).

1.4. Approximate theories for the pair distribution function.

As should be clear by now, the two-particle density, or equivalently the p.d.f. (1.19) plays a central role in the theory of liquids ; moreover it will be shown in the next section that it is a measurable quantity, and it has also been extensively studied by computer simulations (7). For that reason it is not surprising that much theoretical effort has gone into the derivation of approximations applicable to dense liquids. This has resulted in a number of approximate integral equations relating $g^{(2)}$ to the pair potential v and the density $\rho^{(1)} \equiv n$ in a homogeneous fluid (2, 7, 8). Here we only sketch the derivation of one of the most important integral equations, the so-called "hyper-netted chain" (HNC) equation following an elegant procedure due to Percus (12). In a homogeneous fluid translational invariance allows us to single out any one among the N particles and to associate, arbitrarily, the origin O of coordinates with that particle. The N - 1 remaining particles can then be regarded as moving in the "external field" :

$$\varphi(\vec{r}) = v(\vec{r}) \tag{1.51}$$

due to the particle at the origin ; since translational symmetry is broken, the fluid can be considered as inhomogeneous, at least in the vicinity of the origin. The basis of Percus' method lies in the recognition that the one-particle density $\rho^{(1)}(\vec{r})$ of this inhomogeneous fluid can be identified with the quantity $ng_o^{(2)}(\vec{r})$ of the homogeneous fluid. This identification follows in fact directly from equs. (1.15), (1.46) and (1.51). We start from equ. (1.40) describing the effect of switching on continuously the external field, which is here due to the particle at the origin ; in the homogeneous fluid (λ = 0), $\rho_o^{(1)}$ reduces to n and $c^{(1)}$ reduces to n/z. In view of equs. (1.28), (1.51) and of the identification $\rho^{(1)}(\vec{r}) = ng_o^{(2)}(\vec{r})$, , we arrive at the exact expression :

$$\ln[g_o^{(2)}(\vec{r})] = -\beta v(\vec{r}) + \int_o^1 d\lambda \int d\vec{r}'\, c^{(2)}[\rho^{(1)}(\lambda);\vec{r},\vec{r}']\, h_o^{(2)}(\vec{r}') \tag{1.52}$$

where $h_o^{(2)}$ is the pair correlation function of the homogeneous fluid, defined in equ. (1.19). The HNC approximation amounts to replacing $c^{(2)}[\rho^{(1)}(\lambda)]$ by its value $c_o^{(2)}$ for $\lambda = 0$, i.e. for the homogeneous fluid. Thus, dropping the subscript 0 and superscript 2, we arrive at the HNC closure :

$$\ln[g(\vec{r})] = -v(\vec{r}) + n\, c * h \tag{1.53}$$

Exponentiating, and using the OZ relation (1.32), we obtain the equivalent form :

$$g(r) = \exp\{-\beta v(r) + h(r) - c(r)\} \quad \text{(HNC)} \tag{1.54}$$

where restriction was made to the case of isotropic fluids. The closure (1.54) and the OZ relation (1.32) form a closed system which can be solved for g (r), given a pair potential v (r), by Fast Fourier Transform techniques (18). Solving (1.54) for c (r) we obtain :

$$c(r) = -\beta v(r) + h(r) - \ln[1 + h(r)] \quad \text{(HNC)} \tag{1.55}$$

Since $h(r) \to 0$ as $r \to \infty$, equ. (1.55) shows that $c(r) \to -\beta v(r)$ in that limit, in agreemen with the RPA prediction. If the exponential in equ. (1.54) is linearized with respect to its regular part h (r) - c (r), the widely used Percus-Yevick (PY) closure is recovered :

$$g(r) = \exp\{-\beta v(r)\}\,[g(r) - c(r)] \quad \text{(PY)} \tag{1.56}$$

The PY approximation turns out to be most accurate for steeply repulsive, short-range potentials. In the special case of hard spheres of diameter σ, the closure (1.56) reduces to :

$$g(r) = 0 \quad , \quad r < \sigma \tag{1.57a}$$

$$c(r) = 0 \quad , \quad r > \sigma \tag{1.57b}$$

Combined with the OZ relation, the PY closure for hard spheres has been solved analytically (19). Finally, if the potential is separated as in equ. (1.47), the Boltzmann factor in (1.56) can be linearized with respect to the long-range "perturbation" w. Retaining only the dominant asymptotic term ($g - c \simeq 1$) in the coefficient of w, we arrive at the closure (20) :

$$g(r) = \exp\{-\beta v_o(r)\}[g(r) - c(r) - w(r)] \qquad \text{(MSA)} \qquad (1.58)$$

Equ. (1.58) generalizes the "mean spherical approximation" first put foreward in the special case where $v_o(r)$ reduces to the hard sphere potential (21) ; then (1.58) reads :

$$g(r) = 0 \quad , r < \sigma \qquad (1.59a)$$

$$c(r) = -\beta w(r) \quad , r > \sigma \qquad (1.59b)$$

Although not a particularly good approximation, the MSA plays a fundamental role in the theory of ionic and polar liquids (and hence of aqueous solutions !), since it allows analytic solutions to be obtained (cf. lectures 3 and 4).

The various approximate closures suffer from internal thermodynamic inconsistency (the virial energy and compressibility relations do not yield the same equation-of-state). Much recent effort has gone into the derivation of thermodynamically self-consistent closures, e.g. by generalizing the HNC closure (1.54) through the introduction of a "bridge function" B (r) assumed to be "universal" and identified with its hard sphere form for a suitably chosen packing fraction $\eta = \pi n\sigma^3/6$ (22) :

$$g(r) = \exp\{-\beta v(r) + h(r) - c(r) + B_\eta(r)\} \qquad \text{(BHNC)} \qquad (1.60)$$

This scheme has been successfully applied to the "inverse" problem, i.e. extract an unknown pair potential v (r) from a given (measured) g (r) (23). An alternative scheme to obtain self-consistent integral equations amounts to interpolating between the HNC closure (1.54) and either the PY closure (1.56) or the MSA closure (1.58) (24).

Closures similar to those obtained for the pair functions may also be derived for the density profiles $\rho^{(1)}(r)$ of inhomogeneous fluids. Combinat n of equs. (1.28), (1.40) (where $\rho_o^{(1)}$ is taken to be the density n of the homogeneous fluid at the same chemical potential) and use of the same approximation as that leading to (1.53) yields the HNC-like closure :

$$\rho^{(1)}(\vec{r}) = n \exp\{-\beta\varphi(\vec{r}) + c_o^{(2)} * \Delta\rho^{(1)}\} \qquad (1.61)$$

where $c_o^{(2)} = c(r)$ is the d.c.f. of the homogeneous fluid and $\Delta\rho^{(1)} = \rho^{(1)}(\vec{r}) - \rho_o^{(1)} = \rho^{(1)}(\vec{r}) - n$. Given c (r) (as determined in the HNC approximation (1.54) for consistency), (1.61) can be solved to yield the density profile in the external field $\varphi(\vec{r})$.

1.5. Structure factors and linear response.

We now establish the link between the two-body correlation functions considered so far, and the linear response of a homogeneous system to a (weak) external probe. The modulation of the one-particle density induced by an external potential $\varphi(\vec{r})$ is linear in $\varphi(\vec{r})$ if the potential is sufficiently weak :

$$\Delta\rho^{(1)}(\vec{r}) = \rho^{(1)}(\vec{r}) - \rho_o^{(1)} = \int \chi(\vec{r},\vec{r}\,')\,\varphi(\vec{r}\,')d\vec{r}\,' \tag{1.62}$$

where the static density response function is an intrinsic property of the unperturbed system ; if the latter is homogeneous, $\chi(\vec{r},\vec{r}\,') = \chi(\vec{r} - \vec{r}\,')$. Identification with equ. (1.20) shows that :

$$\chi(\vec{r},\vec{r}\,') = -\beta H(\vec{r},\vec{r}\,') \tag{1.63}$$

which is a static version of the "fluctuation-dissipation theorem" (7). Since we restrict our attention to linear response, we may as well consider a single Fourier component of the external potential :

$$\varphi(\vec{r}) = \varphi(\vec{k})\, e^{-i\vec{k}.\vec{r}} \tag{1.64}$$

If the unperturbed fluid is homogeneous, substitution of (1.64) into (1.62) leads to :

$$\Delta\rho^{(1)}(\vec{k}) = \chi(\vec{k})\,\varphi(\vec{k}) \tag{1.65}$$

where $\chi(\vec{k})$ and $\Delta\rho^{(1)}(\vec{k})$ the Fourier transforms of $\chi(\vec{r})$ and $\Delta\rho^{(1)}(\vec{r})$ respectively. According to equs. (1.63) and (1.17) (with $\rho^{(1)}(\vec{r}) = n$) :

$$\chi(\vec{k}) = -\beta n[n\,\hat{h}^{(2)}(\vec{k}) + 1] = -\beta n S(\vec{k}) \tag{1.66}$$

where :

$$\hat{h}^{(2)}(\vec{k}) = \int e^{-i\vec{k}.\vec{r}}\, h^{(2)}(\vec{r})\, d\vec{r} \tag{1.67}$$

and S $(\vec{k})$ is the <u>static structure factor</u> :

$$S(\vec{k}) = \frac{1}{N}\langle\rho_{\vec{k}}\,\rho_{-\vec{k}}\rangle = 1 + n\,\hat{h}^{(2)}(\vec{k}) \tag{1.68}$$

where :

$$\rho_{\vec{k}} = \int e^{-i\vec{k}.\vec{r}}\,\rho(\vec{r})\, d\vec{r} = \sum_{i=1}^{N} e^{-i\vec{k}.\vec{r}_i} \tag{1.69}$$

are the Fourier components of the microscopic density. According to equs. (1.65) and (1.66), S $(\vec{k})$ is a measure of the density response of a system (initially in equilibrium) to a weak external perturbation of wavenumber k. The probe may be, in particular, a beam of photons or neutrons and $S(\vec{k})$ is then directly proportional to the total intensity scattered in a given direction, and hence a measurable quantity. Since, according to equs. (1.67), (1.68) and (1.19) $S(\vec{k})$ is the Fourier transform of the p.d.f. $g^{(2)}(\vec{r})$, the latter quantity can in principle be obtained by inverse Fourier transformation of X - ray or neutron diffraction. data.

Combination of equs. (1.32), (1.42) and (1.68) leads to the familiar long wavelength limit of the structure factor :

$$\lim_{k \to 0} S(k) = \lim_{k \to 0} \left[\frac{1}{1 - n\hat{c}^{(2)}(k)} \right] = \frac{\chi_T}{\chi_T^{(o)}} = \left(\frac{\partial n}{\partial \beta P}\right)_T \qquad (1.70)$$

which is a measure of the "response" of the macroscopic density to a variation of the applied pressure (the "probe").

2. TIME-DISPLACED CORRELATION FUNCTIONS.

2.1. Correlations in space and time.

The first lecture was devoted to static (or equal-time) correlations in space. However, if we are interested in dynamical phenomena in liquids, like the relaxation of an initially perturbed system towards equilibrium, transport phenomena or the inelastic scattering of radiation, we are naturally led to consider time-dependent correlations (7,26). To characterize the decay of a spontaneous (thermal) or induced fluctuation of a dynamical variable A (a function $A(\{\vec{r}_i(t), \vec{p}_i(t)\})$ of the instantaneous positions and (or) momenta of some, or all particles in the system) it is natural to investigate the time-dependence of the projection of its value at time t on its initial value, or on the initial value of another dynamical variable, say B. To that purpose we define the equilibrium time correlation function (TCF) of the two variables A and B as :

$$\begin{aligned} C_{AB}(t',t'') &= \langle A(t')\, B^*(t'')\rangle \\ &= \langle A(t' - t'')\, B^*(0)\rangle \end{aligned} \qquad (2.1)$$

where the angular brackets denote an ensemble average over the initial phase, and advantage has been taken of the stationarity of equilibrium averages. More generally, if the dynamic variables vary in space (like the microscopic density (1.4), and if the fluid is homogeneous (this will be assumed throughout this lecture), the corresponding TCF will also depend on the spatial variable $\vec{r} = \vec{r}' - \vec{r}''$:

$$C_{AB}(\vec{r},t) = \langle A(\vec{r}',t')\, B^*(\vec{r}'',t'')\rangle$$

$$= \langle A(\vec{r},t)\, B^*(\vec{0},0)\rangle \tag{2.2}$$

where t = t' - t'' . Translational invariance implies that only Fourier components of opposite wave-vectors have non-vanishing correlations, so that :

$$C_{AB}(\vec{k},t) = \int e^{-i\vec{k}.\vec{r}}\, C_{AB}(\vec{r},t)\,d\vec{r}$$
$$= \langle A_{\vec{k}}(t)\, B^*_{\vec{k}}(0)\rangle \tag{2.3}$$

The time evolution of any dynamical variable is governed by:

$$\dot{A} = \frac{dA}{dt} = iLA \tag{2.4}$$

where L denotes the Liouville operator :

$$L = \frac{1}{h}[H_N, \ ] \qquad \text{(quantum mechanics)} \tag{2.5a}$$

$$L = i\{H_N, \ \} = i\sum_{j=1}^{N}\left(\frac{\partial H_N}{\partial \vec{r}_j}\cdot\frac{\partial}{\partial \vec{p}_j} - \frac{\partial H_N}{\partial \vec{p}_j}\cdot\frac{\partial}{\partial \vec{r}_j}\right) \tag{2.5b}$$
(classical mechanics)

Equ. (2.4) has the formal solution :

$$A(t) = \exp\{iLt\}A \tag{2.6}$$

where $A \equiv A(0)$ denotes the initial value of the dynamical variable.
We shall only consider dynamic variables of zero mean. Since after a sufficiently long time interval, any dynamic variable will be completely decorrelated from the initial value of the same, or any other, variable, it follows that :

$$\lim_{t\to\infty} C_{AB}(t) = \langle A(t)\rangle\langle B^*(0)\rangle = \langle A\rangle\langle B^*\rangle = 0 \tag{2.7}$$

This property allows us to define the spectrum of any TCF as :

$$\hat{C}_{AB}(\omega) = \frac{1}{2\pi}\int_{-\infty}^{+\infty} e^{i\omega t}\, C_{AB}(t)\, dt \tag{2.8}$$

It is also convenient to introduce the Laplace transform :

$$\tilde{C}_{AB}(z) = \int_{-\infty}^{+\infty} e^{izt}\, C_{AB}(t)\, dt\ ; \quad \mathrm{Im} z > 0$$
$$= i\int_{-\infty}^{+\infty} d\omega\, \frac{\hat{C}_{AB}(\omega)}{z-\omega} \tag{2.9}$$

Inversely :

$$\hat{C}_{AB}(\omega) = \lim_{\varepsilon\to 0} \frac{1}{\pi}\tilde{C}'_{AB}(\omega + i\varepsilon) \tag{2.10}$$

where the prime denotes a real part (a double prime will denote an imaginary part). If A and B are both either even or odd functions of the momenta, their (classical) TCF is an even function of time and the spectrum is an even function of frequency ; $C_{AB}(t)$ then admits a Taylor expansion in even powers of t :

$$\begin{aligned} C_{AB} &= \sum_{n=0}^{\infty} \frac{t^{2n}}{(2n)!} C_{AB}^{(2n)} \quad (t=0) \\ &= \sum_{n=0}^{\infty} \frac{t^{2n}}{(2n)!} (-1)^n \langle A^{(n)}(0) B^{*(n)}(0) \rangle \\ &= \sum_{n=0}^{\infty} \frac{t^{2n}}{(2n)!} \langle (L^n A)(L^n B^*) \rangle \end{aligned} \tag{2.11}$$

where (n) denotes an nth order derivative with respect to time and repeated use was made of the stationarity property :

$$\frac{d}{d\Delta} \langle \dot{A}(t+\Delta) B(\Delta) \rangle = 0 \Rightarrow \langle \dot{A}(t) B \rangle = -\langle A(t) \dot{B} \rangle \tag{2.12}$$

Combination of equs. (2.8) and (2.11) leads to an expression of the frequency moments of the spectrum in terms of the static (equal-time) correlation functions, which are the coefficients in the Taylor expansion (2.11) :

$$\begin{aligned} \Omega_{AB}^{(2n)} &= \int_{-\infty}^{+\infty} \omega^{2n} \hat{C}_{AB}(\omega)\, d\omega \\ &= (-1)^n \; C_{AB}^{(2n)}(0) \end{aligned} \tag{2.13}$$

Substitution of (2.13) into the Hilbert transform (2.9) leads immediately to the high-frequency expansion for the Laplace transform :

$$\tilde{C}_{AB}(z) = \frac{i}{z} \sum_{n=0}^{\infty} \frac{\Omega_{AB}^{(2n)}}{z^{2n}} \tag{2.14}$$

A particularly important class of time correlation functions are the autocorrelation functions (ACF) $C_{AA}(t)$, which are necessarily even functions of time, and which satisfy two important properties :

$$|C_{AA}(t)| \leq C_{AA}(0) \tag{2.15}$$

$$\hat{C}_{AA}(\omega) \geq 0 \tag{2.16}$$

We shall be mostly concerned with <u>local</u> dynamical variables of the general form :

$$A(\vec{r},t) = \sum_{i=1}^{N} a_i(t)\, \delta(\vec{r} - r_i(t)) \tag{2.17}$$

$$A_{\vec{k}}(t) = \int e^{-i\vec{k}.\vec{r}} A(\vec{r},t)d\vec{r} = \sum_{i=1}^{N} a_i(t)\, e^{-i\vec{k}.\vec{r}_i(t)} \tag{2.18}$$

where a_i is any physical quantity, like the mass, the linear or angular momentum, associated with particle i. Important examples of local variables are the microscopic density (1.4) ($a_i = 1$) and the particle current ($a_i = \vec{u}_i(t)$, the velocity of the ith particle) :

$$\vec{j}(\vec{r},t) = \sum_{i=1}^{N} \vec{u}_i(t)\, \delta(\vec{r} - \vec{r}_i(t))$$

$$\vec{j}_{\vec{k}}(t) = \sum_{i=1}^{N} \vec{u}_i(t)\, e^{-i\vec{k}.\vec{r}_i(t)} \tag{2.19}$$

The Fourier components can be separated into their longitudinal (l) and transverse (t) parts which are respectively parallel and orthogonal to the wavevector $\vec{k}$. the longitudinal component $\vec{j}_{\vec{k}l}$ is related to the microscopic density (1.69) via the continuity equation :

$$\dot{\rho}_{\vec{k}}(t) + i\vec{k} \,.\, \vec{j}_{\vec{k}\ell}(t) = 0 \tag{2.20}$$

The ACF which can be constructed from these local dynamical variables are the density ACF (frequently referred to as the "intermediate scattering function") :

$$F(\vec{k},t) = \frac{1}{N} <\rho_{\vec{k}}(t)\, \rho_{-\vec{k}}(0)> \tag{2.21}$$

and the current ACF matrix :

$$C_{\alpha\beta}(\vec{k},t) = \frac{k^2}{N} <j^{\alpha}_{\vec{k}}(t)\, j^{\beta}_{-\vec{k}}(0)>$$

$$= \hat{k}_\alpha \hat{k}_\beta C_\ell(\vec{k},t) + (\delta_{\alpha\beta} - \hat{k}_\alpha \hat{k}_\beta) C_t(k,t)$$

where α, β denote cartesian components and the second equality follows from rotational invariance in an isotropic fluid which implies that the longitudinal and transverse projections of the particle current are uncorrelated ; the $\hat{k}_\alpha$ are the cartesian components of the unit vector $\hat{k} = \vec{k}/k$. Choosing the z - axis along the direction of $\vec{k}$, the longitudinal and transverse current ACF are given by :

$$C_\ell(k,t) = \frac{k^2}{N} <j^z_{\vec{k}}(t)\, j^z_{-\vec{k}}(0)> \tag{2.23}$$

$$C_t(k,t) = \frac{k^2}{N} <j^x_{\vec{k}}(t)\, j^x_{-\vec{k}}(0)> \tag{2.24}$$

Use of the continuity equation (2.20) immediately shows that there are only two independent functions (F and C_t) since :

$$C_\ell(k,t) = \frac{1}{N} \langle \dot\rho_{\vec k}(t)\, \dot\rho_{-\vec k}(0) \rangle$$
$$= -\frac{d^2}{dt^2} F(k,t) \qquad (2.25)$$

The initial (t = 0) values of these ACF are :

$$F(k,t=0) = S(k) \qquad (2.26)$$

$$C_\ell(k,t=0) = C_t(k,t=0) = \frac{k^2 k_B T}{m} \equiv \omega_o^2 \qquad (2.27)$$

Returning to $\vec r$ -space, we define the Fourier transform of F (k,t), called the Van Hove function (25) :

$$G(r,t) = \frac{1}{(2\pi)^3} \int e^{i\vec k.\vec r} F(k,t)\, d\vec k$$
$$= \frac{1}{N} \langle \rho(\vec r,t)\, \rho(\vec 0,0) \rangle$$
$$= \frac{1}{N} \langle \sum_{i=1}^{N} \sum_{j=1}^{N} \langle \delta[\vec r + \vec r_j(0) - \vec r_i(t)] \rangle$$
$$= \frac{1}{N} \langle \sum_{i=1}^{N} \delta[\vec r + \vec r_i(0) - \vec r_i(t)] \rangle$$
$$+ \frac{1}{N} \langle \sum_{i \neq j} \sum \delta[\vec r + \vec r_j(0) - \vec r_i(t)] \rangle$$
$$= G_s(r,t) + G_d(r,t) \qquad (2.28)$$

G_s and G_d are the "self" and "distinct" parts of the Van Hove function, respectively, and a similar separation holds for F(k,t). From a comparison with equs. (1.16) and (1.19) (taken for the case of a homogeneous fluid), it is immediately clear that :

$$G_s(r,t=0) = \delta(\vec r) \qquad (2.29a)$$

$$G_d(r,t=0) = ng(r) \qquad (2.29b)$$

G_d is a natural generalization of the pair distribution function to non-zero time intervals : it is the probability density of finding a particle at time t, at a distance r from the origin, provided <u>another</u> particle was initially at the origin. $G_s(r,t)$, on the other hand, describes the individual ("Brownian") motion of a particle in the fluid ; this function will be reconsidered in the following section.

The spectrum of F(k,t) is called the dynamical structure factor :

$$S(k,\omega) = \frac{1}{2\pi} \int_{-\infty}^{+\infty} e^{i\omega t} F(k,t)dt$$

$$= S_s(k,\omega) + S_d(k,\omega) \tag{2.30}$$

$S(k,\omega)$ and its self part $S_s(k,\omega)$ are directly proportional to the coherent and incoherent cross sections in inelastic neutron scattering experiments, which will be discussed at length by other lecturers. Similarily we define the spectra $\hat{C}(k,\omega)$ and $\hat{C}_t(k,\omega)$ of the longitudinal and transverse current ACF. According to (2.25) :

$$\hat{C}_\ell(k,\omega) = \omega^2 S(k,\omega) \tag{2.31}$$

The two lowest order sum rules obeyed by $S(k,\omega)$ are, according to (2.13), (2.26), (2.27) and (2.21) :

$$\int_-^+ S(k,\omega)\, d\omega = S(k) \qquad \text{("elastic sumrule")} \tag{2.32}$$

$$\int_-^+ \omega^2 S(k,\omega)\, d\omega = -\ddot{F}(k,0) = C(k,0) = \omega_o^2 \tag{2.33}$$

Note that this last sum rule is totally independent of the interactions between particles (ideal gas result). $G(r,t)$ and $S(k,\omega)$ are easily calculated for free particle dynamics. Since positions of different particles are then uncorrelated, $G_d(r,t) \equiv n$ and we are left with the problem of calculating G_s. The probability that a particle of a perfect gas will move at a distance r in time t is proportional to the probability of its having a velocity in the range $(\vec{u}, \vec{u}+d\vec{u})$, where $|\vec{u}| = r/t$, as given by the Maxwell-Boltzmann distribution. Thus :

$$G_s(r,t) = \left(\frac{\beta m}{2\pi t^2}\right)^{3/2} \exp\left\{-\frac{1}{2}\beta m r^2/t^2\right\} \tag{2.34}$$

and Fourier transformation in space and time yields :

$$S_s(k,\omega) = \left(\frac{m\beta}{2\pi k^2}\right)^{1/2} \exp\left\{-\frac{1}{2}\beta m\omega^2/k^2\right\} \tag{2.35}$$

In real liquids the ideal gas limit is reached as $r,t \to 0$, when the particles move as if they were free ; these conditions correspond to the limit $k, \omega \to \infty$. The opposite limit $k, \omega \to 0$, corresponds to long wavelength and time scales ($\lambda \gg \sigma$; $t \gg \tau \simeq \sigma/v_o = (m\sigma^2/k_BT)^{1/2}$), so that the fluid appears as a continuous medium which can be described by the (linearized) equations of hydrodynamics (28). This aspect of large scale collective atomic dynamics is conveniently probed by Rayleigh-Brillouin light scattering experiments which directly measure $S(k,\omega)$ for wavelengths and frequencies comparable to those of light (26, 27). In that regime $S(k,\omega)$ is dominated by collective modes (thermal diffusion and propagating sound waves) which determine its characteristic Rayleigh-Brillouin three-peak structure (29,7,26,27).

Simple examples of such hydrodynamic calculations will be given in sections 2.2 and 2.4.

2.2. Brownian motion.

In this section we investigate the individual ("Brownian") motion of an atom in the liquid, as described by the "self" part of the van Hove function (2.28). At very short times, the atom did not yet have a chance to suffer collisions with its neighbours, and G_s is giver by its free particle limit (2.34). On the contrary after a sufficiently long time interval ($t \gg \tau$), the atom has diffused away from its initial position, undergoing many collisions with other atoms in the liquid throughout. This long time process is well described by the macroscopic diffusion equation :

$$\frac{\partial}{\partial t} G_s(r,t) = D\nabla^2 G_s(r,t) \tag{2.36}$$

where D is the self (or "tagged particle") diffusion coefficient. Equ. (2.36) is easily solved, subject to the initial condition (2.29b), with the result :

$$G_s(r,t) = \frac{1}{(4\pi Dt)^{3/2}} \exp\{-r^2/4Dt\} \tag{2.37}$$

$$F_s(k,t) = \exp\{-Dk^2 t\} \tag{2.38}$$

Equivalently a Fourier-Laplace transform of the diffusion equ. (2.36) yields :

$$\tilde{F}_s(k,z) = \frac{1}{-iz + Dk^2} \tag{2.39}$$

so that a combination of equs. (2.30) and (2.10) leads immediately :

$$S_s(k,\omega) = \frac{1}{\pi} \frac{Dk^2}{\omega^2 + (Dk^2)^2} \tag{2.40}$$

The latter result is valid only in the "hydrodynamic" limit of small k and ω. Note that :

$$D = \lim_{\omega \to 0} \lim_{k \to 0} \Pi \frac{\omega^2}{k^2} S_s(k,\omega) \tag{2.41}$$

In a dilute suspension of large particles in a solvent, $S(k, \omega) \simeq S_s(k, \omega)$ and a light scattering experiment (where scattering of the solvent may be neglected) will directly measure the diffusion coefficient D of the large particles.

Since G_s has a gaussian behaviour both at short (equ. (2.34)) and long (equ. (2.37)) times, it is natural to seek a gaussian representation valid at all times, of the form (30) :

$$G_s(r,t) = \left(\frac{\alpha(t)}{\pi}\right)^{3/2} \exp\{-r^2\alpha(t)\} \tag{2.42}$$

where $\alpha(t) = \beta m/2t^2$ in the free particle limit, and $\alpha(t) = 1/4Dt$ in the hydrodynamic limit. Now the mean square displacement of a particle from its initial position is, by definition :

$$\langle r^2(t)\rangle \equiv \langle|\vec{r}(t) - \vec{r}(0)|^2\rangle = 4\pi\int_0^\infty r^2 G_s(r,t) r^2 dr \tag{2.43}$$

Substitution of the ansatz (2.42) leads to the identification $\langle r^2(t)\rangle = 3/(2\alpha(t))$, and hence to the generalized Gaussian approximation :

$$F_s(k,t) = \exp\{-\frac{k^2}{6}\langle r^2(t)\rangle\} \tag{2.44}$$

Substitution of (2.37) into (2.43) yields the celebrated Einstein relation :

$$\langle r^2(t)\rangle \underset{t\to\infty}{=} 6Dt \tag{2.45}$$

D can also be expressed in terms of the velocity ACF $Z(t)$ defined as the $k \to 0$ limit of the self part of the longitudinal current correlation function (2.23) :

$$\begin{aligned}\lim_{k\to 0}\frac{1}{k^2}C_{\ell s}(k,t) &= -\lim_{k\to 0}\frac{1}{k^2}\frac{d^2}{dt^2}F_s(k,t)\\ &= \langle u_{iz}(t)\, u_{iz}(0)\rangle\\ &= \frac{1}{3}\langle\vec{u}_i(t)\,\vec{u}_i(0)\rangle = Z(t)\end{aligned} \tag{2.46}$$

The spectrum of $Z(t)$ is given by :

$$\begin{aligned}\hat{Z}(\omega) &= \frac{1}{2\pi}\int_{-\infty}^{+\infty} Z(t)\, e^{i\omega t}\, dt\\ &= \frac{\omega^2}{2\pi}\lim_{k\to 0}\frac{1}{k^2}\int_{-\infty}^{+\infty} F(k,t)\, e^{i\omega t}\, dt = \omega^2 \lim_{k\to 0}\frac{1}{k^2}S_s(k,\omega)\end{aligned} \tag{2.47}$$

Hence, according to (2.41) :

$$D = \lim_{\omega\to 0}\pi\hat{Z}(\omega) = \int_0^\infty Z(t)\, dt \tag{2.48}$$

The exact time evolution of the particle velocity $\vec{u}(t)$ is described by equ. (2.6). For practical purposes, a more phenomenological approach is widely used. We start from the familiar Langevin equation for the velocity of a Brownian particle. If u denotes one of its cartesian components, ξ the friction coefficient of the solvent, and $R(t)$ a component of the stochastic random force (due to the collisions suffered by the Brownian particles with the atoms of the solvent), the Langevin equation reads :

$$m\,\dot{u}(t) = -m\xi u(t) + R(t) \tag{2.49}$$

Equipartition of energy, and the assumption that R(t) is orthogonal to u(0), on average , for all t, lead to the following relation between ξ and R (31, 7, 27) :

$$\xi = \frac{1}{mk_BT}\int_0^\infty \langle R(t)\ R(0)\rangle\ dt \qquad (2.50)$$

When the tagged particle is of a size similar to that of the solvent particles, retarded (or non-Markovian) effects become important. In the presence of an external force field X(t), the Langevin equation (2.49) must then be generalized :

$$\dot{u}(t) = -\int_0^t \xi(t-s)\ u(s)\ ds + \frac{1}{m}\left[R(t) + X(t)\right] \qquad (2.51)$$

where $\xi(t)$ is now a non-local friction coefficient. Considering first the case X = 0, projecting both sides of equ. (2.51) onto the initial velocity u(0), and making the usual assumption that :

$$\langle R(t)\ u(0)\rangle = 0 \quad , \quad \forall t > 0 \qquad (2.52)$$

we immediately derive an expression for the Laplace transform (2.9) of the velocity ACF :

$$\tilde{Z}(z) = \frac{Z(t=0)}{-iz + \tilde{\xi}(z)} = \frac{k_BT/m}{-iz + \tilde{\xi}(z)} \qquad (2.53)$$

where the frequency-dependent friction coefficient $\tilde{\xi}(z)$ is given by the following generalization of (2.50) :

$$\tilde{\xi}(z) = \frac{1}{mk_BT}\int_0^\infty \langle R(t)\ R(0)\rangle\ e^{izt}\ dt \qquad (2.54)$$

Taking the zero-frequency limit and remembering (2.48) we recover the Einstein relation :

$$D = \frac{k_BT}{\xi\, m} \qquad (2.55)$$

A direct estimate of the friction coefficient ξ can be obtained from a hydrodynamic calculation of the frictional force on a sphere of diameter σ moving with constant velocity in a viscous fluid. If η is the viscosity of the fluid the result depends on the boundary condition at the surface of the sphere (32,33) :

$$\left.\begin{aligned} \xi &= \frac{3\pi\eta\sigma}{m} \quad &\text{(stick)} \\ &= \frac{2\pi\eta\sigma}{m} \quad &\text{(slip)} \end{aligned}\right\} \qquad (2.56)$$

Combination of equs. (2.55) and (2.56) leads to the familiar Stokes-Einstein law :

$$D\eta = \left.\begin{array}{ll} \dfrac{k_B T}{3\pi\sigma} & \text{(stick)} \\ \dfrac{k_B T}{2\pi\sigma} & \text{(slip)} \end{array}\right\} \quad (2.57)$$

The remarkable aspect of this law is that while it has been derived from purely macroscopic considerations, its approximate validity for self-diffusion of atoms or molecules in liquids is experimentally well-attested, use of the slip boundary condition generally leading to more reasonable values of the effective diameter σ .

The generalized Langevin equation (2.51) can be extended to any dynamical variable, and the generalized "random force" can be given a precise Statistical-Mechanics interpretation in terms of projected time evolution (34, 7, 26). If $A = (A_1, \ldots A_n)$ denotes a set of dynamical variables, the correlation function matrix obeys the memory function equation :

$$\dot{C}_{AA}(t) - i\Omega C_{AA}(t) + \int_0^t M(t-s)\, C_{AA}(s)\, ds = 0 \quad (2.58)$$

where the frequency matrix Ω is defined by :

$$\Omega = \langle \dot{A}A^* \rangle \ (\langle AA^* \rangle)^{-1} \quad (2.59)$$

and the memory function M plays the role of the generalized friction coefficient (t) and is given by the ACF of the set of generalized random forces. If Laplace transforms are taken, equ. (2.58) yields :

$$\tilde{C}_{AA}(z) = C_{AA}(t=0)\ [-iz - i\Omega + \tilde{M}(z)]^{-1} \quad (2.60)$$

which generalizes (2.53). The practical interest of equ. (2.58) or (2.60) lies in the fact that, for a judicious choice of the set of dynamical variables A, the memory functions have a simpler structure (in particular, a faster decay in time) than the corresponding correlation functions, and are hence more readily amenable to simple approximations.

2.3. Linear response theory.

In section 1.5 we examined the static response of a system to a weak external perturbation ; here we generalize these considerations to the dynamic (frequency-dependent) response. First consider the Brownian motion problem described by equ. (2.51), and more specifically the effect of a periodic external force applied to the particle :

$$X(t) = R\, X_o\, e^{-i\omega t} \quad (2.61)$$

The mean velocity at time t (which measures the response of the system to the external field) is given by :

$$\langle u(t)\rangle_{n.e.} = R\mu(\omega)\, X_o\, e^{-i\omega t} \tag{2.62}$$

which defines the frequency-dependent <u>mobility</u> $\mu(\omega)$; an expression for μ is easily derived from equ. (2.51) on the usual assumption that R(t) vanishes in the mean, with the result (cf. (2.53) with $z = \omega$) :

$$\mu(\omega) = \frac{1}{m}\,\frac{1}{-i\omega + \tilde{\xi}(\omega)} = \frac{1}{k_BT}\,\tilde{Z}(\omega) \tag{2.63}$$

In particular, the static mobility $\mu = \mu(\omega=0)$ and the self-diffusion coefficient obey the Einstein relation :

$$\mu = \frac{D}{k_BT} \tag{2.64}$$

$\mu(\omega)$ is the simplest example of a linear response function (or dynamic susceptibility), and equ. (2.63) is a special case of the fluctuation-dissipation theorem (31).

The procedure is easily generalizable to the response of any dynamical variable B to the perturbation due to an external field F(t) which couples to a dynamical variable A, such that the total hamiltonian of the system becomes :

$$H_N = H_N^{(o)} + H_N'(t)$$

$$H_N'(t) = -A\,F(t) = -A\,F_o\, e^{-i\omega t}\, e^{\varepsilon t} \tag{2.65}$$

The time evolution of the non-equilibrium N-particle distribution function $f^{(N)}(\Gamma_N, t)$ is governed by Liouville's equation :

$$\frac{\partial f^{(N)}}{\partial t} = -i\,L\,f^{(N)}(t) = \{H_N, f^{(N)}(t)\}$$

$$= -i\,L_o f^{(N)}(t) - \{A, f^{(N)}(t)\}\,F(t) \tag{2.66}$$

where L_o is the Liouville operator associated with the unperturbed hamiltonian. Since we are interested in the linear response of the system, we write :

$$f^{(N)}(t) = f_o^{(N)} + \Delta f^{(N)}(t) \tag{2.67}$$

where

$$f_o^{(N)} = f^{(N)}(t = -\infty)$$

$$= C\exp\{-\beta H_N^{(o)}\} \tag{2.68}$$

if we assume that the system was in thermodynamic equilibrium in the infinite past, when the external force is applied adiabatically ($\varepsilon \to 0^+$ in equ. (2.65)). Upon linearization of equ. (2.66) we obtain after some straightforward manipulations the following expression for the non-equilibrium expectation value of B at time t (31, 7, 26) :

$$\begin{aligned}\langle \Delta B(t)\rangle_{n.e.} &= \int \Delta f^{(N)}(\Gamma_N,t)\, B(\Gamma_N)\, d\Gamma_N \\ &= \int_{-\infty}^{t} \theta_{BA}(t-s)\, F(s)\, ds \end{aligned} \tag{2.69}$$

where the "after-effect" function θ_{BA} is given by :

$$\begin{aligned}\theta_{BA}(t) &= -\langle\{B(t),A\}\rangle \\ &= \beta \langle B(t)\, \dot{A}\rangle \quad ; \; t \geq 0 \end{aligned} \tag{2.70}$$

The angular brackets denote again equilibrium averages (weighted by (2.68)). The first line in equ. (2.70) allows an immediate extension to the quantum-mechanical case, via the correspondance principle embodied in equ. (2.5). Substitution of the monochromatic form (2.65) of (t) into (2.69) leads to the compact result :

$$\langle \Delta B(t)\rangle_{n.e} = R\, \chi_{BA}(\omega)\, F_o\, e^{-i\omega t} \tag{2.71}$$

where χ_{BA} is the complex dynamic susceptibility or response function :

$$\begin{aligned}\chi_{BA}(\omega) &= \chi'_{BA}(\omega) + i\, \chi''_{BA}(\omega) \\ &= \lim_{\varepsilon \to 0^+} \int_0^{\infty} \theta_{BA}(t)\, e^{i(\omega+i\varepsilon)t}\, dt \end{aligned} \tag{2.72}$$

Clearly $\chi_{BA}(-\omega) = \chi^*_{BA}(\omega)$, so that χ' and χ'' are, respectively, even and odd functions of frequency ; causality implies that the real and imaginary parts are related by the standard Kramers-Kronig relations. Combination of equ. (2.70) and (2.72) yields the fundamental relation :

$$\chi_{BA}(\omega) = [\langle BA\rangle + i\omega\tilde{C}_{BA}(\omega)\,] \tag{2.73}$$

An important case is when B = A. Then, according to (2.10) and (2.73) :

$$\hat{C}_{AA}(\omega) = \frac{k_B T}{\pi\omega}\, \chi''_{AA}(\omega) \tag{2.74}$$

which is a precise statement of the fluctuation-dissipation theorem. Use of the term "dissipation" is connected with the fact that the energy absorbed from the external field and later dissipated as heat is proportional to $\omega\chi''_{AA}(\omega) \geq 0$

As a textbook illustration of linear response theory, we consider the case of electrical conduction in an ionic system. Suppose that a time-dependent electric field is applied to a system containing N mobile charges $q_i = -Z_i e$; then :

$$H'_N(t) = -\vec{M}.\vec{E}(t) \tag{2.75}$$

where M is the total dipole moment :

$$\vec{M} = \sum_{i=1}^{N} q_i \vec{r}_i \tag{2.76}$$

The induced charge current per unit volume is the expectation value of the microscopic current :

$$j_Z(t) = \dot{\vec{M}}(t) = \sum_{i=1}^{N} q_i \vec{u}_i(t) \tag{2.77}$$

Choosing $\vec{E}$ along the x-axis, a straightforward application of the linear response equations (with $A \equiv M_x$ and $B \equiv j_Z^x$ yields :

$$J_Z^x(t) = \frac{1}{V} <j_Z^x(t)>_{n.e.} = R\sigma(\omega) E_o e^{-i\omega t} \tag{2.78}$$

where the complex a. c. conductivity is given by :

$$\sigma(\omega) = \frac{\beta}{V}\int_o^\infty <j_Z^x(t)\; j_Z^x(0)> e^{i\omega t}\, dt$$

$$= \frac{\omega_o^2}{4\pi}\int_o^\infty J_{ZZ}(t)\, e^{i\omega t}\, dt \tag{2.79}$$

where $J_{ZZ}(t)$ is the normalized electric current ACF ($J_{ZZ}(0) = 1$) is the plasma frequency :

$$\omega_p^2 = \frac{4\pi}{V}\sum_{i=1}^{N}\frac{q_i^2}{m_i} = \sum_\nu \frac{4\pi n_\nu q_\nu^2}{m_\nu} \tag{2.80}$$

where the second sum is over ionic species.
From (2.13) , (2.10) and (2.79) it follows that the real part of the conductivity satisfies the sum rule :

$$\int_{-\infty}^{+\infty}\frac{d\omega}{\pi}\sigma'(\omega) = \frac{\omega_p^2}{4\pi} \tag{2.81}$$

An approximate relation between the static conductivity σ ($\omega = 0$) and the mobilities of the various ionic species follows from equs. (2.48), (2.79) and the generalization of (2.64), namely :

$$\mu_\nu = \frac{q_\nu D_\nu}{k_B T} \tag{2.82}$$

where D_ν denotes the self diffusion coefficient of species ν , and the factor q_ν arises because we adopt the convention that the ionic mobility measures the response of a single ion to an applied electric field rather than the corresponding force. If all cross-correlations between the velocities of different ions (of the same or a different species) are neglected, i.e. if one assumes :

$$<\vec{u}_i(t) \cdot \vec{u}_j(0)> = 0 \ \forall t , \ i \neq j$$

then :

$$C_{ZZ}(t) = \frac{1}{N} <j_Z^x(t) \ j_Z^x(0)> = \sum_\nu x_\nu q_\nu^2 Z_\nu(t) \quad (2.83)$$

where $Z_\nu(t)$ denotes the velocity ACF of species ν and $x_\nu = N_\nu/N$ Integration over time then leads to the Nernst-Einstein relation :

$$\sigma = \frac{1}{k_B T} \sum_\nu n_\nu q_\nu^2 D_\nu = \sum_\nu n_\nu q_\nu \mu_\nu \quad (2.84)$$

which is well verified in electrolyte solutions.
As a final illustration of the linear response formalism, we consider the density response function $\chi(k,\omega) \equiv \chi_{\rho\rho}(k,)$ Let $\varphi(k)e^{-i\omega t}$ be a Fourier component of a complex external potential which couples to the particle density $\rho_{\vec{k}}$. The resulting fluctuation in density, denoted by $\Delta\rho(\vec{k},t)$ is then given by a direct application of (2.71) as :

$$\Delta\rho(\vec{k},t) = <\rho_{\vec{k}}(t)>_{n.e.} = \chi(k,\omega)\varphi(\vec{k}) \ e^{-i\omega t} \quad (2.85)$$

which generalizes the static result (1.65). The imaginary part of is related to the dynamic structure factor through the fluctuation-dissipation theorem (2.74), which in this case takes the form :

$$\chi''(k,\omega) = -\pi \beta n \omega S(k,\omega) \quad (2.86)$$

this is the frequency-dependent generalization of (1.66). Note that in view of equs. (2.73) and (1.66), $\chi(k) = \chi(\vec{k},\omega=0)$

2.4. Generalized hydrodynamics and mode coupling.

We finally turn to the question of how to actually calculate time correlation functions explicitly in a given physical situation. The general framework for any attempt in that direction is the memory function formalism summarized in equ. (2.58). Loosely speaking, the memory function can be approximately evaluated by three different approaches :

a) "Generalized hydrodynamics" : the validity of the equations of macroscopic hydrodynamics is extended to microscopic length and time scales by the introduction of wavenumber and frequency-dependent transport coefficients, which play the role of memory functions. Their decay is governed by phenomenological, wavenumber-dependent relaxation times. The correct short-time behaviour is incorporated by imposing exact frequency moment sum rules which are determined by static correlation functions (cf. equ. (2.13)). This procedure, which can be looked upon as an interpolation between the low (k, ω) (or hydrodynamic) and the high (k, ω) (or kinetic) regimes (36,37) will be briefly illustrated in the case of the transverse current ACF (2.24).

b) "Mode coupling" : in this first principles approach, the memory function itself is expressed in terms of products of correlation functions of pairs of conserved dynamic variables associated with hydrodynamic modes (38). This theory also uses static correlation functions as input (to ensure the correct short time behaviour via frequency sum rules) and naturally leads to a self-consistency problem (39). Mode coupling theories correctly predict the slow $(\sim t^{-3/2})$ decay of the correlation functions associated with non-conserved variables, like the velocity of an atom or the components of the stress tensor (40). In particular, the persistence of a "long-time tail" in the velocity ACF $Z(t)$ can be associated with a coupling between the motion of a single atom and the hydrodynamic modes of the system, as materialized by vortex (or backflow) patterns of the velocity field around a moving tagged particle. It is worth pointing out that mode coupling theory has recently been successfully applied to the slow structural relaxation near the glass transition (43).

c) "Kinetic theory" starts from a full phase space (rather than configuration space) description of microscopic dynamics. The time-honoured Boltzmann kinetic equation, which is only applicable to dilute gases, is generalized by introducing a non-markovian collision kernel to cope with non-local effects both in time and in space ; this collision kernel plays the role of a generalized memory function. Generalized kinetic equations have been successfully applied to the microscopic dynamics of the hard sphere fluid which leads to considerable simplifications because of the instantaneous nature of the collisions between particles (26, 41). The formalism can be extended to the case of continuous potentials, by means of some additional approximations (42). It is interesting to note that in order to render the kinetic calculations tractable, a factorization ansatz is generally made (the so-called "disconnected approximation") the physical content of which is very similar to that of mode coupling theories. We conclude this lecture by an illustration of the new insight into microscopic dynamics which may be gained from a "generalized hydrodynamics" approach. The familiar Navier-Stokes equation for the velocity field $\vec{u}(\vec{r},t)$ in a fluid reads, in its linearized form (32) :

$$\rho_m \dot{\vec{u}}(\vec{r},t) + \vec{\nabla} P(\vec{r},t) - \eta \nabla^2 \vec{u}(\vec{r},t) - (\tfrac{1}{3}\eta + \zeta)\vec{\nabla}\vec{\nabla}.\vec{u}(\vec{r},t) = 0 \tag{2.87}$$

where $\rho_m = n\ \ m$ is the macroscopic mass density, P is the local pressure, and η and ζ are the shear and bulk viscosities. Taking the spatial Fourier transform we derive from the transverse projection of equ. (2.87) the following equation for the transverse part of the particle current $\vec{j_k} = n\vec{u}_{\vec{k}}$ (k//0z):

$$\frac{\partial}{\partial t} j^x_{\vec{k}}(t) + \nu k^2 j^x_{\vec{k}}(t) = 0 \tag{2.88}$$

where $\nu = \eta/\rho_m$ is the kinematic viscosity. Multiplying through equ. (2.88) by $j_{-\vec{k}}$, and taking a thermal average, we obtain an equation for C_t (k,t) :

$$\frac{\partial}{\partial t} C_t(k,t) + \nu k^2 C_t(k,t) = 0 \tag{2.89}$$

which is of the same form as the diffusion equation (2.36) : it describes the diffusion of transverse momentum. The solution for the Laplace transform is, remembering (2.27) :

$$\tilde{C}_t(k,z) = \frac{\omega_o^2}{-iz + \nu k^2} \tag{2.90}$$

so that the kinematic viscosity is given by the limit (28) :

$$\nu = \beta m \lim_{\omega \to 0} \lim_{k \to 0} \frac{\omega^2}{k^4} R\, \tilde{C}_t(k,\omega)$$
$$= \pi\beta m \lim_{\omega \to 0} \lim_{k \to 0} \frac{\omega^2}{k^4} \hat{C}_t(k,\omega) \tag{2.91}$$

Proceeding as before, in going from (2.41) to (2.48), equ. (2.91) can be cast in the equivalent Green-Kubo form (7,26):

$$\eta = \frac{\beta}{V} \int_0^\infty \langle \sigma_o^{xz}(t)\, \sigma_o^{xz}(0) \rangle \, dt \tag{2.92}$$

where σ_o^{xy} is the $k \to 0$ limit of an off-diagonal component of the stress tensor.

The hydrodynamic equation (2.89) can now be generalized by incorporating non-markovian (retarded) effects in time, which become important in the kinetic regime (37) :

$$\frac{\partial}{\partial t} C_t(k,t) + k^2 \int_0^t (k,t-s)\, C_t(k,s)\, ds = 0 \tag{2.93}$$

The Kernel $\nu(k, t)$ is a memory function and its Laplace transform $\nu(k,\ )$ plays the role of a generalized kinematic viscosity. Taking the Laplace transform of (2.93) leads to the following generalization of (2.90) :

$$\tilde{C}_t(k,z) = \frac{\omega_o^2}{-iz + k^2\, \tilde{\nu}(k,z)} \tag{2.94}$$

with
$$\nu = \lim_{\omega \to 0} \lim_{k \to 0} \tilde{\nu}(k,\omega) \tag{2.95}$$

A straightforward calculation shows that (7) :

$$\nu(k,t=0) = \frac{1}{\rho m} G_\infty(k) \tag{2.96}$$

where the instantaneous shear modulus $G_\infty(k)$ can be calculated from a knowledge of the pair distribution function. If we make the single relaxation time ansatz :

$$\nu(k,t) = \frac{1}{\rho m} G_\infty(k) \exp\{-t/\tau_t(k)\} \tag{2.97}$$

we obtain immediately from (2.94) that :

$$\hat{C}_t(k,\omega) = \frac{1}{\pi} R\, \tilde{C}_t(k,\omega)$$

$$= \frac{1}{\pi} \frac{\omega_o^2 \omega_{1t}^2 \tau_t}{\omega^2 + \tau_t^2(\omega_{1t}^2 - \omega^2)^2} \tag{2.98}$$

where $\omega_{1t}^2(k) = k^2 G_\infty(k)/\rho m$. It is easy to verify that the spectrum $\hat{C}_t(k,\omega)$ exhibits a peak at non-zero frequency whenever :

$$\omega_{1t}^2(k)\,\tau_t^2(k) > \frac{1}{2} \tag{2.99}$$

In other words (2.99) implies that shear waves will propagate in the liquid for wavenumbers

$$k > k_c = \tau_t^{-1}(k) \left[\frac{\rho m}{2\, G_\infty(k)}\right] \tag{2.100}$$

in agreement with the results of computer simulations (44) These shear waves are characteristic of visco-elastic behaviour of the liquid.

3. IONIC LIQUIDS.

3.1 Models of ionic solutions

An electrolyte solution is, in the simplest cases, a three-component system involving a solvent, usually water, and a solute made up of positive cations and negative anions. These ions carry electric charges, while the solvent is composed of highly polar, hydrogen-bonded molecules. The three species will be labelled by the index $\nu = 0$ (solvent), 1 (cations of charge $q_1 = Z_1 e > 0$) and 2 (anions of charge $q_2 = Z_2 e < 0$) ; the corresponding number densities are $n_\nu = N_\nu/V$ and the total ionic number density will be denoted by $n_i = n_1 + n_2$. Overall charge neutrality requires that :

$$Z_1 n_1 + Z_2 n_2 = 0 \tag{3.1}$$

The total potential energy of the system is the sum of solvent-solvent, solute-solute and solvent-solute terms :

$$V_N = V_{N_o} + V_{N_i} + V_{N_o,N_i} \qquad (3.2)$$

However, a quantitative study of the pure polar solvent already represents a formidable challenge, so that a full first-principles description of the three-component problem has only recently been undertaken in a systematic way. Because of the difficulty of such a description, attempts have been made long ago to reduce the complex three-component system to an effective two-component problem involving only solute coordinates. The reduction can be carried out formally (45) and leads to a solvent-averaged potential energy $\bar{V}_{Ni}$; the price to pay is that $\bar{V}_{Ni}$ is no longer pairwise additive, but contains triplet terms, etc. :

$$\bar{V}_{N_i} = \sum_{i<j}\sum \bar{v}_2(i,j) + \sum_{i<j<k}\sum\sum \bar{v}_3(i,j,k) + \ldots \qquad (3.3)$$

where the $\bar{v}_2$, $\bar{v}_3$, ... are effective pair, triplet,... potentials which are generally temperature and concentration-dependent. In the limit of vanishing solute concentration, we expect that only the pair interaction survives and behaves as :

$$\bar{v}_2(i,j) \simeq \frac{q_i q_j}{\varepsilon_s r_{ij}} \qquad (3.4)$$

when the separation $r_{ij} = |\vec{r}_i - \vec{r}_j|$ between ions goes to infinity ε_s denotes the dielectric constant of the pure solvent, describing the extent to which the bare Coulomb interaction between ions is shielded by the dielectric solvent medium.

In the so-called "primitive model" (PM) of electrolytes, one neglects all higher than pairwise terms in (3.3) and extends the asymptotic form (3.4) of the effective pair potential to finite solute concentrations and interionic distances. In its simplest version the model is one of impenetrable hard spheres ; dropping the bar which symbolizes averaging over solvent effects, we write the PM pair potential as :

$$v_{\nu\mu}(r) = \infty \ ; \ r < \sigma_{\nu\mu} = \frac{1}{2}(\sigma_\nu + \sigma_\mu)$$

$$= \frac{q_\nu q_\mu}{\varepsilon_s r} \ ; \ r > \sigma_{\nu\mu} \ ; \ 1 \leq \nu,\mu \leq 2 \qquad (3.5)$$

More generally $v_{\nu\mu}(r)$ is the sum of a short-range potential $v_{\nu\mu}^{(s)}(r)$ and of the Coulombic term. The PM is a dielectric continum model since the solvent comes in only through the macroscopic dielectric shielding of the Coulomb potential. It cannot cope with specific solvent effects, like ion hydration ; "discrete solvent" descriptions will be the subject of lecture 4.

3.2. Charge response functions.

The Fourier transform of the Coulomb potential :

$$\tilde{v}^{(c)}_{\mu}(k) = \frac{4\pi q_\nu q_\mu}{\varepsilon_s k^2} \qquad (3.6)$$

is singular in the k → 0 limit. This characteristic "Coulomb singularity", which is a manifestation of the infinite range of the Coulomb potential, has a profound influence on the long wavelength response of ionic fluids. This response can be characterized by a charge response function $\chi_{ZZ}(k,\omega)$ which generalizes the density response function (2.85). The dynamical variables of interest are the charge density and current which are the following linear combinations of the partial densities $\rho_{\vec{k}\nu}$ and currents $\vec{j}_{\vec{k}\nu}$ (ν = 1,2) associated with the two ionic species :

$$\begin{aligned} \rho_{\vec{k}Z}(t) &= Z_1\,\rho_{\vec{k}1}(t) + Z_2\,\rho_{\vec{k}2}(t) \\ \vec{j}_{\vec{k}Z}(t) &= Z_1\,\vec{j}_{\vec{k}1}(t) + Z_2\,\vec{j}_{\vec{k}2}(t) \end{aligned} \qquad (3.7)$$

They satisfy the continuity equation (2.20). Note that, while the charge density is a conserved variable in the k → 0 limit, this is not true of the charge current (2.77), due to Ohmic dissipation originating in inter-ionic collisions and solvent friction. In a PM description, solvent friction may be simulated by a stochastic force ("Brownian dynamics" (46)).

The mean induced charge density is linearily related to the external electric potential according to a generalization of (2.85) :

$$\Delta\rho_Z(\vec{k},t) = \langle\rho_{\vec{k}Z}(t)\rangle_{n.e} = \chi_{ZZ}(k,\omega)\; e\,\varphi^{ex}(\vec{k})\; e^{-i\omega t} \qquad (3.8)$$

The external potential derives from an external charge density $\rho_Z^{ex}(k)$ according to Poisson's equation :

$$k^2\,\varphi^{ex}(\vec{k},t) = 4\pi\, e\, \rho_Z^{ex}(k,t) \qquad (3.9)$$

The response to a Fourier component of the external charge density can, alternatively, be described by the inverse (longitudinal) dielectric function :

$$\frac{1}{\varepsilon(k,\omega)} = \frac{\vec{k}.\vec{E}(\vec{k},\omega)}{\vec{k}.\vec{D}(\vec{k},\omega)} = \frac{\rho_Z(\vec{k},\omega)}{\rho_Z^{ex}(\vec{k},\omega)} = 1 + \frac{\Delta\rho_Z(\vec{k},\omega)}{\rho_Z^{ex}(\vec{k},\omega)} \qquad (3.10)$$

where $\vec{E}$ and $\vec{D}$ denote the electric field and displacement vectors, $\rho_Z = \rho_Z^{ex} + \Delta\rho$ is the total (external plus induced) charge density, and use was made of Maxwell's equations. Combination of equs. (3.8), (3.9) and (3.10) leads immediately to the relation :

$$\frac{1}{\varepsilon(\vec{k},\omega)} = 1 + \frac{4\pi e^2}{k^2} \chi_{ZZ}(k,\omega) \tag{3.11}$$

The charge response function is related to the charge dynamic structure factor $S_{ZZ}(k,\omega)$, which characterizes the spontaneous (thermal) fluctuations in the ionic system in equilibrium, by the fluctuation-dissipation relation (2.86).

The functions χ_{ZZ} and ε^{-1} measure the linear response of a fluid of charged particles to an external electric field. This external field polarizes the fluid and thus gives rise to a local internal (Maxwell) field which is the superposition of the field due to the external charge distribution and to the induced charge density ; this local (or screened) electric field is of course the field experienced by the ions. The response of the system to the local electric potential is described by a screened response function χ_{ZZ} according to :

$$\Delta\rho_Z(\vec{k},t) = e\tilde{\chi}_{ZZ}(\vec{k},\omega)[\varphi^{ex}(\vec{k})\, e^{-i\omega t} + \Delta\varphi(\vec{k},t)] \tag{3.12}$$

where $\Delta\varphi$ and $\Delta\rho_Z$ are also related by Poisson's equation (3.9). Confrontation of equs. (3.8) and (3.12) yields the relation between external and screened response functions :

$$\chi_{ZZ}(\vec{k},\omega) = \frac{\tilde{\chi}_{ZZ}(\vec{k},\omega)}{1 - \frac{4\pi e^2}{k^2}\tilde{\chi}_{ZZ}(\vec{k},\omega)} \tag{3.13}$$

while :

$$\varepsilon(\vec{k},\omega) = 1 - \frac{4\pi e^2}{k^2}\tilde{\chi}_{ZZ}(\vec{k},\omega) \tag{3.14}$$

Alternatively, the electric response of an ionic fluid can be measured by the induced electric current. If $\vec{E}(\vec{k},\omega)$ denotes a Fourier component of the local electric field, Ohm's law states that the induced current per unit volume is linearily related to $\vec{E}$:

$$\vec{J}_Z(\vec{k},\omega) = \langle e j_{\vec{k}Z}(\omega)\rangle_{n.e} = \overleftrightarrow{\sigma}(\vec{k},\omega).\vec{E}(\vec{k},\omega) \tag{3.15}$$

where the conductivity tensor can be separated into its longitudinal and transverse parts, in a manner analogous to equ. (2.22). Since the longitudinal current is related to the charge density by the continuity equation, while the longitudinal electric field derives from the local electric potential, it is easily checked that ε and σ_ℓ are related by the fundamental equation :

$$\varepsilon(\vec{k},\omega) = 1 + \frac{4\pi i}{\omega}\sigma_\ell(\vec{k},\omega) \tag{3.16}$$

It must be stressed that σ_ℓ is a screened response function, in the same sense as ε or χ_{ZZ}, since it measures the response to the internal field (47).

We now specialize to the static ($\omega = 0$) response. Generalizing (1.66) we have that :

$$\chi_{ZZ}(\vec{k}) = -\beta n_i \, S_{ZZ}(\vec{k}) \tag{3.17}$$

where $S_{ZZ}(\vec{k})$ is the charge structure factor, which is related to the partial ionic structure factors by :

$$\begin{aligned} S_{ZZ}(\vec{k}) &= \frac{1}{N} \langle \rho_{\vec{k}Z} \, \rho_{-\vec{k}Z} \rangle \\ &= \sum_{\nu} \sum_{\mu} Z_{\nu} Z_{\mu} \, S_{\nu\mu}(\vec{k}) \end{aligned} \tag{3.18}$$

where

$$\begin{aligned} S_{\nu\mu}(\vec{k}) &= \frac{1}{N} \langle \rho_{\vec{k}\nu} \, \rho_{-\vec{k}\mu} \rangle \\ &= x_{\nu} \delta_{\nu\mu} + x_{\nu} x_{\mu} n_i \hat{h}_{\nu\mu}(\vec{k}) \end{aligned} \tag{3.19}$$

In equ. (3.19), $x_\nu = n_\nu / n_i$ is the ionic number concentration of species ν and $\hat{h}_{\nu\mu}(\vec{k})$ is the Fourier transform of the partial correlation function $h_{\nu\mu}(\vec{r}) = g_{\nu\mu}(r) - 1$, which is defined by the multi-component version of equ. (1.19). The OZ relation (1.32) is generalized with the introduction of partial d.c.f. $c_{\nu\mu}(\vec{r})$; in Fourier space this takes the form :

$$\hat{h}_{\nu\mu}(\vec{k}) = \hat{c}_{\nu\mu}(\vec{k}) + \sum_{\lambda} n_{\lambda} \hat{c}_{\nu\lambda}(\vec{k}) \, \hat{h}_{\lambda\mu}(\vec{k}) \tag{3.20}$$

Since $S_{ZZ}(\vec{k})$ is non-negative, combination of equs. (3.14) (with $\omega = 0$) and (3.17) implies the stability condition :

$$\frac{1}{\varepsilon(\vec{k})} < 1 \tag{3.21}$$

An external charge distribution is completely screened by a <u>conducting</u> fluid, which means that the sum of external and induced charge densities must vanish in the long wavelength limit :

$$\lim_{k \to 0} \left[\rho_Z^{ex}(\vec{k}) + \Delta\rho_Z(\vec{k}) \right] = 0 \tag{3.22}$$

According to equ. (3.10) this is equivalent to the assumption that for a conducting fluid :

$$\lim_{k \to 0} \varepsilon(\vec{k}) = \infty \tag{3.23}$$

Equs. (3.11) and (3.17) make it clear that the perfect screening property determines the long wavelength behaviour of the charge structure factor :

$$\lim_{k \to 0} \frac{k_D^2}{k^2} S_{ZZ}(k) = \sum_{\nu} x_{\nu} Z_{\nu}^2 \tag{3.24}$$

where k_D is the so-called Debye wavenumber :

$$k_D^2 = \sum_{\nu} k_{D\nu}^2 = \sum_{\nu} 4\pi n_{\nu} Z_{\nu}^2 \; e^2/k_B T \tag{3.25}$$

It will be shown in the next section that the Debye length $\lambda_D = 1/k_D$ is the distance beyond which the electric potential due to an ion is effectively screened by the local, induced charge density, in dilute electrolyte solutions. Equ. (3.24) implies directly two important sum rules obeyed by the partial pair distribution functions. According to (3.18) and (3.19) :

$$S_{ZZ}(k) = \sum_{\nu} \sum_{\mu} Z_{\nu} Z_{\mu} \left[x_{\nu} \delta_{\nu\mu} + x_{\nu} x_{\mu} 4\pi n_i \int_0^{\infty} \frac{\sin(kr)}{kr} h_{\nu\mu}(r) r^2 dr \right] \tag{3.26}$$

If the $h_{\nu\mu}(r)$ decay sufficiently fast ("clustering property") the Fourier integrals can be expanded to order k^2. A comparison of the coefficients of the zeroth and second powers of k in equs. (3.24) and (3.26) leads then directly to the Stillinger-Lovett (48) sum rules :

$$\sum_{\nu} x_{\nu} Z_{\nu} n_i \int \sum_{\mu} x_{\mu} Z_{\mu} g_{\nu\mu}(r) \, d\vec{r} = -\sum_{\nu} x_{\nu} Z_{\nu}^2 \tag{3.27}$$

$$\sum_{\nu} x_{\nu} Z_{\nu} n_i \int \sum_{\mu} x_{\mu} Z_{\mu} g_{\nu\mu}(r) \, r^2 d\vec{r} = -6\lambda_D^2 \sum_{\nu} x_{\nu} Z_{\nu}^2 \tag{3.28}$$

The first sum rule is nothing but a linear combination of the local electroneutrality conditions :

$$n_i \int \Big[\sum_{\mu} x_{\mu} Z_{\mu} g_{\nu\mu}(r)\Big] d\vec{r} = -Z_{\nu} \tag{3.29}$$

The sum rules (3.27) and (3.28) can in fact be derived rigorously from the Yvon-Born-Green hierarchy for the n-particle densities (7), if appropriate "clustering" assumptions are made (49). Comparing the small k behaviour (3.24) of $S_{zz}(k)$ with the compressibility equation (1.70) for a one-component neutral fluid, we see intuitively that large scale (long wavelength) charge fluctuations are strongly inhibited, compared to the number density fluctuations of a neutral fluid. In fact it has been proved rigorously that the fluctuation of the total charge Q_V contained in a volume V, i.e. $\langle Q_V^2 \rangle - \langle Q_V \rangle^2$, is proportional to the surface S bounding the volume, and not to V itself, as is the case for the number density fluctuations (50).

It is important to realize that the validity of the Stillinger-Lovett conditions is model-independent and also holds within a discrete solvent description. Also note that the limiting behaviour (3.24) of S_{zz} (k) follows from the asymptotic behaviour :

$$c_{\nu\mu}(r) \underset{r \to \infty}{\simeq} -\beta\bar{v}_{\nu\mu}(r) \simeq -\frac{\beta Z_\nu Z_\mu e^2}{\varepsilon_s r} \tag{3.30}$$

of the partial d.c.f..

3.3. Approximations and limiting laws.

The usual approximations for ionic distribution functions are based on the Poisson-Boltzmann equation. The spirit is similar to the derivation of the HNC equation in section 1.4. The charge density around a central ion (looked upon as an "external" charge) is related to the mean electrostatic potential by the Boltzmann distribution, provided correlations between the ions in the "polarization cloud" induced by the central ion are neglected :

$$\rho_Z^\nu(r) = \sum_\mu n_\mu q_\mu \; [g_{\nu\mu}(r) - 1]$$
$$\simeq \sum_\mu n_\mu q_\mu \; [\exp\{-\beta\,[v_{\nu\mu}(r) + q_\mu \varphi^\nu(r)]\} - 1]$$

where $v_{\nu\mu}$ (r) is the total (short-range + Coulomb) potential between ions of species ν and μ, while $\varphi^\nu(r)$ is the mean electric potential of the "polarization cloud" around the central ion. The latter satisfies Poisson's equation :

$$\nabla^2\varphi^\nu(r) = -\frac{4\pi}{\varepsilon_s}\rho_Z^\nu(r) \tag{3.32}$$

Bearing in mind that the Coulomb potential $v_\mu^{(c)}(r) = Z_\mu e/\varepsilon_s r$ is the Green's function of the Laplace operator, we solve (3.32) according to :

$$Z_\mu e\,\varphi^\nu(r) = \int \rho_Z^\nu(r')\, v_\mu^{(c)}(|\vec{r}-\vec{r}'|)\, d\vec{r}'$$
$$= \sum_\lambda n_\lambda h_{\nu\lambda} * v_{\lambda\mu}^{(c)} \tag{3.33}$$

It follows from (3.31) that the p.d.f. satisfy the closed set of the non-linear integral equations :

$$g_{\nu\mu}(r) = \exp\,\{-\beta[v_{\nu\mu}(r) + \sum_\lambda n_\lambda h_{\nu\lambda} * v_{\lambda\mu}^{(c)}]\} \tag{3.34}$$

Comparison with the multi-component generalization of (1 53) shows that the Poisson-Boltzmann approximation can be derived from the HNC closure provided the d.c.f. $c_{\lambda\mu}$ in the latter are replaced by their asymptotic form (3.30).

Linearization of the exponential in equ. (3.34) or in equs. (3.31) and (3.32) leads to equations which are easily solved to yield the well-known Debye-Hückel limiting law In particular the low concentration limit of the osmotic pressure P follows directly from the virial theorem :

$$\frac{\beta P}{n_i} = 1 - \frac{k_D^3}{24\pi n_i} \tag{3.35}$$

where k_D is the Debye wavenumber defined in (3.25). The corresponding charge structure factor is :

$$S_{ZZ}^{DH} = \frac{k^2}{k^2 + k_D^2}\left(\sum_\nu x_\nu z_\nu^2\right) \tag{3.36}$$

so that, according to (3.14) and (3.17) :

$$\varepsilon^{DH}(k) = 1 + \frac{k_D^2}{k^2} \tag{3.37}$$

Replacement in equ. (3.10) clearly shows that an external charge distribution is screened beyond a distance $\lambda_D = 1/k_D$, i.e. the charge distribution function associated with (3.36) decays exponentially beyond $r \simeq \lambda_D$. For <u>concentrated</u> electrolytes, the screening length is renormalized by spatial correlations among ions(51). These correlations are approximately accounted for in the HNC closure (1.53), but a simpler, fully analytic description is obtained from the MSA closure (1.59) for the PM (3.5). For equal size hard spheres carrying opposite charges ("restricted primitive model", or RPM), the MSA equations are :

$$g_{\nu\mu}(r) = 0 \qquad ; \quad r < \sigma$$

$$C_{\nu\mu}^{(s)}(r) = C_{\nu\mu}(r) + \frac{\beta q_\nu q_\mu}{\varepsilon_s r} = 0 \quad ; \quad r > \sigma \tag{3.38}$$

where $c_{\nu\mu}^{(s)}(r)$ denotes the short-range part of the direct correlation function. Equations (3.38) and (3.20) form a closed set which has been solved analytically (52). Since the RPM is symmetric with respect to charge conjugation, $h_{11}(r) \equiv h_{22}(r)$, and the OZ equations (3.20) are easily decoupled by the linear transformation :

$$h^{(1)}(r) = \frac{1}{2}\left[h_{11}(r) + h_{12}(r)\right] \tag{3.39a}$$

$$h^{(2)}(r) = h_{11}(r) - h_{12}(r) \tag{3.39b}$$

The corresponding direct correlation functions $c^{(1)(S)}(r)$ and $c^{(2)(S)}(r)$ turn out to be polynomials of degree 3 (PY solution for neutral hard spheres (19)) and 1 for $r < \sigma$. The excess internal energy per ion (with $Z = Z_1 = - Z_2$) :

$$\frac{U^{ex}}{N_i} = \frac{2\pi Z^2 e^2}{\varepsilon_s} \; n_i \int h^{(2)}(r)\, r\, dr$$
$$= -\frac{Z^2 e^2}{\varepsilon_s \sigma} \; \frac{[\xi + 1 - (1 + 2\xi)^{1/2}]}{\xi} \qquad (3.40)$$

is seen to depend on a single coupling constant, the reduced wavenumber $\xi = k_D\sigma = 4\pi\, n_i^* \beta_i^*$, and not on the reduced density, $n_i^* = n_i \sigma^3$, and on the reduced inverse temperature, $\beta^* = Z^2 e^2/(\varepsilon_s \sigma\, k_B T)$, separately. The equation-of-state is the sum of ideal, Coulomb and "contact" (ionic repulsion) terms:

$$\frac{\beta P}{n_i} = 1 + \frac{\beta U^{ex}}{3N_i} + \frac{2\pi}{3} n_i^* \, g^{(1)}(r=\sigma^+)$$
$$= 1 + \frac{\beta U_i^{ex}}{3N_i} + \frac{2\eta(2-\eta)}{(1-\eta)^2} \qquad (3.41)$$

where $\eta = \pi n_i^*/6$ is the total packing fraction of the ions. In the high temperature or low concentration limit,i.e. for $\xi \ll 1$, the MSA internal energy (3.40) reduces to the Debye-Hückel limiting law :

$$\frac{U_{DH}^{ex}}{N_i} = -\frac{Z^2 e^2}{\varepsilon_s \sigma} \xi = -\frac{k_B T}{4\pi n_i} k_D^3 \qquad (3.42)$$

This law corresponds in fact to the point ion limit ($\sigma \to 0$), where the d.c.f. are identified with their asymptotic values $-\beta v_{\nu\mu}(r)$ for <u>all</u> r, in the spirit of the RPA. In the same limit, the osmotic pressure decreases monotonously with increasing concentration, but at finite concentrations the repulsive contribution in (3.41) leads to a minimum and a subsequent increase of the osmotic pressure. In the high concentration limit ($\xi \gg 1$), the MSA energy (3.40) has a saturation property, in agreement with an exact lower bound due to Onsager (53) :

$$\frac{U^{ex}}{N_i} \geq -\frac{Z^2 e^2}{\varepsilon_s \sigma} \qquad (3.43)$$

The analytic solution of the MSA for the PM has been extended to arbitrary ionic mixtures (i.e. arbitrary charges and diameters) by Blum (54). By adjusting the ion diameters σ_ν, these MSA results can be used to fit the experimental osmotic pressures of many electrolytes over a wide range of concentrations (55). An extension of the PM to non-additive diameters ($\sigma_{12} \lessgtr (\sigma_1 + \sigma_2)/2$) has been analysed within the MSA and HNC approximations to study complex formation by hydrated ions in concentrated aqueous solutions of Cd and N_i phosphate and of Cd sulphate (56). It must however be kept in mind that the (analytic) MSA is less accurate than the numerical solutions of the HNC approximation, particularly at high concentrations. The main defect of the MSA is that it leads to a complete decoupling between density and charge fluctuations, so that the compressibility equation (1.70) (which remains valid for the structure factor associated with the number density) leads to the PY equation-of-state of the underlying uncharged hard sphere fluid, thus stressing the thermodynamic inconsistency of the MSA.

3.4. Primitive models of electric double layers.

The most important example of an inhomogeneous ionic fluid is the electric double layer, an interface region characterized by a separation of opposite charges, giving rise to a charge density profile :

$$\rho_Z^{(1)}(\vec{r}) = \sum_\nu q_\nu \rho_\nu^{(1)}(\vec{r}) \tag{3.44}$$

Common examples are the electrode-electrolyte interface, charge-stabilized colloids or biological membranes. We shall restrict our attention to the planar interface between an impenetrable wall carrying a surface charge σ and an ionic solution. The density profiles are then functions of the normal coordinate, say Z ; the wall is at $Z = 0$. Overall electro-neutrality requires that :

$$\sigma + \int_0^\infty \rho_Z^{(1)}(z)\, dz = 0 \tag{3.45}$$

The mean electrostatic potential $\varphi(z)$ satisfies Poisson's equation :

$$\frac{d^2\varphi(z)}{dz^2} = -\frac{4\pi}{\varepsilon_s}\rho_Z^{(1)}(z) \tag{3.46}$$

which can be integrated with the result :

$$\varphi(z) = \frac{4\pi}{\varepsilon_s}\int_z^\infty (z - z')\, \rho_Z^{(1)}(z')\, dz' \tag{3.47}$$

where use was made of the boundary conditions :

$$\lim_{z\to\infty} \varphi(z) = 0 \tag{3.48a}$$

$$\left.\frac{d\varphi(z)}{dz}\right|_{z=0} = -\frac{4\pi\sigma}{\varepsilon_s} \tag{3.48b}$$

In particular the surface potential $\varphi_0 = \varphi(z=0)$ is given by the dipole moment in the normal direction of the double layer. The local electroneutrality conditions (3.27) for the bulk takes here the form :

$$q_\nu \rho_\nu^{(1)}(z) + \int_{z'>0} d\vec{r}' \sum_\mu q_\mu \rho_\nu^{(1)}(z)\; \rho_\mu^{(1)}(z')\, h^{(2)}(\vec{r},\vec{r}') = 0 \tag{3.49}$$

A similar generalization can be worked out for the perfect screening condition (3.28) (58). The density profiles and mean potential are related by exact equations derived from (1.36) :

$$\frac{d}{dz} \ln \rho_{\nu}^{(1)}(z) = -\beta \frac{d}{dz} v_{o\nu}^{(s)}(z) - \beta q_{\nu} \frac{d}{dz} \varphi(z) - \int d\vec{r}' \sum_{\mu} \rho_{\mu}^{(1)}(z') \frac{\partial}{\partial z'} c_{\nu\mu}^{(s)}(\vec{r},\vec{r}') \quad (3.50)$$

where $v_{o\nu}^{(s)}$ and $c_{\nu\mu}^{(s)}$ are the short-range (non-Coulombic) parts of the wall-ion potentials and of the ion-ion d.c.f.
The mean dipole moment of the excess charge distribution around an ion of species ν at a distance Z from the wall is defined by :

$$\mu_{\nu}(z) = \int (z' - z) \sum_{\mu} q_{\mu} \rho_{\nu}^{(1)}(z) \rho_{\mu}^{(1)}(z') h(\vec{r},\vec{r}') d\vec{r}'$$

It can be proved that this dipole moment vanishes in the bulk ($Z \to \infty$), provided h decays faster than $|\vec{r} - \vec{r}'|^{-3}$ in all directions (59). This is no more true in the double layer region, where (60) :

$$\frac{\partial \rho_{\nu}(z)}{\partial \sigma} = \frac{4\pi\beta}{\varepsilon_s} \mu_{\nu}(z) \quad (3.52)$$

so that $\mu_{\nu}(Z)$ does not vanish, indicating that h must decay as $|\vec{r} - \vec{r}'|^{-3}$ (or slower), in at least one direction. In fact it can be shown that correlations decay as $|\vec{r} - \vec{r}'|^{-3}$ parallel to the surface, as one would intuitively expect in view of the dipole moment created by charge separation ; in other words spatial correlations are screened in the Z-direction, but not parallel to the interface plane (61).

The double layer equivalent of the Poisson-Boltzmann theory for the bulk, sketched in section 3.3, was formulated by Gouy and Chapman (GC) about 15 years before the classic 1923 paper by Debye and Hückel. As in equ. (3.31), GC theory supplements the exact relation (3.46) by the mean field ansatz :

$$\rho_{\nu}^{(1)}(z) = n_{\nu} \exp\{-\beta q_{\nu} \varphi(z)\} \quad (3.53)$$

Linearization (valid for $\varphi < 20$ mV at room temperature) yields an exponential decay of $\varphi(Z)$, with the Debye screening length λ_D (cf. (3.25)). However the non-linear Poisson-Boltzmann equation may be integrated exactly for symmetric electrolytes ($q_1 = -q_2 = q$; $n_1 = n_2 = n$). Combination of (3.46) and (3.53) leads to the non-linear differential equation :

$$\frac{d^2\Phi(z)}{dz} = k_D^2 \sinh \Phi(z) \quad (3.54)$$

where $\Phi = \beta q \varphi$. Multiplying both sides by $d\Phi(Z)/dZ$, we can integrate between Z and ∞ (where Φ vanishes), with the result :

$$\left[\frac{d\Phi(z)}{dz}\right]^2 = 2k_D^2 \left\{\cosh[\Phi(z)] - 1\right\} = 4k_D^2 \sinh^2 [\tfrac{1}{2}\Phi(z)]$$

Setting $Z = 0$, and using (3.48b) we arrive at an important relation between the surface charge σ and the surface potential φ_o, namely :

$$\frac{2\pi\sigma}{\varepsilon_s} = \frac{k_D k_B T}{q} \sinh \left[\frac{q\varphi_o}{k_B T}\right] \tag{3.56}$$

Combination of equs. (3.53) and (3.56) yields the following ionic contact density :

$$\sum_\nu \rho_\nu^{(1)}(0) = \sum_\nu n_\nu + \frac{2\pi\sigma^2}{\varepsilon_s k_B T} \tag{3.57}$$

Taking the square root of (3.55) and integrating once more we obtain the explicit Z-dependence of the reduced potential:

$$\Phi(z) = 4 \tanh^{-1} \left\{\tanh(\tfrac{1}{4}\Phi_o)\, e^{-k_D z}\right\} \tag{3.58}$$

The potential is low when σ is small or the concentration is large (so that k_D is large), and (3.58) reduces then to the simple exponential decay of linearized GC theory. The differential capacitance can be calculated from (3.56) according to :

$$C = \left(\frac{\partial\sigma}{\partial\varphi_o}\right)_{T,n_i} \tag{3.59}$$

Note that in the linear regime the double layer behaves as a condenser, the screening length λ_D playing the role of the distance between plates.

GC theory ignores all ion size effects. Various ways of improving the Poisson-Boltzmann scheme are considered in the excellent review by Carnie and Torrie (57).The simplest correction is to exclude ions (assumed to have all a diameter d) from the region $0 < Z < d/2$ in the vicinity of the wall. According (3.46) and (3.48b) the relation between the surface potential φ and the potential at the plane of closest approach is :

$$\varphi_o = \varphi(d/2) + \frac{4\pi\sigma}{\varepsilon_s} \tag{3.60}$$

while further out the potential of this modified Gouy-Chapman (MGC) theory is simply given by the transformation $\varphi^{GC}(z) \to \varphi^{MGC}(\frac{d}{2} + z)$. The contact relation (3.57) becomes:

$$\sum_\nu \rho_\nu^{(1)}(\tfrac{d}{2}) = \sum_\nu n_\nu + \frac{2\pi\sigma^2}{\varepsilon_s k_B T} \tag{3.61}$$

which follows from an exact result (62) if the osmotic pressure, $P/k_B T$, appearing as the first term in the exact relation, is replaced by its ideal solution value $\sum_\nu n_\nu$

It is much less straightforward task to account for ion-ion correlations due to their finite size and their Coulombic interactions. The most powerful schemes are based on generalizations of the integral equations for bulk p.d.f. (see section 1.4). In particular, the HNC-like closure (1.61) becomes in the present multi-component case :

$$\begin{aligned}\rho_\nu^{(1)}(z) &= n_\nu \exp\left\{-\beta q_\nu \varphi(z) + \sum_\mu c_{\nu\mu}^{(s)} * [\rho_\mu^{(1)} - n_\mu]\right\} \\ &= n_\nu \exp\left\{-\beta q_\nu \varphi(z) + \sum_\mu 2\pi \int_{-\infty}^{+\infty} dz' [\rho_\mu^{(1)}(z') - n_\mu] \right. \\ &\quad \left. \times \int_{|z-z'|}^{\infty} c_{\mu\nu}^{(s)}(r)\, r\, dr\right\} \qquad (3.62)\end{aligned}$$

where $c_{\nu\mu}^{(s)}(r)$ denotes the short-range part (3.28) of the ion-ion d.c.f. in the bulk. Linearization of the exponential in (3.62) leads to an MSA-like closure. In the point-ion limit $c^{(s)} \equiv 0$ and the HNC and MSA-like closures reduce to GC and linearized GC closures respectively. For consistency, the bulk d.c.f. appearing in (3.62) or in the MSA-like closure should be replaced by the functions derived from the corresponding bulk HNC and MSA closures (1.54) and (1.58). The MSA closure for an RPM electrolyte near a wall has been solved analytically (63), but the results present un-physical features. The HNC-like closure, on the other hand, must be solved numerically (64), but leads to much better results in comparison with computer simulations (65). One of the interesting qualitative results is the appearance of oscillatory density profiles for sufficiently high electrolyte concentrations. It must however be stressed that the closures derived from the density functional theory of chapter 1 cannot cope with the problem of image forces which arise from the dielectric discontinuity between the solvent and the surface at $z = 0$ which leads to a polarization of the surface charge. In other words, the integral equations can only treat the somewhat academic situation where the dielectric constants of the wall and of the solvent are equal.

4. POLAR LIQUIDS AND DISCRETE SOLVENT MODELS

4.1. MOLECULAR PAIR DISTRIBUTION FUNCTIONS

Solvents, like water, are made up of polar molecules having both translational and orientational degrees of freedom. The pair-potential between two molecules depends not only on the distance between their centers of mass, but also on their orientations. The distribution functions introduced in the first lecture must be generalized to account for the orientational coordinates. The center of mass (CM) coordinates of a molecule will henceforth be denoted by $\vec{R}_i$, while the angular coordinates in a laboratory fixed frame will be denoted by $\vec{\Omega}_i$. For linear molecules, $\vec{\Omega}_i = (\theta_i, \varphi_i)$ are the usual polar angles, while for non-linear molecules (like H_2O), $\vec{\Omega}_i = (\theta_i, \varphi_i, \chi_i)$ are the Euler angles. The obvious generalization of the microscopic density (1.4) is now :

$$\rho(\vec{R},\vec{\Omega}) = \sum_{i=1}^{N} \delta(\vec{R} - \vec{R}_i)\, \delta(\vec{\Omega} - \vec{\Omega}_i) \tag{4.1}$$

and the one and two-particle densities (1.11) and (1.16) take here the form :

$$\rho^{(1)}(\vec{R},\vec{\Omega}) = \langle\rho(\vec{R},\vec{\Omega})\rangle \tag{4.2}$$

$$\rho^{(2)}(\vec{R},\vec{R}',\vec{\Omega},\vec{\Omega}') = \langle\rho(\vec{R},\vec{\Omega})\, \rho(\vec{R}',\vec{\Omega}')\rangle - \rho^{(1)}(\vec{R},\vec{\Omega})\, \delta(\vec{R}-\vec{R}')\delta(\vec{\Omega}-\vec{\Omega}') \tag{4.3}$$

For future use we change conventions and introduce the more compact notation :

$$1 \equiv (\vec{R}_1,\vec{\Omega}_1) \;\; ; \;\; 2 \equiv (\vec{R}_2,\vec{\Omega}_2)$$

The two-particle density for a homogeneous fluid reads now :

$$\rho^{(2)}(1,2) = \rho^{(2)}(\vec{R}_1 - \vec{R}_2, \vec{\Omega}_1,\vec{\Omega}_2) \tag{4.4}$$

when $R_{12} = |\vec{R}_1 - \vec{R}_2| \to \infty$, the positions and orientations of two molecules are uncorrelated in an isotropic fluid (this excludes, e.g., liquid crystals), and $\rho^{(2)}$ reduces to $(n/\Omega)^2$, where

$$\Omega = \int d\vec{\Omega} = \begin{cases} 4\pi & \text{(linear molecule)} \\ 8\pi^2 & \text{(non-linear molecule)} \end{cases}$$

In analogy with (1.19) we define the molecular p.d.f. :

$$g(1,2) \equiv g^{(2)}(\vec{R}_{12},\vec{\Omega}_1,\vec{\Omega}_2) = \left(\frac{\Omega}{n}\right)^2 \rho^{(2)}(\vec{R}_{12},\vec{\Omega}_1,\vec{\Omega}_2) \tag{4.5}$$

The CM p.d.f. is just the angular average of the molecular p.d.f. :

$$g_c(R) = \frac{1}{\Omega^2}\iint g^{(2)}(\vec{R},\vec{\Omega}_1,\vec{\Omega}_2)\, d\vec{\Omega}_1 d\vec{\Omega}_2 \tag{4.6}$$

A molecular d.c.f. $c(1,2) \equiv c^{(2)}(\vec{R}_{12}, \vec{\Omega}_1, \vec{\Omega}_2)$ can be defined along lines similar to the atomic case, and is related to $h(1,2) = g(1,2)-1$ by a straightforward generalization of the OZ relation (1.32), which must now involve a convolution over CM and angular coordinates ($d3 \equiv d\vec{R}_3\, d\vec{\Omega}_3$):

$$h(1,2) = C(1,2) + \frac{n}{\Omega}\int C(1,3)\, h(3,2)\, d3$$

If the total potential energy V_N of a molecular fluid is pair-wise additive, knowledge of g is sufficient to express the internal energy and the pressure by straightforward generalizations of the corresponding relations for atomic fluids (7). The isothermal compressibility, however, which is solely determined by density fluctuations, is expressible in terms of the angular-averaged p.d.f. (4.6) :

$$\chi_T/\chi_T^{(o)} = 1 + n\int [g_c(R) - 1]\, d\vec{R} \tag{4.8}$$

The difficulty in handling the molecular p.d.f. is that it is a multi-dimensional function depending on 6 and 8 variables for linear and non-linear molecules respectively. Some kind of reduction of the detailed information contained in g(1,2) is needed, and this can be achieved either by projection or by contraction.

Projections of the molecular p.d.f. are obtained by expanding g (1,2) in an infinite set of angle-dependent basis functions. If a laboratory-fixed frame of reference is chosen, the basis functions are the so-called <u>rotational invariants</u> (66) :

$$g(1,2) = \sum_{\substack{mn\ell \\ \mu\nu}} g_{\mu\nu}^{mn\ell}(R)\, \Phi_{\mu\nu}^{mn\ell}(1,2) \tag{4.9}$$

where the sum runs, in the most general case, over 5 indices ; the projections $g_{\nu\mu}^{mnl}$ depend only on the distance R between CM and the rotational invariants depend on the orientational coordinates $\vec{\Omega}_1$ and $\vec{\Omega}_2$, and on the orientation $\hat{R}_{12}$ of the vector joining the CM ; explicitly :

$$\Phi_{\mu\nu}^{mn\ell}(1,2) = \sum_{\mu'\nu'\lambda'} \begin{pmatrix} m & n & \ell \\ \mu' & \nu' & \lambda' \end{pmatrix} D_{\mu\mu'}^{m}(\vec{\Omega}_1)\, D_{\nu\nu'}^{n}(\vec{\Omega}_2)\, D_{o\lambda'}^{\ell}(\hat{R}_{12}) \tag{4.10}$$

where the D are Wigner matrix elements (or generalized spherical harmonics) and the coefficients are the 3-j symbols, familiar from the quantum theory of angular momentum. For linear molecules, the rotational invariants involve only the three indices m, n, l ($\mu = \nu = 0$) and the D reduce to the familiar spherical harmonics :

$$D_{o\mu'}^{m}(\theta,\varphi) = (-1)^m \left(\frac{4\pi}{2m+1}\right)^{1/2} Y_{\mu'}^{m}(\theta,\varphi) \tag{4.11}$$

The Fourier transform of (4.9) (or equivalently

of h(1,2)) with respect to $\vec{R}$ can be calculated by using the standard Rayleigh expansion of $\exp(-i\,\vec{k}.\vec{R})$, with the result :

$$\hat{h}(\vec{k},\vec{\Omega}_1,\vec{\Omega}_2) = \int e^{-i\vec{k}.\vec{R}}\, h(\vec{R},\vec{\Omega}_1,\vec{\Omega}_2)\, d\vec{R}$$

$$= \sum_{\substack{mn\ell \\ \mu\nu}} \hat{h}^{mn\ell}_{\mu\nu}(k)\, \Phi^{mn\ell}_{\mu\nu}(\vec{\Omega}_1,\vec{\Omega}_2,\hat{k}) \qquad (4.12)$$

with

$$\hat{h}^{mn\ell}_{\mu\nu} = 4\pi(i)^{\ell} \int_0^{\infty} h^{mn\ell}_{\mu\nu}(R)\, j_{\ell}(kR)\, R^2 dR \qquad (4.13)$$

where the j_l are the spherical Bessel functions. Through the expansion (4.12) for $\hat{h}$ and $\hat{c}$, the Fourier transform of the OZ relation (4.7) is transformed into an infinite set of coupled algebraic equations relating the $\hat{h}^{mnl}_{\nu\mu}$ and $\hat{c}^{mnl}_{\mu\nu}$.

In practice expansions like (4.9) or (4.12) are useful only if they can be truncated after a limited number of terms without damage to the underlying physics. This problem will be examined in the next section for specific models of the solvent.

The alternative reduction of the molecular p.d.f. is the set of site-site (or atom-atom) p.d.f., which is useful if the intermolecular potential can be accurately represented by an interaction-site model (4,7) :

$$v(1,2) = \sum_{\alpha}\sum_{\beta} v_{\alpha\beta}(|\vec{r}_{1\alpha} - \vec{r}_{2\beta}|) \qquad (4.14)$$

where the site-site potentials are spherically symmetric and depend on the distance between a site (e.g. and atom) on molecule 1 and a site on molecule 2. The simplest example is the "dumbell" model for homonuclear diatomic molecules (like N_2), in which case v (1,2) is the sum of four terms (each atom of one molecule interacts with the two atoms of the other), and all four potentials $v_{\alpha\beta}$ ($1 \leq \alpha,\beta \leq 2$) are identical. Another, more complicated, example is the ST2 model of water (67). The site-site p.d.f. are defined as :

$$n^2 g_{\alpha\beta}(r) = \langle \sum_{i\neq j}\sum \delta(\vec{r}_{i\alpha})\, \delta(\vec{r}_{j\beta} - \vec{r}) \rangle \qquad (4.15)$$

Thus, for a heteronuclear diatomic (like CO) for instance, one would consider 3 site-site p.d.f., namely $g_{CC}(r)$, $g_{CO}(r) \equiv g_{OC}(r)$ and $g_{OO}(r)$. Although the $g_{\alpha\beta}(r)$ can be obtained from the full molecular p.d.f. by integration (and hence correspond to a loss of information) they are sufficient to determine the compressibility, the internal energy (if the intermolecular pair potential is of the form (4.14)) and the structure factors measured by diffraction experiments, since, for instance, neutrons are scattered by the nuclei which are the natural "sites" in a polyatomic molecule.

4.2 Structure and dielectric behaviour of polar solvents

Solvent molecules are characterized by their permanent multipolar moments (dipole $\vec{\mu}$, quadrupole tensor $\overleftrightarrow{Q}$, etc.) and by their polarizability tensor $\overleftrightarrow{\alpha}$, which is frequently isotropic to a good approximation ($\overleftrightarrow{\alpha} \simeq \alpha \overleftrightarrow{I}$, with α a scalar polarizability). Water molecules in the gas phase have $\mu = 1.85 \times 10^{-18}$ esu (=1.85 Debye), $\alpha = 1.44$ Å^3 and a "tetrahedral" quadrupole tensor which is diagonal in a molecular frame with the z-axis along $\vec{\mu}$, with diagonal elements Q_T, $-Q_T$, 0 and $Q_T = 2.5 \times 10^{-26}$ esu.

The simplest model for the pure solvent is one of dipolar hard spheres ($\vec{\mu} \neq 0$, $\overleftrightarrow{Q} = \overleftrightarrow{\alpha} = 0$) characterized by the simple pair potential

$$v(1,2) = \infty \quad : \quad R_{12} \equiv |\vec{R}_1 - \vec{R}_2| < \sigma$$

$$= -\frac{\mu^2}{R_{12}^3} \Phi^{112}(1,2) \quad ; \quad R_{12} > \sigma \tag{4.16}$$

where the rotational invariant $\Phi^{112} \equiv \Phi^{112}_{00}$ is the the angular part of the familiar dipole-dipole interaction :

$$\Phi^{112}(1,2) = 3(\hat{\mu}_1 . \hat{R}_{12})(\hat{\mu}_2 . \hat{R}_{12}) - \hat{\mu}_1 . \hat{\mu}_2 \tag{4.17}$$

This model was first studied in detail by Wertheim (68) who limited the expansion (4.9) to three terms :

$$h(1,2) = h^{000}(R)\, \Phi^{000}(1,2) + h^{110}(R)\, \Phi^{110}(1,2) + h^{112}(R)\, \Phi^{112}(1,2) \tag{4.18}$$

where $\Phi^{000}(1,2) = 1$, $\Phi^{110}(1,2) = \hat{\mu}_1 \cdot \hat{\mu}_2$ and $\Phi^{112}(1,2)$ is defined in (4.17). The reason for this apparently drastic restriction of the infinite set of rotational invariants to just three terms is twofold. First, from a purely mathematical point of view, these three invariants form a closed set with respect to angular convolution : the convolution product of only two of these functions, yields a Φ^{lmn} belonging to the same set and no others. Consequently the expansion (4.18) leads to three coupled OZ relations for the corresponding projections $\hat{h}^{mnl}$ (k) and $\hat{c}^{mnl}$ (k), which are completely decoupled from all other projections. Secondly, a knowledge of the projections h^{000}, h^{110} and h^{112} suffices to determine the thermodynamic and dielectric properties of the model. In particular the compressibility is given by (4.8) (where $g_c(R)-1 \equiv h^{000}(R)$), the excess internal energy is given by :

$$\frac{U^{ex}}{N} = 2\pi n \int g(\vec{R},\vec{\Omega}_1,\vec{\Omega}_2)\, v(\vec{R},\vec{\Omega}_1,\vec{\Omega}_2)\, d\vec{R}d\vec{\Omega}_1 d\vec{\Omega}_2$$

$$= -\frac{4\pi}{3}\mu^2 n \int_0^\infty h^{112}(R)\,\frac{1}{R}\,dR \qquad (4.19)$$

while the dielectric constant ε_s follows directly from Kirkwood's formula (4, 7, 11) :

$$\frac{(\varepsilon_s - 1)(2\varepsilon_s + 1)}{9\varepsilon_s} = y\, g_K \qquad (4.20)$$

where $y = 4\pi\,\mu^2 n/(9k_BT)$ and g_K is Kirkwood's g-factor equal to the mean square fluctuation of the total dipole moment $\vec{M} = \Sigma_i\, \vec{\mu}_i$ of the sample :

$$g_K = \frac{\langle |M|^2\rangle}{N\mu^2} = 1 + \frac{(N-1)}{\mu^2}\langle \hat{\mu}_1 . \hat{\mu}_2\rangle$$

$$= 1 + \frac{n}{3}\hat{h}^{110}\ (k=0) \qquad (4.21)$$

Alternatively it can be shown that :

$$\frac{(\varepsilon_s - 1)^2}{\varepsilon_s} = -3ny\,\hat{h}^{112}\quad (k=0) \qquad (4.22)$$

From the elementary properties of the Hankel transform in eqn (4.13), it follows that :

$$h^{112}(R) \underset{R\to\infty}{\simeq} \frac{1}{4\pi yn}\,\frac{(\varepsilon_s - 1)^2}{\varepsilon_s}\,\frac{1}{R^3} \qquad (4.23)$$

which shows that the orientational correlations decay like the dipole-dipole potential, i.e. there is no screening similar to the ionic case examined in lecture 3.

Starting from the truncated expansion (4.18), Wertheim (68) obtained an analytic solution of the three coupled OZ relations supplemented by the MSA closure :

$$h(1,2) = -1\ ;\ R_{12} < \sigma$$

$$c(1,2) = -\beta v(1,2) = \frac{\beta\mu^2}{R_{12}^3}\,\Phi^{112}(1,2)\ ;\ R_{12} > \sigma \qquad (4.24)$$

The MSA turns out to underestimate considerably the dielectric constant ε_s for large reduced dipole moments $\mu^* = (\mu^2/\sigma^3 k_BT)^{1/2}$. As in the case of ionic correlations in the primitive model, improvement may be sought from the HNC closure :

$$c(1,2) = -\beta v(1,2) + h(1,2) - \ln[1 + h(1,2)] \qquad (4.25)$$

As it stands the closure is useless with respect to the expansions of c and h in rotational invariants, due to the presence of the logarithm. Two ways out of the difficulty have been put foreward and numerically exploited by Patey and co-workers (69,70). In the first scheme, the expansions of h and c are again limited to 3 terms, as in (4.18), and the logarithm in (4.25) is linearized with respect to the projections h^{110} and h^{112}, according to :

$$\ln [1 + h(1,2)] = \ln[g^{000}(R) + h^{110}(R)\Phi^{110}(1,2) + h^{112}(R)\Phi^{112}(1,2)]$$

$$\simeq \ln g^{000}(R) + \frac{h^{110}(R)}{g^{000}(R)}\,\Phi^{110}(1,2) + \frac{h^{112}(R)}{g^{000}(R)}\,\Phi^{112}(1,2) \qquad (4.26)$$

The resulting closure is the so-called linearized hypernetted chain (LHNC) approximation, a set of 3 closure relations between the 3 projections of h and c. The resulting dielectric constant ε_s turns out to be now too large compared to computer simulation results whenever $\mu^*>1$. Another defect which the LHNC closure shares with the MSA approximation is that the spherically symmetric projections h^{000} and c^{000} decouple from the other two projections ; this means that within these approximations, $h^{000}(R)$ is not affected by the strength of the dipolar coupling μ^*, clearly an unphysical feature. These two defects are overcome by recasting the HNC closure (4.25) in a different (but equivalent form) which allows a larger number of rotational invariants to be retained in a consistent way (70). If eqn (4.25) is differentiated with respect to $R = R_{12}$ holding Ω_1 and Ω_2 constant, the result is :

$$\frac{\partial c(1,2)}{\partial R} = -\beta\,\frac{\partial v(1,2)}{\partial R} + \frac{\partial h(1,2)}{\partial R} - \frac{1}{g(1,2)}\,\frac{\partial g(1,2)}{\partial R}$$

$$= -\beta\,\frac{\partial v(1,2)}{\partial R} - h(1,2)\,\frac{\partial w(1,2)}{\partial R} \qquad (4.27)$$

where $W = \beta v + c - h$. Since $c(1,2) \to -\beta v(1,2)$ as $R \to \infty$, integration of (4.27) with respect to R yields the following, equivalent form of the HNC closure :

$$c(1,2) = \int_R^\infty h(1,2)\,\frac{\partial w(1,2)}{\partial R'}\,dR' - \beta v(1,2) \qquad (4.28)$$

in which the product appearing in the integral has an expansion in rotational invariants which is easily derived from the corresponding expansions of c and h, and from the addition theorem of spherical harmonics. The HNC closure (4.28) has been solved (with the corresponding OZ relations) for basis sets involving an increasing number of rotational invariants, and fast convergence was observed (70). The resulting dielectric constant ε_s agrees now quite well with available simulation data, but for values of μ^* appropriate for water ($\mu^* \simeq 2$), the predicted ε_s is significantly larger than the experimental value $\varepsilon_s \simeq 80$. Since the theory is now reasonable accurate, this means that the dipolar hard sphere model is too crude to model water, as one might have expected from the outset. The obvious improvement of the model is to add a point quadrupole to the dipolar hard spheres. The corresponding pair potential now involves additional dipole-quadrupole and quadrupole-quadrupole terms, which introduce further rotational invariants. The corresponding LHNC and full HNC approximations have been solved both for linear (71,72) and for tetrahedral (73) quadrupoles. These solutions show that ε_s is strongly reduced by the presence of even a modest quadrupole moment. Physically, quadrupolar interaction favours T-like molecular pair configurations for which $\hat{\mu}_1 \cdot \hat{\mu}_2 \simeq 0$, leading to a strong reduction of the Kirkwood g-factor (4.21), and hence of ε_s. In fact the reduction of ε_s due to Q_T is too strong, since with H_2O gas phase values of μ and Q_T, the model predicts $\varepsilon_s \simeq 25$ for liquid water at room temperature, well below the experimental value $\varepsilon_s \simeq 80$.

This apparent paradox can be resolved if, in a final step, the polarizability of the molecules is taken into account. The instantaneous dipole moment of molecule i is the sum of its permanent dipole and of the dipole induced by the local electric field $\vec{E}_i$ acting on that molecule :

$$\vec{m}_i = \vec{\mu}_i + \vec{\alpha} \cdot \vec{E}_i \tag{4.29a}$$

$$\vec{E}_i = \sum_{j(\neq i)} \overleftrightarrow{T}_{ij} \cdot \vec{m}_j \tag{4.29b}$$

where the sum runs over the N-1 other molecules, and $\overleftrightarrow{T}_{ij} = \vec{\nabla}_i \vec{\nabla}_j \ (1/R_{ij})$ is the dipolar tensor. Clearly, molecular polarizability introduces a complicated many-body interaction which must be handled by iteration in a computer simulation of polarizable molecules (74). These simulations, as well as theoretical consideration (75), show that, to a good approximation, the properties of a polarizable dipolar fluid can be related to those of a fluid with an <u>effective</u> permanent dipole :

$$\mu_{eff}^2 = \langle |m_i|^2 \rangle \tag{4.30}$$

which may be estimated from a mean field calculation (73) and turns out to be substantially larger than the gas phase value in liquid water : $\mu_{eff} = 2.56$ Debye compared to $\mu = 1.85$ Debye. Inserting this effective dipole moment, together with the quadrupole moment $Q_T = 2.5 \times 10^{-26}$ esu in their LHN calculations, Carnie and Patey (74) find the following values for ε_s in pure water at room temperature ($\sigma = 2.8$ Å, $Q_T^* = (Q_T^2/\sigma^5 k_B T)^{\frac{1}{2}} = 0.9$):

$$\mu = 1.85 \text{ Debye } (\mu^* \simeq 2) : \varepsilon_s = 25$$

$$\mu = \mu_{eff} = 2.55 \text{ Debye } (\mu^* \simeq 2.7) : \varepsilon_s \simeq 80 \simeq \varepsilon_s^{exp}$$

There remains to check whether this surprising agreement with experiment is maintained if the full HNC-closure is used.

4.3 Discrete solvent models of ionic solutions

In lecture 3 we insisted on the shortcomings of primitive model descriptions of ionic solutions. In the preceding section we introduced simple molecular models for the pure solvent. In this concluding section we show how ionic and solvent models can be combined to yield a discrete solvent description of ionic solutions, in which all three components (solvent, anions, cations) are treated on an equal footing. The simplest such model is a mixture of dipolar and charged hard spheres. The solvent is modelled by hard spheres of diameter σ_o carrying a dipole moment $\vec{\mu}$, while the symmetric solute is made up of equal numbers of spheres of diameter $\sigma = \sigma_1 = \sigma_2$ and of charge $q_1 = -q_2 = q$. The pair potentials are of the form (with $r = |\vec{r}_i - \vec{r}_j|$):

$$v_{\nu\mu}(i,j) = v_{\nu\mu}^{000}(r) + v_{\nu\mu}^{101}(r)\,\Phi^{101}(i,j) + v_{\nu\mu}^{011}(r)\,\Phi^{011}(i,j) + v_{\nu\mu}^{112}(r)\,\Phi^{112}(i,j) \; ; \quad 0 \leq \nu, \mu \leq 2 \tag{4.31}$$

where (with the convention $q_o = 0$, $\mu_1 = \mu_2 = 0$) :

$$v_{\nu\mu}^{000}(r) = \infty \; ; \; r < \sigma_{\nu\mu} = \frac{1}{2}(\sigma_\nu + \sigma_\mu)$$
$$= (q_\nu q_\mu)/r \; ; \; r > \sigma_{\nu\mu} \tag{4.32a}$$

$$v_{\nu\mu}^{101}(r) = (\mu_\nu q_\mu)/r^2 \tag{4.32b}$$

$$v_{\nu\mu}^{011}(r) = -(q_\nu \mu_\mu)/r^2 \tag{4.32c}$$

$$v_{\nu\mu}^{112}(r) = -(\mu_\nu \mu_\mu)/r^3 \tag{4.32d}$$

The ion-dipole coupling introduces two new rotational invariants $\Phi^{101}(i,j) = \vec{\mu}_i \cdot \vec{r}_{ij}$ and $\Phi^{011}(i,j) = \vec{\mu}_j \cdot \vec{r}_{ij}$, which, together with those appearing in eqn (4.18), form again a closed set with respect to angular convolution. If the expansion (4.9) is limited to these 5 invariants, it is easy to convince oneself that the pair structure of the symmetric model is entirely described by 7 independent projections, namely :

$$h_{11}^{000}(r) \equiv h_{22}^{000}(r) \quad ; \quad h_{12}^{000}(r) \qquad \text{(ion-ion)}$$

$$h_{00}^{000}(r) \; ; \; h_{00}^{110}(r) \; ; \; h_{00}^{112}(r) \qquad \text{(solvent-solvent)}$$

$$h_{01}^{000}(r) \equiv h_{02}^{000} \quad ; \quad h_{01}^{011}(r) \equiv - h_{02}^{011}(r) \qquad \text{(ion-solvent)}$$

These are related to the corresponding $c_{\nu\mu}^{mnl}(r)$ by 7 coupled OZ relations, which may be solved, subject to approximate closures. For the MSA closure, an analytic solution has been worked out, up to set of coupled, non-linear algebraic equations (76,77). Numerical solutions with the LHNC closure :

$$c_{\nu\mu}(1,2) = -\beta v_{\nu\mu}(1,2) + h_{\nu\mu}(1,2) - \ln g_{\nu\mu}^{000}(r) - \sum_{m,n,\ell} \frac{h_{\nu\mu}^{mn\ell}(r)}{g_{\nu\mu}^{000}(r)} \Phi^{mn\ell}(1,2) \qquad (4.33)$$

with the sum restricted to the 5 invariants enumerated above, have been obtained by Levesque et al. (78). These calculations clearly show that the ionic structure, as characterized by $h_{11}^{000}(r)$ and $h_{12}^{000}(r)$, differs markedly from the predictions of the primitive model of lecture 3. LHNC calculations indicate that both correlation functions exhibit much structure for $r \lesssim 3\,\sigma_i$ at high concentrations, and give strong evidence for the existence of anion-cation pairs separated by exactly one solvent molecule. Under certain physical conditions, such "trimers" may be even more probable than anion-cation "dimers".

In order to make contact with the primitive model of lecture 3, one may define <u>effective</u> ion-ion potentials by reducing the three component solvent-solute mixture to an effective two-component system involving only the ionic species (79). Dropping the superscript 000 for the spherically symmetric ionic correlation functions, this program can be carried out by introducing effective d.c.f. related to the $h_{\nu\mu}$ by generalized OZ equations :

$$\hat{h}_{\nu\mu}(k) = \hat{c}_{\nu\mu}^{eff}(k) + \sum_{\lambda} n_{\lambda}\, \hat{c}_{\nu\lambda}^{eff}(k)\, \hat{h}_{\lambda\mu}(k) \qquad (4.34)$$

where the $\hat{h}_{\nu\mu}$ $(1 \le \nu,\mu \le 2)$ are the exact ion-ion correlation functions calculated from a three-component (i.e. discrete solvent) model. Within the present model of a mixture of charged and dipolar hard spheres, an elementary calculation leads to the following expression for the $\hat{c}^{eff}_{\nu\mu}$ (79,78) :

$$\hat{c}^{eff}_{\nu\mu}(k) = \hat{c}^{000}_{\nu\mu}(k) + n_o \frac{\hat{c}^{000}_{\nu 0}(k)\, \hat{c}^{000}_{\mu 0}(k)}{1 - n_o \hat{c}^{000}_{00}(k)}$$

$$- \frac{n_o}{3} \frac{\hat{c}^{000}_{\nu 0}(k)\, \hat{c}^{000}_{\mu 0}(k)}{1 - (n_o/3)\,(2\hat{c}^{112}_{00}(k) + \hat{c}^{110}_{00}(k))} \tag{4.35}$$

Eqn. (4.35) expresses the fact that the effective ion-ion d.c.f. is determined by ion-ion, ion-solvent and solvent-solvent correlations in the full three-component mixture. Once $c^{eff}_{\nu\mu}$ has been determined, an effective ion-ion potential is derived by inverting the HNC-closure :

$$\beta v^{eff}_{\nu\mu}(r) = h_{\nu\mu}(r) - \ln[g_{\nu\mu}(r)] - c^{eff}_{\nu\mu}(r) \tag{4.36}$$

For large separations, one expects :

$$v^{eff}_{\nu\mu}(r) \underset{r \to \infty}{\simeq} \frac{q_\nu q_\mu}{\varepsilon r} \tag{4.37}$$

where ε ($\neq \varepsilon_s$) is now a solute-dependent dielectric constant. When (4.37) is generalized to all $r > \sigma$, and ε is taken equal to ε_s, the primitive model is recovered. LHNC calculations show that $v^{eff}_{\mu\nu}(r)$ differs markedly from the limit (4.37) when $r \lesssim 3\sigma$, reflecting the structure observed in $g_{\nu\mu}(r)$. $v^{eff}_{\nu\mu}(r)$ should not be confused with the "potential" of mean force :

$$w_{\nu\mu}(r) = - k_B T \ln [g_{\nu\mu}(r)] \tag{4.38}$$

Both potentials coincide only in the limit of infinite dilution ($n_i \to 0$). At finite ionic concentrations, the potentials of mean force between ions exhibit the usual Debye screening :

$$w_{\nu\mu}(r) \underset{r \to \infty}{\simeq} \frac{q_\nu q_\mu}{\varepsilon r} e^{-Kr} \tag{4.39}$$

More interestingly, the ion-dipole and dipole-dipole potentials of mean force exhibit similar exponential screening, due to the presence of free charges (76,80).

It was noted in section 4.2 that the dipolar hard sphere fluid is not a good model for the solvent. For that reason Patey and Carnie (81) refined the discrete solvent model of ionic solutions by solving the LHNC equation for a solution of charged hard spheres in a "waterlike" solvent made up of spheres carrying an effective dipole moment (to account for molecular polarizability) and a tetrahedral quadrupole. Their calculations show that the ion-ion potentials of mean force are less structured and approach their continuum limit more rapidly than those obtained with a purely dipolar solvent yielding the same ε. Ion pairs at contact are also less probable in the "waterlike" fluid which appears to be a "better" solvent than the dipolar fluid since it leads to stronger dissociation.

Recent extensions of the discrete solvent model attempt to account for the non-spherical shape (i.e. the short-range anisotropy) of the solvent molecules, as well as for the extended charge distribution of real water molecules (82). Extensions of the discrete solvent models to inhomogeneous situations (like electric double layers) have also been successfully formulated, in particular within the MSA (83).

REFERENCES

1) J.S.Rowlinson and F.L. Swinton, "Liquids and Liquid Mixtures" (3d edition, Butterworth, London, 1982)

2) E.W.Montroll and J.L.Lebowitz, editors, "The Liquid State of Matter: Fluids, Simple and Complex"(North-Holland, Amsterdam, 1982)

3) J.S.Rowlinson and B.Widom, "Molecular Theory of Capillarity" (Clarendon Press, Oxford, 1982)

4) C.Gray and K.E.Gubbins, "Theory of Molecular Liquids" (Clarendon Press, Oxford, 1984)

5) N.H.March and M.P.Tosi, "Coulomb Liquids" (Academic Press, London, 1984)

6) H.L.Friedman, "A Course in Statistical Mechanics" (Prentice Hall, 1986)

7) J.P.Hansen and I.R.McDonald, "Theory of Simple Liquids" (Academic Press, London, 1976 ; 2d edition, 1986)

8) J.Barker and D.Henderson, Rev. Mod. Phys. 48, 587 (1976)

9) R.Evans, Adv. Phys. 28, 143 (1979)

10) S.L.Carnie and G.Torrie, Adv. Chem. Phys. 56, 141 (1984)

11) P.Madden and D.Kivelson, Adv. Chem. Phys. 56, 467 (1984)

12) J.K.Percus, in "The Equilibrium Theory of Classical Fluids", edited by H.L.Frisch and J.L.Lebowitz (Benjamin, New York, 1964)

13) C. DE Dominicis and P.C.Martin, J. Math. Phys. 5, 14 (1964)
A.J.Young, P.D.Fleming and H.Gibbs, J. Chem. Phys. 64, 3732 (1976)

14) N.D.Mermin, Phys. Rev. 137, A 1441 (1965)

15) C.Ebner, W.F.Saam and D.Stroud, Phys. Rev. A 14, 2264 (1976)

16) W.F.Saam and C.Ebner, Phys. Rev. A 15, 2566 (1977)

17) Ramakrishnan and Yussouff, Phys. Rev. B 19, 2775 (1979)
M.Baus and J.L.Colot, Mol. Phys. 55, 653 (1986)
A.D.J.Haymet and D.W.Oxtoby, J. Chem. Phys. 84, 1769 (1986)
J.L.Barrat, M.Baus and J.P.Hansen, Phys. Rev. Letters 56, 1063 (1986)

18) For particularly efficient algorithms, see M.J.Gillan, Molec. Phys. 38, 1781 (1979)
G.Zerah, J. Comp. Phys. (in Press)

19) M.S.Wertheim, J. Math. Phys. 5, 643 (1964)
E.Thiele, J. Chem. Phys. 39, 474 (1964)
R.J.Baxter, Austr. J. Phys. 21, 563 (1968)

20) L.Blum and A.H.Narten, J. Chem. Phys. 56, 5197 (1972)
J.Chihara, Progr. Theor. Phys. 50, 409 (1973)
W.G.Madden and S.A.Rice, J. Chem. Phys. 72, 4208 (1980)

21) J.L.Lebowitz and J.K.Percus, Phys. Rev. 144, 251 (1966)

22) Y.Rosenfeld and N.W.Ashcroft, Phys. Rev. A 20, 1208 (1979)
F.Lado, S.M.Foiles and N.W.Ashcroft, Phys. Rev. A 28, 2374 (1983)

23) D.Levesque, J.J.Weis and L.Reatto, Phys. Rev. Lett. 54, 451 (1985)

24) F.J.Rogers and D.A.Young, Phys. Rev. A 30, 999 (1984)
G.Zerah and J.P.Hansen, J. Chem. Phys. 84, 2336 (1986)

25) L. Van Hove, Phys. Rev. 95, 249 (1954)

26) J.P.Boon and S.Yip, "Molecular Hydrodynamics", (McGraw Hill, New York, 1980)

27) B.J.Berne and R.Pecora, "Dynamic Light Scattering" (Wiley, New York 1976)

28) L.P.Kadanoff and P.C.Martin, Ann. Phys. (NY) 24, 419 (1963)

29) R.D.Mountain, Rev. Mod. Phys. 38, 205 (1966)

30) B.R.A.Nijboer and A.Rahman, Physica 32, 415 (1966)

31) R.Kubo, Rep. Progr. Phys. 29, 255 (1966)

32) L.D.Landau and E.M.Lifshitz, "Fluid Mechanics" (Pergamon Press, London, 1963)

33) R.Zwanzig and M.Bixon, Phys. Rev. A 2, 2005 (1970)

34) H.Mori, Progr. Theor. Phys. 33, 423 ; 34, 399 (1965)

35) R.G.Gordon, Adv. Mag. Res. 3, 1 (1968)

36) C.H.Chung and S.Yip, Phys. Rev. 182, 323 (1969)

37) N.K.Ailawadi, A.Rahman and R.Zwanzig, Phys. Rev. A 4, 1616 (1971)

38) Y.Pomeau and P.Résibois, Phys. Reports 19, 63 (1975)

39) T.Munakata and A.Igarashi, Progr. Theor. Phys. 60, 45 (1978)
J.Bosse, W.Götze and M.Lücke, Phys. Rev. A 17, 434 (1978)

40) B.J.Alder and T.E.Wainwright, Phys. Rev. A 1, 18 (1970)
D.Levesque and W.T.Ashurst, Phys. Rev. Letters 33, 277 (1974)
J.J.Erpenbeck and W.W.Wood, J. Stat. Phys. 24, 455 (1981) ; Phys. Rev. A 26, 1648 (1982)

41) J.R.Dorfman and H. van Beijeren, in "Statistical Mechanics", part B, edited by B.J.Berne (Plenum, New York, 1977)
S.Yip, Ann. Pev. Phys. Chem. 30, 547 (1979)

42) L.Sjögren and A.Sjölander, J. Phys. C 12, 4369 (1979);
L.Sjögren, Phys. Rev. A 22, 2866 , 2883 (1980)

43) U.Bengtzelius, W.Götze and A.Sjölander, J. Phys. C 17, 5915 (1984)

44) D.Levesque, L.Verlet and J.Kurkijärvi, Phys. Rev. A 7, 1690 (1973)

45) W.G. McMillan and J.E.Mayer, J. Chem. Phys. 13, 276 (1945)

46) P.Turq, F.Lantelme and H.L.Friedman, J. Chem. Phys. 66, 3039 (1977)

47) P.C.Martin, Phys. Rev. 161, 143 (1967)

48) F.Stillinger and R.Lovett, J. Chem. Phys. 49, 1991 (1968)

49) P.A.Martin and C.Gruber, J. Stat. Phys. 31, 691 (1983)

50) P.A.Martin and Yalcin, J. Stat. Phys. 22, 435 (1980)

51) M.Parrinello and M.P.Tosi, Riv. Nuovo Cimento 2, N°6 (1979)

52) E.Waisman and J.L.Lebowitz, J. Chem. Phys. 56, 3086, 3093 (1972)

53) L.Onsager, J. Phys. Chem. 43, 189 (1939)

54) L.Blum, Mol. Phys. 30, 1529 (1975) ; in "Theoretical Chemistry : Advances and Perspectives" (Academic Press, New York 1986)

55) R.Triolo, L.Blum and M.A.Floriano, J. Chem. Phys. 67, 5956 (1978)

56) G.Pastore, P.V.Giaquinta, J.S.Thakur and M.P.Tosi, J. Chem. Phys. 84, 1827 (1986)

57) S.L.Carnie and G.Torrie, Adv. Chem. Phys. 56, 141 (1984)

58) S.L.Carnie and D.Y.C.Chan, Chem. Phys. Lett. 77, 437 (1981)

59) L.Blum, C.Gruber, J.L.Lebowitz and P.A.Martin, Phys. Rev. Letters 48, 1769 (1982)

60) L.Blum, D. Henderson, J.L.Lebowitz, C.Gruber and P.A.Martin, J. Chem. Phys. 75, 5974 (1981)

61) B.Jancovici, J. Stat. Phys. 29, 263 (1982)

62) D.Henderson, L.Blum and J.L.Lebowitz, J.Electroanal Chem. 102, 315 (1979).

63) L. Blum, J. Phys. Chem. 81, 136 (1977)
D. Henderson and L. Blum, J. Chem. Phys. 69, 5441 (1978)

64) D. Henderson, L. Blum and W.R. Smith, Chem. Phys. Lett. 63, 381 (1979). S.L. Carnie, D.Y.C. Chan, D.J. Mitchell and B.W. Ninham J. Chem. Phys. 74, 1472 (1981)

65) G.M. Torrie and J.P. Valleau, J. Chem. Phys. 73, 5807 (1980)

66) L. Blum, J. Chem. Phys. 57, 1862 (1972) ; 58, 3295 (1973)

67) A. Rahman and F.H. Stillinger, J. Chem. Phys. 55, 3336 (1971)

68) M.S. Wertheim, J. Chem. Phys. 55, 4291 (1971)

69) G.N. Patey, Mol. Phys. 34, 427 (1977) ; 35, 1413 (1978)

70) P.H. Fries and G.N. Patey, J. Chem. Phys. 82, 429 (1985)

71) G.N. Patey, D. Levesque and J.J. Weis, Mol. Phys. 38, 219 (1979)

72) J.S. Perkyns, P.H. Fries and G.N. Patey, Mol. Phys. 57, 529 (1986)

73) S.L. Carnie and G.N. Patey, Mol. Phys. 47, 1129 (1982)

74) E.L. Pollock, B.J. Alder and G.N. Patey, Physica A 108, 14 (1981)

75) M.S. Wertheim, Mol. Phys. 25, 211 (1973) ; 26, 1425 (1973) ; 33, 95 (1977) ; 34, 1109 (1977)

76) S.A. Adelman and J.M. Deutch, J. Chem. Phys. 60, 3935 (1974)

77) L. Blum, J. Chem. Phys. 61, 2196 (1974) ; J. Stat. Phys. 18, 451 (1978)

78) D. Levesque, J.J. Weis and G.N. Patey, J. Chem. Phys. 17, 1887 (1980)

79) S.A. Adelman, J. Chem. Phys. 64, 724 (1976)

80) J.S. Høye and G. Stell, J. Chem. Phys. 68, 4145 (1978)

81) G.N. Patey and S.L. Carnie, J. Chem. Phys. 78, 5183 (1983)

82) B.M. Petitt and P.J. Rossky, J. Chem. Phys. 77, 1451 (1982)

83) L. Blum and D. Henderson, J. Chem. Phys. 74, 1902 (1981)

THEORY OF IONIC SOLUTIONS AT EQUILIBRIUM

Harold L. Friedman
Department of Chemistry
State University of New York
Stony Brook, N.Y. 11794, USA

ABSTRACT The pair correlation function theory of ionic interactions in solution is developed using cluster theory methods. Fluctuation theory provides a natural entry to McMillan-Mayer solution theory. It is complemented by a similar operation on the Ornstein-Zernike equation which leads to important conditions on the solvent-averaged interactions among the ions. Certain inconsistencies in the theory which are identified seem to have remarkably little effect on the success of model calculations with solvent-averaged pair potentials, even for electrolyte mixture coefficients which depend on interactions in ion triples and larger clusters, and even for the rate constants of activation-controlled reactions of ionic species.

1. INTRODUCTION

In these lectures we begin with the pair correlation function theory that has been developed for simple fluids and apply it to study the effects of ionic interactions on various properties of solutions at equilibrium. We endeavor to present these developments briefly but at a level in which the logic is clear and persuasive while as many details as possible are suppressed.

Cluster theory methods are employed because they can be partially understood merely as cartoons that map out the various interactions in a way that is concise and clear compared to represention of the same terms as explicit sums of integrals. For full understanding one may consult other reports/1-3/ for the conventions and theorems of the relevant graph theory, including the famous three lemmas.

At this time ionic solution theory, even in the subfield of ion-ion interactions, seems to be ripe for change; some approximations which had seemed to be quite satisfactory are now known to be unrealistic to a significant degree. We expect that some examples of these pressures for change will be developed in other lectures which focus on diffraction experiments or simulation calculations, so we shall describe only one in some detail, namely the problems associated with ϵ_ρ, the thermodynamic dielectric coefficient of an ionic solution.

M.-C. Bellissent-Funel and G. W. Neilson (eds.), The Physics and Chemistry of Aqueous Ionic Solutions, 61–93.

2. PAIR CORRELATION FUNCTION

One way to interpret a pair correlation function is as a special case of the response of a local density to an external field. The local density of particles of species a at coordinate **R** is

$$\hat{\rho} = \Sigma_{i=1}^{N} \delta_{i/a} \, \delta(r_i - R) \tag{2.1}$$

where

$$\begin{aligned} \delta_{i/a} &= 1 \quad \text{if particle } i \text{ belongs to species } a \\ &= 0 \quad \text{otherwise} , \end{aligned} \tag{2.2}$$

and where r_i is the instantaneous location of particle i. The average density field (i.e. measurable local density) of species a is

$$\begin{aligned} \rho_a(R) &= \langle \hat{\rho}_a(R) \rangle \\ &\equiv \int d^{3N}r \, f_N^{\bullet}(r_1, \ldots, r_N) \, \hat{\rho}_a(R) \end{aligned} \tag{2.3}$$

where $f_N^{\bullet}$ is the normalized equilibrium distribution function

$$f_N^{\bullet} \equiv \exp[-\beta(U_N + \Phi_N]/Z_N^{\bullet} \tag{2.4}$$

Here $\beta = 1/k_B T$ while $U_N = U_N(r_1, \ldots, r_N)$ is the potential of the interactions among the N particles and $\Phi_N = \Phi_N(r_1, \ldots, r_N)$ is the potential of their interaction with an external field. The normalization factor, the configuration integral, is

$$Z_N^{\bullet} = \int d^{3N}r \, \exp[-\beta(U_N + \Phi_N)] \tag{2.5}$$

When Φ_N vanishes we drop the • notation and then the equations apply to a homogeneous system in the absence of external fields. Also then the density field reduces to the stoichiometric concentration.

$$\rho_a(R) = \rho_a \equiv N_a/V \quad , \Phi_N = 0 \tag{2.6}$$

A particularly interesting external field is one due to a $N+1^{st}$ particle of species b , possibly the same as a , located at coordinate R′ within the volume occupied by the N-particle system. In this case $U_N + \Phi_N = U_{N+1}$ with $r_{N+1} = R'$ so we may write

$$f_N^{\bullet} Z_N^{\bullet} = f_{N+1} Z_{N+1} \tag{2.7}$$

Also for a large system we have $Z_{N+1} = VZ_N^{\bullet}$ and

$$\begin{aligned} \rho_a(R) &= \int d^{3N}r \, f_N^{\bullet}(r_1, \ldots, r_N) \, \hat{\rho}_a(R) \\ &= V \int d^{3(N+1)}r \, f_{N+1}(r_1, \ldots, r_{N+1}) \, \hat{\rho}_a(R) \, \delta(r_{N+1} - R') \end{aligned}$$

$$\equiv V\langle\hat{\rho}_a(R)\,\delta(r_b-R')\rangle^d \tag{2.8}$$

where $r_b=r_{N+1}$, the location of a particle of species b, and where the superscript d denotes that this is a *distinct* correlation; the particle at R is not the same as the one at **R'**, even if they happen to be of the same species.

A form that exhibits the underlying symmetry follows first from noticing that each of the N_a non-vanishing terms in Eq. (2.1) makes the same contribution to the integral in Eq. (2.8), giving

$$\rho_a(R) = N_a V\langle\delta(\mathbf{r}_a-R)\delta(r_b-R')\rangle^d \tag{2.9}$$

where r_a is the instantaneous location of a particle of species a, and then introducing the function

$$g_{ab}(R,R') \equiv \rho_a(R)/\rho_a = V^2\langle\delta(r_a-R)\delta(r_b-R')\rangle^d \tag{2.10}$$

which has the species interchange symmetry $g_{ab}=g_{ba}$ found in the average $\langle\ldots\rangle$ itself. In the homogenous systems with which we shall be concerned, $g_{ab}(R,R')$ is a function of the distance $r=|R-R'|$ only, so when it is more convenient we shall write the pair correlation function as $g_{ab}(r)$.

These equations for $g_{ab}(r)$ are to be applied in the thermodynamic limit in which V increases without bound while T and every N_a/V are fixed. Therefore these equations, although derived for the canonical N_1, N_2, ..., V, T ensemble, are equally valid for the grand μ_1, μ_2,...,V,T ensemble. The grand ensemble interpretation is exploited in Sections 3.2 and 4.

3. FEATURES OF PAIR CORRELATION FUNCTIONS

3.1 Potential of Average Force

We recall from the theory above that $\rho_a g_{ab}(r)$ is the equilibrium average concentration of species a at a distance r from a point where a particle of species b is located. Equally $\rho_b g_{ab}(r) = \rho_b g_{ba}(r)$ is the equilibrium average concentration of species b at a distance r from a point where a particle of species a is located. It turns out that these local concentrations are just what are needed in a statistical mechanical theory that will replace the activity coefficients, even single ion activity coefficients, in the less satisfactory aspects of thermodynamic solution theory.

Writing the pair correlation function as a Boltzmann factor

$$g_{ab}(r) = \exp[-\beta w_{ab}(r)] \tag{3.1}$$

defines $w_{ab}(r)$, which is called the potential of the average force because it obeys the equation, easily derived from Eq. (3.1) and (2.10),

$$(\partial/\partial R)w_{ab}(R,R') = \frac{\langle(\partial U_N/\partial r_a)\delta(r_a-R)\delta(r_b-R')\rangle^d}{\langle\delta(r_a-R)\delta(r_b-R')\rangle d} \tag{3.2}$$

where $\partial/\partial R$ is the gradient operator. This equation shows that the negative gradient of $w_{ab}(R,R')$ is the equilibrium average of the force on the a particle at **R** due to all of the other particles in the N particle system, one of which is a b particle at **R′**. In the absence of external fields this force is along the **R-R′** direction and its magnitude depends only on $r \equiv |\mathbf{R}-\mathbf{R}'|$. If one expresses the U_N as a sum of pair potential terms together with terms expressing the deviation from pairwise additivity, if any, then in (3.2) the term

$$-\partial u_{ab}(r_a,r_b)/\partial r_a$$

carries the *direct* force on a from b while the rest of the terms carry the indirect shielding and liquid structure effects.

This simple analysis aids in the interpretation of pair correlation data like those in Fig. 1. For example, at the peak of

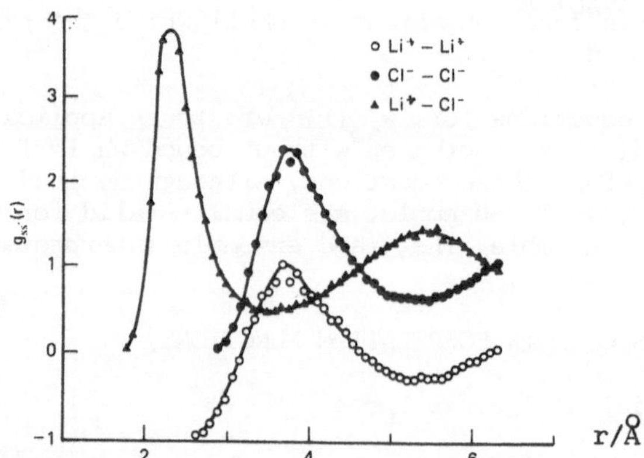

Fig. 1 Pair correlation functions from Monte Carlo simulation for a charged soft sphere model for LiCl liquid at 883K and 28.3cm^{3}mol^{-1} (From McDonald and Singer/4/) The g_{++} curve is shifted down one unit.

the g_{++} curve the ++ force due to the direct interaction, the sum of Coulomb and core repulsions, apparently is balanced by an effective attraction due to many-body forces. Moreover we may conclude that here the core repulsion part of the ++ interaction does not have a dominant effect because the peak of $g_{--}(r)$ is so near that of $g_{++}(r)$ while the respective diameters differ greatly.

3.2 Fluctuation Theory

The theory of concentration fluctuations in the grand ensemble, expressed by the equation/3/

$$\langle N_a N_b\rangle - \langle N_a\rangle\langle N_b\rangle = V\beta^{-1}(\partial\rho_a/\partial\mu_b) \tag{3.3}$$

leads to a connection between the pair correlation functions and thermodynamics. In this equation $\langle\ldots\rangle$ represents a grand ensemble average in a system of specified V, T, μ_a, μ_b, ..., μ_n where n is the number of distinct species of particles in a mixture; thus n concentration variables $\rho_1,\ldots,\rho_n$ determine the composition of the mixture in the thermodynamic sense. By expressing $\langle N_a N_b\rangle$ in terms of g_{ab} and V one arrives at the many-component version of the compressibility equation,

$$\beta(\rho_a\delta_{ab}+\rho_a\rho_b G_{ab}) = M_{ab} \equiv (\partial\rho_a/\partial\mu_b) \tag{3.4}$$

where we define

$$G_{ab} \equiv \int d^3r\,[g_{ab}(r)-1] \quad . \tag{3.5}$$

In view of this fluctuation formula and the observation that diffraction of radiation by a material medium is due to spatial fluctuations in the medium, it is to be expected that the contribution of an ab pair to the diffraction is closely related to $g_{ab}(r)$. Indeed for radiation (x-rays, neutrons, ...) with wave vector k the contribution of the ab species pair to the scattering is proportional to a somewhat generalized form of Eq. (3.5), namely the three dimensional Fourier transform/5/ $H_{ab}(k)$ of

$$h_{ab}(r) \equiv g_{ab}(r)-1 \ . \tag{3.6}$$

$H_{ab}(k)$ is very simply related to the various definitions of the *partial structure factor* encountered in the theory of diffraction. Thus in certain diffraction experiments a single $H_{ab}(k)$ can be determined over some range of k /6/ but more often what is determined is the combination

$$\Sigma_{ab}\, f_a(k) f_b(k)\rho_a\rho_b H_{ab}(k)$$

of correlation functions, concentrations, and scattering lengths $f_s(k)$ which depend on the interaction of the radiation with a particle of species s .

3.3 Other Connections to Thermodynamics

Finally there are several other relations which we shall cite in the respective forms that are exact if the configurational potential U_N is pairwise additive;/3/

$$U_N(r_1,\ldots,r_N) = \Sigma_a\Sigma_b\Sigma_{ij}\,\delta_{i/a}\,\delta_{j/b}\,u_{ab}(ij) \tag{3.7}$$

where the sums are over all distinct pairs of particles and all species while $\delta_{i/a}=1$ if i is a particle of species a and zero otherwise.

The configurational or excess energy is

$$E^{ex} \equiv E - E(\text{of ideal gas in same } \rho_1,\ldots\rho_n, V,T \text{ state})$$
$$= \tfrac{1}{2}V^{-1}\Sigma_a\Sigma_b\, \rho_a\rho_b\int u_{ab}(r)\, g_{ab}(r)\, d^3r \qquad (3.8)$$

The pressure is

$$p = \beta^{-1}\Sigma_a\rho_a - \frac{1}{6}\,\Sigma_a\Sigma_b\rho_a\rho_b \int r(\partial u_{ab}/\partial r)\, g_{ab}(r)\, d^3r \qquad (3.9)$$

The first quantum correction is

$$\beta(A-A^{classical})/V$$
$$= (1/24)(\beta\hbar)^2\Sigma_a\Sigma_b[\rho_a\rho_b/m_{ab}]\int[\nabla_r^2 u_{ab}(r)]\, g_{ab}(r)\, d^3r \qquad (3.10)$$

where

$$m_{ab} \equiv 2m_a m_b/(m_a+m_b)\ .$$

3.4 Cluster Expansions, Graphs

The cluster expansion of g_{ab} is helpful for understanding the integral equation approximations and several other aspects of correlation function theory. It is expressed in terms of graphs in which Mayer f bonds

$$f_{ab}(r) \equiv \exp[-\beta u_{ab}(r)]-1 \qquad (3.11)$$

connect white circles, representing particles at fixed points, and black circles, representing particles at points which are moved about by integration. For example the following graph, with f bonds on two white 1-circles (i.e. unit weights, fixed species, fixed locations) and one black ρ circle (i.e. with weight ρ_s , to be summed over species) represents the function on the right side of the equation/7/

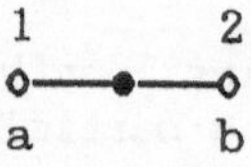

1 2
o——●——o (a, b) $= \Sigma_s\int f_{as}(1\ 3)\ \rho_s\ d(3)\ f_{sb}(3\ 2)$ (3.12)

in a notation in which

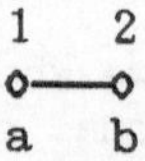

1 2
o——o (a, b) $= f_{ab}(1\ 2) = f_{ab}(r_1, r_2)$ (3.13)

and d(n) means d^3r_n . In this notation we have

$$g_{ab}(1\ 2) = [1+f_{ab}(1\ 2)]y_{ab}(1\ 2)$$

$$= \exp[-\beta u_{ab}(r)]y_{ab}(r) \tag{3.14}$$

where/7/

> $y_{ab}(1\ 2) \equiv 1$ + the sum of all simple, distinct, irreducible graphs of f bonds on two white 1-circles, labelled a and b, and one or more black ρ circles, and with no bond directly connecting the white circles.

$$= 1 + [\text{graph}] + [\text{graph}] + [\text{graph}] + [\text{graph}] + [\text{graph}] + [\text{graph}] \ldots \tag{3.15}$$

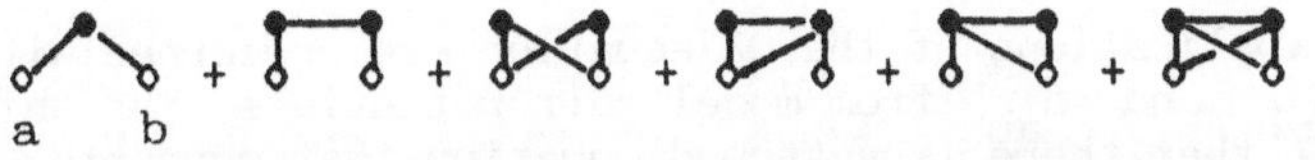

a b

With the help of this cluster expansion it is easy to see how the density expansion of $g_{ab}(r)$ begins, although one may worry whether all of the integrals converge. The corresponding cluster expansion for the potential of the average force between a and b at a separation r also can be written in terms of f-bond graphs,

$$-\beta w_{ab} = -\beta u_{ab} + [\text{graph}] + [\text{graph}] + [\text{graph}] + [\text{graph}] + [\text{graph}] + \cdots \tag{3.16}$$

where the cluster expansion is not quite the same as that for y_{ab} ./3/

3.5 The Direct Correlation Function

Much useful theory is most conveniently cast in terms of the *indirect* correlation function $h_{ab}(r) \equiv g_{ab}(r)-1$ (Eq. (3.6)), a change from g_{ab} that hardly justifies a new name, and the *direct* correlation function $c_{ab}(r)$ which may be given a graphical definition as follows.

We use the term *h-allowed graph* to be any graph in

$$h_{ab} = y_{ab}-1 + f_{ab}y_{ab} \tag{3.17}$$

A *cutting point* in any such graph is a black circle whose removal leaves a disconnected graph in which one can no longer "walk" on the bonds and circles of the graph from the white a circle to the white b circle. The direct correlation function is the subset of h-allowed graphs that have no cutting points. It follows that h_{ab} and c_{ab} are related by the integral equation

$$h_{ab}(1\ 2) = c_{ab}(1\ 2) + \Sigma_s\rho_s\int c_{as}(1\ 3)\ d(3)\ h_{sb}(3\ 2)$$

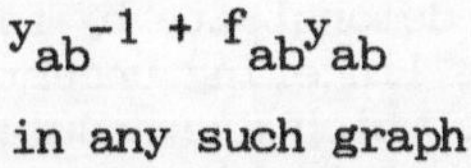

$$= \underset{a\quad b}{\circ\overset{c}{—}\circ} + \circ\overset{c}{—}\bullet\overset{h}{—}\circ \tag{3.18}$$

which is called the Ornstein-Zernike (OZ) equation. It may be represented graphically as shown, where the graph has bonds of the types indicated, where the black circle is a black ρ circle, and where the white 1 circles are separated by a distance $|R_1-R_2|$.

The integral in Eq. (3.18) is of the convolution type, a sort of generalized multiplication which is quite similar to matrix multiplication. The Fourier transform of the OZ equation gives /5/

$$H_{ab}(k) = C_{ab}(k) + \Sigma_s \rho_s C_{as}(k)H_{sb}(k) \tag{3.19}$$

Many applications of the OZ equation are concerned with calculating h_{ab} and c_{ab} from model pair potentials. For this purpose it is important that there is a second equation that connects h and c . It is obtained by classifying the graphs in $w(r)$, the potential of average force. Some will have cutting points; they are also to be found in $h-c$, the sum of all graphs with at least one cutting point. The rest belong to $B(r)$, the sum of bridge graphs, a name suggested by the simplest members in its cluster expansion /8 /

$$B_{ab}(1\ 2) = [\text{graph}] + [\text{graph}] + \ldots \tag{3.20}$$

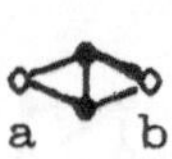

a b

It follows that the required equation, which is exact under pairwise additivity, is

$$c_{ab}(r) = h_{ab}(r) - \ln[1+h_{ab}(r)] - \beta u_{ab}(r) + B_{ab}(r) \tag{3.21}$$

When combined with the OZ equation this equation is the source of many useful approximations. Moreover it can be used to establish the asymptotic large-r behavior of the direct correlation function, namely

$$c_{ab}(r) \rightarrow -\beta u_{ab}(r) \quad \text{as} \quad r\rightarrow\infty \tag{3.22}$$

This very fruitful equation is easy to demonstrate by formal analysis of Eq. (3.21) , but there is room for some lingering uncertainty because the manipulations involve infinite sums which are assumed to converge to their respective leading terms when r is big enough; little is known about the convergence of these series. On the other hand Eq. (3.22) always passes tests in which its consequences are compared with results obtained by other methods, and this has been done for a great variety of systems, the results in Section (3.7) for example.

Various approximate integral equations are based on combining the OZ equation with a truncated form of Eq. (3.21). Most simply we may neglect $B_{ab}(r)$ to get the hypernetted chain (HNC) approximation. This is a non-linear integral equation that may be solved numerically and that proves to be quite accurate for many ionic systems, as judged by comparison with the results of simulation methods applied to the same model pair potentials. In some cases it may be extended by replacing $B(r)$ by $B(r)_{ref}$, the sum of bridge graphs determined for a *reference system* which is hoped to have $B(r)$ that is comparable to the system of interest and which can be treated by some other approximation method which can lead to an estimate of $B(r)_{ref}$; this procedure, the RHNC approximation, is comparable in accuracy to a simulation in some cases./9/

The mean spherical approximation (MSA) is obtained by setting

$$c_{ab}(r)=-\beta u_{ab}(r) \quad \text{if} \quad r>\sigma_{ab} \quad \text{and} \quad h_{ab}(r)=-1 \quad \text{if} \quad r<\sigma_{ab}$$

where σ_{ab} is the diameter of the repulsive core in $u_{ab}(r)$. This is especially neat if $u(r)$ carries a hard core term [$u(r)=\infty$ for $r<\sigma$) plus a "tail" of longer range]. These conditions together with the OZ equation comprise a mathematical structure that can be solved rigorously by analytical mathematics for many tails./11/

We now show how Eqs. (3.18) and (3.22) lead to the Debye-Hückel (DH) theory of dilute electrolytes as well as to the exact Stillinger-Lovett moment conditions.

3.6 The Debye-Hückel approximation.

For simplicity we derive the DH theory for the simple model system in which there is a single species of ion with charge e and concentration ρ embedded in a uniform neutralizing background with the dielectric constant ϵ of free space, thus a one-component plasma (OCP). If ρ is small enough, only interactions at large separation r will be important, so we assume (cf. Eq. (3.22))

$$c(r) = -\beta e^2/\epsilon r \quad , \; 0 \leqslant r \leqslant \infty \tag{3.23}$$

or, in k space,/5/

$$C(k) = -4\pi\beta e^2/\epsilon k^2 \quad , \tag{3.24}$$

an equation that can be more convincingly derived from Poisson's equation for the electrical potential than from Coulomb's law/3/ since the Fourier transform of Eq. (3.23) does not give a definite result. To calculate $h(r)$ from $C(k)$ we notice that in k-space/5/ the OZ equation is (cf. Eq. (3.19))

$$H(k) = C(k) + \rho C(k)H(k) \tag{3.25a}$$

$$= [1-\rho C(k)]^{-1}C(k)$$

$$= -4\pi\beta e^2\epsilon^{-1}/(k^2+\kappa^2) \tag{3.25b}$$

where the last step follows from using Eq. (3.24) together with the one-component version of the Debye κ that is usually defined by the equation

$$\kappa^2 \equiv 4\pi\beta\Sigma_s\rho_s e_s^2 \tag{3.26}$$

where the sum is over species; here we take only the one term ρe^2 of the sum. In r-space Eq. (3.25) becomes

$$h(r) = -\beta e^2 e^{-\kappa r}/\epsilon r \tag{3.27}$$

which is just the one-component version of the DH linearized approximation

$$h_{ab}(r) = -\beta e_a e_b e^{-\kappa r}/\epsilon r \tag{3.28}$$

whose derivation by the above method applied to an electrically neutral mixture of ionic particles is left as an exercise for the reader.

Another result which will be of use below is that the sum of all chains of the "bonds" $-\beta e_s e_{s'}/\epsilon r$ on black ρ circles between white 1 circles labelled a and b and with separation r is

$$\underset{a\quad b}{\circ\!\!-\!\!\circ} + \circ\!\!-\!\!\bullet\!\!-\!\!\circ + \circ\!\!-\!\!\bullet\!\!-\!\!\bullet\!\!-\!\!\circ + \ldots = -\beta e_a e_b e^{-\kappa r}/\epsilon r \tag{3.29}$$

which follows at once from the preceding theory since the OZ equation expresses h as the sum of all chains of c bonds.

It is remarkable that the $e^{-\kappa r}$ shielding factor appears even for the OCP . In this case the familiar Debye-Huckel ionic atmosphere picture does not seem very helpful since in the OCP the interaction between two ions is screened by other ions of the same sign. We conclude that of the two characteristic features of an ionic system, namely the signed potentials and the long range of the interactions, it is the latter that is the seat of the $e^{-\kappa r}$ Debye shielding.

3.7 Stillinger-Lovett Moment Conditions

Although Eq. (3.28) is only exact asymptotically in the limit of *both* low concentrations of the ions and big r , there are other consequences of Eq. (3.22) that are fully exact, namely the moment conditions of Stillinger and Lovett./12/ We first derive them from the approximate equation (3.25b) and then indicate how the derivation can be modified to show that the results are exact and general.

Consider the sum (cf. Eq. (3.4))

$$\Sigma_a e_a M_{ab} = e_b\rho_b + \Sigma_a e_a\rho_a\rho_b[H_{ab}(k)]_{k=0} \tag{3.30}$$

where in place of G_{ab} in Eq. (3.5) we now write $H_{ab}(k)$ at k=0 ./5/ Using Eq. (3.25b) as an estimate for the latter we arrive at the zeroth moment condition

$$\Sigma_a e_a M_{ab} = 0 \tag{3.31}$$

also known as the local electroneutrality conditon.

In the notation in which G_{ab} is the zeroth moment, the second moment of the ab distribution function is

$$\int d^3r\, r^2 h_{ab}(r)$$

which is readily expressed as minus the k=0 limit of $\nabla^2 H_{ab}(k)$, where $\nabla 2$ is the Laplacian operator in k space:

$$\nabla^2 = k^{-2}(\partial/\partial k)k^2(\partial/\partial k) \quad .$$

Again using Eq. (3.25) we easily recover the second moment condition of Stillinger and Lovett/12/

$$\Sigma_a e_a \rho_a \Sigma_b e_b \rho_b \int d^3r\, r^2 h_{ab}(r) = -\Sigma_a e_a \rho_a \Sigma_b e_b \rho_b [\nabla^2 H_{ab}(k)]_{k=0}$$

$$= -3\epsilon/2\pi\beta \tag{3.32}$$

In one method of deriving these results with full generality/3,13/ one expresses $H_{ab}(k)$ in these equations in terms of the exact set of $C_{ab}(k)$ using (3.25a) and the exact version of (3.24),

$$C_{ab}(k) = -4\pi\beta/\epsilon k^2 + O(k^0) \; , \tag{3.33}$$

where the last term stands for a function of k that has a finite value at k=0 ./14/ The proper choice of ϵ here depends on the system; In real systems it is the thermodynamic dielectric coefficient of the mixture./15/ In a model system it is the dielectric coefficient of the medium with the ions removed *if* the potential for any configuration of the ions in the system is pairwise additive. We return to these points in Sections (4.3) and (6.2).

4. SOLUTION THEORIES

4.1 BO-Level

To begin, we explicitly consider certain consequences of viewing g_{ab} as a grand ensemble average,

$$g_{ab}(R,R') = V^2 \langle \delta(r_a - R)\delta(r_b - R') \rangle^d_{\{\mu\},V,T} \tag{4.1}$$

where $\{\mu\} \equiv \mu_1, \mu_2, \ldots$ is the list of the chemical potentials of the species of particles in the system. Then the zeroth moments G_{ab} (Eq. (3.5)) are thermodynamic functions of the independent variables $\{\mu\},V,T$. In Eq. (3.4),

$$\beta(\rho_a \delta_{ab} + \rho_a \rho_b G_{ab}) = M_{ab} = (\partial \rho_a / \partial \mu_b)$$

as elsewhere in this Section, the thermodynamic partial derivatives are to be interpreted in the literal way: they express variation with respect to one of the independent variables with all of the others held fixed. Thus in Eq. (3.4) the derivative $\partial\rho_a/\mu_b$ is the variation of ρ_a due to a change in μ_b while all of the other chemical potentials, the temperature, and the volume are fixed.

We here consider the collection of equations (3.4) for all the species pairs.

$$\underline{\underline{M}} = \beta[\underline{\underline{\rho}} + \underline{\underline{\rho}}\cdot\underline{\underline{G}}\cdot\underline{\underline{\rho}}] \tag{4.2}$$

where $\underline{\underline{\rho}}$ is the $n \times n$ diagonal matrix whose diagonal elements are the stoichiometric concentrations $\rho_1,\ldots,\rho_n$, $\underline{\underline{G}}$ is the $n \times n$ matrix whose elements are the G_{ab} , and $\underline{\underline{M}}$ is the $n \times n$ matrix whose elements are the M_{ab} .

This equation for a mixture can be applied in several ways; here we consider two. In the first, corresponding to what we call the Born-Oppenheimer (BO) level of description of the system,/16,17/ the particles are the molecules and they interact according to a Born-Oppenheimer potential surface obtained (at least in principle) by integrating the Schroedinger Hamiltonian over electronic coordinates. In this case, treated by Kirkwood and Buff,/18/ the matrix inverse $\underline{\underline{L}}=\underline{\underline{M}}^{-1}$ is the matrix whose elements are

$$L_{ab} = \partial\mu_a/\partial\rho_b \tag{4.3}$$

with the independent variables $\rho_1,\ldots,\rho_n$, V, T .

In the case of a two-component system the matrix elements

$$L_{aa} = (\partial\mu_a/\partial\rho_a)_{\rho_b,T}\ , \quad L_{ab} = (\partial\mu_a/\partial\rho_b)_{\rho_a,T}\ , \ \ldots$$

are the derivatives one wants for thermodynamic analysis of systems of specified stoichiometric composition, volume and temperature. Since $\underline{\underline{L}}=\underline{\underline{M}}^{-1}$ the $\partial\mu_a/\rho_b$ can be expressed in terms of the elements of $\underline{\underline{G}}$ to provide a molecular theory of these thermodynamic coefficients. (For condensed matter the L_{ab} are often inaccessible to direct experimental determination but with further thermodynamic manipulation the results may be expressed in terms of more convenient variables, as illustrated by Kirkwood and Buff's explicit results for a two component system./18/)

In the case of ionic systems we have the local electroneutrality condition $\Sigma_a e_a M_{ab}=0$ (cf. Eq. (3.30)) so $\det\underline{\underline{M}}=0$ while $\underline{\underline{L}}$ does not exist. However bulk electrical neutrality imposes the condition $\Sigma_a e_a \rho_a=0$ so the ρ_a are not all independently variable for an ionic system. Choosing n-1 independent concentration variables, just as one does in the thermodynamic theory of ionic systems, we obtain a modified M_{ab} matrix of rank $(n-1)\times(n-1)$ whose inverse does exist, and then proceed as before. This reduction is straightforward for, say, a three component ionic system /19/ but it seems to be very complicated to formulate for an arbitrary mixture./20/ We shall not again mention the $n \times n$ to $(n-1)\times(n-1)$ reduction although it requires attention in any complete model calculation based on the fluctuation theory for ionic systems.

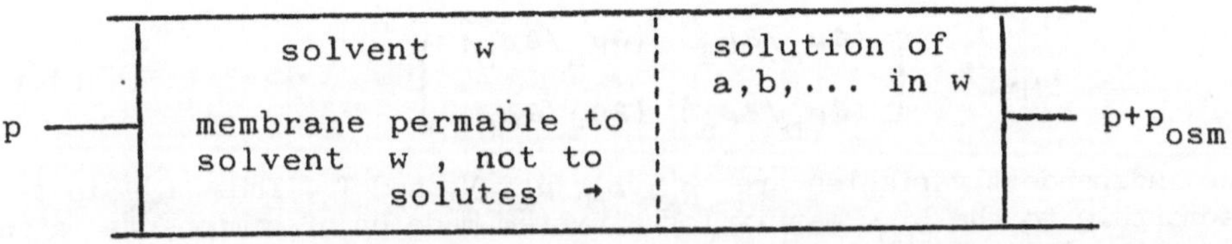

p
solvent w
membrane permable to solvent w , not to solutes →
solution of a,b,... in w
p+p$_{osm}$

Fig. 2 Osmotic pressure experiment related to the McMillan-Mayer theory.

4.2 MM-Level

Another way to interpret Eq. (4.2) is to classify each species as either a solute or a solvent component. For example we consider two solute species (a and b) and one solvent species (w) . We restrict ourselves to a range of states in which μ_W is constant. In this range we need only the solute-solute submatrix of the BO-level matrix

$$\underline{\underline{M}}_{BO} = \begin{pmatrix} M_{aa} & M_{ab} & M_{aw} \\ M_{ba} & M_{bb} & M_{bw} \\ M_{wa} & M_{wb} & M_{ww} \end{pmatrix} \tag{4.4}$$

because for changes in composition within these states we have,

$$d\rho_a = M_{aa}d\mu_a + M_{ab}d\mu_b$$

$$d\rho_b = M_{ba}d\mu_a + M_{bb}d\mu_b$$

These McMillan-Mayer (MM) states are states of a solution that is in osmotic equilibrium with pure solvent that is held at a fixed pressure./21/ (Fig. 2) With μ_w fixed we need only the submatrix

$$\underline{\underline{M}}_{MM} = \begin{pmatrix} M_{aa} & M_{ab} \\ M_{ba} & M_{bb} \end{pmatrix}$$

$$= \begin{pmatrix} (\partial\rho_a/\partial\mu_a) & (\partial\rho_a/\partial\mu_b) \\ (\partial\rho_b/\partial\mu_a) & (\partial\rho_b/\partial\mu_b) \end{pmatrix} \qquad (4.5)$$

which does not involve G_{aw} , G_{bw} or G_{ww} . For the partial derivatives in this matrix the independent variables are μ_a, μ_b, μ_w, V, and T while for the inverse of $\underline{\underline{M}}_{MM}$, namely,

$$\underline{\underline{L}}_{MM} = \begin{pmatrix} (\partial\mu_a/\partial\rho_a) & (\partial\mu_a/\partial\rho_b) \\ (\partial\mu_b/\partial\rho_a) & (\partial\mu_b/\partial\rho_b) \end{pmatrix} \qquad (4.6)$$

the independent variables are ρ_a, ρ_b, μ_w, V, and T . Thus $\underline{\underline{L}}_{MM}$ is isomorphic to the $\underline{\underline{L}}_{BO}$ matrix for a system made up of components a and b ; the solvent is invisible as long as we limit ourselves to the range of states for which μ_w is fixed.

The $\underline{\underline{L}}_{MM}$ matrix carries all of the thermodynamics for changes of state at fixed μ_w . For this range of states the pair correlation functions $g_{ab}(r)$ are functions of ρ_a and ρ_b at the chosen μ_w. It follows that at infinite dilution we have

$$g_{ab} \to \exp[-\beta w_{ab}(r)] \quad , \ \rho_a \to 0 \ , \ \rho_b \to 0 \ , \ \mu_w \text{ constant}$$

$$\equiv \exp[-\beta \bar{u}_{ab}(r)] \qquad (4.7)$$

where the McMillan-Mayer pair potential $\bar{u}_{ab}(r)$ is defined as the limiting value of $w_{ab}(r)$ as the solute concentrations vanish at fixed μ_w .

If N solute particles in any configuration in the pure solvent interact in a way that can be adequately represented as a sum of pair potentials $\bar{u}_{ab}$, then we say that the MM-level potentials are pairwise additive, in analogy to the corresponding BO-level statement (Eq. (3.7)). If both are accurate then we see that the whole theory of the solute-solute correlations at the MM level maps on to the theory of the pair correlation functions at the BO level.

It seems interesting to notice that deviations from Eq. (3.7), i.e. non-additive contributions to the BO potential surface for an N particle system, may be attributed to the prior average over electronic coordinates. In the Schroedinger Hamiltonian pairwise additivity is exact, but averaging over electronic coordinates induces non-pairwise terms on the BO surface. It is expected that averaging over solvent coordinates introduces *additional* non-pairwise additivity in the potential of interaction of N *solute* particles at fixed positions in the solvent, the latter still with the given μ_w . So it is expected that MM-level pairwise additivity is less realistic than pairwise additivity at the BO level.

4.3 MM-BO Correspondence

The BO level thermodynamic relations in Eqs. (3.8) and (3.9) will still hold at the MM level if the sums are restricted to sums over solute particles and u_{ab} is replaced by $\bar{u}_{ab}$. When applied to MM level Eq. (3.8) gives

$$E^{ex} \equiv E\text{-}E(\text{hypothetical ideal solution with same solute composition and same } \mu_w)$$

while Eq. (3.9) gives the osmotic pressure. (But it seems unlikely that Eq. (3.10) can profitably be applied at the MM level; even when the solutes are all much less massive than the solvent particles, the reduced masses of the type m_{aw} would also be small and therefore the omission of terms in g_{aw} could not be justified.)

The concentration dependence of a solute-solute $g_{ab}(r)$ will be given by equations just like (3.14), (3.15) but with f_{ab} replaced by

$$\bar{f}_{ab}(r) \equiv \exp(-\beta\bar{u}_{ab}(r)) - 1 \quad , \tag{4.8}$$

again in cases where the MM-level potentials are pairwise additive. It follows that the rest of Equations (3.11) to (3.20) also apply at the MM level, but with the BO-level pair potential replaced by the MM level solvent averaged pair potential. To give a few examples, in the case of ionic solutions analyzed at the MM level Eq. (3.22) reads

$$c_{ab}(r) \rightarrow -\beta e_a e_b/\epsilon_0 r \tag{4.9}$$

where ϵ_0 is the dielectric coefficient of the pure solvent. In the case of the Stillinger-Lovett moment conditions, they apply as well at MM-level but now the ϵ in Eq. (3.32) and in (3.33) (they must be the same!) is ϵ_0 if MM-level pairwise additivity is assumed. If not, then it is ϵ_ρ, the dielectric coefficient of the mixture. (See Section 6)

4.4 MM Approximation Methods

In the case of BO-level systems some of the most powerful approximation methods are based on the OZ equation as remarked in Section 3. Here we write the OZ equation for a mixture of n species in terms of nxn matrices in k-space (cf. Eq. (4.2))/5/

$$\underline{\underline{H}} = \underline{\underline{C}} + \underline{\underline{C}}\cdot\underline{\underline{\rho}}\cdot\underline{\underline{H}} \tag{4.10}$$

Adding subscripts BO for clarity in the following discussion, and rearranging we obtain

$$[\underline{\underline{1}} - \underline{\underline{\rho}}_{BO}\cdot\underline{\underline{C}}_{BO}]\cdot[\underline{\underline{1}} + \underline{\underline{\rho}}_{BO}\cdot\underline{\underline{H}}_{BO}] = \underline{\underline{1}} \tag{4.11}$$

For the MM-level description of the same mixture, in which we designate one solvent species and n-1 solute species, the corresponding equation is found in terms of $(n-1)\times(n-1)$ matrices in which the indices label only solute species.

$$[\underline{\underline{1}} - \boldsymbol{\varrho}_{MM}\cdot\underline{\underline{C}}_{MM}]\cdot[\underline{\underline{1}} + \boldsymbol{\varrho}_{MM}\cdot\underline{\underline{H}}_{MM}] = \underline{\underline{1}} \tag{4.12}$$

Each matrix element of $\underline{\underline{H}}_{MM}$ is identical to the respective matrix element of $\underline{\underline{H}}_{BO}$ in view of Eq. (3.10); deleting the last colunmn and last row of $\underline{\underline{H}}_{BO}$ gives $\underline{\underline{H}}_{MM}$. The corresponding statement is not true for the relation of $\underline{\underline{C}}_{MM}$ to $\underline{\underline{C}}_{BO}$ as one can easily see by writing out the n=2 equations in detail.

To have an approximation method for the MM system we still need a closure for Eq. (4.12). Formally this is provided by the equation (cf. Eq. (3.21)) relating functions of a given ab pair

$$[c_{ab}]_{MM} = h_{ab} - \ln(1+h_{ab}) - \beta[u_{ab}]_{MM} + [B_{ab}]_{MM} \tag{4.13}$$

where $[B_{ab}]_{MM}$ is the same functional of $\underline{\underline{H}}_{MM}$ as $[B_{ab}]_{BO}$ is of $\underline{\underline{H}}_{BO}$ as implied by Eq. (3.20)./8/ Equation (4/13) follows only under MM-level pairwise additivity; the other case is treated below. Thus when Eq. (4.13) is valid, the pair potential $[u_{ab}]_{MM}$ is independent of the ion concentrations and we have (cf. Eq. (4.7))

$$[u_{ab}(r)]_{MM} = \bar{u}_{ab}(r) \tag{4.14}$$

in full analogy to the BO-level case.

Thus with pairwise additivity for the potential of any N-ion configuration in the solvent, whether as a feature of a model or as an approximation for a real system, the various approximation methods based on the OZ equation together with Eq. (3.21) that may be applied to BO-level systems can equally be applied to MM-level. This simple procedure works reasonably well judging by the comparison of experimental and calculated properties for quite a variety of models, approximation methods, and real solutions./23-29/ A few examples (Fig. 3) are based on the model

$$\bar{u}_{ab}(r) = \gamma_{ab}/r^9 + GUR_{ab}(r) + z_a z_b e^2/\epsilon r \tag{4.15}$$

where the r^{-9} term represents the core repulsion which depends on the ionic radii through γ_{ab}, where the "Gurney" term, which incorporates an adjustable parameter A_{ab}, has a short range and a modest effect but enables us to fit the model to one set of data and then apply it to another, and where we also have the Coulomb term with the ionic charges $z_a e$ and $z_b e$ and the dielectric constant ϵ of the pure solvent.

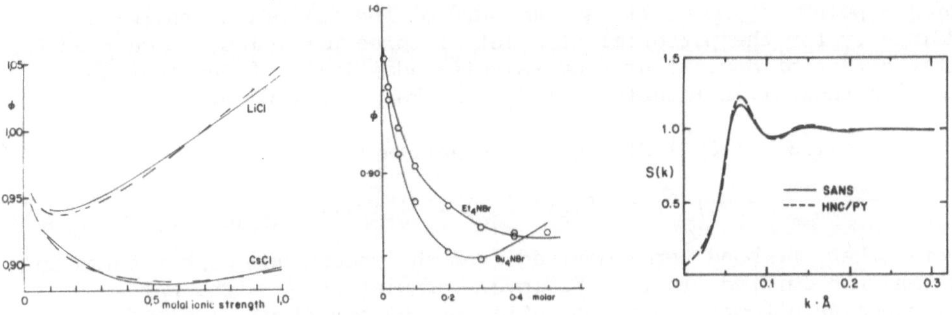

Fig. 3 Comparison of model calculations and experimental data for several real systems. a,b) osmotic coefficient(refs 25,26. c) small angle neutron scattering for a polyelectrolyte $Li_{64}P$. Experimental data for $S_{pp}(k)=1+\rho_p H_{pp}(k)$ compared with an integral equation calculation based on a primitive model./27/

Solvent-averaged pair potentials for aqueous solutions have recently been calculated from BO-level models by simulation/32/ and integral equation methods./30,31/ The $\bar{u}_{ab}(r)$ functions calculated in this way are realistic, judging by comparing the derived osmotic coefficients/30/ and spin relaxation coefficients/31/ with the relevant experimental data. The BO-level-derived solvent-averaged pair potentials show much structure, reflecting the molecular nature of the solvent, compared to the Eq. (4.15) model. It is therefore quite remarkable that calculations based on the latter have been as successful as they have. Thus, in addition to the satisfactory concentration dependence of the calculated results illustrated in Fig. 3, the Gurney parameters determined in the fitting process show reasonable trends when results for whole classes of solutions are compared, and they also accomodate data for mixed electrolytes in a satisfactory way./28,29/

4.5 Non-Pairwise Terms in the Configurational Potential

As an example of what may be expected when the corrections to pairwise additivity are important, we begin with the cluster expansion of the potential of average force for a one-component system at BO level. To Eq. (3.16) we need to add terms such as

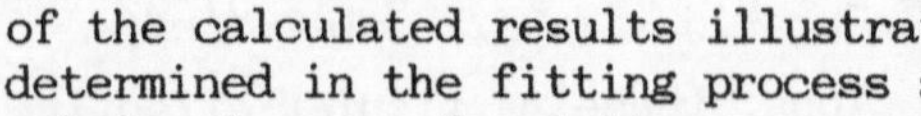 (4.16a)

where we use the notation

$f_3(1\ 2\ 3) \equiv$ 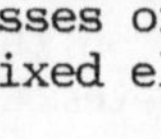 $\equiv \exp\{-\beta[U_3(1\ 2\ 3)-u(1\ 2)-u(2\ 3)-u(3\ 1)]\} - 1$ (4.16b) for a 3-point f bond. It is a measure of the failure of pairwise additivity for the potential of a set of three particles; it carries the first effect of deviations from pairwise additivity of the configurational potential. Evidently in this case Eq. (3.21) becomes

$$c(1\ 2) = -\beta u(1\ 2) + h(1\ 2) - \ln[1+h(1\ 2)] + + B(1\ 2) + \rho\int f_3(1\ 2\ 3)\ d(3) + \ldots \tag{4.17}$$

where other f_3-bond terms derived from the second and higher terms in (4.16a) are omitted. In this BO-level formulation we might well introduce an effective pair potential by the equation

$$-\beta u(1\ 2)_{eff} = -\beta u(1\ 2) + \rho\int f_3(1\ 2\ 3)\ d(3) + \ldots; \tag{4.18}$$

the *effective* pair potential clearly depends on the concentration ρ.

For ionic solutions treated by MM-level theory the f_3 term in Eq. (4.18) is long range, as we shall see. Such a term in the equation for the ion-ion $c_{ab}(r)$ even contributes to the asymptotic big-r behavior of this function.

4.6 Adelman's Solution Theory

We can analyze the system of MM-level equations (4.12) and (4.13) in another way./33,34/ It is sufficient to suppose that there exists an effective N-point solvent averaged potential (cf Eq. (3.7))

$$U_N(1, 2,\ldots,N)_{eff} = \Sigma_{ab}\, \Sigma_{ij}\, \delta_{i/a}\, \delta_{j/b}\, u_{ab}(i\ j)_{eff} \tag{4.19}$$

for any configuration of N ions; these particular effective potentials are expected to be state dependent, depending on the solute concentrations ρ_a, ρ_b, ... as well as on the temperature and the pressure or the solvent chemical potential.

If the effective pair potentials were known we could use them to calculate the solute-solute pair correlations h_{ab} and c_{ab} by solving Eqs. (4.12) and (4.13) together, with some approximation for $B_{ab}(r)$, just as in the discussion of Eq. (4.14). But practically we only know that $[u_{ab}]_{eff}$ is a pair component of an *exactly* pairwise-additive effective potential, which is a simplification compared to $\bar{u}_{ab}$ plus non-pairwise correction terms, but $[u_{ab}]_{eff}$ depends on the solute concentrations, which is a complication compared to $\bar{u}_{ab}$ itself.

The most important feature of $[u_{ab}]_{eff}$ is associated with Eq. (4.13) which we now write in the form

$$[c_{ab}]_{eff} = h_{ab} - \ln(1+h_{ab}) - \beta[u_{ab}]_{eff} + [B_{ab}]_{eff} \tag{4.20}$$

then we can see that

$$[c_{ab}(r)]_{eff} \rightarrow -\beta[u_{ab}(r)]_{eff} \quad \text{as} \quad r \rightarrow \infty \tag{4.21}$$

by the same argument that leads from Eq. (3.21) to (3.22). Because

$$[u_{ab}(r)]_{eff} \rightarrow e_a e_b/\epsilon_\rho r \quad \text{as} \quad r \rightarrow \infty \tag{4.22}$$

may be taken as a definition of the thermodynamic dielectric coefficient of the solution ϵ_ρ and because $[u_{ab}(r)]_{eff}$ depends on the ion concentrations, it follows that ϵ_ρ also depends on the ion concentrations. This analysis leads to the same conclusion as that in Section (6), but by more general arguments.

5. SECOND VIRIAL COEFFICIENTS

5.1 G_{ab} at Infinite Dilution.

In view of Eq. (3.24) we have, in terms of f bonds on two white 1-circles and zero or more black ρ circles,

$$G_{ab} = \int d^3r[\overset{a}{\circ}\!\!-\!\!\!-\!\!\overset{b}{\circ} + \circ\!\!\diagup\!\!\!\bullet\!\!\!\diagdown\!\!\circ + \circ\!\!\diagup\!\!\!\bullet\!\!\!\diagdown\!\!\!-\!\!\circ \ \ldots] \quad . \tag{5.1}$$

In a BO system of two non-ionic components at infinite dilution (ρ_a=0,ρ_b=0) the terms with black ρ circles can be neglected. Then G_{ab} is just the usual second virial coefficient if a=b while it is the corresponding cross interaction in the other case.

In a BO-level system of two ionic components we have $f_{ab}(r) \rightarrow -\beta e_a e_b/\epsilon r$ as r gets big and then each term after the first in Eq. (5.1) leads to a divergent integral. Eq. (3.29) can be applied to analyze these terms and eliminate the divergences if $\kappa>0$ ./3,35/ Then G_{ab} exists at finite concentration but diverges like $1/\kappa$ in the infinite dilution limit, as one may see by substituting Eq. (3.28) in Eq. (3.5) and doing the integration for the $\kappa>0$ case.

In the third case we have a BO-level mixture of a non-ionic substance b , say argon, and a binary electrolyte ca , say NaCl, all in solvent w . One might suppose that as we approach infinite dilution (ρ_a=0, ρ_b=0 , ρ_c=0) we have

$$G_{cb} \rightarrow \int d^3r\, f_{cb}(r) \quad \text{and} \quad G_{ab}(r) \rightarrow \int d^3r\, f_{ab}(r) \tag{5.2}$$

but this is not the case because many of the diagrams in eq. (5.1) give divergent integrals even if b has no electric charge. These complications illustrate the great strength of the correlation induced by Coulombic interactions./29/ However it is easy to see that the complications of the infinite integrals are avoided (i.e. the great range of the Coulombic interactions taken into account) if we look at the coefficient

$$G_{ca,b} \equiv \nu_c G_{cb} + \nu_a G_{ab} \tag{5.3}$$

where ν_c and ν_a are the stoichiometric coefficients of the ca electrolyte. This combination causes a mutual cancellation of all of the most divergent graphs (see Eq. 5.1), in which the second white circle is labelled b and the first white circle, labelled either c or a, is connected to the rest of the graph by a single ion-ion f bond.

5.2 Setchenow Coefficients

In suitable *solutions* with solutes a , b, and c as above, $G_{ca,b}$ is directly measurable . Of course the BO level theory just described is directly convertable to the MM-level solution theory; one has only to change f_{ab} to $\bar{f}_{ab}$, etc.

A relevant experiment is called the measurement of a "salting - out" coefficient, or better, a Setchenow coefficient,/36/ since the effect can equally well have the "salting-in" sign. One measures

$$k_{ca,b} \equiv -(\partial \ln \rho_b/\partial \rho_{ca})_{\mu_b} \tag{5.4}$$

where $\rho_{ca} \equiv \rho_c/\nu_c = \rho_a/\nu_a$. The measurement is feasible if a pure b phase is in equilibrium with the solution. It is most often found that $k_{ca,b}$ is insensitive to the electrolyte concentration over the wide range up to 1mol dm^{-3} , so it often is easy to measure accurately. To get the coefficient of interest we rearrange Eq. (5.4) as follows/37/

$$k_{ca,b} = (\partial \mu_b/\partial \rho_{ca})_{\rho_b} \;/\rho_b(\partial \mu_b/\partial \rho_b)_{\rho_{ac}} \approx -G_{ca,b} \tag{5.5}$$

It is left as an exercise in thermodynamics to show that $k_{ca,b}$ also can be determined if one measures the solubility of a slightly soluble electrolye as a function of the concentration of non-electrolyte./38/ Of course it is not often that a given ca,b solute pair can be investigated in both ways.

The salt:non-electrolyte second virial coefficient $G_{ca,b}$ is potentially very useful in the process of trying to learn what various experimental coefficients tell us about model pair potentials, but it has only rarely been exploited for this purpose. An example of an interesting aspect of alkali metal ion-nonelectrolyte interaction that has been uncovered in this way is shown in Table I. It shows that

$$\int d^3r \, f_{cb}(r)$$

does not vary in the expected monotone way in the series c= $Li^+,\ldots,Cs^+$. In other cases the Setchenow coefficients reveal great regularity in the solute-solute pair interactions. This is especially the case for hydrophobic solute pairs, such as C_6H_6-$(CH_3)_4N^+$ ./23,39/

Table I. Setchenow coefficients, $dm^3 mol^{-1}$ aqueous, $25^{\circ}C$
From ref. 23 compilation.

salt:	LiCl	NaCl	KCl	RbCl	CsCl
Ar	0.226	0.314	0.270		
C_6H_6	0.325	0.449	0.382	0.322	0.203

6. ION-DEPENDENT DIELECTRIC COEFFICIENT

As noted in Section 4, the diagrams in Eq. (4.16) may contribute long-range terms to the effective pair potential. Here we investigate this point by evaluating the leading diagram on the basis of the underlying BO-level theory, i.e. an ionic solution theory in which the solvent molecules appear explicitly./41/ Thus we begin with the potential of average force for a pair of ions in a solution at finite concentration,

$$-\beta w_{ab}(1\ 2) = -\beta w^{0}_{ab}(1\ 2) + \qquad (6.1)$$

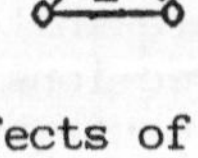

where w^0_{ab} is the function without the effects of deviations from pairwise additivity. The BO-level equation for the same quantity is

$$-\beta w_{ab}(1\ 2) = -\beta u_{ab}(1\ 2) + \quad + \quad + \quad + \quad + \ldots$$

$$+ \quad + \quad + \quad + \quad + \quad + \ldots$$

$$+ \quad + \quad + \ldots \qquad (6.2)$$

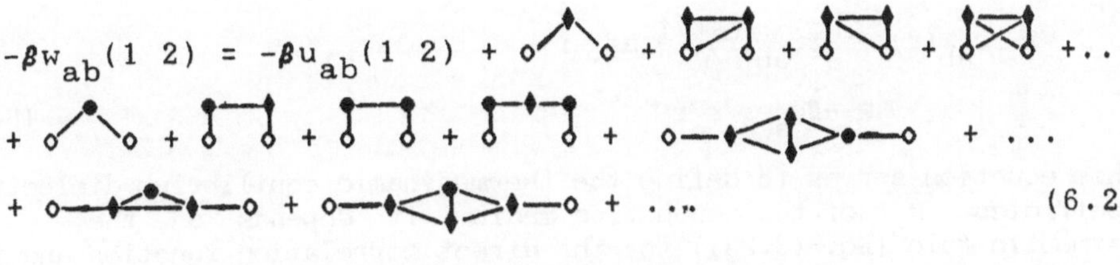

where we have white ion circles ○ labelled a and b , black ρ_{ion} circles ●, and black ρ_w "circles" ♦ .

Real solvents for electrolyte solutions are polyatomic; their pair interactions depend on molecular orientations and, perhaps, distortions, as well as on center of mass distance. Then an f_{aw} or f_{ww} bond depends on the respective coordinates of the particles it connects. Therefore a white solvent circle will depend on the set of relevant molecular coordinates. Blackening the circle implies integrating over all of these coordinates. While the dependence on more than three coordinates per particle makes the evaluation of the cluster integrals relatively difficult, we are here more concerned with the topology of the graphs. Then the problem of actually using the relevant coordinates /3,42-44/ need not concern us.

Equation (6.2) is arranged in such a way that $-\beta u_{ab}$, together with the graphs with only black ρ_w circles combine to give $-\beta \bar{u}_{ab}$, thus $-\beta$ times the MM-level pair potential in Eq. (4.7).

Including graphs like those in the second line, which can be reorganized to become graphs of $\bar{f}_{ab}$ solvent-averaged ion-ion bonds on black ρ_{ion} circles, would give $-\beta w0_{ab}$ in Eq. (6.1) . The characteristic topological feature of the graphs in the second line is that any interconnected cluster of solvent circles ♦ and f_{ww}-bonds alone lies between two ion circles. That is, any solvent cluster can be disconnected from the rest of the graph by deleting two ion circles.

The graphs in the third line violate this condition: They have interconnected clusters of two or more solvent circles that are connected independently to three (or more) ions. It seems clear that the graphs in the third line contribute to deviations from MM pairwise additivity. Indeed if we contemplate the corresponding graphs in which the black *ion* circles are whitened, we can see how the solvent-averaged interaction of the three ions prescribed by each graph is not made up of additive ionic pair contributions./41,43/

The effect of deviations from MM pairwise additivity on the dielectric constant is readily formulated because the McMillan-Mayer level direct correlation function of an electrolyte solution can be written in the form (cf. Eq. 4.9)

$$c_{ab}(r) \to -\beta u_{ab}(r)^{eff} \quad \text{as} \quad r\to\infty$$

$$\equiv -\beta e_a e_b/\epsilon_\rho r \tag{6.3}$$

This equation serves to define the thermodynamic equilibrium dielectric coefficient ϵ_ρ of the conducting medium. It depends on the asympotic rule (Eq. (3.23)) for the direct correlation function and the fact that the direct latter does not carry the exponentially damped distance dependence due to the Debye $\exp(-\kappa r)$ shielding (or its generalization to the regime of non-vanishing ion concentrations./45/) Thus if one actually measured the force between test charges a and b in an ionic medium he would be measuring minus the gradient of the potential of average force $w_{ab}(r)$ of the charges, and would have to devise a correction to cancel the exponential damping and recover the underlying coefficient of the $1/r^2$ dependence of the force. But the theory that shows how the Debye shielding arises (derivation of Eq. (3.27)) also shows that the direct correlation function is not shielded; it is especially suitable as a theoretical probe of the effective dielectric coefficient ϵ_ρ .

By deducing the corresponding term in the direct correlation function from Eq. (6.2) one finds that ϵ_ρ can be related to the structure of the solution as follows./31/

$$\frac{1}{\epsilon_\rho} = \frac{1}{\epsilon_0} - \frac{1}{yV\rho_w}\left[\frac{\epsilon_0^{-1}}{3\epsilon_0}\right]^2 \frac{3}{[8\pi^2]^3} \qquad (6.4)$$

where subscript 0 pertains to the pure solvent, y is the reduced dipole concentration $y=4\pi\beta\rho_w\mu_w^2/9$, the solid bonds in the graph are h bonds connecting particles in the pure solvent, the dotted "bond" connecting the two solvent circles is $\hat{\mu}_1 \cdot \hat{\mu}_2$, the product of the unit dipoles in the solvent molecules, and each black circle implies integration over the orientation (three Eulerian angles) as well as location. The correction term is linear in ρ_{ion} *via* the black ρ ion circle in the graphs; higher terms in ρ_{ion} are neglected here. Also the diagram shown is only the first of an infinite set, but of these it is the only one that would be included in the HNC approximation applied to the relevant BO-level model./41,43/

The experimental situation is complicated by the fact that ϵ_ρ is not directly measurable. What is accessible is $\epsilon(\omega)$ measured at relatively high frequency ω and extrapolated to $\omega=0$, yielding the sum $\epsilon_{sol} \equiv \epsilon_\rho + \epsilon_{KDD}$, where the kinetic dielectric decrement ϵ_{KDD} is a dynamical contribution whose effect does not vanish at $\omega=0$. Using the theory of ϵ_{KDD} due to Onsager, Hubbard, and Wolynes/46/ together with his BO-level models for aqueous electrolytes, Patey and coworkers have calculated both ϵ_ρ and ϵ_{KDD} and find substantial agreement with experimental data for ϵ_{sol}./47/ Their results are at least roughly consistent with the equation $\epsilon_{sol}=80/(1+I/3)$ where I is the molar ionic strength.

7. MIXED ELECTROLYTES

7.1 Mixing Coefficients for Common Ion Mixtures

In Section 5 we formulated the theory of the experimental coefficients that can be determined by varying the electrolyte concentration in solutions with a non-electrolyte at a fixed chemical potential. A somewhat analogous theory is described here; it concerns the thermodynamic coefficients that characterize the process of mixing two electrolytes having a common ion.

Here the composition variations which are most interesting are at fixed ionic strength. With this constraint the cluster integrals, which have been modified by repeated application of the operation in Eq. (3.29) to replace bare Coulomb interactions by shielded, hence κ-dependent interactions, are independent of the remaining composition variable(s). So we recover a bit of the simplicity of the theory of non-ionic systems.

When solutions of electrolytes A (ion species 1 and 3) and B (ion species 2 and 3) having the same molal ionic strength I , are mixed at fixed pressure and temperature, the change in the excess Gibbs function G^{ex} per unit of solution may be written in the form/40/

$$\Delta_m G^{ex}(y,I) \equiv G^{ex}(y,I)-yG^{ex}(1,I)-(1-y)G^{ex}(0,I)$$

$$= I^2 RTy(1-y)\Sigma_{n=0}\, g_n(I)(1-2y)^n \qquad (7.1)$$

where y is the fraction of I coming from the ions of electrolyte A, and g_n is the *nth order mixing coefficient*, a combination of the modified cluster integrals mentioned above. By convention $|z_1|\geq|z_2|$; then it may be shown that g_1 is positive at small I for unsymmetrical mixtures./40/ The unit of solution here is the quantity of solution containing a kilogram of solvent.

The enthalpy of mixing $\partial(\Delta_m G^{ex}/T)/\partial(1/T)$ has a similar expansion in which the coefficients are

$$h_n \equiv -T(\partial g_n/\partial T)$$

The study of the n=1 coefficients was inspired by the measurements of $h_1(I)$ for mixtures of Na_2SO_4 with NaCl and mixtures of $BaCl_2$ with NaCl by Cassel and Wood./48/ When combined with the cluster theory prediction that $h(I) \sim \sqrt{I}$ at small enough I their data lead to a picture of $h_1(I)$ for each system that is qualitatively similar to the calculated $g_1(I)$ function shown in Fig. 4; in particular there is a quite sharp peak in h_1 in the low I-range in which the Debye-Hückel linearized theory is accurate for many purposes. (But see ref. 40 for remaining inconsistencies.)

In more familiar cases the thermodynamic coefficients of electrolytes exhibit especially bland I-dependence in the low I range. So it is interesting to see whether this behavior of $h_1(I)$ is consistent with what is known about ionic solution theory and to ask what might be learned about specific ion-ion interactions from comparison of theory and experiment for h_1 or for g_1 . The latter function is more directly accessible in a theoretical study.

7.2 Mixture Limiting Laws

The cluster theory sketched below/22,40,49,51/ shows that g_0 is governed by one limiting law

$$\ln[g_0(I)/g_0(0)] = (z_1^2\lambda^{3/2}/2\pi)(N_{AV}/V_w)^{1/2}\, I^{1/2} \qquad (7.2)$$

if $z_1=z_2$ and another

$$g_0(I) = (z_1-z_2)^2(\lambda^3/96\pi^2)(N_{AV}/V_w)\ \ln I \qquad (7.3)$$

otherwise. In these equations $\lambda \equiv 4\pi\beta e^2/\epsilon$, where e is the electronic charge, ϵ the solvent dielectric constant. Thus λ is 8π times the

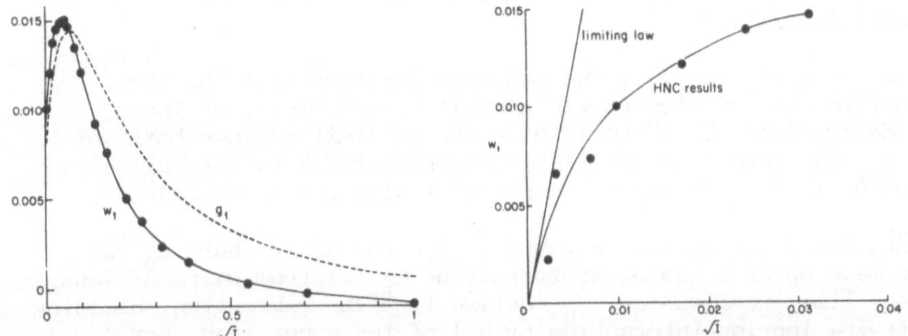

Fig. 4. a) First-order mixing coefficients calculated by the HNC approximation applied to MM-level models. •, $w_1(I)$ data points. ——, fitted w_1 curve. - - -, $g_1(I)$ from integrating w_1 according to Eq. 7.9). b) Enlarged view near I=0 to show the approach to the limiting law for w_1 . From ref. 40.

Bjerrum length for a 1-1 electrolyte./50/ Also N_{AV} is Avogadro's number, and V_w is the volume, in the same units as $\lambda 3$, occupied by a kilogram of solvent in the same P,T state as the solution. These limiting laws are distinct from the familiar Debye-Hückel single electrolyte limiting law for G^{ex} and its thermodynamic derivatives.

The origins of the limiting laws can be described in terms of the cluster expansion/3,22,35,51/ for the extensive excess Helmholtz free energy A^{ex} in a McMillan-Mayer state (specified temperature, volume V, solvent activity, and concentrations $\rho_1,\ldots,\rho_s,\ldots$ of ions of various species) of the solution

$$-A^{ex}/VkT = \kappa^3/12\pi + \Sigma_s\Sigma_t\rho_s\rho_t B_{st}(\kappa) + \Sigma_s\Sigma_t\Sigma_u\rho_s\rho_t\rho_u B_{stu}(\kappa) + \cdots$$

$$= \kappa^3/12\pi + \bullet\!\!-\!\!\bullet + \triangle + \square + \boxslash + \boxtimes + \ldots \quad (7.4)$$

Here $B_{stu\ldots}(\kappa)$ is the κ-dependent nth-order virial coefficient resulting from the Mayer resummation of chains of Coulombic interaction (Mayer renormalization). In the second line we represent the same series in terms of graphs of f' bonds on black ρ_{ion} circles. The bond function $f'_{ab}(r)$ is given by

$$f'_{ab}(r) = \exp[-\beta u'_{ab}(r)] - 1 \quad (7.5)$$

where $u'_{ab}(r)$ is obtained from the full McMillan-Mayer $\bar{u}_{ab}(r)$ by inserting the shielding factor $\exp(-\kappa r)$ in the Coulomb term $e_a e_b/\epsilon r$. (Eq. 3.29) The graphical representation in Eq. (7.4) is incomplete because well-known corrections for the overcounting of whole classes of interactions have been suppressed. This simplification seems desirable for our purpose. Of course the actual calculations have been made with the fully correct version of the cluster expansion or with the HNC approximation./22,40/

In terms of Eq.(7.4), the ordinary limiting laws for the osmotic coefficient and electrolyte activity coefficients of ionic solutions derive from the κ^3 term while the g_0 limiting laws both derive from the $B_{st}(\kappa)$ terms. The available experimental data for $g_0(I)$ and its thermodynamic derivatives are consistent with the theory./48,49/

Analyzing the cluster expansion also shows/51/ that g_0 at low I is made up of a sum of contributions $B_{st}(\kappa)$ from pairs of ions, and that the limiting laws for g_0 derive from the Debye-like shielding of the pairwise ion-ion interaction by all of the other ions. For example, if $z_1=z_2$ we find as $I\to 0$

$$g_0(I) \to 2\frac{\int_0^\infty(2f^s_{12}-f^s_{11}-f^s_{22})\exp[-z_1^2\lambda e^{-\kappa r}/4\pi r]\ 4\pi r^2 dr}{z_1^2(z_1-z_3)^2} \tag{7.6}$$

where f^s_{ab} is the cluster function for the short-range part of the pair potential that remains after the Coulombic term is removed. Thus g_0 is dominated by contributions from close ion pairs with the common ion excluded. The data for g_0 are available for many common-ion mixtures of two 1-1 electrolytes. It is found that Eq. (7.6) applied to the primitive model gives g_0 that is only a fraction of that determined experimentally; with the Eq. (4.15) type of model, however, agreement with experiment is readily obtained./19,51/

In the unsymmetrical case, distant pairs of all charge types determine g_0 ./22,51/

7.3 The g_1 Mixing Coefficient

The coefficient g_1 derives mostly from the $B_{stu}(\kappa)$, thus from clusters of three ions, with a minor contribution from the first and third B_4 diagrams in Eq. (7.4), all with the corrections mentioned above. In view of the structure of the f' bonds in Eq.(7.4) there is Debye screening of the Coulombic parts of the interactions in these clusters. On this basis a molecular explanation for the strong maximum in g_1 (Fig. 4) in dilute solutions may be tentatively suggested:

Among the triple ions which may be found in a mixture of electrolytes A and B, are some with mixed charges, say two positive charges and a negative charge. Such an ion triple, presumably mostly in a linear +-+ configuration, is stabilized by reducing I , thus reducing the Debye shielding of the two +- interactions. However, as I is reduced even more, we reach a regime in which the secular concentration dependence [law of mass action ~ explicit concentration factors in Eq. (7.4)] is the dominant factor in the stability of the complex. The biggest such effect may be expected for the ion triple with the strongest +- interactions. This picture is analogous to the molecular interpretation of the minimum in $[\phi(I)-1]/\sqrt{I}$ in aqueous solutions of single electrolytes of the $CuSO_4$ charge type/52,53/ (ϕ is the osmotic coefficient).

All of the above equations except (7.4) are in the Lewis and Randall thermodynamic system of independent variables (T, P, and molalities)/22,54/ which are most directly applicable to analyzing the thermodynamic data. On the other hand the statistical-mechanical calculations are made in the framework of the McMillan-Mayer theory and are most simply based on the change in Helmholtz free energy when A and B are mixed at fixed Debye κ (hence fixed *molar* ionic strength), fixed temperature, and fixed solvent activity. The subtle aspects of the conversion from the resulting McMillan-Mayer system to the Lewis-Randall system are well known/54/ but are of little consequence at low I . Therefore we simply overlook these complications here.

Among the free energy functions that characterize the common-ion electrolyte mixture the following are the most straightforward to determine, whether by experiment or by model calculations.

$\phi(y,I)$ is the osmotic coefficient, given by

$$(1-\phi)m_t RT = [\partial(G^{ex}/m_t)/\partial(1/m_t)]_y \tag{7.7}$$

where m_t is the total molality of ions

$$m_t = m_1+m_2+m_3 = 2I[y/|z_1z_3| + (1-y)/|z_2z_3|] \;.$$

The way the osmotic coefficient changes in the mixing process is given by a new set of mixing coefficients

$$\Delta_m\{m_t(y,I)[\phi(y,I)-1]\} = y(1-y)I^2(w_0+w_1(1-2y)+\ldots) \tag{7.8}$$

Thermodynamic consistency leads to various relations among the mixing coefficients of given order n . The one we need is

$$w_n = \partial(Ig_n)/\partial I \tag{7.9}$$

for the n=1 case.

7.4 Model and HNC approximation

Beginning with model solvent-averaged pair potentials (Eq. (4.15)) which earlier had been fitted to thermodynamic data for a $MnCl_2$-LiCl mixture, the HNC approximation was used to calculate the set of model ion-ion pair correlation functions $g_{ab}(r)$ for given y, I ./40/ The osmotic coefficient ϕ was then calculated *via* the MM-level osmotic pressure version of Eq. (3.9), in which a and b are limited to solute species. Then $\phi(y,I)$ over a range of y at fixed I was fitted to Eq. (7.8) to extract $w_1(I)$ with the results shown in Fig. 4.

The results for w_1 and g_1 shown in Fig. 4 are found to be accurate enough to satisfy self-consistency tests except at the lowest concentrations shown. The peak in g_1 is a result of deviations from the g_1 limiting law; it is sensitive to short-range interactions. This was shown "experimentally" by varying the Gurney parameter for the $Cl^-:Mn^{2+}$ interaction.

The resulting effect on g_1 is noticeable, but less directly interpretable than the effect shown in Fig. 5 for the I-dependence of reduced pair correlation functions formed by dividing by a Debye-Hückel type of estimate to remove the strongest part of the ionic strength dependence. (cf. Eq. 7.5)

$$\hat{g}_{ab}(r) \equiv g_{ab}(r)/\exp[-\beta u'_{ab}(r)] \qquad (7.10)$$

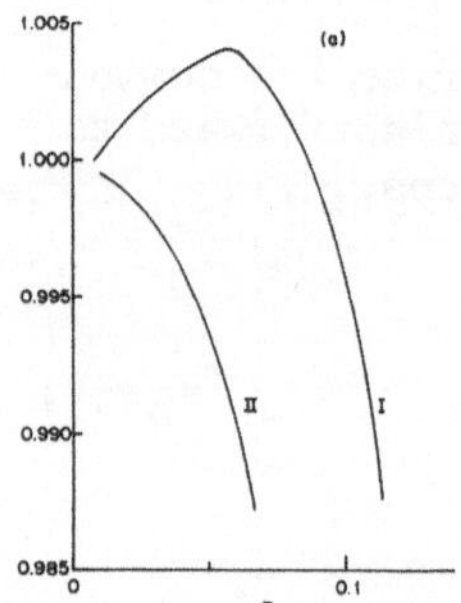

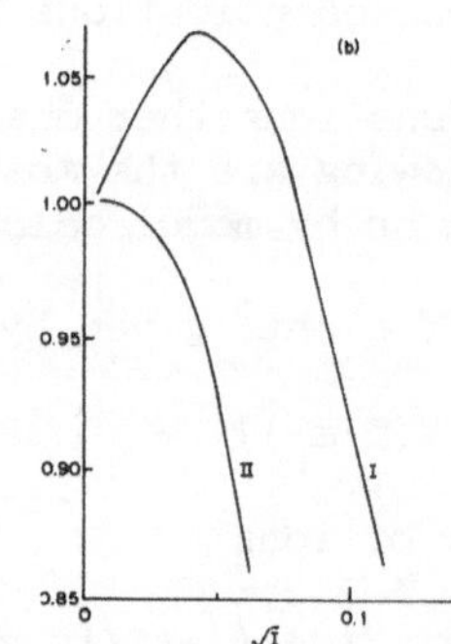

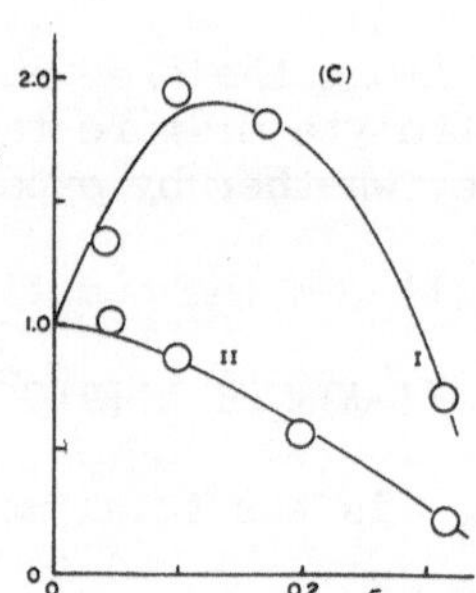

Fig. 5. (a,b) Ionic strength dependence of the reduced pair correlation function $\hat{g}_{ab}(4Å)$) for $ab=Li^+Mn^{2+}$ in (a) and $ab=Mn^{2+}Mn^{2+}$ in (b). In each case curve I is for the model fitted to thermodynamic data while curve II is for the same model except that the Gurney parameter $A_{Mn,Cl}$ is increased by 0.1RT. (c) The ordinate is $g_{ab}(R)$ for the pair $Ru(NH_3)_4bpy^{2+}$, $Ru(NH_3)_4bpy^{3+}$ calculated from the electron exchange rate data of Brown and Sutin./55/ For curve I the medium is $HClO_4$ while for curve II it is CF_3SO_3H.

In Fig. 5c is an *experimental* coefficient that is closely related to the theoretical coefficients discussed above. Its origin is described below; here we conclude by pointing out that taken together these figures indicate that the sharp peak in g_1 may well be related to certain features of the medium-dependence of the rates of activation-controlled reactions between highly charged ions of the same sign.

8. MEDIUM EFFECTS IN ACTIVATION-CONTROLLED REACTIONS

Ionic solution theory has long been applied to the study of medium effects on the rate constants of activation-controlled chemical reactions. Thus for the chemical reaction

$$A + B \rightarrow AB \rightarrow \text{products}$$

which involves solute species A and B and a transition complex AB, we ask how the rate constant k_{ab} is affected by changes in the composition of the solution. If the rate constant is activation controlled, it is insensitive to the diffusive dynamics by which the reactants form the transition complex. The medium effects traditionally have been discussed in terms of activity coefficients of A, B, and the AB complex./56/ To apply the pair correlation function theory to this problem we write

$$k_{ab} = \int g_{ab}(r)\, \hat{k}_{ab}(r)\, d^3r \tag{8.1}$$

where $\hat{k}_{ab}(r)$ is the *local* rate constant, the rate constant for the unimolecular process in which the AB pair with the center to center separation r passes over to products. (In the usual case in which A and B are polyatomic, Eq. (8.1) implies that part of the barrier to this process is associated with attaining the reactive orientations of A and B within the AB pair at separation r.)

Certain spin relaxation processes can also be cast in the form of Eq. (8.1), in which case the local rate constant may be well known. For example, consider the nuclear relaxation of lithium ion due to collisions with the paramagnetic ion Ni^{2+} in solution, i.e. the "reaction"/57/

$$^{7}Li^{+}(m) + Ni^{2+} = {}^{7}Li^{+}(m') + Ni^{2+} \tag{8.2}$$

where $m \rightarrow m'$ denotes a change in the nuclear spin state of the lithium. For this process the rate constant k_{ab} is the term in the spin-lattice relaxation rate law of $^{7}Li^{+}$ that is linear in Ni^{2+} molarity while the local rate constant is a suitable generalization of the Solomon-Bloembergen form

$$\hat{k}_{ab}(r) = C_{IS}[H_1(r)/\hbar]^2/[1/T_{1S} + 1/\tau(r)] \tag{8.3}$$

where C_{IS} is a known combination of spin factors and where $H_1(r) \sim 1/r3$ is the spin Hamiltonian term due to the interaction of the $^{7}Li^{+}$ nuclear dipole with the Ni^{2+} electronic dipole. The effect of the $H_1^{\,2}$ term is to make this experimental probe of $g_{ab}(r)$ sensitive to small r relative to the range probed by thermodynamic coefficients. Ni^{2+} is selected for the paramagnetic ion because its spin-lattice relaxation time T_{1S} is so small (3 ps at zero magnetic field) compared to $\tau(r)$, a characteristic time for the poorly known modulation of the dipolar interaction by the relative diffusion of Li^+ and Ni^{2+} when they are close together.

One learns from this experiment that $g_{ab}(r)$ is significantly larger than zero for r-values that are small compared to 0.69nm, the sum of the radii of the hydration complexes $M(H_2O)_6^{z+}$, a result also found for the Al^{3+}-Ni^{2+} interaction by a similar method. Indirect confirmation of these results is provided by a refined neutron diffraction study of aqueous $NiCl_2$. It shows/6/ that two hexaaquonickel +2 ions in solution come much closer together than one might expect from the diameter of the spherical envelope around the complex ion.

Eqs. (8.1) and (8.3) were recently applied to interpret experimental spin relaxation rates $1/T_1$ for the protons in $(CH_3)_4N^+$ ions due to collisions with the paramagnetic nitrilodisulfonate ion NDS^{2-} ion in aqueous solutions./58/ An integral equation approximation method led from a BO-level model [in which the water molecules and the ions are represented as hard spheres variously decorated with dipoles, quadrupoles and polarizability (water) or just with charges (ions)] to the solvent-averaged ion-ion correlation functions. Although the calculation of $1/T_1$ is governed by Eq. (8.3) we now are in a very different regime, both because the electron spin relaxation of the S=1/2 ion NDS^{2-} is slow compared to the time $\tau(r)$ that characterizes the diffusive motion, and because the Coulombic attraction tends to lengthen the ion-ion encounters. The spin relaxation data are in excellent agreement with the model calculations, an impressive accomplishment since the BO-level models have no structural parameters that are adjusted to fit *solution* data.

Equation (8.1) has also been applied to model calculations for the rate of a particular chemical reaction, the ferrous-ferric electron exchange in aqueous solution./59/ In this case the calculation of the local rate constant is far more difficult and fraught with uncertainty, but with the best current estimate it is clear that the important range of r is well below 0.69nm, the distance at which the envelopes of hexaaquo ferrous and ferric ions would touch, in accord with the recent results for Li^+-Ni^{2+}, Al^{3+}-Ni^{2+}, and Ni^{2+}-Ni^{2+} interactions mentioned above. Furthermore the calculated ionic strength dependence of the rate constant k_{23} for the ferrous-ferric electron exchange and its temperature derivative (expressed as the entropy of activation $\Delta S^{\ddagger}$) are both much more realistic than has been obtained with various forms of the Debye-Hückel theory./59/

Modelling the I-dependence of k_{23} can be done easily when the sensitivity of the local rate constant to changes in the ionic composition of the medium can be neglected, and when the integrand in Eq. (8.1) is strongly peaked at some separation R . Then we have

$$k_{23}(I)/k_{23}(I=0) = g_{23}(R,I)/g_{23}(R,I=0) \tag{8.5}$$

where I is the ionic strength to which the coefficient applies. Thus in suitable cases the experimental rate constant data can be used to derive the medium effect on the relevant pair correlation function. To compare with the results in Fig. 5a,b we notice that, with $z_2z_3=6$ the reduced pair correlation function defined in Eq. (7.10) is

$$\hat{g}_{23}(R,I) = [k_{23}(I)/k_{23}(I=0)]/\exp[q_{23}(R,I)-q_{23}(R,I=0)]$$
$$\approx [k_{23}(I)/k_{23}(I=0)]/\exp(6\kappa\lambda/4\pi) \qquad (8.6)$$

As an example we use data/55/ for the $Ru(NH_3)_4bpy^{2+/3+}$ electron exchange. The reduced pair correlation functions in Fig. 5c were obtained, showing the sensitivity of these +2,+3 correlations to the anion, even though both anionic species involved here are noted for their low degree of specific interaction with cations. The similarity to the reduced correlation functions in Fig. 5a,b shows that only *very* weak differences in cation-anion interaction are required to account for the observations while the need for more directly applicable model calculations is clear, since the amplitudes of the peaks of the reduced correlation functions are so sensitive to the charges.

9. CONCLUSIONS

We have emphasized some of the difficulties that are developing with the MM-level of description which, by the way, can be extended into the linear response dynamic regime rather directly with the use of the N-body Smoluchowski equation, thus making the MM theory of pair correlation functions applicable to measurable transport coefficients./60/ It may be argued that it is worth the effort to master these difficulties because the MM-level of description provides an economical basis for understanding a variety of solution phenomena, even in complex systems, as we have tried to illustrate. Of course the difficulties could be avoided by using the BO level of description for all purposes, which is now becoming possible by integral equation approximations/30,31/ and, no doubt, soon by simulation techniques.

Another point of view is that only the most basic and complete level of description has lasting value, in which case we should start with the Schroedinger Hamiltonian for the solution. Even this program is becoming feasible, and in some sense is only an extension of what is already done in studies based on BO-level *pair* potentials derived from S-level computations. It may be the only reliable way to formulate the distinctive properties of solutions with open-shell ions, especially when 'crystal field' effects are very large as in the case of Ni^{2+}.

REFERENCES AND FOOTNOTES

1. J. P. Hansen and I. R. McDonald, *Theory of Simple Liquids,* Academic Press, New York, 1976.
2. G. Stell in *Graph Theory and Theoretical Physics*, Academic Press, New York, 1967. editor F. Harary

3. H. L. Friedman, *A Course in Statistical Mechanics*, Prentice-Hall, Inc., Englewood Cliffs, N.J., 1985.
4. I. R. McDonald and K. Singer, *Chem. Britain*, 9, 544 (1973).
5. For any function f(r) we denote the Fourier transform and its inverse as, respectively,

$$F(k) = \int f(r)\, e^{ik\cdot r}\, d^3r \quad ; \quad f(r) = (2\pi)^{-3} \int F(k)\, e^{-k\cdot r}\, d^3k$$

We make much use of the convolution theorem: The Fourier transform of $\int f(r-R) d^3R s(R)$ is the product of the transforms $F(k)S(k)$.
6. G. W. Neilson and J. E. Enderby, *Proc. Roy. Soc.*(London)A390, 353 (1983).
7. The details involving graphical notation and conventions, symmetry numbers, and the three lemmas which help with analytical manipulations, are needed to take full advantage of the power of the graphical methods. See references 1-3 , for example.
8. It is sometimes useful to notice that the prescription for B(r) in terms of f bonds may be replaced by one in terms of h bonds. See for example, R. J. Bacquet and P. J. Rossky, *J. Chem. Phys.* 79, 1419 (1983).
9. D. M. Ceperley and G. V. Chester, *Phys Rev.* A15, 755 (1977).
F. Lado, S. M. Foiles, and N. W. Ashcroft, *Phys. Rev.* A28, 2374 (1983).
10. E. Thiele, *J. Chem. Phys* 39, 474 (1963).
M. S. Wertheim, *Phys. Rev. Lett.* 10, 321 (1963).
11. L. Blum in *Theoretical Chemistry* Academic Press, New York, 1980, D. Henderson, editor, Vol. 5, Chapter 1.
R. J. Baxter, *J. Chem. Phys.* 52,4559 (1968).
E. Waisman and J. L. Lebowitz *J. Chem. Phys.* 56, 3086, 3093 (1972).
12. F. H. Stillinger and R. Lovett, *J. Chem. Phys.* 48, 3858 (1968).
13. D. J. Mitchell, D. A. McQuarrie, A. Szabo, and J. Groeneveldt *J. Stat. Phys.* 17, 15 (1977).
14. G. Stell, J. S. Høye, and G. N. Patey, *Adv. Chem. Phys.* 48, 183 (1981).
15. F. H. Stillinger and R. Lovett *J. Chem. Phys.* 49, 1991 (1968).
16. H. L. Friedman and W.D.T.Dale, in *Modern Theoretical Chemistry*,Plenum Press, New York, Vol. 5:*Statistical Mechanics*, p. 85. B. J. Berne, editor.
17. H. L. Friedman, *Faraday Disc.* 64, 7 (1977).
18. J. G. Kirkwood and F. P. Buff, *J. Chem. Phys.* 19, 744 (1951).
19. H. L. Friedman and P. S. Ramanathan, *J. Phys.Chem.* 74, 3756 (1970).
20. R. L. Perry and J. P. O'Connell, *Mol. Phys.* 11, 1 (1984).
21. W. G. McMillan and J. E. Mayer, *J. Chem. Phys.* 13, 276 (1945).
22. H. L. Friedman, *Ionic Solution Theory*, Wiley, New York, 1962.
23. C. V. Krishnan and H. L. Friedman, *J. Solution Chem.* 3, 727 (1974).
24. R. Triolo, L. Blum and M. A. Floriano, *J. Phys. Chem.* 82, 1368 (1978).
24. W. H. Streng and W. Y. Wen, *J. Solution Chem.* 3, 65 (1974).
25. P. S. Ramanathan and H. L. Friedman, *J. Chem. Phys.* 54, 1086 (1971).
26. P. S. Ramanathan, C. V. Krishnan, and H. L. Friedman, *J. Solution Chem.* 1, 237 (1972)
27. D. Bratko, H. L. Friedman, and E. C. Zhong, *J. Chem. Phys.* 85, 377 (1986).

28. H. L. Friedman, C. V. Krishnan, and L. P. Hwang in *Structure of Water and Aqueous Solutions*, Verlag Chemie Gmbh, 1974. Editor W. Luck
29. H. L. Friedman, C. V. Krishnan, and C. Jolicouer, *Ann. New York Academy Sci.* 264 ,79 (1973).
30. B. M. Pettitt and P. J. Rossky *J. Chem. Phys.* 84, 5836 (1986).
31. G. N. Patey and S. L. Carnie, *J. Chem. Phys,* 78, 5183 (1983).
32. M. Berkowitz, O. A. Karim, A. McCammon, and P. J. Rossky, *Chem. Phys. Lett.* 105 577 (1984).
33. S. A. Adelman, Chem. Phys. Lett. *38, 567 (1976).*
34. S. A. Adelman *J. Chem. Phys.* 64, 724 (1976).
35. J. E. Mayer, *J. Chem. Phys.* 18, 1426 (1950).
36. M. Setchenow, *Ann. chim. phys.*,[6] 25, 226 (1892).
37. An exact expression for $k_{ca,b}$ in terms of $G_{ca,b}$ or even G_{ab} (!) can be obtained from Eqs. (3.4), (3.31), and (5.5) ./29/ The form given here is only accurate if the solution is dilute enough so that $k_{ca,b}$ is independent of solute concentrations, in which case it reduces to a combination of second virial coefficients./3/
38. G. R. Haugen and H. L. Friedman, *J. Phys. Chem.* 60, 1363 (1956).
39. P. J. Rossky and H. L. Friedman, *J. Phys. Chem.* 84 ,587 (1980).
40. T. K. Lim, E. C. Zhong, and H. L. Friedman, *J. Phys. Chem.* 90, 144(1986).
41. H. L. Friedman, *Kinam* 3, 101 (1981).
42. L. Blum and A. J. Torruella, *J. Chem. Phys.* 56, 303 (1972).
43. H. L. Friedman, *J. Chem. Phys.* 76, 1092 (1982).
44. C. G. Gray and K. E. Gubbins, *Theory of Molecular Fluids*, Oxford, 1984.
45. G. Stell and J. S. Høye, *Faraday Discussion.* 64, 17 (1977).
46. J. B. Hubbard and L. Onsager, *J. Chem. Phys.* 68, 1649 (1977).
J. B. Hubbard, P. Colonomos, and P. G. Wolynes, *J. Chem. Phys.* 71, 2652 (1979).
47. P. G. Kusalik and G. N. Patey, *J. Chem. Phys.* 79, 4468 (1983).
48. R. B. Cassel and R. H. Wood, *J. Phys. Chem.*78, 1924 (1974).
49. H. L. Friedman and and C. V. Krishnan, *J. Phys. Chem*,78, 1927 (1974).
50. H. S. Harned and B. B. Owen, *The Physical Chemistry of Electrolyte Solutions*, 3rd edition, Reinhold, New York, 1972.
51. H. L. Friedman, *J. Chem. Phys.* 32, 1134 (1960),
52. H. L. Friedman and B. Larsen, *J. Chem. Phys.* 70, 92 (1979).
53. H. L. Friedman and B. Larsen,*Pure and Applied Chem.* 51, 2147 (1979).
54. H. L. Friedman, *J. Solution Chem.* 1, 387, 418 (1972).
55. G. Brown and N. Sutin, *J. Am. Chem. Soc.* 101, 883 (1979).
56. K. Laidler, *Chemical Kinetics*, McGraw-Hill Book Co. New York, 1965.
57. F. Hirata, H.L. Friedman, M. Holz, and H. G. Hertz, *J. Chem. Phys.* 73 , 6031 (1980).
58. P. H. Fries, N. R. Jaganathan, F. G. Herring, and G. N. Patey, *J. Chem. Phys.* 80, 6267 (1984).
59. B. L. Tembe, H. L. Friedman, and M. D. Newton, *J. Chem. Phys.* 76 1490 (1982).
60. A. R. Altenberger and H. L. Friedman, *J. Chem. Phys.*, 78, 4162 (1983).

20. [illegible] and [illegible], [illegible] in [illegible], [illegible]
[illegible], [illegible] (19[illegible]).
21. [illegible], Mol. Phys. [illegible] (19[illegible]).
[illegible]
22. [illegible], J. Chem. Phys. [illegible] (19[illegible]).
23. [illegible], [illegible], J. Chem. Phys. [illegible] (1966).
[illegible], [illegible], [illegible], and [illegible], [illegible], Mol. Phys. [illegible]
(19[illegible]).
[illegible] Chem. Phys. Lett. [illegible] (19[illegible]).
24. J. G. Kirkwood, J. Chem. Phys. 2, 351 (1934).
25. L. Onsager, J. Am. Chem. Soc. 58, 1486 (1936).
26. H. Fröhlich, [illegible] (19[illegible]).
[illegible] (the) [illegible], [illegible] (1) [illegible]
[illegible] from [illegible] and [illegible] (1.[illegible]) [illegible]
[illegible] of the [illegible] [illegible]
[illegible] (19[illegible]).
28. [illegible], [illegible] and [illegible], J. Chem. Phys. 50, [illegible] (1969).
39. [illegible], [illegible] and [illegible], [illegible] (19[illegible]).
40. [illegible], [illegible] and [illegible], J. Chem. Phys. [illegible]
(19[illegible]).
41. [illegible], [illegible], [illegible] (19[illegible]).
42. [illegible] and [illegible], J. Chem. Phys. [illegible] (19[illegible]).
43. [illegible], J. Chem. Phys. [illegible] (19[illegible]).
44. C. G. Gray and K. E. Gubbins, Theory of Molecular Fluids, [illegible]
[illegible]
45. [illegible] and [illegible], [illegible] (19[illegible]).
46. [illegible] and [illegible], [illegible] (19[illegible]).
47. [illegible], [illegible], J. Chem. Phys. [illegible]
(19[illegible]).
48. [illegible] and [illegible], J. Chem. Phys. [illegible] (19[illegible]).
49. [illegible] and [illegible], J. Phys. Chem. [illegible] (19[illegible]).
50. [illegible] and [illegible], [illegible] (19[illegible]).
51. [illegible] and [illegible], The Physical Chemistry of Electrolytic
Solutions, 3rd edition, Reinhold, New York, 1958.
[illegible] (19[illegible]).
52. [illegible] and [illegible], J. Chem. Phys. 70, [illegible] (1979).
53. [illegible] and [illegible], [illegible] (19[illegible]).
54. [illegible], J. [illegible] (1972).
55. [illegible] and [illegible], [illegible] (1979).
56. [illegible], [illegible], [illegible], New York, 19[illegible].
57. [illegible], [illegible] and [illegible], J. Chem. Phys.
[illegible] (19[illegible]).
58. [illegible], [illegible], [illegible] and [illegible],
[illegible] (19[illegible]).
59. [illegible], [illegible] and [illegible], J. Chem. Phys.
[illegible] (19[illegible]).
60. [illegible] and [illegible], J. Chem. Phys. [illegible]
(19[illegible]).

NON-EQUILIBRIUM THEORIES OF ELECTROLYTE SOLUTIONS

Joseph B. Hubbard
Thermophysics Division
National Bureau of Standards
Gaithersburg, Md. 20899 USA

ABSTRACT. Dynamic aspects of electrolyte solution theory are explored through a generalized Langevin approach as well as through a van Hove/Smoluchowski description. The classical theory of ion-ion dynamical interactions is presented at the Debye-Falkenhagen-Onsager level, while the non-equilibrium ion-solvent interaction is analyzed from both a microscopic and a continuum viewpoint. Emphasis is placed on understanding the physics of simple models on which explicit calculations can be performed. These include electrical conductance, ionic friction coefficients, space-time correlation functions, and laser light scattering spectra for simple electrolytes and charged macromolecules.

1. INTRODUCTION

If these lectures were given fifteen or twenty years ago, they would consist of a derivation of the Onsager reciprocal relations along with the fluctuation - dissipation theorem and generalized susceptibility, followed by a detailed analysis of the concentration dependence of the conductance, diffusion, and viscosity of electrolyte solutions in the spirit of Debye and Falkenhagen. While these fundamentals are the working equipment of any electrolyte theorist worth his salt, more recent developments, both theoretical and experimental, have prodded me to revise the classical format so as to incorporate some of the machinery of modern statistical physics. Photon and neutron scattering experiments are supplying us with detailed structural and dynamical information on electrolytes, computer simulations are becoming sophisticated enough to deal with these complex systems, and theorists are beginning to develop a sense for which approximations work best (and why), and how they should be extended so as to process predictive as well as explanatory power.

Electrolyte dynamics was traditionally a playground (or swamp) for a few ambitious thinkers who wanted to construct an exact solution to a dynamical many-body problem with long-range interactions. This was one of the very few instances in which a transport property (conductance) could be calculated as accurately as it could be measured, and the measurements were of the highest quality. The basic assumptions of this theory were quite simple: Brownian motion for the ions, a viscous dielectric continuum for the solvent, and a two-point probability conservation law for the distribution funtions. It turns out that this description is strictly valid only for dilute, strong electrolytes, and it has proved to be extremely difficult, if not impossible, to generalize it rigorously to higher concentrations.

M.-C. Bellissent-Funel and G. W. Neilson (eds.), The Physics and Chemistry of Aqueous Ionic Solutions, 95–128.

With the evolution of modern statistical physics, emphasis has shifted away from classical extensions of the Debye-Falkenhagen-Onsager (DFO) theory, and a much richer, albeit complex, set of questions is now being raised: for instance, "What does the velocity autocorrelation function of a lithium ion in water look like?" "What are the dynamics of a solvated ion complex?" What is the lifetime of a closely associated ion pair in water or methanol?" "How should one describe the motion of a charged macroion in an external field, if the zeta potential of the ion is fluctuating?" "What is the role of a polar solvent in charge transfer reactions?" The search for answers to such questions is both intriguing and frustrating: intriguing, because of the novelty, relative to simple liquids, of solvated ions and the possibility of discovering some "universal" aspect of ion dynamics, as in the DFO theory; frustrating, because the high specificity of such questions tends to preclude a general analysis, and there appears to be no single optimal theoretical approach.

We have just heard Professors Friedman and Hansen, in their eloquent lectures, outline several systematic statistical mechanics approaches to electrolyte solutions. Partly because of time constraints but mostly because of personal taste, I shall take an "anti-Kirkwoodian" approach: rather than attempt to derive correlation functions and transport coefficients directly from fundamental statistical mechanics, I shall, in most instances, begin with phenomenological theories such as the generalized Langevin equation, proceed to shamelessly (though not recklessly) combine microscopic and macroscopic concepts, and attempt to describe electrolyte dynamics in terms of relatively "accessible" quantities. While this leaves me vulnerable to the purists' charge of "uncontrolled, premature calculation," it will hopefully enable the non-professional user of statistical mechanics acquire a firmer grasp of the physics, along with a sense of the power as well as the limitations of the methods employed.

2. FLUCTUATIONS AND LIGHT SCATTERING IN STRONG ELECTROLYTES

This example provides the simplest, most instructive introduction to electrolyte dynamics. In many ways this is the prototype self-consistent field theory, and the astute observer will immediately recognize the very strong resemblance to the famous DFO treatment. Since most of you are quite familiar with the basic lore of dynamic light scattering,/1/ I shall skip the preliminaries and begin by considering the binary electrolyte (A,B) which completely dissociates as

$$A_p B_q \rightarrow pA + qB \tag{1.1}$$

If the electric charge on A is z_1, and that of B is z_2, then bulk neutrality requires

$$pz_1 + q z_2 = 0, \tag{1.2}$$

or in terms of the equilibrium number density n^o,

$$z_1 n_1^o + z_2 n_2^o = 0 \tag{1.3}$$

Suppose the ions obey a Langevin equation of the form

$$m_i \frac{d\mathbf{v}_i}{dt} = - \xi_i \mathbf{v}_i + z_i \mathbf{E} + \mathbf{F}(t) \qquad i = (1,2) \tag{1.4}$$

where m_i is the ion mass, $\mathbf{E}(\mathbf{r},t)$ is the local electric field, $\mathbf{F}(t)$ is the random force, and ξ_i is assumed to be a constant. The average velocity at time t, conditional on the initial velocity being v_o, is then given by

$$<\mathbf{v}(t)>_{\text{cond } v_o} = \mathbf{v}_o \exp - \left(\frac{\xi_i t}{m_i}\right)$$

$$+ \frac{z_i E}{m_i} \left[1 - \exp - \left(\frac{\xi_i t}{m_i}\right)\right] \tag{1.5}$$

which evolves to $z_i E/\xi_i$ after a time ~ m_i/ξ_i. If we express the friction coefficient in terms of the single ion diffusion coefficient D_i via the Einstein relation

$$D_i = \frac{k_B T}{\xi_i} = \beta^{-1} \xi_i^{-1}, \tag{1.6}$$

we have, after an inertial time m_i/ξ_i has elapsed,

$$\mathbf{v}_i (\mathbf{r},t) = \beta D_i z_i \mathbf{E} (\mathbf{r},t). \tag{1.7}$$

Since A and B are separately conserved, the following continuity condition applies:

$$\frac{\partial n_i(\mathbf{r},t)}{\partial t} + \nabla \bullet \mathbf{J}_i (\mathbf{r},t) = 0 \ (i=1,2) \tag{1.8}$$

where the number density (or equivalently, probability density) flux may be written

$$\mathbf{J}_i(\mathbf{r},t) = - D_i \nabla_i n_i(\mathbf{r},t) + \beta D_i z_i n_i(\mathbf{r},t) \mathbf{E}(\mathbf{r},t)$$

$$n_i = n_i^o + \delta n_i(\mathbf{r},t) \tag{1.9}$$

The field $\mathbf{E}$ can be expressed in terms of the dielectric constant ϵ_o of the solvent and the ion densities through Poisson's equation:

$$\nabla \bullet \mathbf{E} = \frac{4\pi}{\epsilon_o} \sum_i z_i n_i \, (\mathbf{r},t)$$

$$= \frac{4\pi}{\epsilon_o} \sum_i z_i \, \delta n_i(r,t) \tag{1.10}$$

where the zeroth order term vanishes because of electroneutrality. When Eqns (1.9) and (1.10) are substituted into (1.8), the result is two coupled linear diffusion equations whose spatial Fourier transform may be written

$$\frac{\partial \delta n_i(\mathbf{k},t)}{\partial t} = -\,(k^2 + \kappa_i^2)\, \delta n_i(\mathbf{k},t) + \left|\frac{z_j}{z_i}\right| \kappa_i^2 \, \delta n_j(\mathbf{k},t)$$

$$(i,j) = 1,2 \qquad i \neq j \tag{1.11}$$

with

$$\kappa_i^2 = \frac{4\pi}{\epsilon_o} \beta \, z_i^2 \, n_i^o . \tag{1.12}$$

The density fluctuation correlation functions are now easily constructed by Laplace transformation (variable s) with respect to time, multiplying by $\delta n^* (\mathbf{k},0)$ ($* \rightarrow$ complex conjugate), and averaging over an equilibrium ensemble. After a bit of algebra, the elements of the dynamic structure factor matrix emerge as/1/

$$\tilde{S}_{11}(\mathbf{k},s) = \frac{1}{\Delta(s)} \left\{ [s + (k^2 + \kappa_2^2)\, D_2] \; S_{11}\,(\mathbf{k},0) + \left|\frac{z_2}{z_1}\right| \kappa_1^2 \, D_1 \, S_{12}\,(\mathbf{k},0) \right\}$$

$$\tilde{S}_{21}\,(\mathbf{k},s) = \frac{1}{\Delta(s)} \left\{ [s + (k^2 + \kappa_2^2)\, D_2] \; S_{21}\,(\mathbf{k},0) + \left|\frac{z_2}{z_1}\right| \kappa_1^2 \, D_1 \, S_{22}\,(\mathbf{k},0) \right\}$$

$$\tilde{S}_{12}\,(\mathbf{k},s) = \frac{1}{\Delta(s)} \left\{ [s + (k^2 + \kappa_1^2)\, D_1] \; S_{12}\,(\mathbf{k},0) + \left|\frac{z_1}{z_2}\right| \kappa_2^2 \, D_2 \, S_{11}\,(\mathbf{k},0) \right\}$$

$$\tilde{S}_{22}\,(\mathbf{k},s) = \frac{1}{\Delta(s)} \left\{ [s + (k^2 + \kappa_1^2)\, D_1] \; S_{22}\,(\mathbf{k},0) + \left|\frac{z_1}{z_2}\right| \kappa_2^2 \, D_2 \, S_{21}\,(\mathbf{k},0) \right\} \tag{1.13}$$

where the determinant

$$\Delta(s) = [s + (k^2 + \kappa_1^2) D_1][s + (k^2 + \kappa_2^2) D_2] - \kappa_1^2 D_1 \kappa_2^2 D_2. \tag{1.14}$$

The $\tilde{S}_{ij}(\mathbf{k},s)$ are the Laplace transforms of $<\delta n_i^*(\mathbf{k},0)\ \delta n_j\ (\mathbf{k},t)>$, and the S_{ij} are the equilibrium structure factors $<\delta n_i^*\ (\mathbf{k},0)\ \delta n_j\ (\mathbf{k},0)>$. Note that $S_{12} = S_{21}$ but that the matrix $\tilde{S}_{ij}$ is not obviously symmetric.

The usual procedure at this point is to factor the dispersion determinant as

$$\Delta(s) = (s - s_+)\ (s - s_-) \tag{1.15}$$

and identify the roots $s_\pm$ with the relaxation times for the correlation functions $<\delta n_i^*(\mathbf{k},0)\ \delta n_j\ (\mathbf{k},t)>$. In other words, the correlation functions are a sum of two exponentials with the decay times s_+, s_-. The general expression for the roots is messy, but matters are simplified if we realize that in a typical light scattering experiment k $\sim 10^5\ \text{cm}^{-1}$, whereas D $\sim 10^{-5}\ \text{cm}^2/\text{sec}$, and κ_i for a typical dilute 1-1 electrolyte is $10^8 \sim 10^9\ \text{cm}^{-1}$. Thus $k^2 D_i \ll [(k^2 + \kappa_1^2) D_1 + (k^2 + \kappa_2^2) D_2]$. In this approximation we can write, after a couple of lines of algebra

$$\begin{matrix} s_+ \\ s_- \end{matrix} = \begin{cases} -k^2 D_0\ (k) \\ -[(k^2 + \kappa_1^2) D_1 + (k^2 + \kappa_2^2) D_2] + k^2 D_0(k) \end{cases} \tag{1.16}$$

with

$$D_0(k) = \frac{(k^2 + (\kappa_1^2 + \kappa_2^2))\ D_1 D_2}{(k^2 + \kappa_1^2)\ D_1 + (k^2 + \kappa_2^2)\ D_2}, \tag{1.17}$$

and so we see that the "effective diffusion coefficient" measured in an electrolyte scattering experiment is the k-dependent quantity $D_0(k)$. The spectral densities of the density fluctuation correlation functions will therefore consist of a superposition of two Lorentzian profiles, a narrow line of width $k^2 D_0$ and a broader line of width $|s_-| \sim \kappa_1^2 D_1 + \kappa_2^2 D_2$ which we can identify as the reciprocal of the Maxwell time

$$\tau_M = \frac{1}{\kappa_1^2 D_1 + \kappa_2^2 D_2} \tag{1.18}$$

for the decay of charge density fluctuations.

Even this simple model of electrolyte dynamics leads to some interesting predictions. For example, consider the small wave vector limit of the apparent (effective) diffusion coefficient $D_0(k)$ for a "macroion" of charge $z = |z_1|$ with counterion valence $|z_2| = 1$:

$$D_0(0) = \frac{(\kappa_1^2 + \kappa_2^2)\, D_1 D_2}{\kappa_1^2 D_1 + \kappa_2^2 D_2} \cong \frac{(1 + z^2)\, D_1 D_2}{D_2 + z^2 D_1} \tag{1.19}$$

Since a macroion is much larger than a counterion, we expect $D_1 \ll D_2$, and so

$$D_0(0) \cong (1 + z^2)\, D_1 > D_1 \tag{1.20}$$

On the other hand, we also expect $z \gg 1$, and so for given D_1 and D_2,

$$D_0(0) \leqq D_2 . \tag{1.21}$$

Charge density fluctuations therefore enhance the Brownian motion of macroions while inhibiting that of the counterions.

As another example, consider the decay of a localized charge density fluctuation $\delta\rho$:

$$\delta\rho = \frac{\epsilon_o}{4\pi} (\nabla \bullet \mathrm{E}) = z_1\, \delta n_1 + z_2\, \delta n_2 \tag{1.22}$$

If we assume $D_1 = D_2 = D$, then our preceding analysis yields the following equation for $\delta\rho(\mathrm{r},t)$:

$$\frac{\partial\, \delta\rho(\mathrm{r},t)}{\partial\, t} = D\, [\nabla^2 \delta\rho - \kappa^2\, \delta\rho] \tag{1.23}$$

where κ is the inverse Debye screening length

$$\kappa^2 = \kappa_1^2 + \kappa_2^2 = \frac{4\pi\beta}{\epsilon_o} \left[\sum_i z_i^2\, n_i^o\right] . \tag{1.24}$$

For a permanent, spherically symmetric charge distribution centered around the origin (an ion, for instance), we have the stationary solution

$$\delta\rho(\mathrm{r},o) = \frac{A\, e^{-\kappa r}}{r} . \tag{1.25}$$

If the charge at r = 0 were to be instantaneously extinquished at t = 0, then Eqn. (1.23) says that the atmospheric charge density decays as

$$\delta\rho(r,t) = \left(\frac{A\, e^{-\kappa r}}{r}\right) \frac{1}{\sqrt{\pi}} \int_{\sqrt{\tau} - \kappa r/2\sqrt{\tau}}^{\infty} dx\, e^{-x^2}$$

$$\tau = \frac{t}{2\tau_M^o}, \qquad \tau_M^o = \frac{1}{2D\kappa^2} \tag{1.26}$$

Note that at long times $\delta\rho$ does not decay as a simple exponential, and in fact $\delta\rho$ relaxes as $\tau^{-1/2}\exp(-\tau)(\tau \to \infty)$. This qualitative difference between micro- and macroscopic relaxation is a common occurence in statistical physics, and it should be borne in mind when constructing models for molecular-scale processes.

3. THE DEBYE-FALKENHAGEN-ONSAGER THEORY

A most important concept in the DFO theory, and in statistical physics in general, is that of a time dependent pair correlation function. We define the pair correlation function g_{ab} so that $n_a\, g_{ab}(\mathbf{r}_a,\mathbf{r}_b)\, dV_a$ is the number of ions of type a in a volume dV_a about the point $\mathbf{r}_b$, given that there is one ion of type b at the point $\mathbf{r}_b$; the a and b ions may be the same or different. It is obvious that

$$g_{ab}(\mathbf{r}_a,\mathbf{r}_b) = g_{ba}(\mathbf{r}_b,\mathbf{r}_a) \tag{2.1}$$

and that $g_{ab} \to 1$ as $|\mathbf{r}_b - \mathbf{r}_a| \to \infty$. The correlation functions satisfy a continuity equation in the configuration space of two particles (six dimensions):/2,3/

$$\frac{\partial g_{ab}}{\partial t} + \mathrm{div}_a\, \mathbf{J}_a + \mathrm{div}_b\, \mathbf{J}_b = 0 \tag{2.2}$$

where $\mathbf{J}_a$ (b) are the probability fluxes for the a(b) particles. The flux $\mathbf{J}_a$ is given by

$$\mathbf{J}_a = \omega_a^o\, z_a\, g_{ab}\, (\mathbf{E} - \nabla_a\phi_b) - \beta^{-1}\, \omega_a^o\, \nabla_a\, g_{ab} \tag{2.3}$$

with $\mathbf{J}_b$ the same with a and b interchanged./4/ Here ω_a^o is the ion mobility at infinite dilution where all ion-ion interactions vanish, $\mathbf{E}$ is the external field and ϕ_b is the potential due to a b ion located at $\mathbf{r}_b$. Note the similarity between Eqn (2.3) and (1.9). As before, ϕ_b

satisfies Poisson's equation

$$\nabla_a^2 \phi_b (\mathbf{r}_a, \mathbf{r}_b) = \frac{-4\pi}{\epsilon_o} \left[\sum_c z_c n_c h_{cb} + z_b \delta(\mathbf{r}_a - \mathbf{r}_b)\right]$$

$$h_{ab} = g_{ab} - 1, \tag{2.4}$$

to first order in h, where we have invoked electroneutrality as in section 1. In the same approximation, we write the probability flux as

$$\mathbf{J}_a = \omega_a^o \left[- \beta^{-1} \nabla_a \bullet h_{ab} + z_a (1 + h_{ab}) \mathbf{E} - z_a \nabla_a \phi_b\right] \tag{2.5}$$

Taking **E** to be constant in space and time, and making use of the symmetry $\nabla_a h_{ab} = - \nabla_b h_{ab}$, we now substitute Eqns (2.5) and (2.4) into (2.2) with $\partial h_{ab}/\partial t = 0$, and arrive at

$$\beta^{-1} (\omega_a^o + \omega_b^o) \nabla^2 h_{ab}(\mathbf{r}) + z_a \omega_a^o \nabla^2 \phi_b (\mathbf{r}) + z_b \omega_b^o \nabla^2 \phi_a(-\mathbf{r}) = (z_a \omega_a^o - z_b \omega_b^o) \mathbf{E} \bullet \nabla h_{ab}(\mathbf{r}) \tag{2.6}$$

where derivatives are with respect to $\mathbf{r} = \mathbf{r}_b - \mathbf{r}_a$. If we set $\mathbf{E} = 0$, then the equilibrium potential ϕ^o must be an even function of r, and since h and ϕ must vanish as $r \to \infty$, we recover

$$\beta^{-1} (\omega_a^o + \omega_b^o) h_{ab}^o + (z_a \omega_a^o \phi_b^o + z_b \omega_b^o \phi_a^o) = 0 \tag{2.7}$$

By requiring solutions to Eqn (2.7) of the form

$$h_{ab}^o = z_a z_b h^o, \quad \phi_a^o = - \beta^{-1} z_a h^o \tag{2.8}$$

we see that (2.7) becomes an identity and (2.4) reverts to the equilibrium linearized Poisson Boltzmann equation, whose solution may be written in terms of h^o as

$$h^o(r) = \frac{-\beta}{\epsilon_o} \frac{e^{-\kappa r}}{r} \tag{2.9}$$

where κ is again the inverse Debye length. Since **E** is assumed to be small, we write

$$\phi_a = \phi_a^o + \delta \phi_a \qquad h_{ab} = h_{ab}^o + \delta h_{ab} \tag{2.10}$$

where the perturbations $\delta\phi$ and δh are now odd functions of $\mathbf{r}$. If we restrict our attention to a binary electrolyte, then we need consider only one function $\delta h_{12}(\mathbf{r}) = -\delta h_{21}(\mathbf{r})$, and Poisson's equation (2.4) becomes

$$\nabla^2 \delta\phi_2(\mathbf{r}) = -\frac{4\pi z_1}{\epsilon_o} n_1 \delta h_{12}(\mathbf{r})$$

$$\delta\phi_1(\mathbf{r}) = \delta\phi_2(\mathbf{r}). \tag{2.11}$$

Eqn (2.6) then becomes, to lowest order in the perturbations

$$\beta^{-1}(\omega_1^o + \omega_2^o)\nabla^2 \delta h_{12}(\mathbf{r}) + (z_1\omega_1^o - z_2\omega_2^o)\nabla^2 \delta\phi_2(\mathbf{r})$$

$$= (z_1\omega_1^o - z_2\omega_2^o) z_1 z_2 \mathbf{E} \bullet \nabla h^o(\mathbf{r}) \tag{2.12}$$

which is readily solved by Fourier transforms. After substituting Eqn (2.11) into (2.12) and noting that

$$h^o(k) = -\left(\frac{\beta}{\epsilon_o}\right)\frac{4\pi}{k^2 + \kappa^2} \tag{2.13}$$

it is easily found that the Fourier components of $\delta\phi$ satisfy

$$\delta\phi_2(\mathbf{k}) = 4\pi z_1 z_2 \beta\kappa^2 q\, \epsilon_o^{-1}\left[\frac{i\,\mathbf{k}\bullet\mathbf{E}}{k^2(k^2+\kappa^2)(k^2+q\kappa^2)}\right] \tag{2.14}$$

where

$$q = \frac{\omega_1^o z_1 - \omega_2^o z_2}{(z_1 - z_2)(\omega_1^o + \omega_2^o)}, \qquad 0 < q < 1 \tag{2.15}$$

What we want is the perturbed electric field acting on ion 2, which is given by

$$\delta E_2(\mathbf{r})\Big|_{\mathbf{r}=0} = -\nabla\,\delta\phi_2(\mathbf{r})\Big|_{\mathbf{r}=0}$$

$$\delta E_2(\mathbf{k}) = -i\,\mathbf{k}\;\delta\phi_2(\mathbf{k}) \tag{2.16}$$

The inverse Fourier transform of $\delta E_2(\mathbf{k})$ with $\mathbf{r}=0$ then gives us

$$\delta E_2(0) = \frac{-i}{(2\pi)^3}\int \mathbf{k}\,\delta\phi_2(\mathbf{k})\, d^3\mathbf{k}, \tag{2.17}$$

and so from Eqn (2.14), we must evaluate the integral

$$I = \frac{1}{(2\pi)^3} \int \frac{\mathbf{k}\,(\mathbf{k} \bullet \mathbf{E}) \quad d^3 k}{k^2\,(k^2 + \kappa^2)(k^2 + q\,\kappa^2)} \ . \tag{2.18}$$

The angular average over the directions of **k** results in the replacement of **k** (**k** • **E**) by $(1/3)(k^2\mathbf{E})$, and the integral over $|\mathbf{k}|$ is performed via the calculus of residues (poles at $k = i\,\kappa,\ i\sqrt{q}\,\kappa$):

$$I = \frac{1}{12\pi(1 + \sqrt{q})\kappa} \ \mathbf{E} \ . \tag{2.19}$$

Finally, the total electric field acting on ion 2 is

$$\mathbf{E} + \delta\mathbf{E}_2(0) = \left[1 - \frac{|z_1 z_2|\ q\ \beta\ \kappa}{3\ \epsilon_o\ (1 + \sqrt{q})} \right] \mathbf{E} \ , \tag{2.20}$$

which is the famous atmospheric relaxation correction./5/ Note that the identical result is obtained for $\delta\mathbf{E}_1(0)$. Since $\omega z\mathbf{E}$ is the velocity of an ion of mobility ω in the external field, we may interpret Eqn (2.20) as an effective reduction in ion mobility:

$$\delta\omega = - \frac{\omega^o |z_1 z_2|\ q\ \beta\ \kappa}{3\ \epsilon_o\ (1 + \sqrt{q})} \ , \ \kappa^2 = \frac{4\pi\beta}{\epsilon_o} \sum_i z_i^2\, n_i \tag{2.21}$$

where $\delta\omega$ is proportional to the square root of the electrolyte concentration through the factor κ.

Let's see how the relaxation effect behaves for a macromolecular ionic solution as a complement to the analysis of section 1. Consider the q-factor for a solution with macroions (1) and counter-ions (2) such that

$$\frac{\omega_1^o}{\omega_2^o} \to 0 \quad |z_2| = 1,\ |z_1| = z \gg 1,\ q \to \frac{1}{z} \tag{2.22}$$

This implies that $|\delta\omega/\omega| \sim (zn_2)^{1/2}$, where we have used the neutrality condition $zn_1 = n_2$. This represents a considerable enhancement of the relaxation effect relative to the case of a simple electrolyte with, say, $|z_1| = |z_2| = 1$, for which $q = 1/2$ and $|\delta\omega/\omega| \sim (n_1 + n_2)^{1/2}$.

The electrophoretic correction to the ionic mobility turns out to be as significant, both conceptually and numerically, as the relaxation effect which we have just calculated. But in marked contrast to the latter contribution, whose origin is the direct electrical interaction between ions, electrophoresis is an indirect effect which arises from the hydrodynamic momentum transfer between ions and solvent. As in the case of Coulomb interaction the electrophoretic force is long ranged, inasmuch as the flow field far from a moving object decays as the inverse of the distance; however, the physical reason for this is that the solvent possesses a finite viscosity, and most of the viscosity in a dense fluid may be attributed to the short-ranged, predominantly repulsive, intermolecular forces. In view of the microscopic complexities of these phenomena, it is truly remarkable that the macroscopic shear viscosity of the solvent in a classical hydrodynamic description yields exact results, even in the limit of high dilution.

The problem may be formulated as follows. Consider a particular ion in solution together with its atmosphere, the electrical charge density of which is

$$\delta\rho = \sum z_a \, \delta n_a, \; \delta n_a = n_a - n_a^0 \tag{2.23}$$

where δn_a is the difference between the local and average density of ion a. In the presence of the field **E**, this charge density experiences a force $\mathbf{f} = \mathbf{E}\,\delta\rho$ which induces motion in the surrounding liquid, and this motion is then transmitted to the central ion. In short, the central ion moves in a counter-current produced by the interaction of its atmosphere with the external field. δn_a is related to the local potential ϕ:

$$\delta n_a = - z_a \, \phi \, n_a^o \, \beta,$$

$$\phi = \frac{z_b \, e^{-\kappa r}}{r} \tag{2.24}$$

where we have linearized the exponential in the Boltzmann factor and assumed spherical symmetry since the field is weak (these approximations may be rigorously justified)./2,3/ The atmospheric charge density may be written

$$\delta\rho = - \frac{z_b}{4\pi} \kappa^2 \frac{e^{-\kappa r}}{r} \tag{2.25}$$

and with the further assumption (easily justified) of slow, steady, incompressible hydrodynamic flow, we can formulate a Navier-Stokes equation

$$\eta \, \nabla^2 \mathbf{v} - \nabla p + \mathbf{E}\,\delta\rho = 0 \tag{2.26}$$

where p is a hydrodynamic pressure and η is the shear viscosity of the solvent. Taking the Fourier components in Eqn (2.26), we arrive at

$$- \eta \, k^2 \, \mathbf{v}(\mathbf{k}) - i \, \mathbf{k} \, p(\mathbf{k}) + \mathbf{E} \, \delta\rho(\mathbf{k}) = 0 \tag{2.27}$$

which reduces to

$$\mathbf{v(k)} = \frac{\delta\rho(\mathbf{k})}{\eta} \left[\frac{k^2 \ \mathbf{E} - \mathbf{k}\ (\mathbf{k} \bullet \mathbf{E})}{k^4} \right] \tag{2.28}$$

since $\mathbf{k} \bullet \mathbf{v(k)} = 0$ because of incompressibility. The Fourier component of the charge density $\delta\rho(\mathbf{k})$ is

$$\delta\rho(\mathbf{k}) = - z_b/(\kappa^{-2} k^2 + 1) \tag{2.29}$$

What we now require is the velocity at the origin $\mathbf{v}(0)$, which is simply given by the inverse Fourier transform with $\mathbf{r}$=0:

$$\mathbf{v}(0) = \frac{1}{(2\pi)^3} \int \mathbf{v(k)}\ d^3\ \mathbf{k}$$

$$= \frac{- z_b}{(2\pi)^3\ \eta} \left(\frac{8\pi}{3}\right) \mathbf{E} \int_0^\infty \frac{dk}{[\kappa^{-2} k^2 + 1]} \tag{2.30}$$

where we have again integrated over the directions of $\mathbf{k}$. The integral in (2.30) is trivial, the result being

$$\mathbf{v}(0) = - \frac{z_b\ \kappa}{6\ \pi\ \eta}\ \mathbf{E} \tag{2.31}$$

and, as in the relaxation effect, we may express this in terms of an electrophoretic correction to the ion mobility

$$\delta\omega = \frac{-\kappa}{6\pi\eta} \tag{2.32}$$

which is the same for all ions. The total correction to the mobility is then given by the sum

$$\delta\omega_i = - \frac{(\omega_i^o)\,|z_1 z_2|\,q\beta\kappa}{3\epsilon_o(1 + \sqrt{q})} - \frac{\kappa}{6\pi\eta} \qquad i = 1,2 \tag{2.33}$$

which is sometimes referred to as the "limiting law for electrical conductance"./2,3/ Its derivation and experimental verification was a major scientific achievement.

4. BROWNIAN DYNAMICS WITH FLUCTUATIONS

In dealing with the Brownian motion of ions we have noted the close correspondence between particle flux and probability flux. This analogy can be made precise, as you recall from the lectures of Professors

Enderby and Hansen, by introducing the van Hove correlation function $G(\mathbf{r},t)$ which describes the probability density, at a point $\mathbf{r}$ in three dimensional space, of a set of N particles at positions $\mathbf{r}_1 \ldots \mathbf{r}_N$ at time t, given the distribution at t=0./6/ The Fourier transform of G in space-time is the dynamic scattering function $S(\mathbf{k},\omega)$. Here we shall only be interested in the "self-part" of G, G_s, defined as the ensemble average

$$G_s(\mathbf{r},t) = N^{-1} < \sum_j \delta\ [\mathbf{r} + \mathbf{r}_j(0) - \mathbf{r}_j(t)]>$$

$$G_s(\mathbf{r},0) = \delta(\mathbf{r}), \qquad \int G_s(\mathbf{r},t)d^3\mathbf{r} = 1 \tag{3.1}$$

Knowledge of G_s, which reflects single particle dynamics, can be obtained from incoherent neutron or light scattering experiments. I would like to introduce a diffusion-level calculation of a closely related correlation function as a first order model for solvated ion dynamics, and for electrophoretic laser light scattering from a macromolecular solution. The method, though somewhat novel, is quite simple, and is based on a straightforward extension of the Smoluchowski equation so as to include stochastic, as well as deterministic, forces. As far as I am aware, Professor P. Wolynes and I were the first to per-form an actual calculation using this technique in 1978;/7/ at that time we were interested in developing a theory of molecular reorientation and dielectric friction, as so we considered the rotational diffusion of dipoles with stochastic torque fluctuations due to polarization fluctuations in the surrounding medium. The motivation for all this is that we can, in some instances, expect G_s and the force or torque fluctuations to evolve on the same time scale, which implies that the Einstein relation between diffusion and friction coefficient will break down at finite frequencies ($D(\omega) \neq k_BT/\xi(\omega)$). Furthermore, this anomalous behavior should be observable in certain spectroscopic and scattering experiments./7/

The theory as applied to electrophoretic light scattering is as follows./1,8/ Imagine a dilute solution of spherical macromolecules, such as a globular protein in a salt-containing aqueous solution, and suppose that a laser light scattering experiment is performed in the presence of a constant electric field. Also assume that the velocity of the macroion in the external field is μE, where μ is called the electrophoretic mobility, and is related to the zeta-potential across the protein-solvent interface./8/

Now take

$$\mu(t) = \mu_o + \delta\mu(t) \tag{3.2}$$

where μ_o is the average mobility and $\delta\mu$ is a fluctuation such that its equilibrium ensemble average vanishes:

$$<\mu(t)> = \mu_o \quad , \quad <\delta\mu(t)> = 0 \tag{3.3}$$

Here inertial effects have been neglected, so that the molecule instantaneously attains the electrophoretic velocity appropriate to $\mu(t)$ (this constraint can be relaxed without excessive complication).

Define a modified van Hove G_s as follows. Let $\psi(\mathbf{r},t|\mu(o), \mu(t))$ be the probability that a molecule initially at $\mathbf{r}=0$ with mobility $\mu(o)$ will be located at position $\mathbf{r}(t)$ with mobility $\mu(t)$. Note that ψ is a function of the stochastic variable μ as well as the deterministic variables $\mathbf{r},t$. Now write (suppressing the μ-dependence for brevity of notation)

$$\begin{aligned} \psi = < \psi(\mathbf{r},t) > + \delta\psi(\mathbf{r},t) \\ < \delta\psi > = 0 \end{aligned} \tag{3.4}$$

and the analogy between $<\psi>$ and G_s as defined by Eq. (3.1) becomes obvious, and in fact the two are identical if we neglect the mobility fluctuations.

Suppose that ψ satisfies a stochastic Smoluchowski equation

$$\frac{\partial\psi}{\partial t} = - \nabla\bullet\mathbf{J}$$

$$\mathbf{J} = - D\nabla\psi - \mu\mathbf{E}\psi \qquad (t>0) \tag{3.5}$$

If we reverse the sign of time, then

$$\frac{\partial\psi}{\partial t} = - D\,\nabla^2\psi + \mu(\mathbf{E}\bullet\nabla)\,\psi \qquad (t<0) \tag{3.6}$$

Note that the convective term is invariant under $t \to - t$. If the origin of $\delta\mu$ had been assumed to be "internal" or "dissipative" noise, then $\mu(\mathbf{E}\bullet\nabla)\psi$ would change sign under time reversal./7/ With the assumption of convective fluctuations, we have

$$\frac{\partial<\psi>}{\partial t} = D\nabla^2 <\psi> + \mu_o(\mathbf{E}\bullet\nabla)\, <\psi> + < \delta\mu(\mathbf{E}\bullet\nabla)\delta\psi > \qquad (t>0) \qquad (a)$$

$$\frac{\partial<\psi>}{\partial t} = - D\,\nabla^2<\psi> + \mu_o(\mathbf{E}\bullet\nabla)\, <\psi> + <\delta\mu(\mathbf{E}\bullet\nabla)\delta\mu> \qquad (t<0) \qquad (b) \tag{3.7}$$

and, from Eqn (3.4),

$$\begin{aligned} \frac{\partial\,\delta\psi}{\partial t} = - D\nabla^2\delta\psi + \mu_o(\mathbf{E}\bullet\nabla)\delta\psi + (\delta\mu\mathbf{E}\bullet\nabla)<\psi> \\ + (\delta\mu\mathbf{E}\bullet\nabla)\delta\psi - < \delta\mu\mathbf{E}\bullet\nabla\delta\psi> \qquad (t<0) \end{aligned} \tag{3.8}$$

The idea is to linearize (3.8) in $\delta\psi$, $\delta\mu$, solve for $\delta\psi$ and substitute back into Eqn (3.7a) to obtain a closed expression for $<\psi>$. In other words, $<\psi(t)>$ depends on $\delta\psi(s)$ and $\delta\mu(s)$, $s<t$, and in turn, $\delta\psi(s)$ depends on $<\psi(\tau)>$ and $\delta\mu(\tau)$ at previous times $\tau<s$. The linearized version of Eqn (3.8) is readily solved with a Green's function:

$$\delta\psi = E \int d^3r' \int_{-\infty}^{t} ds\ G(\mathbf{r}|\mathbf{r}',\ t-s) \left[\delta\mu(s) \frac{\partial}{\partial x'} \ \langle\psi(\mathbf{r}',s)\rangle\right]$$

$(t \gtrless 0)$ (3.9)

where the electric field is in the x-direction and

$$G(\mathbf{r}|\mathbf{r}',t-s) = \left[4\pi D(t-s)\right]^{-3/2} \exp - \left[\frac{(x-x'+\mu_o E(t-s))^2+(y-y')^2+(z-z')^2}{4D(t-s)}\right]$$

(3 .10)

Substitute Eqn (3.9) into (3.7a) to obtain

$$\frac{\partial \langle\psi(\mathbf{r},t)\rangle}{\partial t} - D\,\nabla^2\langle\psi\rangle - \mu_o E \frac{\partial \langle\psi\rangle}{\partial x} = Q\,\langle\psi\rangle$$

$$= E^2 \int d^3r' \int_{-\infty}^{t} ds \left[\langle\delta\mu(t)\ \delta\mu(s)\rangle \frac{\partial}{\partial x} G(\mathbf{r}|\mathbf{r}',t-s) \bullet \frac{\partial}{\partial x'} \langle\psi(\mathbf{r}',s)\rangle\right]$$

(3.11)

The exact solution to (3.11) may be written in the Gaussian form

$$\langle\ \psi(\mathbf{r},t)\rangle = C(t)\ G_o(\mathbf{r},t)$$

$$G_o(\mathbf{r},t) = [4\pi Dt]^{-3/2} \exp - \left[\frac{(x+\mu_o Et)^2 + y^2 + z^2}{4\ Dt}\right] \qquad (3.12)$$

and an equation for C(t) can be obtained by taking Fourier transforms (wavevector **k**) in Eqn (3.11):

$$\frac{dC(t)}{dt} = -\ E^2\ k_x^2 \int_{-\infty}^{t} ds\ C(s)\ \langle\delta\mu(t)\ \delta\mu(s)\rangle$$

$$= -\ E^2\ k_x^2 \int_{o}^{t} ds\ C(s)\ \langle\delta\mu(o)\ \delta\mu(t-s)\rangle \qquad (3.13)$$

where use has been made of the time displacement invariance symmetry of an equilibrium correlation function. Eqn (3.13) is readily solved by the convolution theorem of Laplace transforms (variable s) to yield

$$\tilde{C}(s) = \frac{1}{s + E^2 k_x^2 \tilde{\Gamma}(s)} \tag{3.14}$$

$$\tilde{\Gamma}(s) = \int_0^\infty d\tau e^{-s\tau} \langle \delta\mu(o)\ \delta\mu(\tau) \rangle$$

It is instructive to see what C(t), and therefore, by Eqn (3.12), $\langle \psi(\mathbf{r},t) \rangle$, looks like for various forms of the mobility fluctuation correlation function. Suppose the fluctuation spectrum is taken to be white noise:

$$\langle \delta\mu(o)\delta\mu(\tau) \rangle = \gamma\ \delta(\tau) \tag{3.15}$$

$$\gamma \text{ constant}$$

In this case C(t) is given by

$$C(t) = \exp - \left[\frac{E^2 k_x^2\ \gamma t}{2} \right] \tag{3.16}$$

which implies that the diffusion tensor has been transformed from isotropic to the anisotropic form:

$$\begin{bmatrix} D & 0 & 0 \\ 0 & D & 0 \\ 0 & 0 & D \end{bmatrix} \rightarrow \begin{bmatrix} D + \delta D_{xx} & 0 & 0 \\ 0 & D & 0 \\ 0 & 0 & D \end{bmatrix}$$

$$\delta D_{xx} = \frac{1}{2} E^2\ \gamma \tag{3.17}$$

Alternatively, one could take the singular limit of the Green's function in Eqn (3.11), $G \rightarrow \delta(\mathbf{r}-\mathbf{r}')$, and arrive at

$$\delta D_{xx} = E^2 \int_0^\infty d\tau\ \langle \delta\mu(o)\ \delta\mu(\tau) \rangle \tag{3.18}$$

In these cases, the diffusion coefficient is anisotropically enhanced because the mobility fluctuations give the molecule rapidly varying random kicks in the E-direction. An observer unaware of these fluctuations might conclude that the molecule's shape is distorted by the external field.

Suppose that the mobility correlation function decays exponentially:

$$\langle \delta\mu(o)\ \delta\mu(\tau) \rangle = M(o)\ \exp(-a\tau). \tag{3.19}$$

we now have

$$\tilde{C}(s) = \frac{s + a}{(s-R_1)(s-R_2)} \qquad \left.\begin{matrix} R_1 \\ R_2 \end{matrix}\right\} = \frac{-a \pm(a^2-4b)^{1/2}}{2}$$

$$b = E^2\, k_x^2\, M(o) \tag{3.20}$$

and, after inverting the Laplace transform,

$$C(t) = - \frac{(R_1+a)}{R_2-R_1} \exp(R_1 t) - \frac{(R_2+a)}{R_1-R_2} \exp(R_2 t) \tag{3.21}$$

Since R_1 and R_2 are in general complex, C(t) decays as the superposition of two oscillating exponentials. Note that if we take a=0, which corresponds to an infinite correlation time, we recover a diffusion model for two very slowly interconverting species at equal concentration./8/ In this case

$$\left.\begin{matrix} R_1 \\ R_2 \end{matrix}\right. = \pm\, i \sqrt{b} \quad , \; C(t) = \cos(\sqrt{b}\, t) \tag{3.22}$$

and the periodicity of C(t) gives the amplitude of the mobility fluctuations. The total correlation function (in complex form) then becomes

$$B(t) = (\text{const}) \cos (\sqrt{b}\, t) \exp[-i\, k_x \mu_o E t] \exp[-k^2 D t] \tag{3.23}$$

and the scattered laser light intensity (heterodyne detection) at frequency ω and wavevector **k** is the sum of two Lorentzians:

$$S(\mathbf{k},\omega) = (\text{const}') \left\{ \frac{Dk^2}{(\omega - k_x E(\mu_o - \sqrt{M(0)}))^2 + (Dk^2)^2} + \frac{Dk^2}{(\omega - k_x E(\mu_o + \sqrt{M(0)}))^2 + (Dk^2)^2} \right\} \tag{3.24}$$

Inasmuch as we expect electrophoretic mobility fluctuations to be correlated with "structural" fluctuations in the protein-solvent interface, electrophoretic light scattering could prove to be a useful probe of dynamics in the outer regions of solvated proteins.

5. THE MEMORY FUNCTION

Eqn (3.13) is a form that is familiar to most of you. It gives the time rate of change of a correlation function in terms of a superposition of products of that correlation function and a memory function evaluated at two previous times. In that example, the memory kernel was assumed to be independent of the correlation function. Here the general case will be presented.

Suppose we have a normalized autocorrelation function $\psi(t)$ of some phase space function $f(p,q)$:

$$\psi(t) = \langle f(o) f(t) \rangle, \quad \psi(0) = 1 \tag{4.1}$$

with

$$\langle f(t) \rangle = 0 \tag{4.2}$$

Now recall from Prof. Hansen's lectures that the time evolution of f is described by

$$\frac{df}{dt} = i \mathcal{L} f, \qquad \exp(i t \mathcal{L}) \dot{f}(0) = \dot{f}(t) \tag{4.3}$$

where $\mathcal{L}$ is the Liouville operator defined by

$$\mathcal{L} f = i (\mathcal{H}, f) \quad (\ \ ) = \text{Poisson Bracket} \tag{4.4}$$

and where $\mathcal{H}$ is the system Hamiltonian. Differentiate $\psi(t)$ in Eqn (4.1) twice with respect to time and use the (easily demonstrated) Hermitian property of $\mathcal{L}$ to arrive at/9,10/

$$\begin{aligned} \frac{d^2\psi}{dt^2} &= - \langle i \mathcal{L} f \exp(i t \mathcal{L}) \, i \mathcal{L} f \rangle \\ &= - \langle \dot{f}(0) \, \dot{f}(t) \rangle \end{aligned} \tag{4.5}$$

with the initial condition

$$\langle f(0) \, \dot{f}(0) \rangle = 0 \quad . \tag{4.6}$$

Laplace transformation (variable s) of (4.5) yields

$$\begin{aligned} &s^2 \, \tilde{\psi}(s) - s = -\tilde{\phi}(s) \\ &\phi(t) = \langle \dot{f}(0) \, \dot{f}(t) \rangle \end{aligned} \tag{4.7}$$

or equivalently

$$s \, \tilde{\psi}(s) - 1 = - \left[1 - \frac{\tilde{\phi}(s)}{s} \right]^{-1} \tilde{\phi}(s) \, \tilde{\psi}(s). \tag{4.8}$$

When this is inverted using the convolution theorem of Laplace transforms, we arrive at

$$\frac{d\psi}{dt} = - \int_0^t d\tau \, K(\tau) \, \psi(t-\tau), \tag{4.9}$$

$$\tilde{K}(s) = \left[1 - \frac{\tilde{\phi}(s)}{s}\right]^{-1} \tilde{\phi}(s),$$

which means that the memory function is completely determined by $<\dot{f}(0)\dot{f}(t)>$.

Eqn (4.9) is exact, but this fact offers little advantage unless the memory kernel $K(\tau)$ lends itself to simple approximations: happily, this turns out to be the case. I won't go into the details, but it is a straightforward matter to prove that $K(\tau)$ is an even, real function of time./9,10/ With these constraints and a fair amount of intuition, plausible forms of memory functions for various processes can be constructed./11/ One popular example is the pathological but useful $K(\tau) \sim \delta(\tau)$, which implies simple exponential relaxation for ψ. Another is

$$K(\tau) = a \exp[-b|\tau|], \quad a,b>0 \tag{4.10}$$

which, by following the procedure in the previous section, leads to

$$\psi(t) = \frac{1}{R_1 - R_2}\left\{R_1 e^{R_2 t} - R_2 e^{R_1 t}\right\}$$

$$\begin{matrix} R_1 \\ R_2 \end{matrix} = \left\{ -\frac{b}{2}\left[1 \mp \left(1 - \frac{4a}{b^2}\right)^{1/2}\right]\right. \tag{4.11}$$

which is again the sum of two oscillating exponentials. For sufficiently large values of a/b^2, R_1 and R_2 are complex, and ψ can be written in the form

$$\psi(t) = \exp(-\frac{bt}{2}) \left[\frac{b}{2\gamma} \sin \gamma t + \cos \gamma t\right]$$

$$\gamma = -\frac{b}{2}\left(1 - \frac{4a}{b^2}\right) > 0. \tag{4.12}$$

For reasonable choices of a and b, it turns out that Eqn. (4.12) does a remarkably good job of describing the velocity autocorrelation functions observed in molecular dynamics simulations of simple liquids./11/

In condensed media where the collision frequency is large compared to other characteristic frequencies, a Langevin or generalized Langevin description is frequently preferred to the exact Eqn. (4.9). In terms of single particle motion, the generalized Langevin equation is usually written/9,10,11/

$$\frac{dv(t)}{dt} = -\int_0^t K(t-\tau)v(\tau)d\tau + F(t) \tag{4.13}$$

in which the random force F(t) has the following properties

$$\langle \mathbf{F}(t-\tau)\bullet\mathbf{F}(\tau)\rangle = \frac{3\,k_BT}{m}\,K(t) \qquad \text{(a)} \qquad (4.14)$$

$$\langle \mathbf{F}(t)\bullet\mathbf{v}(0)\rangle = 0 \qquad \text{(b)}$$

The first is a consequence of the so-called second fluctuation-dissipation theorem discussed by Professor Hansen, while the second is an orthogonality condition which says that the random force is uncorrelated with the velocity of the Brownian particle. Note that since

$$\left\langle \frac{d\mathbf{v}(t)}{dt}\bullet\mathbf{v}(0)\right\rangle = \frac{d}{dt}\langle \mathbf{v}(t)\bullet\mathbf{v}(0)\rangle \,, \qquad (4.15)$$

multiplication of (4.13) by $\mathbf{v}(0)$ followed by Laplace transformation gives us the familiar

$$\tilde{C}(s) = [s + \tilde{K}(s)]^{-1} \quad , \qquad (4.16)$$

where $\tilde{C}(s)$ is the Laplace transform of the normalized velocity autocorrelation function. As in the ordinary Langevin formulation, the time integral of the random force autocorrelation function in Eqn. (4.14), or equivalently, the time integral of K(t), can be identified with the friction coefficient of the Brownian particle. In the case of a Brownian ion, it is reasonable to suppose that the "short range" ion-solvent force relaxes on a much shorter time scale than the long range attractive force. If one further assumes that these so-called "hard" forces and "soft" forces are uncorrelated, then the friction coefficient takes the approximate form

$$\xi = \xi_o + \frac{1}{3k_BT}\int_o^{\infty} \langle \mathbf{F}_s(0)\bullet\mathbf{F}_s(t)\rangle\, dt \qquad (4.17)$$

where ξ_o arises solely from the hard force correlations, and $\mathbf{F}_s$ is determined by the long range ion-solvent dipole interaction. Eqn (4.17) is the decoupling ansatz made by P. Wolynes in his theory of solvated ion dynamics./12/ The Wolynes theory will be discussed in more detail later on.

The conventional Langevin description of a stochastic process deals with additive random noise having prescribed properties. There is also another possibility; namely, multiplicitive noise,/13/ wherein the "Langevin variable" gets multiplied by a stochastic variable, as in

$$\frac{dv(t)}{dt} + \xi(t)\,v(t) = 0$$

$$v(t) = \langle v(t)\rangle + \delta v(t)$$

$$\xi(t) = \xi_o + \delta\xi(t), \quad \xi_o > 0$$

$$\langle \delta v\rangle = \langle \delta\xi\rangle = 0 \qquad (4.18)$$

Consider v(t) to be one component of the velocity of a Brownian particle and take $\xi(t)$ to be a fluctuating friction coefficient whose equilibrium ensemble average is ξ_o. The generalization to three dimensions or to other interpretations of the equations is straightforward. The particle therefore experiences a fluctuating resistance which is determined by fluctuating environmental dynamics, and this is in addition to the systematic resistance due to the "average structure" of the medium. We can use the same method employed in the previous section to analyze the properties of a Smoluchowski equation with fluctuations. First, note that for positive times, we can write

$$\frac{d\langle v(t)\rangle}{dt} + \xi_o\langle v(t)\rangle = -\langle\delta\xi(t)\ \delta v(t)\rangle \qquad t>0 \tag{4.19}$$

or in terms of a normalized velocity autocorrelation function:

$$\langle\dot{\psi}\rangle + \xi_o\langle\psi\rangle = -\langle\delta\xi(t)\ \delta\psi(t)\rangle \qquad t>0 \tag{4.20}$$

As before, we reverse the sign of time to obtain an expression for $\delta\psi(t)$, but since we are assuming internal or dissipative noise, these terms change sign under $t \rightarrow -t$. The linearized result is

$$\delta\psi(t) = e^{\xi_o t}\int_{-\infty}^{t} ds\ e^{-\xi_o s}\ \delta\xi(s)\ \langle\psi(s)\rangle \qquad t \gtrless 0 \tag{4.21}$$

which, when inserted into Eqn (4.20), yields

$$\langle\dot{\psi}\rangle + \xi_o\ \langle\psi\rangle = -\int_0^t ds\ \Big\{\exp[\xi_o(t-s)]\ \langle\delta\xi(0)\ \delta\xi(t-s)\rangle \bullet\langle\psi(s)\rangle\Big\} \qquad t>0 \tag{4.22}$$

In terms of Laplace transforms,

$$\tilde{\psi}(s) = \frac{1}{s+\xi_o + \tilde{\Gamma}(s-\xi_o)}$$

$$\tilde{\Gamma}(s) = \int_0^\infty dt\ e^{-st}\ \langle\delta\xi(0)\ \delta\xi(t)\rangle \tag{4.23}$$

For $\Gamma(t) \sim \delta(t)$ we recover a renormalized friction coefficient $\xi_o+\Delta\xi_o$, and $\langle\psi(t)\rangle$ decays as a simple exponential. Consider the case

$$\Gamma(t) = a\, e^{-bt} \qquad a,b>0 \tag{4.24}$$

for which

$$\tilde{\psi}(s) = \frac{(s - \xi_o + b)}{(s - \xi_o + b)\,(s + \xi_o) + a} \tag{4.25}$$

The roots of the denominator are

$$\begin{matrix} s_+ \\ s_- \end{matrix} = \left\{ -\frac{b}{2}\left[1 \pm \left(1 - \frac{4\gamma}{b^2}\right)^{1/2}\right]\right.$$

$$\gamma = \xi_o\,(-\,\xi_o + b) + a \tag{4.26}$$

which can be compared with Eqns (4.10) and (4.11). Indeed, for $\xi_o=0$ we recover the velocity autocorrelation in Eqn (4.11), while for $\xi_o=b$, the roots (Eqn (4.26) and (4.11)) are identical. From Eqns (4.25) and (4.26), we see that $\langle\psi(t)\rangle$ decays as the sum of two oscillating exponentials, which, because of the systematic friction ξ_o, turns out to be even more accurate than Eqn (4.12) at describing computer simulated velocity autocorrelation functions. At this time it is still an open question as to how well these simple theories (models) describe the dynamics of solvated ions.

6. MICROSCOPIC THEORY OF ION-SOLVENT DYNAMICS

Up to this point we have assiduously avoided the introduction of detailed physical models for ion-solvent dynamics. This approach has the advantage of generality and places a limit on conceptual blunders, but we have yet to grapple with such questions as "What makes ion-solvent dynamics different from the dynamics of ordinary liquids?" or "How does the long-range ion-solvent interaction affect the ionic friction coefficient?" From Professors Neilson and Enderby we have been hearing about the structure of water around ions, and it is natural to ask how the temporal behavior of this structure is reflected in a transport coefficient such as the ionic conductance.

A few years ago Peter Wolynes formulated a theory of ionic friction which, though heuristic, is physically appealing, thought provoking, and fits in nicely with these lectures./12/ Furthermore, it is the only microscopic treatment in which explicit calculations of ionic friction coefficients have been performed. Wolynes took the view that the main difference between the dynamics of simple liquids (argon, methane) and solvated ion dynamics is that the attractive forces between an ion and a solvent molecule, being stronger than van der Waals forces in simple liquids, do play a major role in ionic motion. While this is a trivial observation for the physical chemist, it poses an exceedingly difficult problem for the physicist who wants to approach the problem from fundamental kinetic theory, inasmuch as it renders void any attempt to treat the ion-solvent molecule collisions in the binary approximation: the ion interacts simultaneously with many solvent molecules. Of course, even for van der Waals liquids the binary collision approximation breaks down badly, and some time ago Allnatt and Rice came up with a simple remedy:/14/ decompose the intermolecular potential into a steep, repulsive, "hard" part V_H and a gradually sloping, longer-ranged "soft" V_s, treat the V_H interaction as binary collisions to calculate a friction coefficient ξ_H, and leave the remaining terms for the next generation of Dutch postdocs. Wolynes applied this decoupling idea to the ion-solvent interaction where ξ_H was calculated from hydrodynamics, and the correction term ξ_s was calculated from a simple "dielectric friction" picture. According to this model, an appreciable part of the resistance experienced by a migrating ion arises from the delayed reorientation of solvent dipoles in the electrostatic field of the ion, so this is a direct analogue of the ion atmosphere relaxation effect which we calculated in Section 2.

We begin with our Langevin result from the previous section:

$$\xi = \frac{1}{3k_BT}\int_0^\infty dt\langle \mathbf{F}(0)\bullet\mathbf{F}(t)\rangle \tag{5.1}$$

which relates the friction coefficient to the random force autocorrelation associated with a fixed ion. The Wolynes ansatz may now be written

$$\langle \mathbf{F}(0)\bullet\mathbf{F}(t)\rangle = \langle \mathbf{F}_H(0)\bullet\mathbf{F}_H(t)\rangle + \langle \mathbf{F}_H(0)\bullet\mathbf{F}_s(t)\rangle + \langle \mathbf{F}_s(0)\bullet\mathbf{F}_H(t)\rangle + \langle \mathbf{F}_s(0)\bullet\mathbf{F}_s(t)\rangle$$

$$\langle \mathbf{F}(0)\bullet\mathbf{F}(t)\rangle \cong \langle \mathbf{F}_H(0)\bullet\mathbf{F}_H(t)\rangle + \langle \mathbf{F}_s(0)\bullet\mathbf{F}_s(t)\rangle \tag{5.2}$$

where $\mathbf{F}_H$ is the force associated with V_H and similarly for $\mathbf{F}_s$. Eqn (5.2) assumes that the hard force - soft force cross correlation function is negligible. Eqn (5.2) may also be written

$$\xi = \xi_o + \xi_s = \xi_o + \frac{1}{3k_BT}\int_0^\infty dt\langle \mathbf{F}_s(0)\bullet\mathbf{F}_s(t)\rangle \tag{5.3}$$

Since $\mathbf{F}$ ($\mathbf{F}_s$) is a dynamical variable, the time rate of change of $\mathbf{F}_s$ can be written in terms of the linear and angular velocities of each solvent molecule in the system:

$$\dot{\mathbf{F}}_s = \sum_i\left\{\mathbf{v}_i\bullet\nabla_i\,\mathbf{F}_s + \mathbf{\Omega}_i\bullet\mathbf{L}_i\,\mathbf{F}_s\right\} \tag{5.4}$$

where $\mathbf{v}_i$ and $\mathbf{\Omega}_i$ are the linear and angular velocities of the i th solvent molecule and $\mathbf{L}_i$ is the orientational gradient $\mathbf{r}_i\mathrm{x}\nabla_i$. At this point another assumption is made; namely, that after a short induction time during which $<F_s(0)\bullet F_s(t)>$ decays quadratically, the relaxation may be described by a simple exponential

$$<F_s(0) \bullet F_s(t)> = <F_s(0)^2>\exp\left[-\frac{|t|}{\tau_F}\right] \qquad (5.5)$$

The next step is to relate the soft force correlation time τ_F to the correlation function associated with $\dot{F}_s$ in Eqn (5.4). First, note that because of time displacement invariance of an equilibrium correlation function, we have

$$\frac{d}{dt} <F_s(0) \bullet F_s(t)> = <F_s(0) \bullet \dot{F}_s(t)> ,$$

$$-\frac{d}{dt} <F_s(0) \bullet F_s(t)> = - \frac{d}{dt} <F_s(-t) \bullet F_s(0)>$$

$$= - <\dot{F}_s(-t) \bullet F_s(0)> = - <\dot{F}_s(0) \bullet F_s(t)> \qquad (5.6)$$

and therefore

$$\frac{d^2}{dt^2} <F_s(0) \bullet F_s(t)> = - <\dot{F}_s(0) \bullet \dot{F}_s(t)> \qquad (5.7)$$

Integrate (5.7) over a short time interval from 0 to t, and arrive at

$$- \tau_F^{-1} <F_s(0)^2> \exp(\frac{-t}{\tau_F}) + \frac{d}{dt} <F_s(0)^2>$$

$$= - \int_0^t <\dot{F}_s(0) \bullet \dot{F}_s(t')> dt' \qquad (5.8)$$

where we have substituted from Eqn. (5.5) for times that are not too short. Since the time derivative of an equilibrium correlation function vanishes identically, the second ℓ.h.s. term in Eqn (5.8) is zero, and we arrive at

$$<F_s(0)^2> \tau_F^{-1} = \int_0^\infty < \dot{F}_s(0) \bullet \dot{F}_s(t)> dt \qquad (5.9)$$

where we have assumed that $t<<\tau_F$, and also that t is much larger than the correlation time of $<\dot{F}_s(0) \bullet \dot{F}_s(t)>$. This implies that $\dot{F}_s$ relaxes much faster than F_s itself.

It is instructive to go back to the exact expression for the memory function in Eqn (4.9) in order to ascertain which form of $<\dot{F}(0) \bullet$

$\dot{F}_s(t)$> can produce an exponential force autocorrelation function, as in Eqn (5.5). First, recall that the memory kernel $K(t) \sim \delta(t)$ gives a correlation function $\psi \sim \exp(-at)$. Since

$$\tilde{K}(s) = \left[1 - \frac{\tilde{\phi}(s)}{s}\right]^{-1} \tilde{\phi}(s)$$

$$\tilde{\phi}(s) = \int_0^\infty dt\, e^{-st} \langle \dot{F}_s(0) \bullet \dot{F}_s(t) \rangle \tag{5.10}$$

it is easy to see that the highly pathological inverse transform of

$$\tilde{\phi}(s) = \frac{cs}{s+c} \qquad c \text{ constant} \tag{5.11}$$

which resembles a closely spaced pair of positive and negative "spikes" of equal amplitude, does indeed lead to an exponential form for ψ. An increasing force acting on the ion at any given time therefore tends to be cancelled by a decrease at the following instant.

From Eqn (5.4), we derive the formidable-looking expression

$$\langle \dot{F}_s(0) \bullet \dot{F}_s(t) \rangle = \sum_i \sum_j \Big\langle (\nabla_i F_s)(\nabla_j F_s) : \langle v_i(0) v_j(t) \rangle^*$$

$$+ (\nabla_i F_s)(L_j F_s) : \langle v_i(0) \Omega_j(t) \rangle^*$$

$$+ (L_i F_s)(\nabla_j F_s) : \langle \Omega_i(0) v_j(t) \rangle^*$$

$$+ (L_i F_s)(L_j F_s) : \langle \Omega_i(0) \Omega_j(t) \rangle^* \Big\rangle \tag{5.12}$$

where (:) denotes a double scalar product and the * superscript over the velocity ensemble averages signify that the averages are to be taken over the velocities of all the particles but with the position of particles i and j kept fixed. The forces are to be taken at the same (initial) time. In a sense, the decoupling approximation in Eqn (5.12) implies that the solvent is frozen in some amorphous state on a velocity relaxation timescale. With the approximations Eqns (5.5) and (5.9), the ionic friction coefficient may be derived from Eqn (5.3):

$$\xi = \xi_o + \frac{\langle F_s(0) \rangle^2}{3k_B T} \tau_F$$

$$\xi = \xi_o + \frac{\langle F_s(0)^2 \rangle^2}{3k_B T} \left[\int_0^\infty dt\, \langle \dot{F}_s(0) \bullet \dot{F}_s(t) \rangle \right]^{-1} \tag{5.13}$$

where the integrand is given by Eqn (5.12).

We have finally arrived at an explicit but seemingly intractable expression for the ionic friction coefficient. Not only do we have to decide what form the equilibrium soft force should take, but we have to calculate velocity cross-correlation functions that involve three fixed particles (two solvent molecules and the ion). If one or both solvent molecules are near the ion, we expect the velocity correlation functions to deviate considerably from their form in the pure solvent. Moreover, very little is known about the structure of these dynamic cross-correlations, even for pure liquids. Nevertheless, Eqns (5.12) and (5.13) represent a conceptual breakthrough, inasmuch as the role of translational and rotational degrees of freedom of the solvent dipoles is clearly displayed, and one can easily imagine testing the theory at this level by a molecular dynamics simulation of an ion in a polar liquid. As we have heard from several speakers, such computer experiments are being planned for the near future. I won't go into any detail, but Wolynes and co-workers have succeeded in calculating friction coefficients using a physically crude though conceptually elegant hydrodynamic approximation for the cross-correlation functions, together with a simple model for the equilibrium ion-solvent structure./12/ They obtained impressive agreement with experimental results for lithium and sodium ions in water and methanol. At this time it seems as if the Wolynes theory or perhaps a similar hybrid will be the basis for future theoretical advances in this area.

7. CONTINUUM THEORIES

Historically speaking, a good deal of the current interest in solvated ion dynamics arose from the development of continuum models of the ion-solvent interaction. Inasmuch as there exists a close analogy between these approaches and the equilibrium Born model or the enormously successful DFO theory, such notables as M. Born, R. Zwanzig, and L. Onsager apparently felt justified in investing in the construction of a continuum theory of ionic friction which would help to explain the observed trends in single ion mobilities: small ions in polar liquids tend to possess smaller mobilities, and hence, larger friction coefficients, than somewhat larger ions. This is, of course, counter to purely hydrodynamic reasoning, which says that the friction coefficient should be proportional to the product of solvent viscosity and ion radius ($\xi \sim \eta R$). The chemists' explanation of this "anomaly" was clear but crude: small ions must carry a tightly bound cluster of solvent molecules with them as they migrate, the extra baggage implies a larger hydrodynamic size, and therefore an enhanced friction coefficient. Although this is probably correct for most divalent cations in water, there is ample evidence, some of which has been presented at these meetings, that ion-solvent complexes should generally be regarded as dynamic, rather than static, entities. The question is "How is the dynamics manifested in a transport coefficient, such as ion mobility?"

Rather than subject you to a baptism (immersion stylè) in continuum electrodynamics and hydrodynamics, I have opted for a poor man's treatment of dielectric relaxation in a non-uniform electric field, a brief discussion of polarization relaxation around an ion where the amplitude of the ion charge oscillates, followed by a physicist's view of energy transfer from a moving ion to a dielectric medium with loss, and finally, a few comments on how one goes about constructing a self-consistent continuum theory of ionic friction.

Consider a polar liquid in which there initially exists a spatial gradient in the distribution of molecules having a specified orientation. This distribution may be defined as follows. At time t, let the number of polar molecules located at $\mathbf{r}$ in $d^3\mathbf{r}$ with their dipole vector pointing in $d^2\Omega$ (solid angle) be given by

$$\psi(t,\ \mathbf{r},\ \Omega)\ d^3\mathbf{r}\ d^2\Omega \tag{6.1}$$

and the number density at $\mathbf{r}$ without regard to orientation is

$$n(t,\mathbf{r}) = \int_0^{\pi}\int_0^{2\pi} \psi(t,\mathbf{r},\Omega)\ d^2\Omega \tag{6.2}$$

At equilibrium, ψ and n become constants ($\psi_o = n_o/4\pi$). As the system relaxes to equilibrium, we expect both translational and rotational motion of dipoles. In the absence of molecular reorientation, the translational migration smoothes out the initial orientation inhomogeneity in the system, but this does not affect the net polarization. Focus on a single dipole, and assume the following for its equation of motion;/15/

$$m\ddot{\mathbf{r}} = \nabla(\mu\bullet\mathbf{E}) - \xi_o\dot{\mathbf{r}} \tag{6.3}$$

where m is the mass, μ the dipole moment, ξ_o is a friction coefficient which we take to be constant, and $\mathbf{E}$ is the local field acting on the dipole. By discarding inertial effects, we have

$$\dot{\mathbf{r}} = \xi_o^{-1}\ \nabla(\mu\bullet\mathbf{E}) \tag{6.4}$$

and the corresponding "microscopic flux" is

$$\mathbf{j}_{trans}(\text{electric}) = \frac{1}{\xi_o}\ \psi\ \nabla(\mu\bullet\mathbf{E})\ d^2\Omega \tag{6.5}$$

To this field-driven flux we add the diffusive flux

$$\mathbf{j}_{trans}(\text{diffusion}) = -\ D\ \nabla\psi\ d^2\Omega \tag{6.6}$$

where D is the self diffusion coefficient. We now invoke the Einstein relation ($D = k_BT/\xi_o$) to write the net translational flux as

$$\mathbf{j}_{trans} = -\ D\left[\nabla\psi - \beta\ \psi\ \nabla(\mu\bullet\mathbf{E})\right]\ d^2\Omega \tag{6.7}$$

Suppose that the non equilibrium part of ψ is due to a small field gradient. Then, to first order in $\nabla\mathbf{E}$, we can replace ψ in the second term on the r.h.s. of Eqn (6.7) by $\psi_o = n_o/4\pi$ and write

$$\left.\frac{\partial n}{\partial t}\right|_{trans} = -\nabla\cdot\mathbf{j}_{trans}$$

$$\left.\frac{\partial\psi}{\partial t}\right|_{trans} = D\,\nabla^2\psi - \left(\frac{Dn_o\mu}{4\pi k_B T}\right)\nabla^2[\boldsymbol{\mu}\cdot\mathbf{E}] \qquad (6.8)$$

Consider now the rate of change of ψ due to rotational diffusion. A dipole oriented at an angle $\boldsymbol{\theta}$ to the field $\mathbf{E}$ experiences a driving torque $\boldsymbol{\mu}\times\mathbf{E}$ and a frictional torque $\xi_{rot}\dot{\boldsymbol{\theta}}$. If rotational inertia is neglected, we have

$$\dot{\boldsymbol{\theta}} = \xi_{rot}^{-1}\,(\boldsymbol{\mu}\times\mathbf{E}) \qquad (6.9)$$

The rotational flux associated with this torque is therefore

$$\mathbf{j}_{rot}(\text{electric}) = \frac{1}{\xi_{rot}}\,(\boldsymbol{\mu}\times\mathbf{E})\,\psi\,d^2\Omega \qquad (6.10)$$

Now add the diffusive flux associated with the angular gradient ($\mathbf{L} = \mathbf{r}\times\nabla$) of ψ, and the net rotational flux into $d^2\Omega$ becomes

$$\mathbf{j}_{rot} = -D_{rot}\left[\mathbf{L}\,\psi - \beta\psi\,(\boldsymbol{\mu}\times\mathbf{E})\right]d^2\Omega \qquad (6.11)$$

where $D_{rot} = k_B T/\xi_{rot}$. As in Eqn (6.8), we can write

$$\left.\frac{\partial n}{\partial t}\right|_{rot} = -\mathbf{L}\cdot\mathbf{j}_{rot}$$

$$\left.\frac{\partial\psi}{\partial t}\right|_{rot} = D_{rot}\,L^2\psi - \left(\frac{D_{rot}}{k_B T}\right)\mathbf{L}\cdot\left[(\boldsymbol{\mu}\times\mathbf{E})\psi\right] \qquad (6.12)$$

The connection between this and the continuum notion of polarization relaxation is established as follows. Define the macroscopic orientation polarization by

$$\mathbf{P} = \mu\int_\Omega \psi\,\hat{\boldsymbol{\mu}}\,d^2\Omega \qquad (6.13)$$

where $\hat{\boldsymbol{\mu}}$ is a unit vector in the $\boldsymbol{\mu}$ direction. Now multiply Eqn (6.8) by $\boldsymbol{\mu}$ and integrate over angles to obtain

$$\frac{\partial}{\partial t}\int \psi\,\hat{\mu}_\alpha\,d^2\Omega = D\,\nabla^2\int_\Omega \psi\,\hat{\mu}_\alpha\,d^2\Omega$$

$$- \frac{D n_o \mu^2}{4\pi k_B T} \nabla^2 E_\beta \int_\Omega \hat{\mu}_\alpha \hat{\mu}_\beta \, d^2\Omega \tag{6.14}$$

Identify $n_o \mu^2/3k_B T$ with the dielectric susceptibility χ and use the identity

$$\int_\Omega \hat{\mu}_\alpha \hat{\mu}_\beta \, d^2\Omega = \frac{4\pi}{3} \delta_{\alpha\beta} \tag{6.15}$$

together with Eqns (6.13) and (6.14) to obtain

$$\left.\frac{\partial \mathbf{P}}{\partial t}\right|_{trans} = D \nabla^2 (\mathbf{P} - \chi \mathbf{E}) \tag{6.16}$$

A similar procedure with Eqn (6.12) yields

$$\frac{\partial}{\partial t} \int_\Omega \mu \psi \, d^2\Omega = D_{rot} \int_\Omega \mu L^2 \psi \, d^2\Omega$$

$$- \frac{D_{rot}}{k_B T} \int_\Omega \mu \, \mathbf{L} \bullet (\mu \times \mathbf{E} \psi) \, d^2\Omega \tag{6.17}$$

Expand the first term on the r.h.s. of Eqn (6.17) in spherical harmonics to give

$$\int_\Omega \mu L^2 \psi \, d^2\Omega = - 2 \int_\Omega \mu \psi \, d^2\Omega \tag{6.18}$$

Use the identity $\mathbf{L} \bullet (\mu \times \mathbf{E}) = - 2\mu \bullet \mathbf{E}$ together with Eqns (6.13) and (6.15), then linearize Eqn (6.17) in **E** to arrive at

$$\left.\frac{\partial \mathbf{P}}{\partial t}\right|_{rot} = - 2 D_{rot} (\mathbf{P} - \chi \mathbf{E}) = - \tau^{-1} (\mathbf{P} - \chi \mathbf{E}) \tag{6.19}$$

which is the famous Debye equation for dielectric relaxation. The net result is the sum of Eqns (6.19) and (6.16):/15/

$$\frac{\partial \mathbf{P}}{\partial t} = - \frac{1}{\tau} (\mathbf{P} - \chi \mathbf{E}) + D \nabla^2 (\mathbf{P} - \chi \mathbf{E}) \tag{6.20}$$

where τ is called the Debye relaxation time.

An alternative method of arriving at an expression like Eqn (6.20), and which provides some additional insight, is simply to define a macroscopic polarization flux tensor $\tilde{J}$ such that

$$\left.\frac{\partial P}{\partial t}\right|_{trans} = - \nabla \cdot \tilde{J} \tag{6.21}$$

For an isotropic medium, $\tilde{J}$ can be written

$$\begin{aligned} J_{\alpha\beta} &= (D_1 - \tfrac{2}{3} D_2)(\nabla \cdot P^*)\, \delta_{\alpha\beta} \\ &+ D_2(\nabla_\alpha P^*_\beta + \nabla_\beta P^*_\alpha) \\ &+ D_3(\nabla_\alpha P^*_\beta - \nabla_\beta P^*_\alpha) \\ P^* &= \chi E - P \end{aligned} \tag{6.22}$$

where the D's are phenomenological diffusion coefficients. Note that $\tilde{J} = 0$ for P^* constant. The net polarization current can then be written/15/

$$\frac{\partial P}{\partial t} = \frac{1}{\tau} P^* - D_L \nabla(\nabla \cdot P^*) + D_T \nabla \times (\nabla \times P^*) \tag{6.23}$$

$$D_L = D_1 + \frac{4}{3} D_2 \qquad D_T = D_2 - D_3$$

which reduces to Eqn (6.20) for $D_L = D_T = D$. Eqn (6.23) says that the Debye term acts as a sink, while the non-equilibrium polarization diffuses in both longitudinal and transverse modes.

I'll skip the details, but it is easy to show that Eqn (6.23) is equivalent to deriving a frequency and wavenumber dependent dielectric constant

$$\epsilon(k,\omega) = \frac{\epsilon_o - i\omega\tau + \epsilon_o D_L \tau k^2}{1 - i\omega\tau + D_T \tau k^2} \tag{6.24}$$

where the hydrodynamic-like k^2 terms arise from translational motions of dipoles. This expression will be used in our discussion of the energy loss of an ion moving through a dielectric medium.

Eqns (6.23) or (6.20) can be used to glean information about the polarization dynamics around a microscopic charge distribution, such as a cavity dipole or cavity ion, where "cavity" refers to a molecular size vacuum in which the charges are located. The idea is that electric field inhomogeneities near such a particle can result in a qualitatively different picture of polarization relaxation than that obtained from a local description, such as the Debye theory. Since we are in a bulk charge-free, quasi-static regime, we may write

$$\mathbf{D} = \mathbf{E} + 4\pi\mathbf{P}, \quad \nabla\bullet\mathbf{D} = 0, \quad \nabla \times \mathbf{E} = 0 \tag{6.25}$$

for the medium, with the boundary conditions

$$\hat{\mathbf{n}}\bullet\mathbf{D} = \hat{\mathbf{n}}\bullet\mathbf{E}_c \qquad \hat{\mathbf{n}} \times \mathbf{E} = \hat{\mathbf{n}} \times \mathbf{E}_c \tag{6.26}$$

where $\hat{\mathbf{n}}$ is a unit normal to the surface and "c" denotes the vacuum cavity. We also have a conservation law for the polarization flux; namely

$$\int d^3r\, \nabla\bullet\tilde{\mathbf{J}} = 0 \tag{6.27}$$

which, according to the theorems of Gauss and Stokes, together with Eqns (6.23) and (6.25), may be transformed into the additional boundary conditions

$$\hat{\mathbf{n}} \times (\nabla \times \mathbf{D}) = 0 \qquad \hat{\mathbf{n}}(\nabla\bullet\mathbf{E}) = 0 \tag{6.28}$$

Similarly, elementary electrostatics says that Eqn (6.23) may be written in the Fourier component forms

$$(1 - i\omega\tau - D_T\tau\nabla^2)\, \mathbf{D}(\mathbf{r},\omega) = (\epsilon_o - i\omega\tau - \epsilon_o D_L\tau\nabla^2)\, \mathbf{E}(\mathbf{r},\omega) \quad \text{(a)}$$

$$(\epsilon_o - i\omega\tau - \epsilon_o D_L\tau\nabla^2)\, \nabla^2\, \phi(\mathbf{r},\omega) = 0 \quad \text{(b)} \tag{6.29}$$

where $\mathbf{E} = -\nabla\phi$. Thus, we have a fourth order equation for the potential with the four boundary conditions Eqns (6.26) and (6.28).

van der Zwan and Hynes, in what amounts to a tour de force performance in "cavity field polarization dynamics", have solved these equations analytically for a variety of models, including that of a charge moving along the axis of a hollow cylinder embedded in a dielectric, where the long axis is supposed to represent the reaction coordinate for a charge transfer process./16/ Their paper is equivalent to six problems from Landau and Lifshitz.

For the simplest case of a bare point charge in a dielectric, they obtain

$$\mathbf{P}(\mathbf{r},\omega) = \frac{q(\omega)}{4\pi}\left(\frac{\hat{\mathbf{r}}}{r^2}\right)\left[1 - \frac{1}{\epsilon(\omega)} + \exp(-D_\epsilon r) \bullet (1 + D_\epsilon r)\left(\frac{1}{\epsilon(\omega)} - \frac{1}{\epsilon_o}\right)\right]$$

$$\epsilon(\omega) = \frac{\epsilon_o - i\omega\tau}{1 - i\omega\tau} \tag{6.30}$$

for the oscillating charge distribution

$$\rho(\mathbf{r},\omega) = q(\omega)\, \delta(\mathbf{r}) \tag{6.31}$$

with

$$D_\epsilon^2 = \frac{(\epsilon_o - i\omega\tau)}{\epsilon_o D_L \tau} \tag{6.32}$$

They also derive, for the same charge embedded in a spherical cavity of radius R

$$P(r,\omega) = \frac{q(\omega)}{4\pi}\left(\frac{\hat{r}}{r^2}\right)\left\{1 - \frac{1}{\epsilon(\omega)} + \left(\frac{1}{\epsilon(\omega)} - \frac{1}{\epsilon_o}\right)\right.$$
$$\left. \bullet \frac{(1 + D_\epsilon r) \quad \exp[-D_\epsilon(r-R)]}{1 + D_\epsilon R + (1/2)(D_L/D_T)\, D_\epsilon^2 R^2}\right\} \tag{6.33}$$

Note that an oscillating point charge excites only longitudinal modes, and that for a finite size charge, **P** reverts back to the Debye form if D_T=0. Also, note the existence of a dynamic Debye screening effect due to the "polarization charge atmosphere", even though there is no ionic atmosphere.

From Maxwell's equations it is readily demonstrated that the rate of energy loss of a charge q moving in a straight trajectory **v** through a dielectric medium is given by the volume integral

$$\dot{W} = \frac{1}{4\pi}\int d^3\mathbf{r}\left(\mathbf{E} \bullet \frac{\partial D}{\partial t}\right)$$
$$= -\frac{q^2}{2\pi^2}\int d^3\mathbf{k}\left[\frac{\mathbf{k} \bullet \mathbf{v}}{k^2}\right] \mathrm{Im}\left[\frac{1}{\epsilon(\mathbf{k},\mathbf{k}\bullet\mathbf{v})}\right] \tag{6.34}$$

Upon substituting Eqn (6.24) with $D_T = D_L = D$ into (6.34) and integrating over the directions of **k**, we obtain/17/

$$\dot{W} = \frac{2q^2}{\pi}\left(\frac{\epsilon_o - 1}{\tau}\right)\int_o^{k_o} dk\; p(k)\left[1 - \frac{p(k)}{\alpha k}\tan^{-1}\frac{\alpha k}{p(k)}\right]$$
$$p(k) = 1 + \tau D k^2 \qquad \alpha = \frac{v\tau}{\epsilon_o} \tag{6.35}$$

Here the infinite limit in Eqn (6.34) has been replaced by a cut-off k_o such that

$$k_o \lesssim \min\left\{\frac{1}{a}, \frac{\omega_\infty}{v}\right\} \tag{6.36}$$

where a is some average intermolecular spacing and ω_∞ is a frequency beyond which Debye dispersion is invalid. In the fast ion regime with $\alpha k_o/p(k_o) \gg 1$, Eqn (6.35) is approximated accurately by

$$\dot{W} = \frac{2q^2}{\pi} (\epsilon_o - 1) k_o \left(\frac{1}{\tau} + \frac{1}{3} Dk_o^2 + \dots\right) \tag{6.37}$$

At high velocity $k_o \sim v^{-1}$ and $\dot{W}$ vanishes, which means that the charge is moving too fast for the ambient dielectric to dissipate energy. Note that at high velocity the translational motion of polar molecules leads to an increase in the energy loss. At low velocity with $\alpha k_o/p(k_o) \ll 1$, Eqn (6.35) becomes

$$\dot{W} = \frac{2}{9\pi} q^2 \left(\frac{\epsilon_o - 1}{\epsilon_o^2}\right) \tau v^2 k_o^3 \left[1 - \frac{3}{5}\tau Dk_o^2 + \dots\right] \tag{6.38}$$

for $k_o(\tau D)^{1/2} \ll 1$, and

$$\dot{W} = \frac{2}{3\pi} q^2 \left(\frac{\epsilon_o - 1}{\epsilon_o^2}\right) \frac{v^2}{D} \left[k_o - \frac{\pi}{2} (\tau D)^{-1/2} + \dots\right] \tag{6.39}$$

if $k_o(\tau D)^{1/2} \gg 1$. As $v \to 0$, k_o becomes independent of v, and if we identify k_o with the inverse ion radius and write

$$\dot{W} = \xi_D v^2 \tag{6.40}$$

then Eq(6.38) says that $\xi_D \sim R^{-3}$. In other words, the dielectric friction coefficient increases as the ion size decreases. In this case translational diffusion diminishes the frictional drag. If we take the residual drag to be given by Stokes' law, then, with Eqn (6.38), the total ionic friction coefficient takes the form

$$\xi = \alpha\eta R + \gamma q^2 \tau R^{-3} + \dots \tag{6.41}$$

Eqn (6.41) has some of the features of experimental friction coefficient vs. ion size plots.

It turns out that Eqn (6.34), with its implicit assumption of no ion-induced hydrodynamic motion in the surrounding dielectric, is inadequate as a self-consistent continuum model for ionic friction in a polar fluid. Such a theory was in fact developed back in 1977 by Lars Onsager and myself./18/ We came up with a method for combining Navier-Stokes hydrodynamics with dielectric relaxation in an arbitrary electric field to produce a theory of "electrohydrodynamics". In order to achieve this, we had to postulate a principle of "material frame indifference" or "Poincare invariance" which, though just an ansatz, has proved to be very successful in explaining the viscoelastic behavior of polymer solutions and melts./19/ This principle states that both the relaxation equations and the equations of motion must be invariant under the class of rigid body translations and rotations of the medium. The predictions of this continuum dielectric friction theory have recently been very thoroughly tested by the experimental investigations of Professor M. Nakahara and co-workers, who claim that it successfully describes several key aspects of single ion transport. These include temperature, pressure, solvent isotope, and even mixed solvent effects for small ions such as lithium./20/

References and Footnotes

Inasmuch as a complete set of references would be far too lengthy, for a general background I refer the reader to P. Wolynes' excellent review article in Ann. Rev. Phys. Chem. 31, 345 (1980).

1. B.J. Berne and R. Pecora, Dynamic Light Scattering, Wiley & Sons, New York (1976).
2. H. Falkenhagen, Electrolytes, Oxford, New York (1934).
3. R.M. Fuoss and F. Accascina, Electrolytic Conductance, Interscience, New York (1959).
4. In this derivation I take the value of the electronic charge to be unity.
5. A similar Fourier transform derivation may be found in E.M. Lifshitz and L.P. Pitaevskii, Physical Kinetics, Pergamon, New York (1981).
6. L. van Hove, Phys. Rev. 95, 249 (1954).
7. J.B. Hubbard and P.G. Wolynes, J. Chem. Phys. 69, 998 (1978).
8. B.R. Ware, Adv. Coll. Int. Sci. 4, 1 (1974).
9. R. Zwanzig, Ann. Rev. Phys. Chem. 16, 67 (1965).
10. D.A. McQuarrie, Statistical Mechanics, Harper and Row, New York (1976).
11. B.J. Berne, J.P. Boon and S.A. Rice, J. Chem. Phys. 45, 1086 (1966).
12. P.G. Wolynes, J. Chem. Phys 68, 473 (1978); P. Colonomos and P.G. Wolynes, J. Chem. Phys. 71. 2644 (1979).
13. N.G. van Kampen, Stochastic Processes in Physics and Chemistry, North-Holland, New York (1981).
14. S.A. Rice and A.R. Allnatt, J. Chem. Phys., 34, 2144 (1961); A.R. Allnatt and S.A. Rice, J. Chem. Phys. 34, 2156 (1961).
15. P.J. Stiles and J.B. Hubbard, Chem. Phys. 84, 431 (1984).
16. G. van der Zwan and J.T. Hynes, Physica 121A, 227 (1983).
17. J.B. Hubbard and P.J. Stiles, Chem. Phys. Lett. 114, 121 (1985).
18. J.B. Hubbard and L. Onsager, J. Chem. Phys. 67, 4850 (1977); J.B. Hubbard, J. Chem. Phys. 68, 1649 (1978).
19. R.B. Bird, R.G. Armstrong and O. Hassager, Dynamics of Polymeric Liquids Vol. 1, Wiley & Sons, New York (1977).
20. Recent references are: K. Ibuki and M. Nakahara, J. Chem. Phys. 84, 2776 (1986); K. Ibuki and M. Nakahara, J. Chem. Phys. 84, 6979 (1986); M. Nakahara and K. Ibuki, J. Phys. Chem. 90, 3026 (1986).

DIFFRACTION STUDIES OF AQUEOUS IONIC SOLUTIONS

J. E. Enderby
Directeur-Adjoint
Institut Laue-Langevin
156X, 38042 Grenoble, France

ABSTRACT. A description of neutron diffraction and the method of differences as applied to aqueous ionic solutions is given. It is shown that the method yields precise information about the structure of a range of aquo-ions and their solvation. The possibility of using non-neutron methods, particularly laboratory based X-ray diffraction is discussed.

1. INTRODUCTION

The pair correlation function, $g_{\alpha\beta}(r)$, measures the probability of finding a β-type particle at a distance r from an α-type particle placed at the origin. In order to explain this idea in a quantitative fashion, let us consider an α-type particle at the origin and ask what is the average number of β-type particles which occupy a spherical shell of radius r and thickness dr at the same instant of time. That number is given by

$$dn_r = 4\pi\rho_\beta g_{\alpha\beta}(r)r^2dr \tag{1}$$

where $\rho_\beta = N_\beta/V$ and N_β is the number of β species contained in the sample of volume V. In Figure 1 we sketch a hypothetical g(r) for a simple liquid which contains just one chemical species. Let us focus attention on the parameters indicated on the sketch. The chance of finding two particles separated by a distance less than r_1 is negligible. Thus, r_1 measures the closest distance of approach of two particles in the system. On the other hand, $\bar{r}$ allows us to define the most probable separation of two atoms, and r_2 tells us the range over which near neighbour interactions are likely to be important. It follows from the definition of g(r) given in Equation (1) that the value of integral

$$4\pi\rho \int_0^{r_s} g(r)r^2dr$$

is the running co-ordination number, i.e. the number of neighbours within a spherical shell of radius r_s for one particle chosen to be at

M.-C. Bellissent-Funel and G. W. Neilson (eds.), The Physics and Chemistry of Aqueous Ionic Solutions, 129–145.

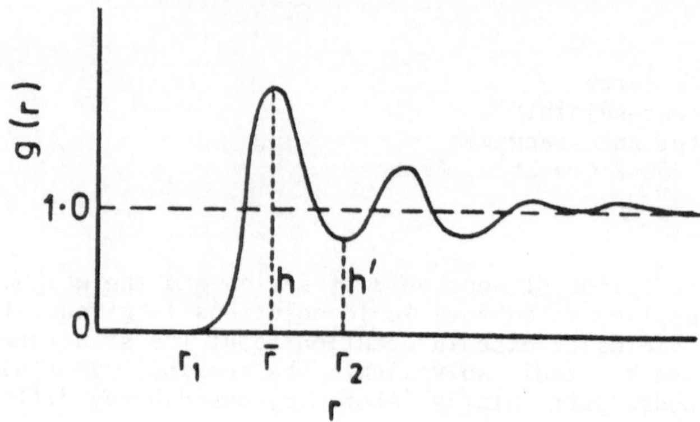

Figure 1. A hypothetical radial distribution function g(r) for a simple liquid which contains just one chemical species.

the origin. If r_s is chosen as r_2, this value of the running coordination number is usually referred to as "the co-ordination number", ($\bar{n}$ for a pure liquid, $\bar{n}_{\alpha\beta}$ for a liquid containing more than one chemical species), and when combined with a value of $\bar{r}$ allows us to build up a chemically plausible picture of the short-range order. The ratio h'/h is typically ~1/4 for simple liquids and tends to zero if well-defined local order persists for $t \geq 10^{-11}$s.

In order to determine $g_{\alpha\beta}(r)$ experimentally, we must link this quantity with the results of diffraction theory. If neutrons or X-rays are incident on a liquid containing several chemical species, a measure of the amplitude of the scattered waves is given by

$$\sum_{\alpha} b_{\alpha} \sum_{i(\alpha)} \exp\,[i\mathbf{k} \cdot \mathbf{r}_i(\alpha)] \tag{2}$$

where b_α is the neutron coherent scattering length in neutron scattering (or the X-ray form factor in X-ray scattering, usually written f_α) and $r_i(\alpha)$ denotes the position of the ith nucleus of the α type. We assume here that there is no multiple scattering. Thus, in Eq. 2 the second sum looks after the phase relationships of the waves scattered from the nuclei at different positions. The sum over the β values, on the other hand, takes account of the different scattering amplitudes for the different kinds of nuclei or atoms. The mean intensity is the square modulus of the amplitude and is given by

$$\mathbf{I(k)} = \overline{\sum_{\alpha}\sum_{\beta} b_{\alpha} b_{\beta} \sum_{i(\alpha)} \sum_{j(\beta)} \exp\left\{i\mathbf{k} \cdot [\mathbf{r}_j(\beta) - \mathbf{r}_i(\alpha)]\right\}} \tag{3}$$

and can in principle be obtained from experimental setups shown in Figures 2 and 3. The quantity **k** is the scattering vector whose modulus, k, for elastic scattering (i.e. $|\underset{\sim}{k}_0| = |\underset{\sim}{k}_1|$) (see Figure 2b) is given by

$$k = 2k_o \sin \theta$$

or, since $k_o = 2\pi/\lambda_o$, $k = (4\pi \sin \theta/\lambda_o)$ where θ is half the scattering angle. To do diffraction experiments an intense source of neutrons or X-rays is required. For neutrons this is normally a high-flux nuclear reactor, although pulsed sources based on nuclear spallation will play a major role in the future. X-rays can be derived from conventional laboratory sources or, if very high fluxes or tunability are required, synchrotron radiation.

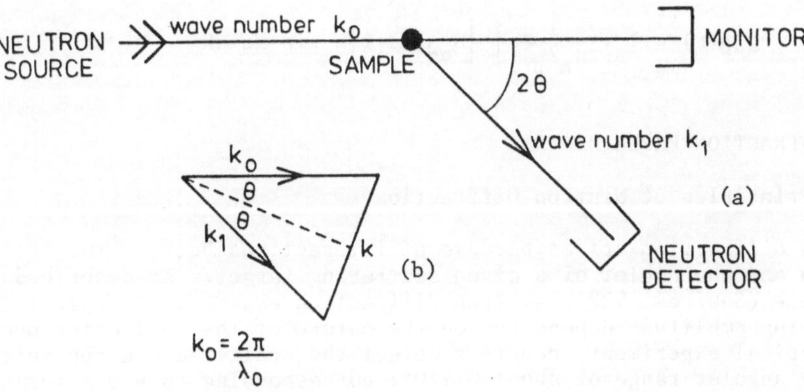

Fig. 2. The conventional arrangement for neutron diffraction studies.

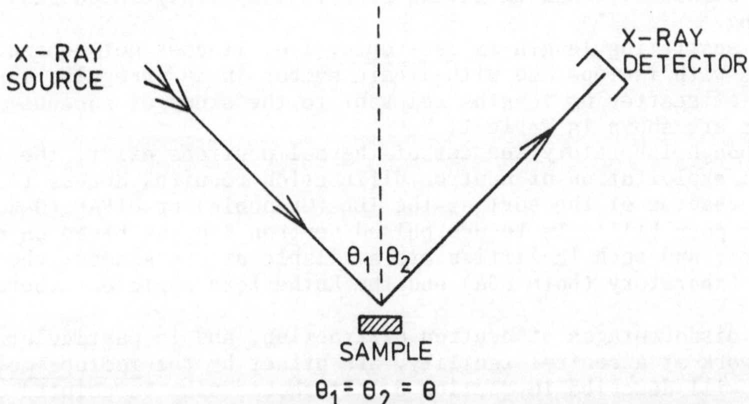

Fig. 3. Conventional arrangement for X-ray diffraction studies on liquids.

Elementary manipulation of equation 3 yields an expression of the form

$$I(k) = N\left[\sum_{\alpha} c_{\alpha} b_{\alpha}^{2} + F(k)\right] \tag{4}$$

where c_α is the atomic fraction of the α species and F(k) is a weighted average of the <u>partial structure factors</u>, $S_{\alpha\beta}(k)$, whose Fourier transform yields $g_{\alpha\beta}(r)$. Explicitly

$$F(k) = \sum_{\alpha}\sum_{\beta} c_{\alpha}c_{\beta}b_{\alpha}b_{\beta}\left[S_{\alpha\beta}(k) - 1\right] \tag{5}$$

and

$$g_{\alpha\beta}(r) = 1 + \frac{1}{2\pi^{2}\rho r}\int\left[S_{\alpha\beta}(k)-1\right]k \sin kr \, dk \tag{6}$$

2. DIFFRACTION EXPERIMENTS

2.1. Principles of Neutron Diffraction

Neutron diffraction arises because of interactions between the incident neutron and the nuclei of a given scattering target. As described elsewhere (Squires 1982), neutron diffraction experiments exploit the scattering amplitude dependence on the nature of the scattering nucleus. In a typical experiment, counters detect the scattered neutron intensity over an angular range of about 0-120°, corresponding to a scattering vector in the range 0.2-16$Å^{-1}$, for a neutron wavelength of about 0.7Å. The data reduction required to obtain F(k) consists of absorption, multiple scattering and Placzek corrections, the latter to account for inelastic effects when neutrons are scattered by light elements. The reduced scattered intensity is put on an absolute scale by means of a vanadium standard, vanadium giving rise to completely incoherent scattering.

The scattering length is <u>isotropic</u>, i.e. it does not depend on k. It varies with isotope and with atomic number in an irregular manner. Examples of scattering lengths relevant to the study of aqueous solutions are shown in Table I.

Although laboratory sources of thermal neutrons exist, the practical exploitation of neutron diffraction requires access to a research reactor of the sort at the ILL (Grenoble) or HIFAR (Oak Ridge). A further possibility is to use pulsed neutron sources based on nuclear spallation, and such facilities are available at Los Alamos, the Argonne National Laboratory (both USA) and the Rutherford Appleton Laboratory (UK).

The disadvantages of neutron diffraction, and in particular the need to work at a central facility, are offset by the <u>isotope method</u> which we will describe in section 3. It should also be pointed out that, since b does not depend on k, the weighted correlation function G(r) obtained by taking the Fourier transform of F(k) is a <u>linear combination</u> of $g_{\alpha\beta}(r)$. Explicitly

$$G(r) = 1 + \frac{1}{2\pi^2 r\rho} \int F(k)k \sin krdk = \sum_\alpha \sum_\beta c_\alpha c_\beta b_\alpha b_\beta \left[g_{\alpha\beta}(r)-1\right] \quad (7)$$

2.2. Principles of X-ray Diffraction

X-ray scattering has been used extensively to study aqueous solutions (see, for example Marques and Cabaço 1986; Magini et al 1982; Johansson and Ohtaki 1973) carried out on an X-ray diffractometer (Figure 3), typically using MoKα (λ = 0.711Å). Commercial θ-θ diffractometers which are suitable for liquid studies are available from several

TABLE I. Examples of Coherent Scattering Lengths (fm*)

Element or isotope	b	Element or isotope	b
H	-3.72	Fe	9.51
D	6.70	^{54}Fe	4.2
^{6}Li	1.8	^{56}Fe	10.1
^{7}Li	-2.1	^{57}Fe	2.3
N	9.36	Ni	10.3
^{14}N	9.37	^{58}Ni	14.4
^{15}N	6.44	^{60}Ni	2.82
K	3.67	^{62}Ni	-8.7
^{41}K	2.58	^{64}Ni	-0.37
Cl	9.58	Cu	7.689
^{35}Cl	1.7	^{63}Cu	6.7
^{37}Cl	2.9	^{65}Cu	11.1
Ca	4.9	Zn	5.686
^{40}Ca	4.8	^{64}Zn	5.5
^{44}Ca	1.8	^{68}Zn	6.7

* 1 fm = 10^{-15}m

suppliers with a k range of 1-15Å^{-1}. Measured intensities are corrected for polarisation and absorption in the sample, and for back-ground. The coherent X-ray intensity (in electron units), conventionally written $I^{coh}_{eu}(k)$ is given by an expression equivalent to Equation (4):

$$I^{coh}_{eu}(k) = F(k) - \sum_\alpha \sum_\beta c_\alpha c_\beta f_\alpha(k) f_p(k)$$

$$= \sum_\alpha \sum_\beta c_\alpha c_\beta f_\alpha (k) f_\beta(k) S_{\alpha\beta}(k) \quad (8)$$

where $f_\alpha(k)$ and $f_\beta(k)$ are the k-dependent X-ray atomic form factors and f(k=0) increases with atomic number in a linear way. In principle

$f_\alpha(k)$ depends on the wavelength of the incident X-rays but this effect is only significant near an absorption edge (see section 5). The total radial distribution function, $G_X(r)$, is usually expressed for X-rays as

$$G_X(r) = 1 + \frac{\Sigma(\bar{K}_m)^2}{(\Sigma\bar{K}_m)^2 2\pi^2 \rho r}\int_0^\infty ki(k)\sin kr\, dk \tag{9}$$

where $$i(k) = \frac{I_{eu}^{coh}(k)}{\sum_m f_m^2(k)} - 1$$

and $\bar{K}_m$ is the effective number of electrons in atom m. It follows that $G_X(r)$ is not a linear combination of $g_{\alpha\beta}(r)$, since k dependent form factors remain inside the integral of equation 9 and this may lead to a 'convolution broadening' of some of the structural features.

3. THE METHOD OF DIFFERENCES

F(k) is, as we have seen, a weighted average of several partial structure factors; for aqueous solutions of the form MX_n in D_2O, ten structure factors enter into F(k). The difficulty in interpreting F(k) in terms of ion-water and ion-ion correlation functions, even for concentrated solutions, can be appreciated by reference to the bar chart shown in Figure 4. For practical purposes these patterns can yield information only about $S_{DD}(k)$ and $S_{OD}(k)$ or their transforms $g_{DD}(r)$ and $g_{OD}(r)$.

The neutron 'first order' difference method (Soper et al, 1977; Neilson & Enderby, 1978; Enderby & Neilson, 1981) allows one to gain direct information about the detailed arrangement of the water molecules around the ions in aqueous solutions. The quantity that is central to the method is the difference in F(k) between two samples that are identical in all respects except that the isotopic state of the cation, M, (or the anion X) has been changed; this quantity, denoted $\Delta_M(k)$ or $\Delta_X(k)$ is the sum of four partial structure factors $S_{\alpha\beta}(k)$ weighted in such a way that only those relating to ion-water correlations are significant. Explicitly:

$$\Delta_M(k)=A_M(S_{MO}(k)-1)+B_M(S_{MD}(k)-1)+C_M(S_{MX}(k)-1)+D_M(S_{MM}(k)-1)\ )$$

$$\Delta_X(k)=A_X(S_{XO}(k)-1)+B_X(S_{XD}(k)-1)+C_X(S_{MX}(k)-1)+D_X(S_{XX}(k)-1)\ )$$

where

$$A_M = 2c_Mc_Ob_O(b_M-b'_M); \quad A_X = 2c_Xc_Ob_O(b_X-b'_X);$$

$$B_M = 2c_Mc_Db_D(b_M-b'_M); \quad B_X = 2c_Xc_Mb_D(b_X-b'_X);$$

$$C_M = 2c_Mc_Xb_X(b_M-b'_M); \quad C_X = 2c_Xc_Mb_M(b_X-b'_X);$$

$$D_M = c_M^2(b_M^2-(b'_M)^2); \quad D_X = c_X^2(b_X^2-(b'_X)^2)$$

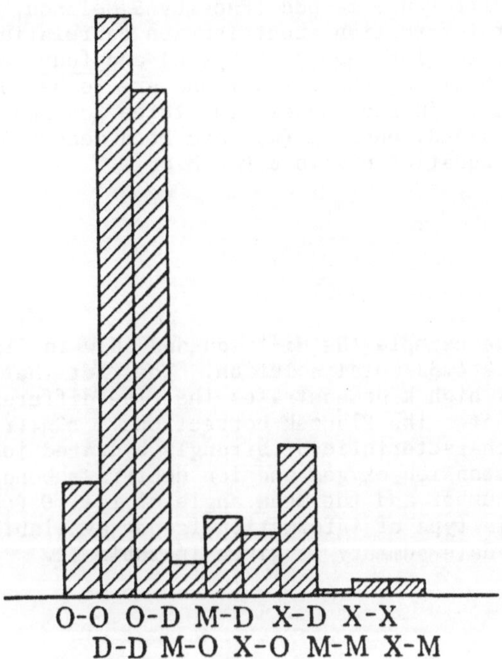

Figure 4. The weighting of the various contributions to a total neutron pattern for a 4.35 molal solution of $NiCl_2$ in D_2O.

and b_O and b_D are the neutron coherent scattering amplitudes for oxygen and deuterium and b_M, b'_M, b_X and b'_X are the mean scattering amplitudes for the isotopic states used in producing the salt MX_n.

The properties of $\Delta(k)$ have been discussed in detail elsewhere (Soper et al, 1977) and need not be enlarged on here. The crucial property, apart from the fact that $A,B > C,D$, is that Placzek distortions are essentially eliminated so that a difference function G(r) can be determined directly from

$$\Delta G(r) = \frac{V}{2\pi^2 Nr} \int \Delta(k)\, k \sin(kr)\, dk$$

In terms of the correlation functions, $g_{\alpha\beta}$, it follows at once that

$$\Delta G_M(r) = A_M(g_{MO}-1) + B_M(g_{MD}-1) + C_M(g_{MX}-1) + D_M(g_{MM}-1)$$

and

$$\Delta G_X(r) = A_X(g_{XO}-1) + B_X(g_{XD}-1) + C_X(g_{MX}-1) + D_X(g_{XX}-1).$$

Since A and B are much greater than C and D, the method yields a high resolution measurement of an appropriate combination of g_{MO} and g_{MD} or g_{XO} and g_{XD}.

The 'second order' difference method (Enderby & Neilson, 1981) allows one to gain direct information about ion-ion correlations. The method requires three samples for $S_{MM}(k)$ or $S_{XX}(k)$ and four samples for $S_{MX}(k)$ and formulae for obtaining these functions are to be found in Enderby and Neilson (1981). In real space, the three ion-ion correlation functions can be obtained, once $S_{\alpha\beta}(k)$ have been measured, by numerical integration of equation 6 with α, β = M or X.

4. RESULTS

4.1. Cationic hydration

We take as an illustrative example the Ni^{2+} ion and show in Figure 5 and 6 $\Delta_{Ni}(k)$ and $\Delta G_{Ni}(r)$ for a 4.32 molal solution. The fact that $\Delta_{Ni}(k)$ is well behaved at high k demonstrates that the difference method effectively eliminates the Placzek corrections. $\Delta G_{Ni}(r)$ shows the twin peak structure characteristic of strongly hydrated ions and allows $\bar{r}_{MO}$ and $\bar{r}_{MD}$ (the mean ion-oxygen and ion-deuterium bond lengths), the hydration number and the mean angle of tilt θ (see Figure 6) to be determined. This type of information is now available for some ten cations and an up-to-date summary is given in Table II.

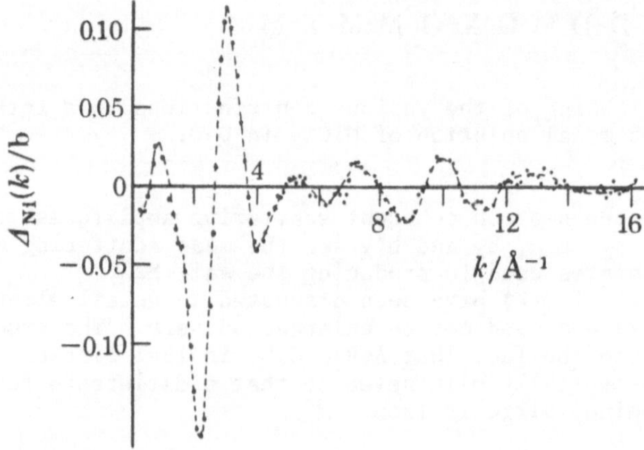

Figure 5. $\Delta Ni(k)$ for a 4.35 molal solution of $NiCl_2$ in D_2O. 1b (barn) = $10^{-28}m^2$

4.2. Anionic hydration

Some years ago, Cummings et al (1980) found that form of $\Delta G_{Cl}(r)$, for which Figure 7 is a typical example, was remarkably insensitive to the nature of the cation. This finding has been confirmed by many subsequent investigations and a summary of the present experimental situation is given in Table III.

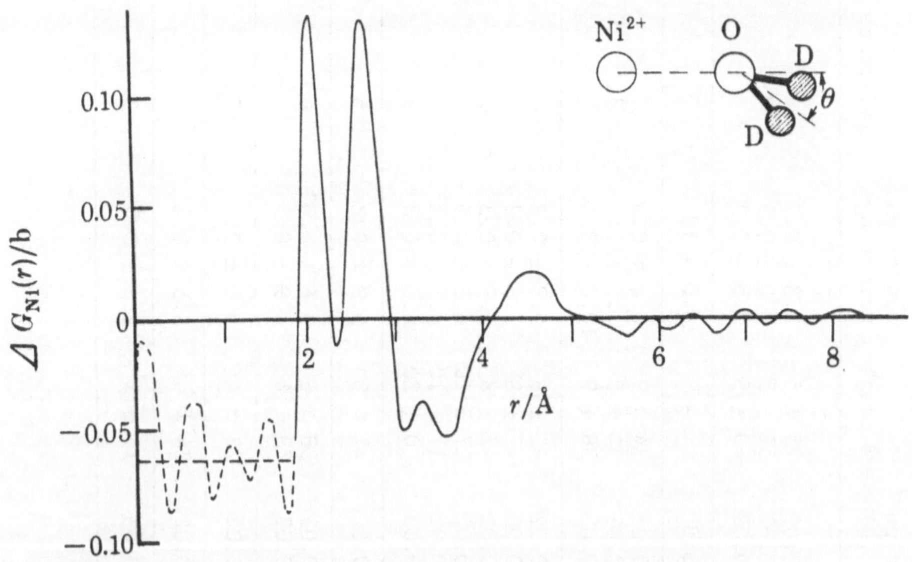

Figure 6. $\Delta G_{Ni}(r)$ for a 4.35 molal solution of $NiCl_2$ in D_2O. 1b (barn) $= 10^{-28}$ m^2

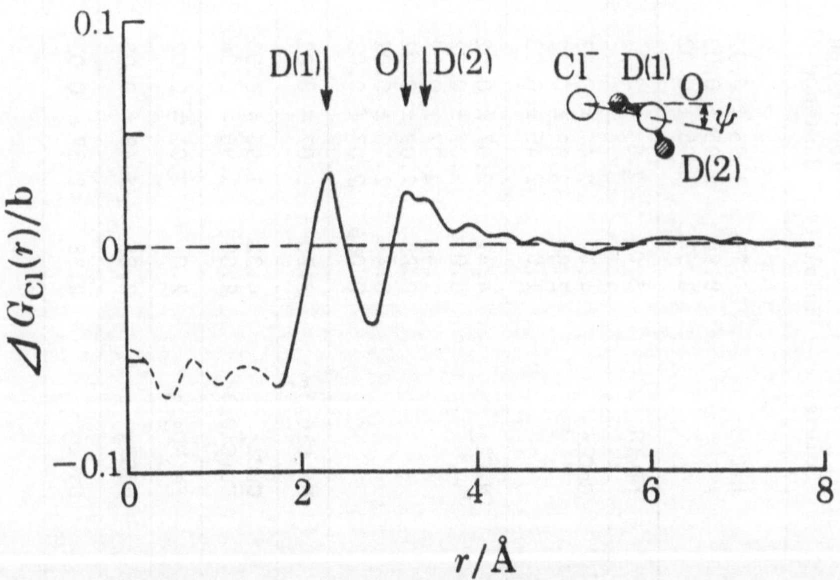

Figure 7. $\Delta G_{Cl}(r)$ for a 4.35 molal solution of $NiCl_2$ in D_2O. 1 b (barn) $= 10^{-28}$ m^2

TABLE II. Cation hydration determined by neutron diffraction

Ion	Solute	Molality	Ion-oxygen distance (Å)	Ion-deuterium distance (Å)	θ (deg)	Hydration number $\bar{n}_M$	Reference
Li^+	LiCl	27.77	1.95 ± 0.02	2.31 ± 0.02	75 ± 5	2.3 ± 0.2	
		9.95	1.95 ± 0.02	2.50 ± 0.02	52 ± 5	3.0 ± 0.5	1
		3.57	1.95 ± 0.02	2.55 ± 0.02	40 ± 5	5.5 ± 0.3	
ND_4^+	ND_4Cl	5.0	2.8 – 3.2	3.4 – 3.8	–	10.0 – 12.0	2
Ca^{2+}	$CaCl_2$	4.49	2.41 ± 0.03	3.04 ± 0.03	34 ± 9	6.4 ± 0.3	
		2.80	2.39 ± 0.02	3.02 ± 0.03	34 ± 9	7.2 ± 0.2	3
		1.0	2.46 ± 0.03	3.07 ± 0.03	38 ± 9	10.0 ± 0.6	
Ni^{2+}	$NiCl_2$	4.41	2.07 ± 0.02	2.67 ± 0.02	42 ± 8	5.8 ± 0.2	
		3.05	2.07 ± 0.02	2.67 ± 0.02	42 ± 8	5.8 ± 0.2	
		1.46	2.07 ± 0.02	2.67 ± 0.02	42 ± 8	5.8 ± 0.3	4
		0.85	2.09 ± 0.02	2.76 ± 0.02	27 ± 10	6.6 ± 0.5	
		0.46	2.10 ± 0.02	2.80 ± 0.02	17 ± 10	6.8 ± 0.8	
		0.086	2.07 ± 0.03	2.80 ± 0.03	0 ± 20	6.8 ± 0.8	
Ni^{2+}	$Ni(ClO_4)_2$	3.80	2.07 ± 0.02	2.67 ± 0.02	42 ± 8	5.8 ± 0.2	5
Cu^{2+}	$CuCl_2$	4.32	1.96 ± 0.03	2.58 ± 0.03	38 ± 6	3.6 ± 0.03	6
Cu^{2+}	$Cu(ClO_4)_2$	2.00	1.96 ± 0.04	2.58 ± 0.03	38 ± 6	4.9 ± 0.3	
Fe^{3+}	$Fe(NO_3)_3$	2.0	2.01 ± 0.02	2.67 ± 0.02	22 ± 4	5.0 ± 0.2	7
Nd^{3+}	$NdCl_3$	2.85	2.48 ± 0.02	3.13 ± 0.02	24 ± 4	8.5 ± 0.2	8
Dy^{3+}	$DyCl_3$	2.38	2.37 ± 0.03	3.04 ± 0.03	17 ± 3	7.4 ± 0.5	9

TABLE III. Chloride hydration determined by neutron diffraction

Ion	Solute (reference)	Molality	X-D(1) (Å)	X-O (Å)	X-D(2) (Å)	ψ(deg)	Coordination number	Reference
Cl^-	LiCl	14.9	2.24 ± 0.02	3.25 ± 0.03	3.50–3.60	0	4.4 ± 0.3	1
		9.95	2.22 ± 0.02	3.29 ± 0.04	3.50–3.68	0	5.3 ± 0.2	
		3.57	2.25 ± 0.02	3.34 ± 0.05	3.50–3.7	0	5.9 ± 0.2	
	NaCl	5.32	2.26 ± 0.04	3.20 ± 0.05	–	0–20	5.5 ± 0.4	2
		2.99	2.24 ± 0.04	3.25 ± 0.05	–	0–15	6.0 ± 0.3	
		1.49	2.24 ± 0.04	3.25 ± 0.05	–	0–15	6.0 ± 0.3	
	RbCl	4.36	2.26 ± 0.04	3.20 ± 0.05	–	0–20	5.8 ± 0.3	1
	$CaCl_2$	4.49	2.25 ± 0.02	3.25 ± 0.04	3.55–3.65	0–7	5.8 ± 0.2	1
	$NiCl_2$	4.35	2.29 ± 0.02	3.20 ± 0.04	3.40–3.50	5–11	5.7 ± 0.2	1
	$NiCl_2$	3.00	2.23 ± 0.04	3.25 ± 0.04	3.40–3.50	0–8		
	$CuCl_2$	4.32	2.27 ± 0.02	3.25 ± 0.05	–	0–3	3.3 ± 0.4	3
	$ZnCl_2$	45.1 (100°C)	2.25 ± 0.05				1.0 ± 0.8	4
		19.0	2.24 ± 0.04	3.40 ± 0.2	3.7–3.9	0–7	1.9 ± 0.2	
		4.9	2.25 ± 0.06	3.25 ± 0.1	3.7–3.9	0–3	3.7 ± 0.2	
	$NdCl_3$	2.85	2.29 ± 0.02	3.45 ± 0.04	–	0	3.9 ± 0.2	5

References for Table II.

1. Newsome et al (1980) and Ichikawa et al (1984).
2. Hewish & Neilson (1981).
3. Hewish et al (1982).
4. Neilson and Enderby (1978)
5. Newsome et al (1981)
6. Salmon (1985)
7. Herdman, J. and Neilson, G.W. (1986) private communication
8. Narten and Hahn (1982)
9. Annis et al (1985)

References for Table III.

1. Cummings et al (1980) and Copestake et al (1985).
2. Soper (1977).
3. Salmon (1985).
4. Biggin (1986).
5. Biggin et al (1984).

As the molten salt (high salt molality) region is approached, the hydration number will fall. From the data shown in Table III, it is evident that in some solutions the hydration number achieves the value of 6 provided the H_2O/Cl^- ratio is itself ≥ 6. For other solutions, notably $ZnCl_2$ and $CuCl_2$, the value of 6 is not achieved until the water/chloride ratio is ≥ 12. This 6 reflects the importance of inner-sphere complexing by the chloride ion and will influence the form of the ion-ion distribution functions.

4.3. Ion-Ion distribution functions

Neilson and Enderby (1983) determined $g_{NiNi}(r)$, $g_{NiCl}(r)$ and $g_{ClCl}(r)$ for 4.35 molal solution of $NiCl_2$. A summary of the results, together with $g_{ClCl}(r)$ for two other solutions is given in Table IV. These functions have been discussed in detail elsewhere, but six points are worth emphasising.

(i) The cut-off at r = 4.1Å for the Ni-Ni interaction is remarkably insensitive to errors in the raw data. A comparable distance to this has been calculated theoretically by Tembe et al (1982) for iron chloride solutions. However, $\bar{r}_{NiNi}$ is subject to considerable uncertainty and is in the range 4.7 - 5.5Å (see 5. below).

(ii) There is no evidence for substantial inner-sphere complexing of Cl^- by Ni^{2+}. The r_c value of 3.9Å (Table IV) suggests that Cl^- is effectively excluded from the inner sphere, a conclusion supported by the recent NMR work of Hunt and Friedman (1983). On the other hand, Weingartner and Hertz (1979) previously suggested that the concentration of the inner-sphere complex $[Ni(H_2O)_5Cl]^+$ is about 2.5 molal, but this is not consistent with the experimental results.

(iii) There is a strong pairing between Ni^{2+} and Cl^- as shown by the peak centred around 4.6Å. The co-ordination number for Cl^- around Ni^{2+} of 5.8 ± 0.3 suggests that two-thirds of the available sites associated with the water in the $[Ni(H_2O)_6]^{2+}$ complex are occupied.

(iv) The ion pairing is mediated by the inner-sphere water and will therefore have a lifetime of 5×10^{-12}s or less (i.e. the lifetime of $Cl^- - D_2O$). This time is comparable with the binding time between water molecules themselves so that in spite of a well-defined peak in $g_{NiCl}(r)$ the ion pair must not be thought of as a stable entity. Chloride ions exchange rapidly with the hydrated nickel ion although the interaction is both directional and specific.

(v) The form of $g_{ClCl}(r)$ in concentrated $NiCl_2$ solution is a direct consequence of the geometrical stability of the $[(Ni(D_2O)_6]^{2+}$ species because the separation of two chloride ions is determined by their co-ordination to the hydrated ion.

(vi) For very concentrated solutions near to the molten salt limit, direct cation-anion interactions will become more important and will lead to a short $\bar{r}_{ClCl}$ distance (the "molten salt" value) and a long $\bar{r}_{ClCl}$ distance (the "aqueous solution" value). This effect has actually been observed for 14.9 molal LiCl (see Figure 8).

TABLE IV. Ion-ion properties in aqueous solution

Solute	Molality	Pair correlation function	Cut-off distance (r_c)(Å)	Position of first maximum	Coordination number	Ref.
$NiCl_2$	4.35	Ni-Ni	4.1 ± 0.1	4.7 - 5.7	-	
		Ni-Cl	3.9 ± 0.1	4.6 ± 0.2	5.8 ± 0.3 (Cl^- around Ni^{2+})	1
		Cl-Cl	5.0 ± 0.1	6.1 ± 0.2	8.5 ± 0.3	
LiCl	14.9	Cl-Cl	2.95 ± 0.05	3.75 ± 0.05	2.3 ± 0.3	2
$ZnCl_2$	5.0	Cl-Cl	3.3 ± 0.1	3.8 ± 0.1	3.0 ± 1.0	3

References. (1) Neilson & Enderby (1983); (2) Copestake et al (1985); (3) Biggin (1986).

5. OTHER STRUCTURAL METHODS FOR AQUEOUS SOLUTIONS

It is clear from what has been said already that the neutron method of differences is formally exact and is capable of yielding detailed structural information for aqueous solutions with precision. There are three disadvantages in its practical implementation, however. First, it relies on the use of separated isotopes which are invariably expensive and sometimes not actually available in sufficient quantity for neutron work. Secondly, the experiments have to be performed at a central facility which can provide intense beams of thermal neutrons. The user pressure on these facilities is often so great that important experiments may have to wait several months or even years before they can be tackled. Moreover, it is often difficult to justify, to the relevant allocation committee, a series of experiments which, by their

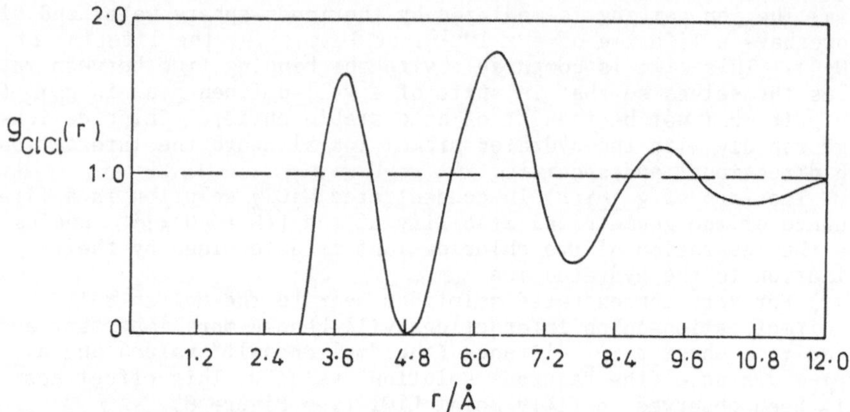

Figure 8. The pair correlation function $g_{ClCl}(r)$ for a 14.90 molal solution of LiCl in D_2O. The two principal peaks occur at 3.75 ± 0.03Å and 6.38 ± 0.03Å, the former corresponding to the separation of chloride ions which are in direct contact with Li^+.

nature, are time-consuming. Experiments in which the concentration of the additive is changed over a wide range in a systematic way would, if time were allocated, occupy unacceptably long periods of machine time to the exclusion of other users.

Finally, the variation (Δb) of b which arises from isotopic substitution is ~20 fm in the most favourable case and is often considerably less (Table I). It follows that the second-order difference method developed to obtain ion-ion correlation functions is, with current technology, limited to relatively few solutions at concentrations in excess of ~3 molal.

There are three other methods which have been proposed. The modulation of the X-ray absorption coefficient within a few hundred eV beyond the absorption edge (the Kronig oscillation) produces the so-called EXAFS signal. EXAFS contain information to be determined about the local order associated with a particular chemical species and is, therefore, a powerful experimental method for structures. So far, little work has been carried out on solutions, but in an interesting paper Lagarde et al (1980) used EXAFS to establish the existence in solution of inner sphere complexing of the bromide ion in $ZnBr_2$ solutions. A full account of the method can be found in Lagarde (1985).

A second alternative is based on the anomalous X-ray scattering near an absorption edge. In the elementary theory of X-ray scattering described in 2.2. above, the form factor f depends only on k. However, near an absorption edge, f changes its character and can be written as

$$f(k,E) = fo(k) + f'(E) + i f''(E)$$

where E is the incident energy of the X-ray and $f^\circ(k)$ is the conventional form factor. Thus a change (Δf) in f can be induced by varying the wavelength of the X-rays close to an absorption edge characteristic of one of the chemical species in the system. The changes induced are small ($\Delta f/f \sim 1$) and the method is not yet fully developed (Waseda 1984) but it clearly offers considerable potential for the future.

Both anomalous scattering and EXAFS involve intense tunable sources of X-rays and these are normally available only at a central facility. The method we have developed to extract the required information makes use of laboratory-based X-ray diffraction (as in Figure 3) and has three novel features. First it relies on isomorphic rather than isotopic substitution. In other words, $f_\alpha(k)$ is changed to $f_{\alpha'}(k)$ by replacing one element with another on the assumption that the set of structure factors $S_{\alpha\beta}(k)$ is unaffected. This is, of course, not new and has been tried by other workers, notably Bol et al (1970) for a range of aqueous solutions. The second feature concerns the way the difference function is treated. Suppose that an isomorphic substitution has been satisfactorily made. Then the X-ray difference function $\tilde{\Delta}_\alpha(k)$ will be of the form

$$\tilde{\Delta}_\alpha(k) = c_\alpha \Delta f_\alpha(k)\{2 \sum_{\beta\neq\alpha} c_\beta f_\beta(k)[S_{\alpha\beta}(k)-1]+c_\alpha^2(f_\alpha^2(k)-f_{\alpha'}^2(k)\}(S_{\alpha\alpha}(k)-1)$$

where β labels all the chemical species other than one isomorphically substituted and the tilde is used to differentiate the X-ray case from the neutron case. As in the neutron method, the structural effects due to the water are not present in the difference function and the systematic corrections are greatly simplified. In order to eliminate the convolution broadening due to the k dependence of $f_\alpha(k)$, let us extract from the sum the term arising from a particular chemical species denoted by γ. Then

$$\begin{aligned}\tilde{\Delta}_\alpha(k) &= 2c_\alpha c_\gamma \Delta f_\alpha(k) f_\gamma(k)(S_{\alpha\gamma}(k)-1) + \\ &+ 2c_\alpha \Delta f_\alpha(k)\{\sum_{\beta\neq\gamma\neq\alpha} c_\beta f_\beta(k) S_{\alpha\beta}(k)-1)\} + \\ &+ c_\alpha^2(f_\alpha^2(k)-f_{\alpha'}(k))(S_{\alpha\alpha}(k)-1)\end{aligned}$$

The Fourier transform of $\tilde{\Delta}(k)/2c_\alpha c_\gamma \Delta f_\alpha f_\gamma(k)$, $\Delta\tilde{G}(r)$, is of the form

$$\Delta\tilde{G}_\alpha(r) = g_{\alpha\gamma}(r) + \sum_{\beta\neq\gamma} a_\beta \int H_\beta(|\underset{\sim}{r}-\underset{\sim}{r}'|) g_{\alpha\beta}(\underset{\sim}{r}')d\underset{\sim}{r}'$$

where $H_\beta(|\underset{\sim}{r}-\underset{\sim}{r}'|)$ is a convolution function associated with the k-dependence of the form factor and a_β are constants which can be evaluated for any pair of isomorphs. Provided that the first peak in $g_{\alpha\gamma}(r)$ is well separated from those of $g_{\alpha\beta}(r)$ ($\beta\neq\gamma$), a situation which normally obtains for well co-ordinated ions, the isomorphic method, when treated in this way, yields an unbroadened pair correlation function, the choice of which is determined by γ. For example, if α represents a metal ion in solution and γ refers to oxygen, $\bar{r}_{MO}$ and $\bar{n}_{MO}$ can be measured.

The third feature of the new method is that one of the isomorphs should also show a neutron isotopic effect. This enables an exact $g_{\alpha\gamma}(r)$ to be obtained via neutron scattering and a once-for-all comparison made between it and the $g_{\alpha\gamma}(r)$ derived from the X-ray method in the neighbourhood of the first peak. Non-isomorphism will necessarily produce broadening in the X-ray distribution function and comparison with the neutron measurement enables the deviation from ideal isomorphic behaviour to be quantified. Once isomorphism has been established, the time-consuming but scientifically significant studies involving changes in concentration, temperature etc. can all be carried out in the laboratory without recourse to central facilities.

It is the combination of these three features which makes our approach to the study of aqueous solution by isomorphic substitution novel and represents a significant advance on what has been tried before. Furthermore, the economic advantages of the method are clearly very substantial.

Skipper et al (1986) have applied the method to an aqueous solution of $NiCl_2$ paired with $MgCl_2$. It was shown that Ni^{2+} and Mg^{2+} are isomorphic to within the limits set by the neutron method (± 0.01Å). They then performed a second-order difference experiment and deduced the Ni-Ni pair correlation function for the first time by X-ray methods, with an accuracy higher than was obtained by neutrons (see Table V). Studies are in progress on a wide range of candidates for isomorphic substitution and a systematic procedure for identifying them will be published in due course.

TABLE V. Ni-Ni correlation functions by the isomorphic method (Skipper et al 1986).

Solute	Molality		Cut-off distance (Å)	Position of 1st max (Å)
$NiCl_2$ ($MgCl_2$)	4.20	Ni-Ni (Mg-Mg)	4.1 ± 0.1	6.4 ± 0.8

6. ACKNOWLEDGEMENTS

This paper summarises the work of the Bristol aqueous solution group and I wish to thank George Neilson, Philip Salmon, John Herdman, Sue Biggin and Neil Skipper for allowing me access to unpublished data.

The group wishes to thank the SERC for continued support of its solution programme and the staff of the Institut Laue-Langevin, and in particular Dr. Pierre Chieux, for assistance with the experiments.

7. REFERENCES

Annis, B.K., Hahn, R.L. and Narten, A.H. (1985), J. Chem. Phys. **82**, 2806.

Biggin, S. (1986), ISAS report 16/86/CM, Trieste.

Biggin, S., Enderby, J.E., Hahn, R.L. and Narten, A.H. (1984), J. Phys. Chem. **88**, 3634.

Bol, W., Gerrits, G, and Eck, C.LvP. (1970), J. App. Cryst. **3**, 486.

Copestake, A., Neilson, G.W. and Enderby, J.E. (1985), J. Phys. C: Solid State Physics, **18**, 4211-16.

Cummings, S., Enderby, J.E., Neilson, G.W., Newsome, J.R., Howe, R.A., Howells, W.S. and Soper, A.K. (1980), Nature, **287**, 714.

Enderby, J.E. and Neilson, G.W. (1981), Rep. Prog. Phys. **44**, 593.

Hewish, N.A. and Neilson, G.W. (1981), Chem. Phys. Lett. **84**, 425.

Hewish, N.A., Neilson, G.W. and Enderby, J.E. (1982), Nature, **297**, 138.

Hunt, J.P. and Friedman, H.L. (1983), Prog. Inorg. Chem. **30**, 359.

Ichikawa, K., Kameda, Y., Matsumoto, T. and Misawa, M. (1984), J. Phys. C: Solid State Physics **17**, L725.

Johansson, G. and Ohtaki, H. (1979), Acta. Chem. Scand. **A33**, 305.

Largarde, P. (1985), Amorphous Solids and the Liquid State (eds. N.H. March et al, Plenum Press) p.365.

Largarde, P., Fontaine, A., Raoux, D., Sadoc, Migliardo, P. (1980), J. Chem. Phys. **72**, 3061.

Magini, M., Paschina, G. and Piccaluga, G. (1982), J. Chem. Phys. **76**, 2.

Marques, M.A. and Cabaço, M.I. (1986), Chem. Phys. Lett. **123**, 73.

Narten, A.H. and Hahn. R.L. (1982) Science, **217**, 1249.

Neilson, G.W. and Enderby, J.E. (1978), J. Phys. C: Solid State Phys., **11**, L625.

Neilson, G.W. and Enderby, J.E. (1983), Proc. R. Soc., **A 393**, 353.

Newsome, J.R., Neilson, G.W. and Enderby, J.E. (1980), J. Phys. C: Solid State Phys., **13**, L923.

Newsome, J.R., Neilson, G.W., Enderby, J.E. and Sandström, M. (1981), Chem. Phys. Lett., **82**, 399.

Skipper, N.T., Cummings, S., Neilson, G.W. and Enderby, J.E. (1986), Nature **321**, 52.

Soper, A.K., Neilson, G.W., Enderby, J.E. and Howe, R. (1977), J. Phys. C: Solid State Phys., **10**, 1793.

Squires, G. (1978), Introduction to the Theory of Thermal Neutron Scattering (CUP, Cambridge).

Tembe B.L., Friedman H.L. and Newton, M.D. (1982), J. Chem. Phys. **76**, 1490.

Waseda Y. (1984), Novel Applications of Anomalous (Resonance) Scattering for Structural Characterisation of Disordered Materials (Springer-Verlag, Berlin).

Weingartner H. and Hertz H.G. (1979), J. Chem. Soc. Faraday Trans. II, **75**, 2700.

References

[illegible], [illegible], R.J., and [illegible] (19[illegible]), [illegible]
[illegible] (19[illegible]), [illegible]
[illegible], and [illegible] (194[illegible]), [illegible] Phys. [illegible]
[illegible] (1979), [illegible]
[illegible], and [illegible] (19[illegible]), [illegible] Phys. [illegible]
[illegible] Physics, [illegible]
[illegible] (1980), [illegible]
[illegible] and [illegible], G.V. (1984), Am. Min., [illegible]
[illegible] and [illegible], G.V. (19[illegible]), Chem. Phys. [illegible]
[illegible] (19[illegible]), Phys. [illegible]
[illegible] (19[illegible]), [illegible]
[illegible] (19[illegible]), [illegible] State Physics [illegible]
[illegible] (19[illegible]), Acta Cryst. [illegible]
[illegible] (19[illegible]), [illegible] and the [illegible] State [illegible]
[illegible] (19[illegible]), [illegible]
[illegible] (19[illegible]), [illegible] J. Chem. Phys. [illegible]
[illegible] (1986), [illegible]
[illegible] (19[illegible]), Science [illegible]
[illegible] (19[illegible]), [illegible] and Structure [illegible]
[illegible] (19[illegible]), [illegible]
[illegible] (19[illegible]), [illegible] State Phys. [illegible]
[illegible] (19[illegible]), [illegible]
[illegible] (19[illegible]), [illegible]
[illegible] (19[illegible]), [illegible]
[illegible] (1978), Introduction to the theory of thermal neutron scattering (CUP, Cambridge).
[illegible] (19[illegible]), J. Chem. Phys. [illegible]
[illegible] (198[illegible]), Novel applications of anomalous (resonance) x-ray scattering for structural characterization of disordered materials [illegible]
[illegible] and [illegible] (197[illegible]), J. Chem. [illegible]

NEUTRON AND NMR SPECTROSCOPY OF AQUEOUS IONIC SOLUTIONS

A. J. Dianoux
Institut Laue-Langevin
Avenue des Martyrs
156 X
38042 Grenoble Cedex, France

ABSTRACT

These two lectures give a short presentation of Neutron and NMR spectroscopies and review the pertinent works in the field of aqueous ionic solutions.

The first chapter deal with the problem of motions in liquids and presents some models for molecular motions which can be probed by spectroscopic measurements.

The second chapter gives a very brief overview of the neutron scattering technique. Quasi-elastic scattering is shortly described and the concept of Elastic Incoherent Structure Factor (EISF) is introduced. Some selected examples are presented. The informations contained in the inelastic spectra are sorted out. The method to extract, from the spectra, the generalized frequency distribution is outlined, and some applications are given.

The third chapter begins with a brief presentation of Nuclear Magnetic Resonance. Some generalities on spin-lattice relaxation time are given, together with the application of the technique to measure correlation times in liquids. Some examples of proton and deuteron spin lattice relaxation time measurements in electrolyte solutions are analyzed. The spin-echo measurement of self-diffusion is presented, with some results obtained in aqueous ionic solutions.

As a conclusion, a short comparison is done between these two spectroscopic techniques and their importance to test realistic models for atomic and ionic interactions is stressed.

M.-C. Bellissent-Funel and G. W. Neilson (eds.), The Physics and Chemistry of Aqueous Ionic Solutions, 147–179.

1. MOLECULAR MOTIONS IN LIQUIDS

1.1. General

It is outside the scope of these two lectures to present a comprehensive treatment of Neutron and NMR spectroscopies applied to the problem of molecular motions in liquids. I will give a short presentation of the methods and review the pertinent work in the field of aqueous ionic solutions.

Experimentally, the study of molecular motions in liquids can be approached in two complementary ways : by Molecular Dynamics simulations and by spectroscopic experiments.

Since there is an excellent presentation of Molecular Dynamics Simulation in these proceedings [1] I will just give here some of its limitations. A Molecular Dynamics Simulation can be considered as an "experiment" having a well defined starting point, but is the potential used in the simulation the real one ? It is pointed out in these proceedings that "up to now no model has been able to describe satisfactorily all physical - chemical properties of a liquid as complex as liquid water" [1]. Another limitation comes from the fact that the calculations are restricted to classical mechanics. This is valid only when calculating frequency responses corresponding to an energy less than k_BT ($\sim$ 26 meV at room temperature). This means that MD simulations can give only an approximate treatment of the effects of the vibrational motion of the solvent molecules on the overall dynamics. There is also a limitation of the time scale : long range and long time phenomena cannot be treated. Nevertheless MD simulation is a very fruitful approach into ion dynamics since it can be used as a predictive tool or can produce results which can be accessed experimentally only with great difficulties (e.g. : the dynamics of the water in the first two coordination shells of an ion).

The more frequent approach for the study of molecular motions is to perform spectroscopic experiments. The idea is to prepare a probe in a well specified state (for example, a monochromatic neutron beam) and to make it interact with the system under study. The interaction with the degrees of freedom of the system will change the state of the probe (for example, change the energy of the incident neutron beam). One can relate this change to some correlation functions which describe the dynamical state of the system. Depending upon the technique one can have access to dynamical quantities (e.g. : NMR), to spatial information (e.g. : X-Ray or Neutron diffraction), or to both (quasi-elastic and inelastic neutron scattering).

1.2. Models for molecular motions

There are three kinds of degrees of freedom :

. Vibrational motion : needs a quantum description

. Rotational motion } can be described classically

. Translational motion } or quantum mechanically

- Vibrational motion : Molecular vibrations are localized. They will appear as an attenuation factor in the energy spectrum (e.g. Debye-Waller factor in quasielastic neutron scattering). Sometimes there can be a direct coupling with other degrees of freedom leading to vibrational relaxation (for example in Raman scattering)

- Rotational motions : Many models have been put forward. They can be classified as inertial and stochastic models. The inertial models are valid for low density systems : the molecules are essentially rotating but suffer random collisions which change their dynamical states. The stochastic model apply for dense systems : the motion occur by rapid rotational jumps over barriers (for example in plastic crystals). However, when the mean jump time τ is of the order of the free rotational time $(k_BT/I)^{1/2}$, one should combine the two aspects as in the Langevin rotational model.

- Translational motion : The self motion of the molecular centre of mass is often described by the Langevin model, where the particle is submitted to a viscous and a random force. In the limit of strong viscosity one obtains the simple translational diffusion model which is described by only one parameter, D_t, the translational self-diffusion coefficient (for an isotropic medium).

1.3. Spectroscopic measurements of molecular motions

We consider Figure 1, which is a sketch of a spectroscopic experiment.

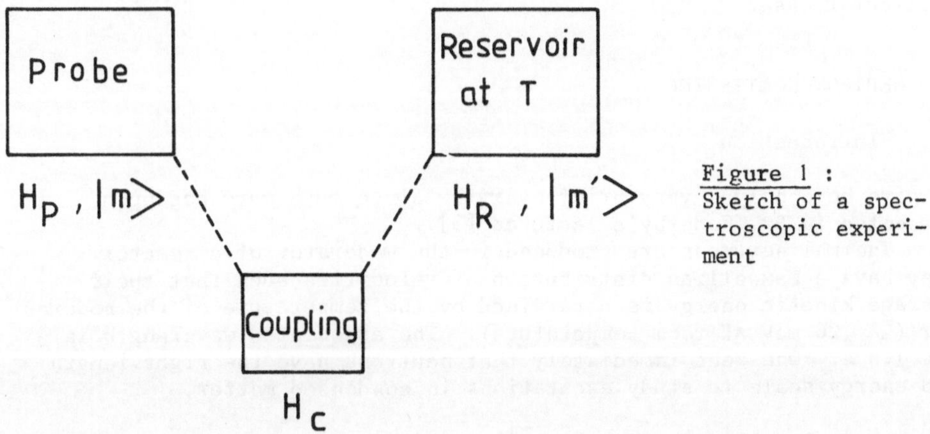

Figure 1 : Sketch of a spectroscopic experiment

The system under study, at thermal equilibrium T, constitutes the reservoir R described by the Hamiltonian H_R. The probe, which is described by the Hamiltonian H_p, interacts with the reservoir via a coupling described by the Hamiltonian H_c. We are following here the presentation by Volino [2] to which we refer the reader for details.

When the interaction is switched on, the probe changes its state $|m>$ to a final state $|n>$. In the linear approximation (H_c small compared to H_p or H_R), this change is characterized by a probability per unit time W_{nm}. The principle of a spectroscopic experiment is to measure a quantity proportional to W_{nm} as a function of $|n>$ or $|m>$. Since W_{nm} is a function of the dynamical variables of the reservoir R this measurement will yield information about the molecular motions in our system.

Using the Fermi golden rule, one obtains [2] :

$$W_{nm} = \frac{2\pi}{\hbar^2} C_{\bar{H}_c \bar{H}_c}(\omega) \quad ; \quad \hbar\omega = E_m - E_n \qquad (1)$$

where $C_{\bar{H}_c \bar{H}_c}(\omega)$ is the Fourier transform (spectral density) of the autocorrelation function of $\bar{H}_c$. $\bar{H}_c$ is the matrix element of the coupling Hamiltonian H_c between the initial and final states of the probe :

$$\bar{H}_c = <n|H_c|m> \qquad (2)$$

For a classical system, one writes

$$C_{\bar{H}_c \bar{H}_c}(t) = <\bar{H}_c(o) \; . \; \bar{H}_c(t)> \qquad (3)$$

where <...> is a thermal average and the operators in $\bar{H}_c$ are replaced by their classical expressions.

The relationship between W_{nm} and a measurable quantity depends upon the type of experiment. This has to be established for each particular case.

2. NEUTRON SCATTERING

2.1. Introduction

We give here only a very brief overview since much more has been presented in Pr. Enderby's lectures [3].

Thermal neutrons are produced in the moderator of a reactor. They have a Maxwellian distribution of velocities such that their average kinetic energy is determined by the temperature of the moderator ($\bar{E} \simeq 26$ meV at room temperature). The associated wavelength is $\bar{\lambda} = 1.8$ Å. One sees immediately that neutrons have the right length and energy scale to study excitations in condensed matter.

(Recall : 1 meV $\sim$ 8 cm^{-1})

Dynamical studies are mostly done by using incoherent scatterers. Neutron-nuclei interaction is very short range ($\sim 10^{-12}$ cm compared to the size of an atom $\sim 10^{-8}$ cm). It can be written as the Fermi pseudo potential

$$V(\vec{r}) = \frac{2\pi\hbar^2}{m_n} b_i \ \delta(\vec{r} - \vec{r}_i) \tag{4}$$

b_i is the scattering amplitude of the nuclei i : it is independent of the neutron energy (in the thermal and sub-thermal range).
For an assembly of nuclei, having different isotopes and spins, one defines the scattering lengths :

$$\begin{aligned} &- \text{coherent} \quad b_i^{coh} = \langle b_i \rangle \\ &- \text{incoherent} \quad b_i^{inc} = [\langle b_i^2 \rangle - \langle b_i \rangle^2]^{1/2} \end{aligned} \tag{5}$$

The cross-sections are defined by :

$$\begin{aligned} \sigma^i_{coh} &= 4\pi \ b_i^{2\ coh} \\ \sigma^i_{inc} &= 4\pi \ b_i^{2\ inc} \end{aligned} \tag{6}$$

We list below these cross-sections in the case of Hydrogen and Deuterium where the differences are enormous.

	spin	σ_{coh}	σ_{inc}
H	1/2	1.76	79.7
D	1	5.6	2.0

Table 1

The unit is in barn (1 barn = 10^{-24} cm^2). H has a very large incoherent scattering cross-section which is more than twenty times bigger than for other nuclei.

Besides the nuclear interaction with the nuclei, the neutrons interact with the nuclear and electronic spins via dipole-dipole interaction. This permits studies of magnetic structures or magnetic excitations (spin waves).

Using the Fermi pseudo-potential (4), the coupling Hamiltonian for an assembly of nuclei i is given by :

$$H_c = \frac{2\pi\hbar^2}{m_n} \sum_i b_i \ \delta(\vec{r} - \vec{r}_i) \tag{7}$$

The neutron states before and after scattering are defined by the wave-vectors $\vec{k}_o$ and $\vec{k}$. Using Eq. 2, one obtains :

$$\bar{H}_c = \langle k|H_c|k_o\rangle \; \alpha \; \sum_i b_i \exp(\vec{Q}.\vec{r}_i) \tag{8}$$

with $\vec{Q} = \vec{k} - \vec{k}_o$ the momentum transfer. The correlation function (Eq. 3) is directly :

$$C_{\bar{H}_c\bar{H}_c}(t) \; \alpha \; \sum_{i,j} \langle b_i b_j \exp i\,\vec{Q}.[\vec{r}_i(t) - \vec{r}_j(0)]\rangle \tag{9}$$

The scattered intensity in a solid angle $d\Omega$ and an energy window $d\omega$ is given by :

$$I = I_o N \frac{k}{k_o} S(\vec{Q},\omega)\, d\Omega d\omega \tag{10}$$

for an incident neutron flux I_o on N scattering centres.

The scattering law, $S(\vec{Q},\omega)$ is the Fourier transform of the correlation function (Eq. 9) (often called the intermediate scattering law)

$$S(\vec{Q},\omega) = \frac{1}{2\pi}\int_{-\infty}^{+\infty} C(t)\, e^{-i\omega t}\, dt \tag{11}$$

A complete derivation is given in reference [2].

2.2. Quasi-elastic scattering

2.2.1. General

We will restrict ourselves to incoherent scattering. The correlation function called the intermediate scattering law, writes :

$$I_s(\vec{Q},t) = \frac{1}{N}\sum_i b_i^{2\,inc} \langle \exp i\,\vec{Q}\,[\vec{r}_i(t) - \vec{r}_i(0)]\rangle \tag{12}$$

Writing :

$$\vec{r} = \vec{d} + \vec{\rho} + \vec{u}, \tag{13}$$

where $\vec{d}$ specifies the centre of mass position, $\vec{\rho}$ defines the position of the scattering centre with respect to the c.o.m and $\vec{u}$ stands for a small vibrational displacement around the average position. For uncoupled motions, we can write with obvious notations :

$$I_s(\vec{Q},t) = I_s^{trans} \,.\, I_s^{rot} \,.\, I_s^{vib} \tag{14}$$

By Fourier transform, one obtains :

$$S_s(\vec{Q},\omega) = S_s^{trans} \otimes S_s^{rot} \otimes S_s^{vib} \tag{15}$$

where $\otimes$ stands for the convolution product.

- Vibrational part : When we restrict to the quasi-elastic region, the characteristic times we are interested in are much longer than the reciprocal of the vibrational frequencies ($\tau_c \gg \frac{1}{\omega_{vib}}$). I_s^{vib} is a simple Debye-Waller factor :

$$S_s^{vib} = I_s^{vib} = e^{-Q^2 \langle u^2 \rangle} \tag{16}$$

where $\langle u^2 \rangle$ is a mean square amplitude of vibrations.

- Translational part : for the simple translational diffusion model [4], characterized by the self-diffusion coefficient D_t, valid at low Q-values ($Q\,a \ll 1$, where a is a molecular radius) one has :

$$I_s^{trans}(Q,t) = e^{-D_t Q^2 |t|} \tag{17}$$

which gives by Fourier transform

$$S_s^{trans}(Q,\omega) = \frac{1}{\pi} \frac{D_t Q^2}{(D_t Q^2)^2 + \omega^2}$$

The full width at half maximum (FWHM) is given by $\Delta E = 2D_t Q^2$ (fig. 2)

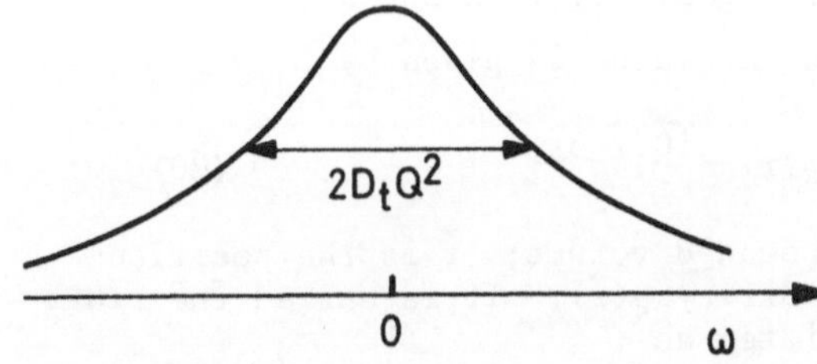

Figure 2. Theoretical incoherent neutron quasi-elastic scattering spectrum for isotropic translational diffusion.

- Rotational part : In the case of the rotational diffusion on a sphere of radius ρ, one obtains [5],

$$S_s^{rot}(Q,\omega) = j_o^2(Q\rho)\delta(\omega) + \sum_{\ell=1}^{\infty} (2\ell+1)\; j_\ell^2(Q\rho) \frac{1}{\pi} \frac{\ell(\ell+1)D_r}{[\ell(\ell+1)D_r]^2 + \omega^2} \tag{18}$$

Using Eqs. 11 and 12 it is easy to show that :

$$I_s(\vec{Q},0) = \int_{-\infty}^{+\infty} S_s(\vec{Q},\omega)\, d\omega \equiv 1 \tag{19}$$

Eq. 18 must verify this, which gives the sum rule for spherical Bessel functions :

$$\sum_{\ell=0}^{\infty} (2\ell+1)\; j_\ell^2(Q\rho) \equiv 1$$

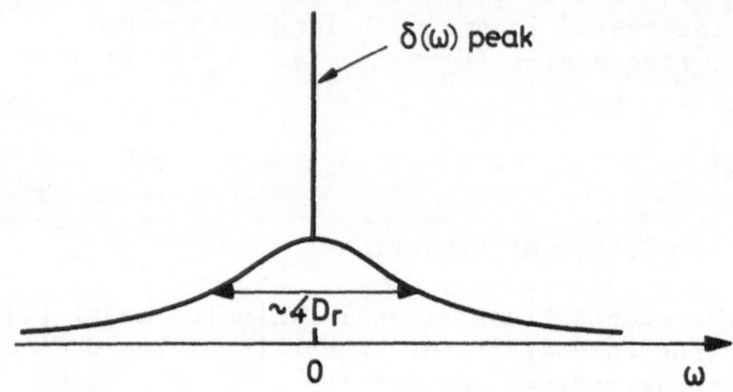

Figure 3. Theoretical incoherent neutron quasi-elastic scattering spectrum for a purely rotational diffusion motion

On Figure 3, we have sketched the scattering law Eq. 18. In an actual experiment, the elastic peak reproduces the shape of the resolution function.

2.2.2. Elastic Incoherent Structure Factor (EISF)

In the preceding paragraph, we have seen an example of a scattering law with a $\delta(\omega)$ peak, whose intensity is a function of $\vec{Q}$. This is quite general for any bounded motion : for $t \to \infty$, $I_s(\vec{Q},t)$ does not decay to zero since there will be always a finite probability to find the particle inside this volume. Since for ergodic systems one has :

$\vec{r}(t=0) \equiv \vec{r}(t \to \infty)$, this constant value is given by :

$$A_o(\vec{Q}) = |\langle e^{i\vec{Q}.\vec{r}}\rangle|^2 = \left|\int p(r)\ e^{i\vec{Q}.\vec{r}} d^3r\right|^2 \quad (20)$$

for one particle moving inside a bounded volume; $\vec{r}$ is the position vector of this particle with probability $p(\vec{r})$. It is called the EISF. The scattering law can thus be written as :

$$S_s^{bounded}(\vec{Q},\omega) = A_o(\vec{Q})\ \delta(\omega) + \sum_n \text{broadened terms} \quad (21)$$

Experimentally, the EISF is given by :

$$A_o(\vec{Q}) = \frac{I_e}{I_e+I_q} \quad (22)$$

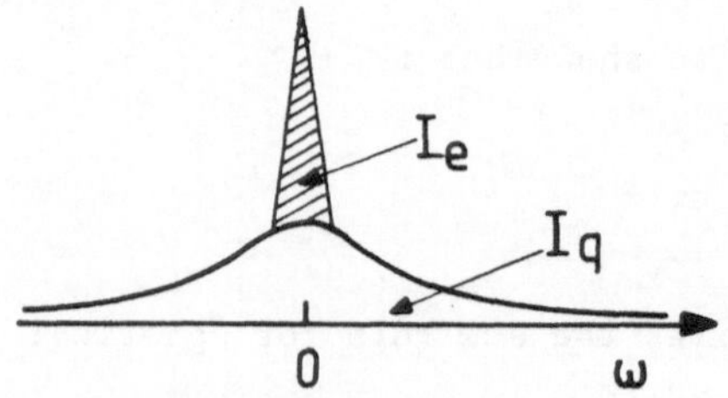

Figure 4. Experimental determination of the Elastic Incoherent Structure Factor (EISF)

This is a model independent quantity

2.2.3. Selected examples

a) Low Q-high resolution studies of aqueous solutions :

In a series of experiment using the backscattering instrument IN10 at the ILL (resolution ∿ 1 μeV FWHM), Enderby and co-workers [6] have shown that for ions for which the primary hydration shell is in slow exchange with the remaining water, the frequently used two-state model is not appropriate. Their experiments provide the evidence that the dynamics of water molecules other than these in the first shell are affected by the presence of ions. This point has been recently verified by MD simulations [1].

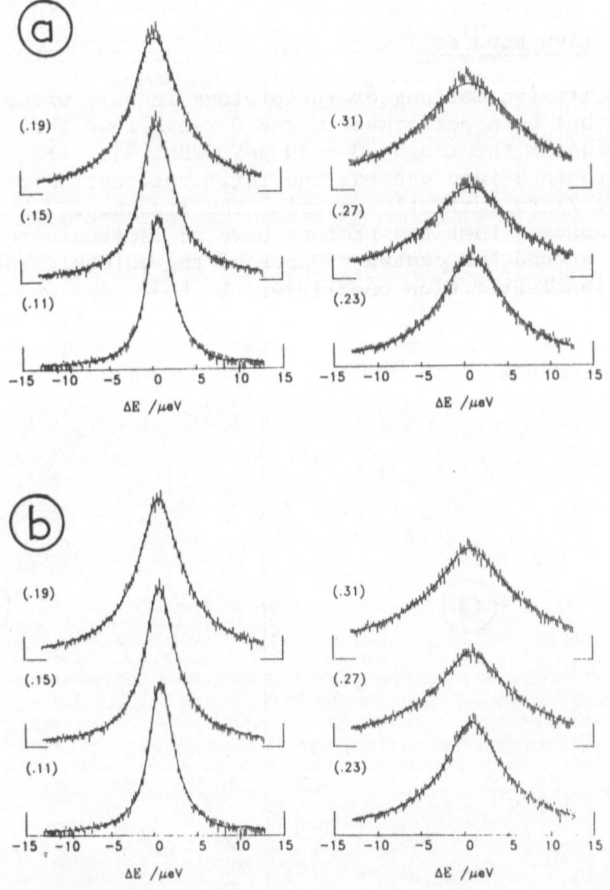

Figure 5. High resolution data on 2.0 m $NiCl_2$ aqueous solution; (a) : fit with the fast exchange scattering law; (b) : fit with the slow exchange scattering law with n_h = 6.

This work has been extended by P.S. Salmon [7] and we present below some of his results. More is to be found in Pr. Enderby's lectures [3]. The 2m $NiCl_2$ spectra presented in Figure 5 show that a single proton population inadequately describes the system. The slow exchange scattering law [7], with a hydration number $n_h = 6$ gives a good fit to the data. On Figure 6 are reported the results of two fits using the slow exchange scattering law : one with $D_1 = D_{Ni}$ determined by tracer diffusion, the other with D_1 free to vary. The results are $D_1 = (0.45 \pm 0.02)\ 10^{-9}\ m^2\ s^{-1}$ at 300°K. Note that the fitted D_1 value is only 7% larger than the tracer diffusion D_{Ni} value. One can conclude that if n_h is known, quasi-elastic neutron scattering can give reliable value of the self-diffusion coefficient of ions, when the water in the first shell is in slow exchange.

b) Medium resolution studies :

A study of the diffusive motions of the protons in pure water and $ZnCl_2$ aqueous solution has been performed in the Q-range less than 1.7 Å^{-1} and with resolution in the range 50 - 70 μeV FWHM [8]. It is shown that it is essential to take into account the rotational motion of the water molecules (Figure 7) to correctly describe the lineshape. For simplicity, it has been assumed that the protons have an isotropic rotational diffusive motion around the centre of mass of the molecule and characterized by a rotational diffusion coefficient D_r [5]. One can define

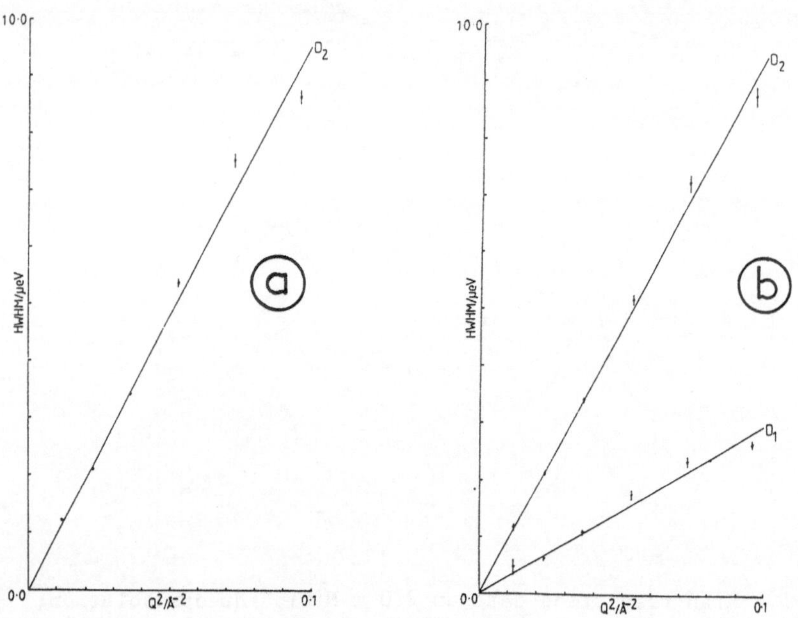

Figure 6. HWHM versus Q^2 for the slow exchange scattering law fit to the 2.0 m $NiCl_2$ solution; (a) : fit with $D_1 = D_{Ni}$; $n_h = 6$. (b) : D_1 free to vary; $n_h = 6$.

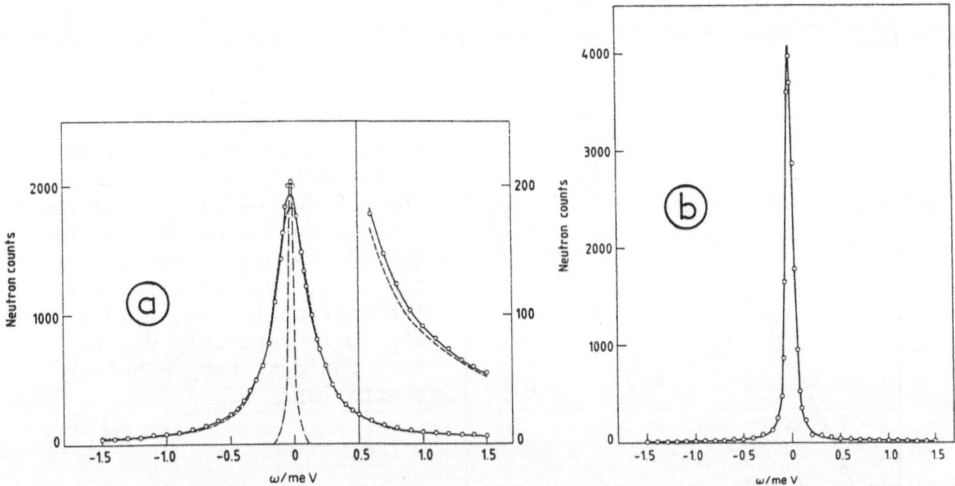

Figure 7. Quasi-elastic spectra at 298 K for $Q = 1.13\ \text{Å}^{-1}$. (a) O : experimental points for pure water. --- single Lorentzian fit; - : fit with $D_r = 0.13$ meV (insert X10); -.-.- : Vanadium data (FWHM = 48.8 μeV) (b) O : experimental points for saturated $ZnCl_2$ solution; - : fit with $D_r = 0.005$ meV.

a characteristic relaxation time $\tau_1 = \frac{1}{6D_r}$. At room temperature, this time varies from ∿ 1 ps for pure water to 20 ps for the saturated solution. Having determined D_r, it is possible to extract the translational width ΔE_T (HWHM) versus the momentum transfer Q. This is shown in Figure 8, for pure water and three concentrations of $ZnCl_2$. For pure water and the concentrated solution, where it is assumed that there is only one dynamical state for the water molecules, this width has been fitted with the Random Jump Diffusion model [9] :

$$\Delta E_T = \frac{DQ^2}{1+DQ^2\tau_o} \qquad (23)$$

where D is the self-diffusion coefficient and τ_o is a residence time.

A characteristic jump length can be defined by $L = (6D\tau_o)^{1/2}$. For pure water, τ_o is very near τ_1 at room temperature ($\tau_o \sim 1.5$ ps). For the saturated solution and for pure water, one obtains $L \sim 1.6$ Å which is close to the distance between protons in the H_2O molecule.

Another study has been confined to pure water, but in the supercooled range [10], down to -20°C. This is achieved by using an array of capillaries having an inner diameter of 0.3 mm. The linewidth of the translational component is presented in Figure 9. One can see that the residence time τ_o, defined by Eq. 23, increases dramatically when the temperature decreases. On the contrary, the rotational correlation time τ_1 increases less rapidly and follows an Arrhenius behaviour. This is

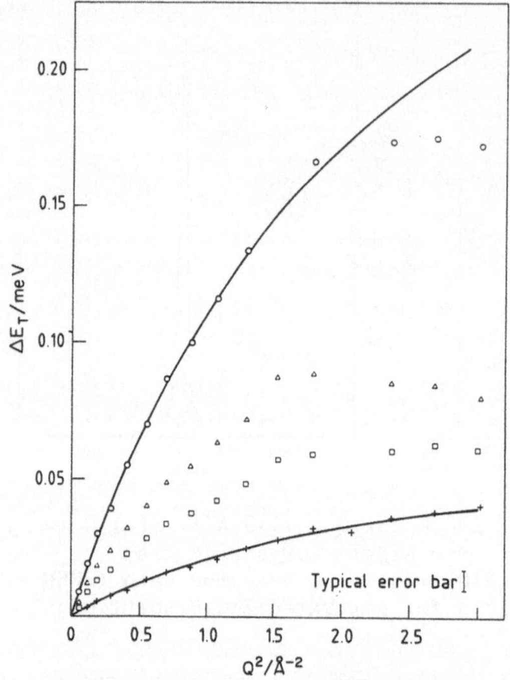

Figure 8. ΔE_T versus Q^2 at T = 298 K. O, H_2O (D_r = 0.13 meV); Δ, 3M $ZnCl_2$ solution; , 6M $ZnCl_2$ solution; +, saturated $ZnCl_2$ solution (D_r = 0.005 meV); ——, random jump diffusion model for pure H_2O and saturated $ZnCl_2$ solution; I, typical error bar. The systematic deviation at high Q is certainly due to some residual sample holder scattering.

shown on Figure 10. An explanation has been given for this non-Arrhenius behaviour of the residence time τ_o. Another interesting feature is the nearly constant value with temperature of the vibrational amplitude $\langle u^2 \rangle^{1/2} \sim 0.5$ Å. This value corresponds to an amplitude of vibration around 25°, when according to M.D. simulations, the hydrogen bond is broken. More on this subject can be found in the proceedings of a Workshop which was held at the ILL in April 1984 [11].

c) Water mobility in an ionic membrane

This experiment is an excellent example of the power of quasi-elastic neutron scattering to provide both dynamical and structural information [12]. Water mobility in an acid Nafion® membrane containing 15% water by weight has been studied by using different resolutions ranging from 18.5 to 9 μeV FWHM. An example of a spectrum is given on figure 11. One sees that it consists of a sharp peak superimposed on a broad component having about 100 μeV FWHM.

By studying carefully the shape of this broadened component, one finds that it deviates systematically from a single Lorentzian (Fig. 12). Furthermore the width of the best fitted Lorentzian is practically constant at low Q and then increases with Q. This is shown on Figure 13. This behaviour is consistent with a model of diffusion inside a sphere [13]. We cannot report here the mathematical details, but this model depends upon two parameters : a self-diffusion coefficient D and the

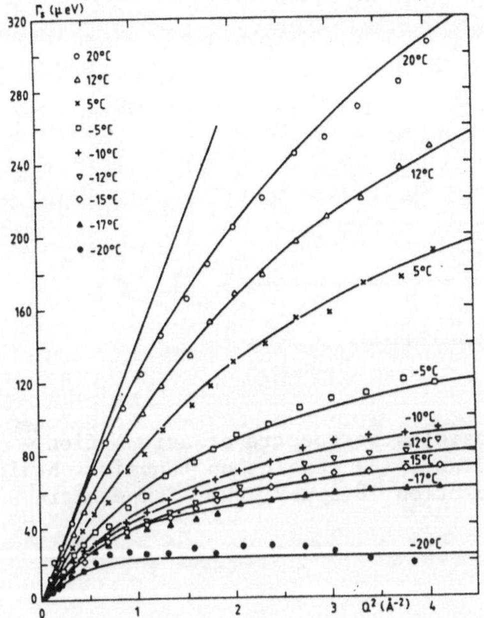

Figure 9. Linewidth Γ of the translational component of the spectrum vs Q^2. Note that a one-Lorentzian fit, as commonly done, gives much larger linewidths with a different Q dependence. —— : best fit using Eq. (23). The straight line gives the self-diffusion constant at 20°C.

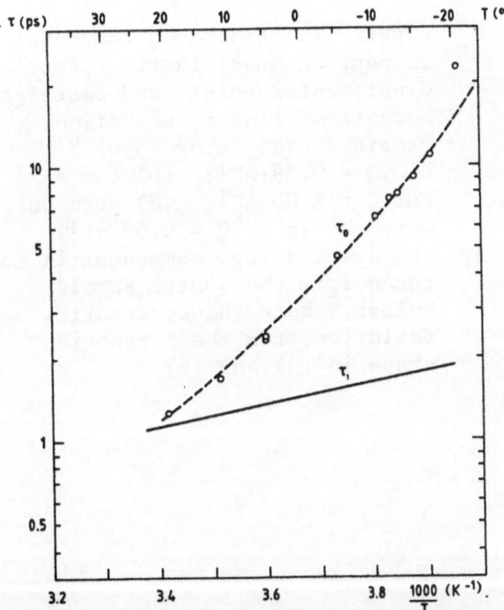

Figure 10. Residence time τ_0 of the jump diffusion and an Arrhenius plot of the relaxation time τ_1.

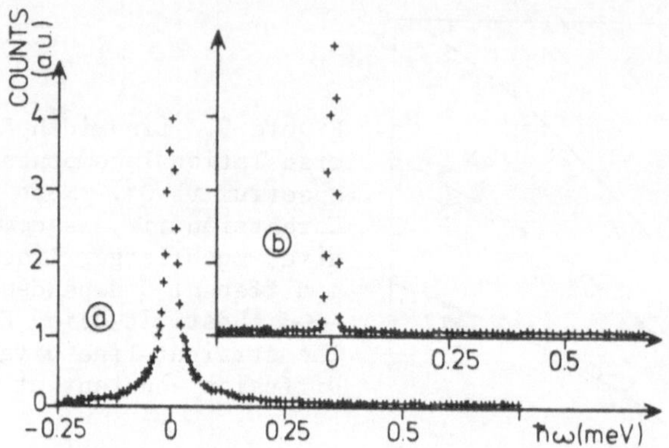

Figure 11. Examples of neutron quasielastic spectra of acid Nafion® membrane for $Q = 0.59\ \text{Å}^{-1}$. (a) Membrane, (b) resolution function obtained with a vanadium sample (energy resolution 18.5 μeV FWHM, temperature ca. 25°C).

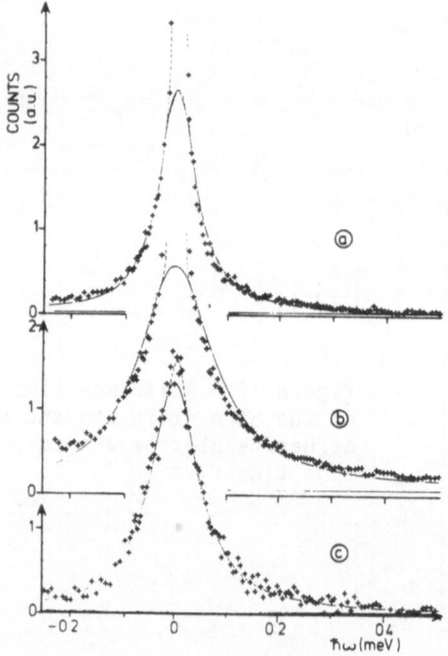

Figure 12. Broadened component of neutron quasielastic spectra : experimental points and best-fit Lorentzian line : (a) Nafion® membrane containing 15 wt % H_2O, $Q = 0.59\ \text{Å}^{-1}$; (b) the same but $Q = 1.05\ \text{Å}^{-1}$; (c) pure bulk water at 28°C, $Q = 0.59\ \text{Å}^{-1}$. The small narrow component in (c) comes from the quartz sample holder. Note the systematic deviation from the Lorentzian shape in (a) and (b).

radius of the sphere a. By applying this model to the measured spectra, the values of D and a have been extracted and are reported on figure 14.

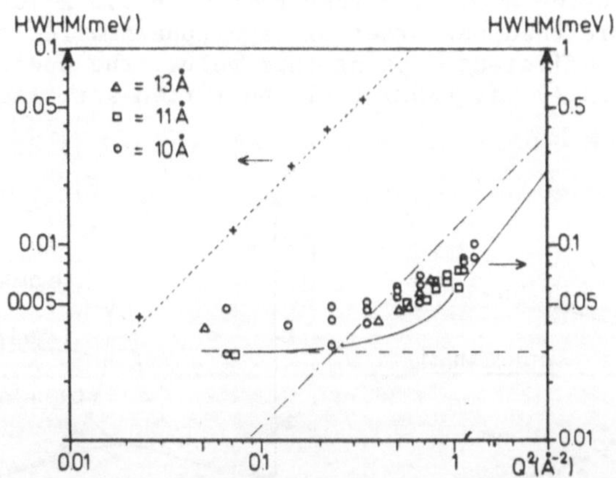

Figure 13. Half-width at half-maximum of broadened component of neutron quasielastic spectra obtained from acid Nafion® membrane containing 15 wt% water. The points are the widths of the best-fit Lorentzian curves: (O) from spectra obtained with incident wavelength 10 Å, (□) 11 Å, (Δ) 13 Å. The full line is the theoretical width predicted by the model with diffusion in a sphere with D = 1.8 x 10^{-5} cm^2/s and a = 4.25 Å. The two theoretical asymptotes for $Q \to 0$ and $Q \to \infty$ are also shown. (+) half-width at half-maximum of the best-fit Lorentzian lines to spectra obtained from bulk water at 28°C (incident wavelength 10 Å). The straight line passing through the points (+) is the theoretical width predicted by the simple self-diffusion model with D_t = 2.5 x 10^{-5} cm^2/s. Note the different vertical scales for the Nafion® and bulk-water samples.

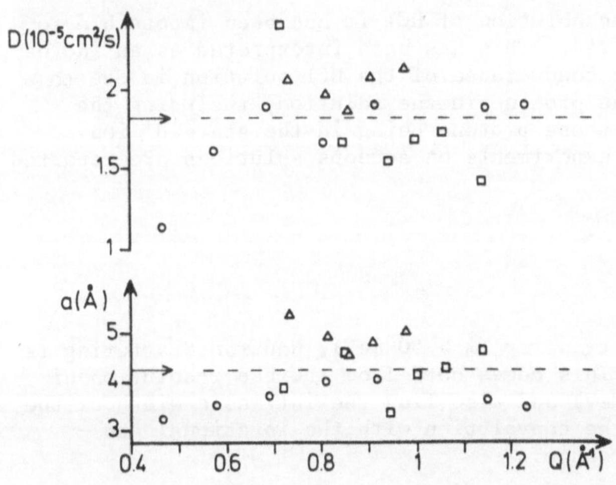

Figure 14. Values of diffusion coefficient D and radii of spheres a obtained by fitting the theoretical scattering law to all the quasi-elastic spectra with the central part excluded : (O) from spectra obtained with incident wavelength 10 Å, (□) 11 Å, (Δ) 13 Å. The mean values of D and a are also indicated.

One sees that the water diffuses nearly as fast as in pure water but in a volume of less than 10 Å in diameter.

The long-range diffusion coefficient of water has been measured by tracer diffusion using HTO. The result is $D_t = 1.6 \times 10^{-6}$ cm^2/s at 25°C. This is more than one order of magnitude smaller than the short range diffusion coefficient. Using this value, the spectra obtained with the best resolution (9 μeV FWHM) could be fitted satisfactorily (Fig. 15b).

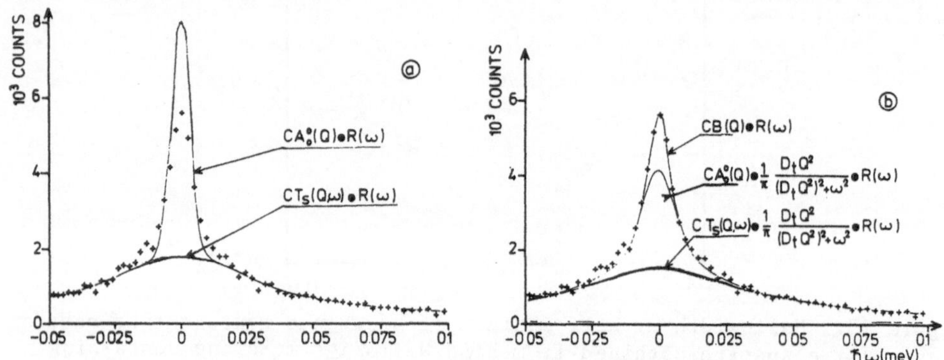

Figure 15. Quasielastic spectrum for Q = 0.635 $Å^{-1}$ obtained with an incident wavelength of 13 Å, and best-fit curves using $D = 1.8 \times 10^{-5}$ cm^2/s, a = 4.25 Å (a) or $D = 1.8 \times 10^{-5}$ cm^2/s, a = 4.25 Å, and $D_t = 1.6 \times 10^{-6}$ cm^2/s (b). The separation between the various components of the theoretical spectra is also indicated. In (a) the best fit gives B(Q) = 0.

d) Other studies : Very few studies have been performed on aqueous solutions, using quasi-elastic neutron scattering. The main reason is that outside the ILL, the available instruments are lacking the necessary resolution.

In a 15.5 molal aqueous solution of HCl it has been impossible to detect any fast H^+ motion [14]. This has been interpreted as an indication that the high electric conductance of the HCl solution is due to a collective motion of all the protons in the solution, including the water protons, not only of those protons being in the state H^+ ion.

Some lower resolution experiments on aqueous solutions are reported in references 15 to 17.

2.3. Inelastic scattering

2.3.1. General

In liquids, at high energy transfer (ω > 50 meV), Neutron Scattering is often resolution limited. This comes both from the bad resolution of the spectrometers (4 - 10 meV) and also from the intrinsic width of the vibrational bands, due to the convolution with the rotational and translational degrees of freedom.

However, Neutron Inelastic Scattering has a distinct advantage over all the other spectroscopic techniques : the spectra result from a direct coupling with the positions of the scattering particles (see Eq. 9). There is no selection rules so that one can reveal vibrational bands which are neither Infrared nor Raman active : this kind of studies are usually done by using a Beryllium-filter spectrometer. Another advantage for assignment of bands, is the possibility of using selective deuteration, due to the big difference in cross-sections between Hydrogen and Deuterium (see Table I). At low frequency (ω < 50 meV) neutron scattering will reveal the whole molecule vibrational motion (c.o.m. motion). For that purpose one needs a spectrometer with a good energy resolution : this can be achieved by using a time-of-flight (TOF) spectrometer with cold neutrons (λ > 4 Å). However for this kind of instrument, the resolution degrades rapidly for $\omega > E_o$ (incident energy).

2.3.2. Generalized Frequency distribution

We present here a very brief account of the derivation leading to the generalized frequency distribution. A full derivation can be found in ref. [18]. It is straightforward to show that for isolated or coupled harmonic oscillators, the intermediate incoherent scattering law (Eq. 12) takes the form :

$$I_{inc}(Q,t) = e^{-\frac{1}{2} Q^2 \gamma(t)} \tag{24}$$

where $\gamma(t)$ is called the width function. For an harmonic oscillator, in the classical limit ($\hbar\,\omega_o \ll k_BT$) one has :

$$\gamma(t) = \frac{2\,k_BT}{M\,\omega_o^2}(1 - \cos\omega_o t) \tag{25}$$

$\gamma(t)$ is related directly to the mean square displacement

$$\gamma(t) = \frac{1}{3} \langle r^2(t)\rangle \tag{26}$$

In the Gaussian approximation, one writes for any scattering law :

$$S_{inc}(Q,\omega) = \frac{1}{2\pi}\int_{-\infty}^{\infty} dt\; e^{i\omega t}\; e^{-\frac{1}{2} Q^2\,\gamma(t)} \tag{27}$$

one can show [18] that $\gamma(t)$ is related to the Fourier transform of the frequency distribution $g(\omega)$. A practical way to extract this quantity is the extrapolation procedure due to Egelstaff and Schofield [19].

Putting $\alpha = \frac{\hbar^2 Q^2}{2Mk_BT}$ and $\beta = \frac{\hbar\omega}{k_BT}$

the frequency distribution is obtained as :

$$g(\beta) = 2\beta \sinh(\beta/2) \lim_{\alpha \to o} \frac{S(\alpha,\beta)}{\alpha} \qquad (28)$$

This expression is nearly equal to :

$$g(\omega) \sim \omega^2 \lim_{Q \to o} \frac{S(Q,\omega)}{Q^2} \qquad (29)$$

An example of a vibrational density of states is the one predicted by the Debye model :

$$g(\omega) = \frac{3}{\omega_D^3}\,\omega^2 \quad \text{for} \quad \omega \leqslant \omega_D$$

$$= 0 \quad \text{for} \quad \omega > \omega_D$$

This generalized frequency distribution can be compared to the one obtained by Raman scattering. It will be shown that while in the neutron case one obtains a quantity which can be directly compared to MD simulations, in the Raman case, the low frequency part is quite insensitive to the whole molecule vibrations.

2.3.3. Selected examples

a) - earlier studies : Very few inelastic neutron scattering studies on ionic solutions have been reported in the literature. Several systems were investigated at the end of the sixties in view of quantifying the ion-water interactions through the variation of the observed intermolecular frequencies. These studies can be found in refs. [16] and [17]. Although the resolution of the spectrometer was quite poor, the authors could classify the different ions studied as acting as "positive" or "negative hydrators".

b) - low- and very low-frequency dynamics in $ZnCl_2$ aqueous solutions
A comprehensive investigation on the low (ω < 50 meV) and very low (0.02 < ω < 5 meV) frequency vibrational dynamics in water and aqueous solutions of $ZnCl_2$ (for concentrations up to saturation) has been done by TOF neutron spectroscopy and Raman scattering [20]. The reduced depolarized Raman intensity $g^R(\omega)$ and the generalized frequency distribution deduced from neutron data $g^N(\omega)$ are compared in figure 16.

One can see that for pure water the strong peak in $g^N(\omega)$ around 7 meV ($\sim$ 60 cm^{-1}) is only apparent as a small shoulder in $g^R(\omega)$. In a very drastic approximation [21] one can write :

$$g^R(\omega) = C^R(\omega) \;.\; g(\omega)$$

where $C^R(\omega)$ is the electron-vibration coupling function. It can be obtained by forming the ratio $g^R(\omega)/g^N(\omega)$ and is shown in figure 17 for pure water and three concentrations of $ZnCl_2$.

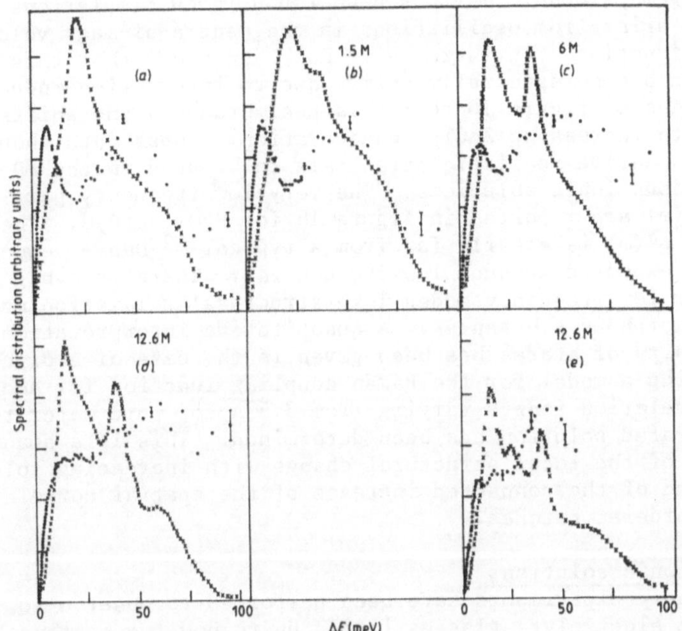

Figure 16. Reduced depolarised Raman intensity $g^R(\omega)$ (crosses) and generalised frequency distribution $g^N(\omega)$ (dots) for pure D_2O (a) and solutions of $ZnCl_2$ in D_2O (b–d). The reduced spectra for the saturated $ZnCl_2$ solution in H_2O are also shown for comparison (e).

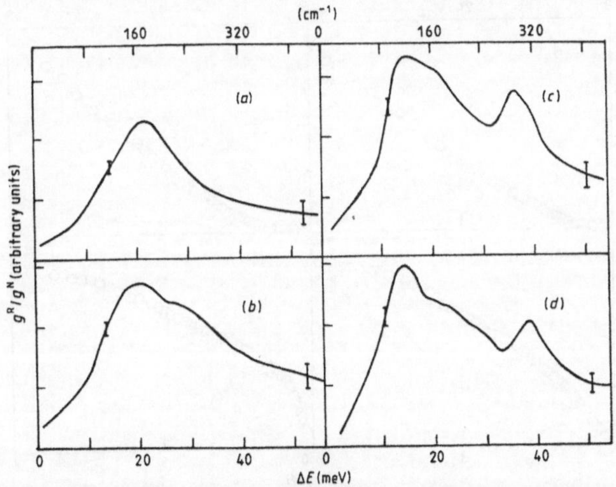

Figure 17. Concentration dependence of the experimentally obtained Raman coupling function $C^R(\omega) = g^R(\omega)/g^N(\omega)$: pure water (a), 3 M $ZnCl_2$ in H_2O (b), 6 M $ZnCl_2$ in H_2O (c) and 12.6 M $ZnCl_2$ in H_2O (d).

The peak in $g(\omega)$ around 7 meV has been found in MD simulations [1], [22] and shown to arise from oscillations in the centre-of-mass velocity correlation function. Furthermore it has been shown that this translational motion become spectrally active due to interaction-induced dipoles [22]. One sees on figure 16 that this peak broadens and shifts to higher frequency with increasing $ZnCl_2$ concentration. These data should be of some value to derive realistic potentials to be used in the MD simulations of aqueous ionic solutions. The very-low frequency part of $g^N(\omega)$ and $g^R(\omega)$ are reported in figure 18 for H_2O and D_2O. The observed behaviour of $g^N(\omega)$ is clearly far from a typical ω^2 Debye dependence which can be expected at such low frequencies. There is a big contribution of other low frequency modes like structural relaxation modes observed in all amorphous substances. A quantitative interpretation of this enhanced density of states has been given in the case of amorphous silica [23]. By using a model for the Raman coupling function for acoustic modes, a correlation length varying from 3.5 Å for pure water to 7.5 Å for the saturated solution has been determined. This is a quantitative confirmation of the local structural change with increasing solute concentration and of the connected increase of the spatial correlation of the locally ordered patches.

c) - Glassy ionic solutions

Some preliminary experiments have been performed to observe low-frequency vibrations in electrolyte glasses [24]. Up to now two systems have been investigated : D_2O - 14% LiCl and D_2O - 13% $BeCl_2$. A low frequency peak is observed in both cases in the vicinity of 4 meV in the scattering law. However the form of the density of states in these two cases seems to be qualitatively different as shown on figures 19 and 20.

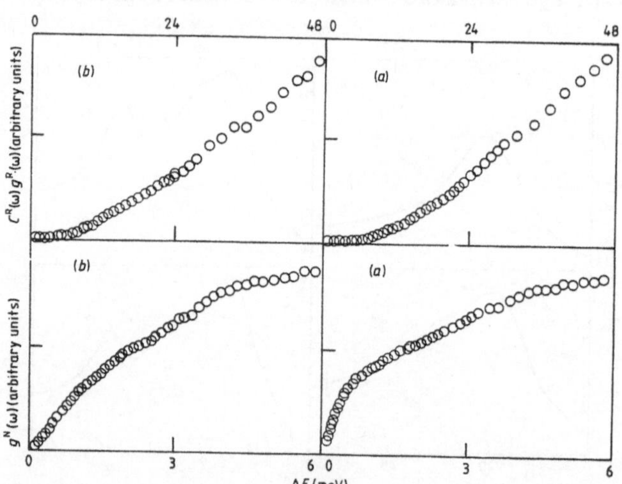

Figure 18. Very-low frequency spectrum of the generalised distribution $g^N(\omega)$ and of the reduced Raman intensity after subtraction of the quasi-elastic contribution for H_2O (a) and D_2O (b).

With LiCl it appears that there is a sudden increase in the density of states above a ω^2 dependence for ω bigger than 3.5 meV. This is not seen in the $BeCl_2$ system.

A currently fashionable explanation for the form of the density of states of amorphous materials is based on "fractons" which are excitations on a fractal network [25]. Originally developed for polymeric materials, it has been extended to other glasses on the basis that all amorphous materials show a characteristic length scale above which the solid is essentially homogeneous. In order to test this "fracton" model we need to know the sound velocities in these glasses.

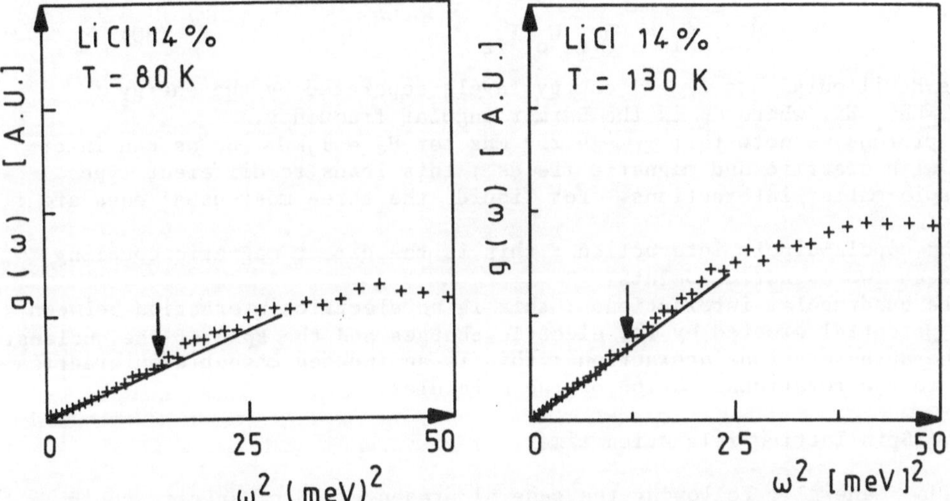

Figure 19 : Generalised frequency distribution for the glassy electrolyte. D_2O - 14% LiCl for two temperatures. The arrows indicate a positive deviation above the ω^2 dependence.

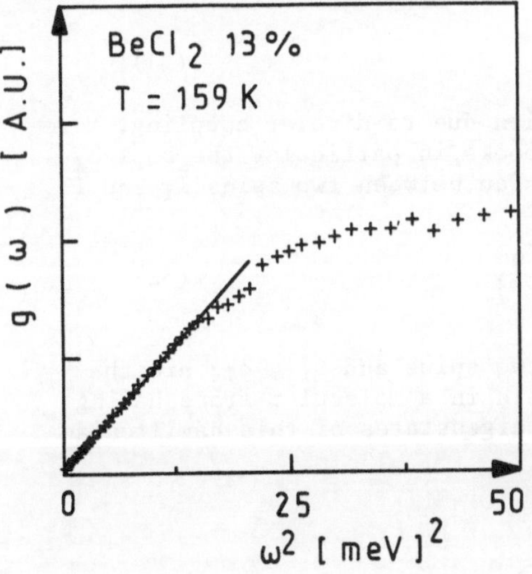

Figure 20. Generalized frequency distribution for the glassy electrolyte D_2O - 13% $BeCl_2$. There is no evidence of a positive deviation above the ω^2 dependence.

3. NUCLEAR MAGNETIC RESONANCE

3.1. Introduction

We will present very briefly how nuclei which have a spin can be used to study molecular motion. A complete derivation can again be found in ref. [2].

A particle i whose nucleus possesses a spin $\vec{I}_i$ has a magnetic moment $\vec{\mu}_i = \hbar\,\gamma_i\,\vec{I}_i$. γ_i is called the gyromagnetic ratio. This magnetic moment interacts with a magnetic field $\vec{H}_o$ through the Zeeman Hamiltonian :

$$H_z = -\vec{H}_o \cdot \vec{\mu}_i = -\hbar\gamma_i H_o I_{iz} \tag{30}$$

This Hamiltonian has 2I + 1 energy levels separated by the energy : $\hbar\omega_L = \hbar\gamma_i H_o$, where ω_L is the Larmor angular frequency.
For protons we note that ω_L = 4.257 MHz for H_o = 1 kG. Spins can interact with electric and magnetic fields : this leads to different types of spin-matter interactions. For liquids the three most usual ones are :

- the dipole-dipole interaction : this is the direct magnetic coupling between two magnetic moments.
- the quadrupolar interactions : this is an electric interaction between the potential created by the electric charges and the spin of the nucleus.
- the spin-rotation interaction : this is an induced magnetic interaction due to the rotational motion of the molecules.

3.2. Spin lattice relaxation time

3.2.1. General : Following the general presentation of spectroscopic methods (see § 1.3), the system considered is a liquid of N particles i possessing a spin $\vec{I}_i$, and placed in a strong magnetic field $\vec{H}_o$. The probe is directly made up by the spin magnetization. The total nuclear magnetization is :

$$\vec{M} = \sum_i \hbar\,\gamma_i\,\vec{I}_i \tag{31}$$

We will consider here only the relaxation due to dipolar coupling. More details can be found in standard text books, in particular the book by Abragam [26]. The interaction hamiltonian between two spins $\vec{I}_i$ and $\vec{I}_j$ is :

$$H_d^{ij} \propto \frac{1}{r_{ij}^3} Y_2^m(\theta_{ij}, \phi_{ij})\,\vec{I}_i \cdot \vec{I}_j \tag{32}$$

where r_{ij} is the distance between the two spins and θ_{ij}, ϕ_{ij} are the polar angles of the static magnetic field in a molecular frame having $\vec{r}_{ij}$ as z-axis. For two spins 1/2, the eigenstates of this hamiltonian are noted :

$$|++\rangle \quad |+-\rangle, \quad |-+\rangle \quad \text{and} \quad |--\rangle.$$

The two central eigenstates are degenerate, so one has three levels separated by the energies $\hbar\omega_L$ and $2\hbar\omega_L$. Since H_d (Eq. 32) has non-zero matrix elements between these three levels, the relaxation of the longitudinal magnetization M_z will follow a characteristic time $T_1{}^d$ given by :

$$\frac{1}{T_1{}^d} = \sum_{m=-2}^{+2} A_m J_m(\omega_L) + B_m J_m(2\omega_L) \tag{33}$$

$T_1{}^d$ is called the <u>spin lattice relaxation time</u> and $J_m(\omega)$ is the spectral density of the correlation function of $1/r_{ij}^3 \cdot Y_2^m(\theta_{ij}, \phi_{ij})$.

There are a priori two contributions :

- the intramolecular contribution : it is determined by the rotational motion only through the time variation of θ_{ij}, ϕ_{ij}

- the intermolecular contribution : in addition to rotational contribution, there is an influence of translational motion through the time variation of r_{ij}.

The separation between these two contributions is a priori difficult and one needs models for the motions.

Looking at Eq. 33, one sees that in principle we need measurements at different Larmor frequencies ω_L (corresponding to different strengths of the magnetic field H_o) to probe the spectral density in a wide range.

More frequently, one follows the variation of the spin-lattice relaxation time with temperature. This variation is sketched on figure 21. If one obtains a minimum of T_1 (BPP minimum) the scale of the relevant <u>correlation time</u> is directly fixed by :

$$\omega_L \tau_c \sim 1 \tag{34}$$

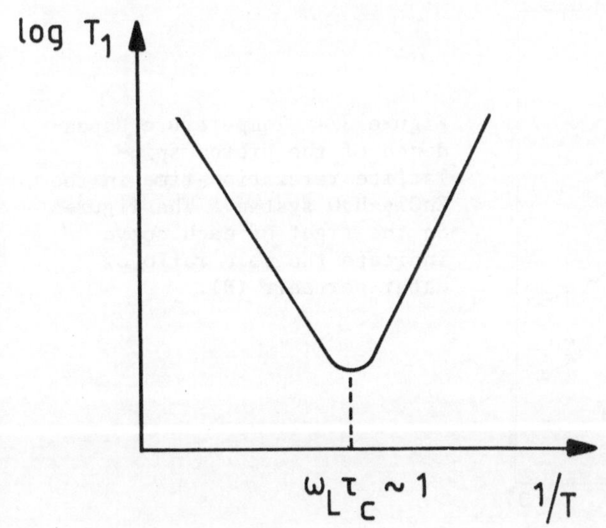

Figure 21. Variation of the spin-lattice relaxation time versus reciprocal temperature, showing the well-known BPP minimum.

where the exact value depends upon the specific model. The slope of the straight lines defines the activation energy of the motion.

For normal liquids, even in strong magnetic fields, one is always in the extreme narrowing limit, $\omega_L \tau_c \ll 1$. In this case one has

$$1/T_1 = A\tau_c \tag{35}$$

where the constant A depends upon the specific model for the motion.

3.2.2. Selected examples : The literature on the application of NMR in the field of aqueous ionic solutions is much more extended than the one on Neutron Scattering. We will just report here some specific examples and give references to other works.

a) Proton spin-lattice relaxation time in concentrated aqueous solutions Concentrated aqueous $ZnCl_2$ solutions have been measured as a function of salt concentration and temperatures [27]. Figure 21 shows this dependence. One sees that it is difficult to have an estimate of the magnitude of the relevant correlation time.

On the solution with R = 4.1 the measurements were extended deeply into the super-cooled region down to -55°C. Figure 23 gives the result and a minimum is obtained at about -40°C. The analysis shows that in this temperature range, the intramolecular contribution, which is influenced by the rotational motion only, is important. The value of the rotational correlation time τ_{intra} can thus be calculated. Since the authors have also measured the viscosity of the solution, they could compared this value with the one calculated by using the Debye equation : they found that this equation does not apply in these concentrated solutions.

Another measurements of T_1 in concentrated $Ca(NO_3)_2$ solutions have been analyzed in term of intramolecular and intermolecular contributions [28]. The authors show that the simple BPP analysis using the Debye

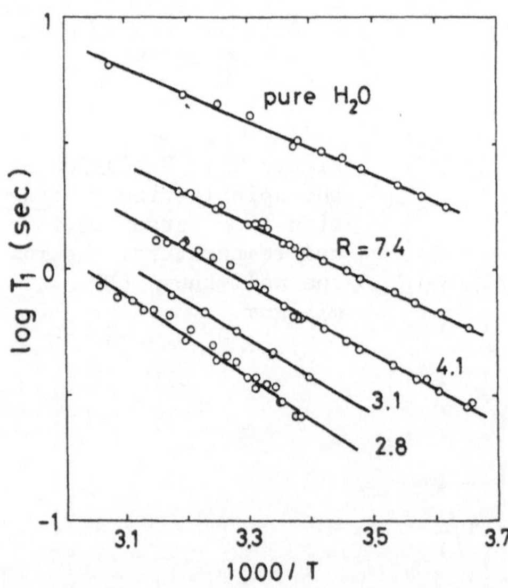

Figure 22. Temperature dependence of the proton spin-lattice relaxation time in the $ZnCl_2$-H_2O system. The figures on the right of each curve indicate the mole ratio of water per salt (R).

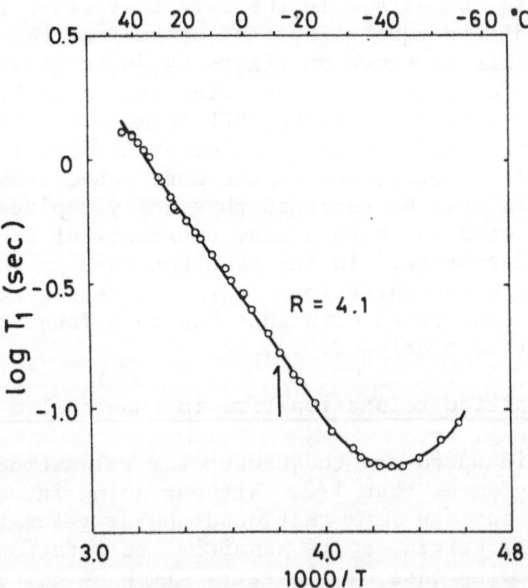

Figure 23. Temperature dependence of the proton spin-lattice relaxation time of the solution with R = 4.1. The arrow indicates the freezing point.

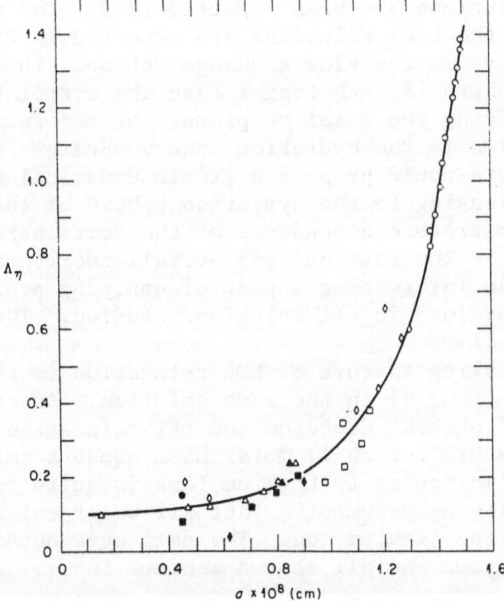

Figure 24. Ratio of rotational to intermolecular correlation time and effective molecular radius corresponding to forced fit of the BPP equation.

theory cannot explain the data; furthermore they state that the apparent agreement for dilute aqueous systems at ambient temperature appears to be fortuitous. This is shown on figure 24 which gives the parameter $\Lambda = 9\ \tau_{rot}/\tau_{inter}$ as a function of the effective molecular radius a which corresponds to forced fit of the BPP formula. On this figure, the points on the right correspond to water at different temperatures while those at the left correspond to the most concentrated solutions. It is apparent that a must be assigned physically implausible values if one assumes the BPP equation with a near constancy of Λ, except for water at ambient temperature. In any case the results show, that with increasing electrolyte concentration, τ_{inter} increases much more rapidly than τ_{rot}. This is consistent with what has been found by quasi-elastic neutron scattering (see section 2.3.3. b).

b) <u>Deuteron spin-lattice relaxation time in supercooled electrolyte solutions</u>

Nothing has been said up to now to quadrupolar relaxation which arises when the spins are greater than 1/2. Without going into any details (see [2] and [26]), one can state that quadrupolar relaxation has a purely intramolecular origin, s_o is sensitive to rotational motions only. In the extreme narrowing case, T_1 is still given by Eq. 35, but with the constant A depending upon the quadrupolar coupling constant and its asymmetry.

A detailed study of the pressure, temperature and concentration dependence of deuterium relaxation times in supercooled LiCl -, NaCl - and $MgCl_2$ - D_2O solutions has been reported [29]. The results for the 225 MPa isobars of the LiCl solutions are reported in Figure 25. This system is always in the slow exchange region, since the lifetime of the hydration sphere is much longer than the correlation times of the water molecules. Using the detailed picture of the geometrical arrangement of the molecules in the hydration sphere obtained by neutron diffraction, the authors could propose a simple dynamical model of the water molecules belonging to the hydration sphere of the Lithium ion, and extract the temperature dependence of the correlation times τ_i and τ_r which characterize the internal and overall reorientations of the water molecules. An interesting aspect of applying pressure is that a minimum of T_1 is obtained in all solutions studied. The results are tabulated in Table II.

Another interesting feature of NMR relaxation is that it can be applied to different nuclei in the same solution. An example is provided by the study of proton, deuteron and ^7Li relaxation times as a function of temperature for an 11 molar LiCl aqueous solution [30]. The ratio of water molecules to Lithium ions is quite low (R = 4.9), allowing the simplifying assumption that all water exist in the first hydration shell of the Lithium ion. The data were obtained down to the supercooled region and all show a minimum in T_1. Figure 26 presents the results.

From these data, the authors could extract the various correlation times and the self-diffusion coefficient of water. A consistency check of the analysis is provided on Figure 27, where the same correlation time explains both the 7 MHz proton and ^7Li data.

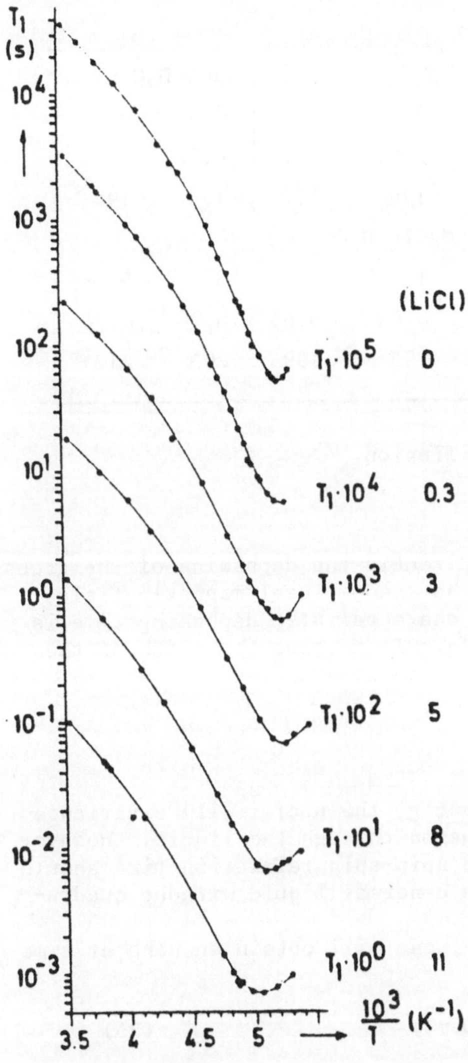

Figure 25. 225 MPa Isobars of Deuteron-T_1 in D_2O /LiCl Solutions
(LiCl) ≡ Molality LiCl

c) Other studies

A recent review can be found in ref. [31]. Several studies are aiming at obtaining detailed information concerning the structure and the interactions in ionic solutions [32 - 34], through the measurement of spin-lattice and spin-spin relaxation times. Recent experiments are using ^{17}O enriched solutions, in view of obtaining direct measurements of the rotational motion of the hydrated ions [35,36].

TABLE II

	$LiCl-D_2O$					$NaCl-D_2O$	
C(molal)	0.3	3	5	8	11	0.1	3
T_1(ms) ± 10%	0.54	0.62	0.65	0.67	0.70	0.58	0.54
T_{min}(K) ± 2	192	192	192	194	197	192	194
				$MgCl_2-D_2O$			
C(molal)	0.1	0.3	0.6	1	2	3	5
T_1(ms) ± 10%	0.5	0.54	0.58	0.63	0.65	0.70	0.74
T_{min}(K) ± 2	192	192	194	195	199	203	217

3.3. Spin-echo measurement of self-diffusion

3.3.1. General

In a pulsed experiment, one observes directly the dephasing of the transverse component of the spins, due to the random local magnetic fields arising from molecular motions. This characteristic dephasing time is called the spin-spin relaxation time T_2. It governs the width of the absorption spectrum :

$$T_2 \propto \frac{1}{\Delta\omega}$$

where $\Delta\omega$ is the linewidth.

By imposing a static field gradient g, the nuclei will experience different magnetic fields due to diffusion through the liquid. However in order to follow this diffusion, the spin-spin relaxation time should not be too short. This is the case in a normal liquid without quadrupolar relaxation : $T_2 \sim T_1 \sim$ 1- 10 sec.
After a 90°(o)- 180°(τ) pulse sequence, one will obtain an echo at time 2τ given by [26] :

$$A(2\tau) = A_o \exp\left(-\frac{2}{3}\gamma_I^{\,2}\, g^2\, D\, \tau^3\right) \qquad (36)$$

where A_o is governed by T_2 :

$$A_o \propto \exp(-t/T_2) \qquad (37)$$

It is easy to see in Eqs. 36 - 37, that T_2 should be long enough so that the echo will not die out too rapidly. Usually the value of the field gradient g is determined by using a standard sample of known self-diffusion coefficient, as pure water. Another NMR method for measuring the self-diffusion coefficient of liquids is the pulsed gradient method, that we will not detail here.

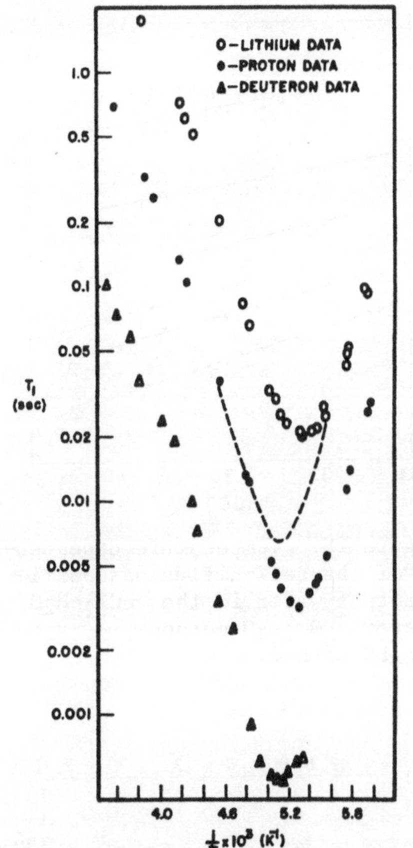

Figure 26. T_1 data vs. reciprocal temperature. The resonant nuclei are as indicated. Both the 4- and 7-MHz deuteron data are plotted with one symbol since the 4-MHz data do not go to a low enough temperature to show frequency dependence. The proton and lithium data are at 7 MHz. The 14-MHz proton data are not shown to avoid clutter, but the minimum region is indicated by the dotted line.

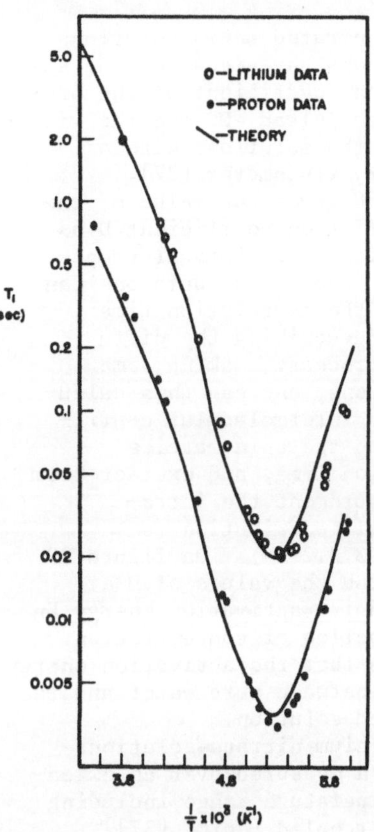

Figure 27. Comparison of $(R_A + R_B)^{-1}$ (lower curve) with the 7-MHz proton data (dots). The upper curve is a fit at the ^{7}Li minimum.

3.3.2. Selected examples

On concentrated $ZnCl_2$ solutions, the authors have measured the self-diffusion coefficient of the protons with pulsed NMR and the viscosity of the solutions with a capillary viscometer [27]. Figure 28 gives the value of the self-diffusion coefficient D as a function of reciprocal temperature. From these data one can extract the correlation time τ_{inter}, describing the diffusional process. Making some assumptions, one can thus calculate the intermolecular contribution to the spin-lattice relaxation time, and extract from the measurement the intramolecular contribution (see section 3.2.2.-a). On figure 29 is plotted the values of D at 25°C and its activation energy E_D as a function of concentration. One sees that the activation energy doubles between pure water and the saturated solution.

Calcium nitrate solutions have been measured over an extensive temperature range including the supercooled regime [37]. The data are covering 2.5 orders of magnitude in the self-diffusion constant (see figure 30). Their temperature dependence is described by a modified Arrhenius (VTF) relationship, which provides a convenient means of comparing the temperature dependences of different transport processes.

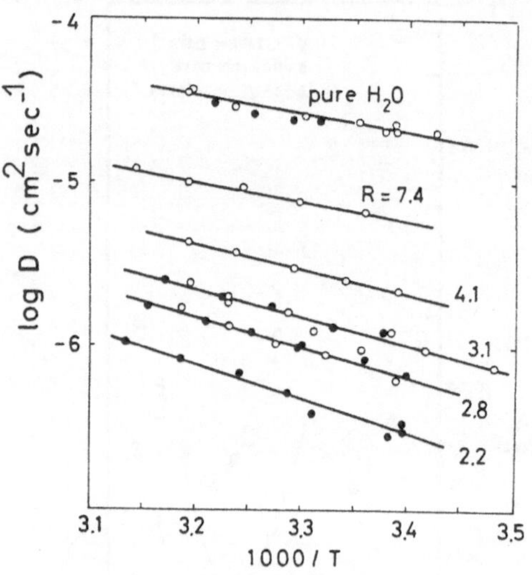

Figure 28. Temperature dependence of the self-diffusion coefficient of water in the $ZnCl_2$-H_2O system. O : SG method, ● : PG method.

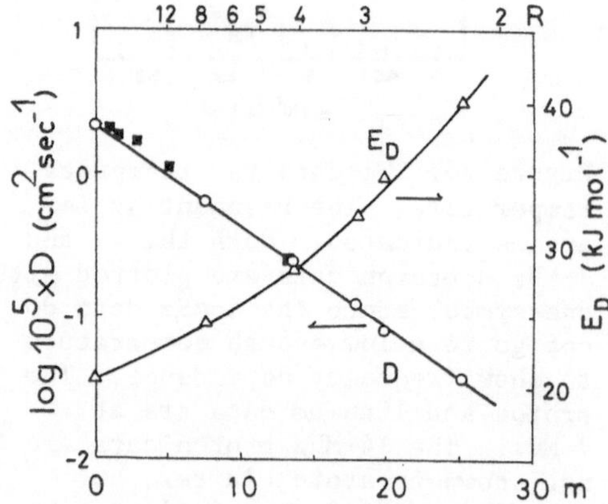

Figure 29. Composition dependence of the self-diffusion coefficient (D) of water at 25°C and its activation energy (E_D). O, Δ : Present work, ■ : McCall and Douglass (23°C).

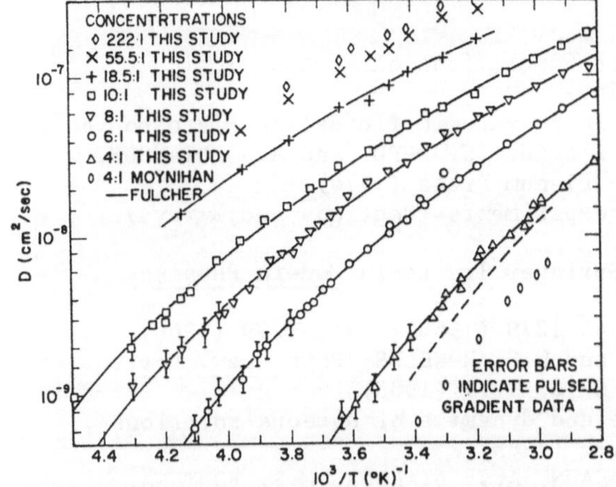

Figure 30. Diffusion constant vs inverse temperature in aqueous calcium nitrate solutions for various concentration.

4. CONCLUSIONS

We have presented here a short review of the application of two powerful spectroscopic techniques to the study of the dynamics of aqueous ionic solution.

The Neutron Scattering technique (quasi-elastic or inelastic scattering) measures directly a correlation function of the displacement vector of the particle. Different kinds of motions can be separated through the momentum transfer variation of the spectrum. It is very well adapted to the study of hydrogeneous samples due to the high value of the incoherent scattering cross-section of the hydrogen. The main drawback is the low flux of neutron scattering and its high cost which makes it available only at selected places around the world.

Nuclear Magnetic Resonance, on the contrary, measures a complicated correlation function, only at the Larmor frequency. The mixing of translational and rotational contributions in the case of molecular liquids is always difficult to disentangle. But NMR is a widely used technique due to its relatively low cost; it has also a very high sensitivity and selectivity.

It is the purpose of spectroscopic techniques to provide data which can, in principle, be understood by a careful Molecular Dynamics simulation. Furthermore, this kind of comparison, on the same systems, will lead to realistic models for the atomic and ionic interactions.

5. REFERENCES

[1] P. BOPP, 'Molecular Dynamics Simulations of Aqueous Ionic Solutions', these proceedings

[2] F. VOLINO, 'Local dynamics in polyatomic fluids' in Microscopic Structure and Dynamics of Liquids, J. DUPUY and A.J. DIANOUX, Eds. NATO-ASI series B33 - Plenum Press (1978)

[3] J.E. ENDERBY, 'Scattering experiments (Neutrons and X-rays)', these proceedings

[4] See eg. : T. SPRINGER in Springer Tracts in Modern Physics 64, (1972)

[5] V.F. SEARS, Can J. Phys. 44, 1279 (1966); 44, 1299 (1966)

[6] N.A. HEWISH, J.E. ENDERBY and W.S. HOWELLS, Phys. Rev. Lett. 48, 756 (1982); J. Phys. C 16, 1777 (1983)

[7] P.S. SALMON 'The structure and dynamics of aqueous solutions', Ph.D. Thesis - Bristol, G.B. (1985)

[8] M.C. BELLISSENT-FUNEL, R. KAHN, A.J. DIANOUX, M.P. FONTANA, G. MAISANO, P. MIGLIARDO and F. WANDERLINGH, Mol. Phys. 52, 1479 (1984)

[9] See eg. : P.A. EGELSTAFF, 'An introduction to the liquid state', Academic Press (1967)

[10] J. TEIXEIRA, M.C. BELLISSENT-FUNEL, S.H. CHEN and A.J. DIANOUX, Phys. Rev. A31, 1913 (1985)

[11] Workshop on Water, 'Structure and dynamics of water and aqueous solutions : anomalies and their possible implications in Biology', J. Physique-Colloque, C7 (1984)

[12] F. VOLINO, M. PINERI, A.J. DIANOUX and A. DE GEYER, J. Pol. Science, Pol. Phys. Ed., 20, 481 (1982)

[13] F. VOLINO and A.J. DIANOUX, Mol. Phys. 41, 271 (1980)

[14] H. BERTAGNOLLI, P. CHIEUX and H.G. HERTZ, Z. Phys. Chem., 135, 125 (1983)

[15] T. SAKUMA, S. HOSHINO and Y. FUJII, J. Phys. Soc. Japan, 46, 617 (1979)

[16] Symposium on structures of Water and Aqueous Solutions, J. Phys. Chem., 74, 3677 (1970)

[17] G.J. SAFFORD, P.S. LEUNG, A.W. NAUMANN and P.C. SCHAFFER, J. Chem. Phys. 50, 4444 (1969)

[18] H. BOUTIN and S. YIP, 'Molecular spectroscopy with neutrons', MIT Press (1968)

[19] P.A. EGELSTAFF and P. SCHOFIELD, Nucl. Sci. Eng. 12, 260 (1962)

[20] G. MAISANO, P. MIGLIARDO, M.P. FONTANA, M.-C. BELLISSENT-FUNEL and A.J. DIANOUX, J. Phys. C 18, 1115 (1985)

[21] R. SHUKER and R.W. GAMON, Phys. Rev. Lett., 25, 222 (1970)

[22] P.A. MADDEN and R.W. IMPEY, Chem. Phys. Lett., 123, 502 (1986)

[23] A.J. DIANOUX, U. BUCHENAU, M. PRAGER and N. NÜCKER, Physica, 138 B, 264 (1986)

[24] A. ELARBY-AOUIZERAT, J.F. JAL, P. CHIEUX, J. DUPUY and A.J. DIANOUX, to be published

[25] S. ALEXANDER, C. LAERMANS, R. ORBACH and H.M. ROSENBERG, Phys. Rev. B 28, 4615 (1983)
B. DERRIDA, R. ORBACH, KIM-WAH-YU, Phys. Rev. B 29, 6645 (1984)

[26] A. ABRAGAM, 'The principles of magnetic resonance', Oxford University Press (1960)

[27] Y. NAKAMURA, S. SHIMOKAWA, K. FUTAMATA and M. SHIMOJI, J. Chem. Phys., 77, 3258 (1982)

[28] C. GIRARD, J. BRAUNSTEIN, A.L. BACARELLA, B.M. BENJAMIN and L.L. BROWN, J. Chem. Phys. 67, 1555 (1977)

[29] E. LANG, W. FINK and H.D. LÜDEMANN in ref. [11], p. 173

[30] J.F. HARMON and E.J. SUTTER, J. Phys. Chem., 82, 1938 (1978)

[31] H.G. HERTZ, J. Chimie Physique, 82, 557 (1985)

[32] N. BODEN and M. MORTIMER, J. Chem. Soc., Faraday II, 74, 353 (1978)

[33] I.C. BAIANU, N. BODEN, D. LIGHTOWLERS and M. MORTIMER, Chem. Phys. Lett., 54, 169 (1978)

[34] F. HIRATA, H.L. FRIEDMAN, M. HOLZ and H.G. HERTZ, J. Chem. Phys. 73, 6031 (1980)

[35] Y. MASUDA, M. SANO and H. YAMATERA, J. Chem. Soc., Faraday I, 81, 127 (1985)

[36] J.R.C. VAN DER MAAREL, D. LANKHORST, J. DE BLEIJSER and J.C. LEYTE, Chem. Phys. Lett., 122, 541 (1985), J. Phys. Chem., 90, 1470 (1986)

[37] P.M. GAMMEL and R. MEISTER, J. Chem. Phys., 64, 4287 (1976)

LIGHT SCATTERING AND RELATED TOPICS

P.A. Madden
Physical Chemistry Laboratory
South Parks Road
Oxford OX1 3QZ
U.K.

§1. INTRODUCTION

The objective of this article is to present a description of the light scattering spectra of water in such a form that the information on molecular motion gained from the spectra, may be usefully compared with information from other sources. The 'other sources' include not only real experiments (quasi- and inelastic neutron scattering and n.m.r.) and computer simulation "experiments" on water itself, but also the ideas gained from studies of other liquids - from which water is physically indistinguishable at the macroscopic level, after all.

The natural formalism for a description which will embrace a number of experiments is expressed in terms of time correlation functions of molecular properties. Whilst the focus of the article is light scattering, closely related problems occur in analysing infrared/dielectric spectra within this formalism and these cases will also be considered, parenthetically.

The time correlation function formalism has not normally been used to analyse light scattering and infrared spectra of water.[1-3] A major reason for this is undoubtedly the great complexity of the spectra, which are dominated by many remarkable features not found in the spectra of non-hydrogen bonded fluids. Classical analyses of water spectra have therefore sought to associate these features with specific hydrogen-bond structures:- "..... the band at - is due to the broken H-bond, that at ....". These analyses may very well be correct and it is not my purpose to dispute them. Rather I wish to emphasise that the focus on the <u>bond</u> between molecules makes the information gained from light scattering and infrared on water difficult to compare with that gained from other experiments, where the observables are naturally expressed as properties of <u>molecules</u>, and with what is known about other fluids, for which a <u>molecular</u> description is appropriate.

A spectral analysis within the correlation function framework consists of two distinct stages. Firstly, the fluid property which the experiment senses (the polarizability or dipole density for light scattering or infrared) must be expressed as a function of the positions and orientations of the molecules. This is an <u>electrodynamic</u> development since it involves a molecular description of the

M.-C. Bellissent-Funel and G. W. Neilson (eds.), The Physics and Chemistry of Aqueous Ionic Solutions, 181–216.

field-matter interaction. Secondly, the amplitude and relaxation characteristics of the fluctuations of this function, as seen in the behaviour of its correlation function, must be understood; these involve the structure and dynamics of the molecules in the fluid. For "normal" fluids, like N_2 or CS_2, the first of these developments is reasonably well established and correlation functions of well-defined functions of molecular positions and orientations have been characterised.[4] For water, however, one might expect severe electrodynamic effects (in the form of additional contributions to the polarizability) from H-bond formation. The required function of molecular positions might then be difficult, or impossible, to obtain and that which has been learnt from studies of normal fluids of little relevance. One can also expect marked effects of the H-bond on the molecular dynamics and, thus, for the water molecule correlation functions to show unique features. In the classical "bond" analysis of water spectra it is unnecessary to make this separation but the price paid is that the spectral features are never identified with well-defined functions of molecular coordinates which might be compared with other experiments.

In this article I shall adopt the (optimistic) viewpoint that there are no specific H-bonding effects on the electrodynamic aspects of the problem and I will produce a certain amount of evidence to corroborate this view. The correlation functions which appear are therefore those of the same well defined functions of molecular properties as occur for normal fluids. On the other hand, the characteristics of the correlation functions themselves are very strongly affected by H-bonding and differ markedly from the "normal" fluid case. I will not be concerned with explaining this remarkable structure here, this undoubtedly requires a theory of H-bond dynamics. Rather, I will try to show where the same dynamical features are seen in other experiments and to contrast the information available from different sources. In order to tie together these different strands I will be heavily dependent on results from computer simulation studies, in this my article will be complementary to that of Bopp.[5]

§1.2 BASICS

A diagram which illustrates the possible types of light scattering experiment is shown in figure 1. In the geometry in which the polarization of the incident and scattered fields are both vertical (VV) "polarized" scattering is observed. Of the other "depolarized" geometries, VH and HV are completely equivalent and they almost always give the same information as HH. The remaining variable is the scattering angle θ; this and the wavelength of the light determine the scattering vector

$$\underline{K} = \underline{K}_i - \underline{K}_s \tag{1}$$

In general,[6] the light scattered at a given angle should be described as the result of diffraction of the incident beam by a periodic disturbance in the polarizability of the fluid characterised by the vector $\underline{K}$ (as illustrated). In practice, the properties of the scattered light only depend on $\underline{K}$ when the frequencies of the incident and scattered light are almost equal (Rayleigh scattering); in this case

$$|\underline{K}_i| = |\underline{K}_s| = \frac{2\pi}{\lambda_i} \tag{2}$$

$$|K| = \frac{4\pi}{\lambda_i} \sin \theta/_2 \tag{3}$$

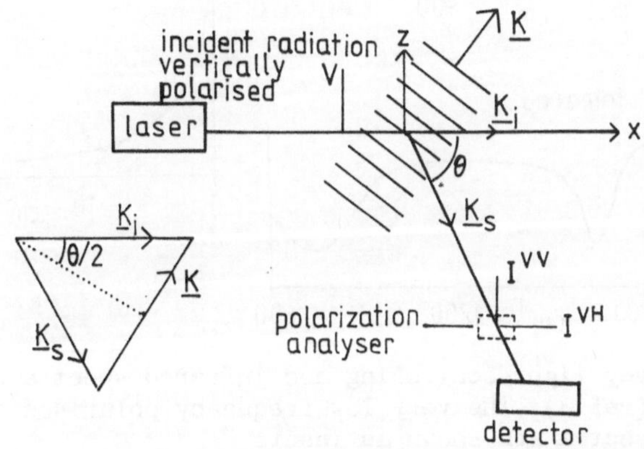

Figure 1 : Light scattering geometry : the vertical direction is parallel to z and $\underline{K}_i$, $\underline{K}_s$ and $\underline{K}$ lie in the xy plane. Light is diffracted from a wave characterised by the wavevector $\underline{K}$.

The shortest wavelength disturbance which may be observed is therefore $\lambda_i/2$, in backscattering. For blue light this is 2500Å, which should be contrasted with the 1Å range of typical neutron scattering experiments.

Survey spectra[3] of pure water in the frequency régime from 0-4000 cm^{-1} are shown in figure 2. (Neutron scatterers may note that 100 cm^{-1} is 12.4meV, n.m.r. practitioners that 1cm^{-1} is 30 GHz and the rest that K_BT at room temperature is 208 cm^{-1}). Except in the very low frequency régime (< 1cm^{-1}) the polarized and depolarized spectra show substantially the same structure, and I have only shown the polarized. The low frequency spectra are shown as insets[7,8] on an expanded frequency scale (these are conventionally referred to as Rayleigh spectra with the rest of the spectrum described as Raman scattering, though the dividing line is really arbitrary).

The low frequency polarized spectrum, shown in the inset[7], is the only region of the spectra of water which shows any dependence on the scattering vector $\underline{K}$, so far as I know. (Some organic liquids show $\underline{K}$-dependence in the depolarized Rayleigh spectrum too[6]). The shift of the two "Brillouin" peaks is proportional to $|K|$ and the widths of the

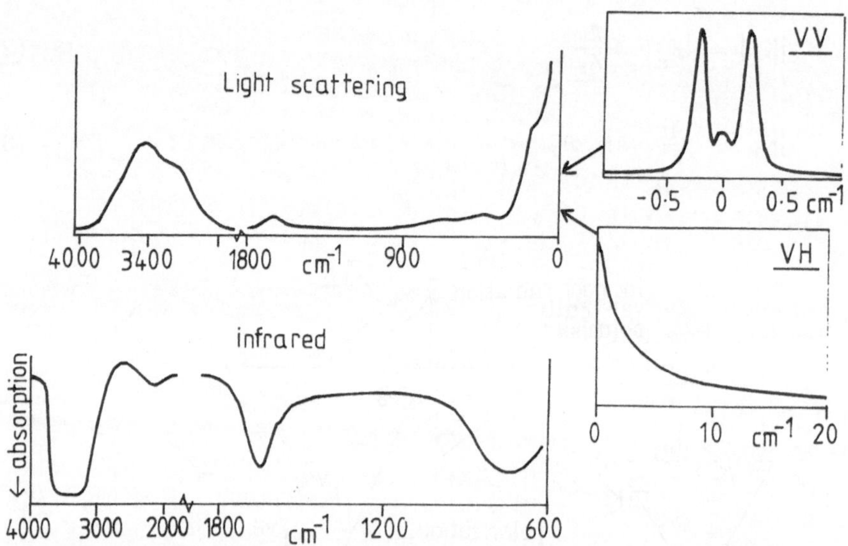

Figure 2 : Survey light scattering and infrared spectra of pure H_2O are shown (ref.3). The very low frequency polarised[7] and depolarised[8] spectra are shown as insets.

lines are proportional to K^2. What this means is that the molecular motions required to relax the periodic disturbance of the polarizability as seen in the <u>other</u> regions of the spectrum do not depend on the wavelength of the disturbance. They must involve local motions of single molecules or groups of molecules over distances small compared to $|K|$. On the other hand, the low frequency polarized spectrum is attributable to a periodic <u>density</u> fluctuation in the fluid. For this to relax, molecules must move from the peaks of the density wave to the troughs and the rate of relaxation depends upon the distance they must travel, i.e. $(K/2\pi)^{-1}$.

Figure 2 also shows the infrared absorption coefficient[3] over the same frequency range as the light scattering spectrum, the distribution of intensity in the two spectra is somewhat different. However, if one were to divide the infrared absorption at frequency ω (in cm^{-1}) by $\omega[1-\exp(-h\omega c/K_BT)]$ - for reasons to be discussed below - one would end up with a spectrum quite similar to the depolarised light scattering one. The information contained in the two spectra is complementary.

The link between the spectral shape and molecular motion is embodied in the formal relationship.[9,10]

$$I(\omega)\ \alpha \left\{\frac{(\omega_L-\omega)^4}{1+\exp(-h\omega/K_BT)}\right\} \mathrm{Re}\int_0^\infty dte^{i\omega t} < \Pi_a(\underline{K},t)\Pi_a^*(\underline{K},0) > \qquad (4)$$

where $I(\omega)$ is the intensity of scattered light with polarization a, at frequency $\omega_L - \omega$ and ω_L is the laser frequency (here in sec^{-1}). $\Pi(\underline{K},t)$ is the spatial Fourier transform of the polarization density.

$$\Pi_a(\underline{K},t) = \int \underline{dr}\ \Pi_a(\underline{r},t)\exp\left[i\underline{K}.\underline{r}\right] \qquad (5)$$

The polarization density is the density of dipoles induced in the sample by the incident electric field. The important part of this expression is that not in curly brackets, the spectrum is related to a time correlation function. This type of relationship is very general (see ref. 11); for example, the infrared absorption coefficient is given by

$$A(\omega)\ \alpha \left\{\frac{h\omega(1-\exp(-h\omega/K_BT))}{n(\omega)[1+\exp(-h\omega/K_BT)]}\right\} \mathrm{Re}\int_0^\infty dte^{i\omega t} < \underline{M}_T(\underline{K},t)\cdot\underline{M}_T^*(\underline{K},0) > \qquad (6)$$

where $\underline{M}_T$ is a component of the fourier transforms of the dipole density (and $n(\omega)$ is the refractive index).

For the rest of the article I will be discussing the time correlation functions but before doing so I would like to mention the factors in curly brackets, for completeness.

The $(\omega_L-\omega)^4$ factor in equation 4 is responsible for the blue appearance of the sea and sky (in Corsica, at least!). The scattered light may be viewed as the electromagnetic radiation emanating from oscillating electric dipoles induced in the sample by the incident electric field and the dependence of the radiation intensity on the oscillation frequency gives rise to this factor (see e.g. ref.10, chapter 1).

$[1 + \exp(-h\omega/K_BT)]^{-1}$ is a "detailed balance" factor. Notice that this leads to a difference of $\exp(-h\omega/K_BT)$ between the intensity of light scattered at $\omega_L + \omega$ ("anti-Stokes scattering"), to the high frequency of the laser, compared to the "Stokes" scattering at $\omega_L-\omega$. In Stokes scattering the radiation field gives up energy $h\omega$ to the fluid or, in quantum terms, the fluid makes a transition from some initial state to one of energy $h\omega$ higher. In anti-Stokes scattering the reverse transition from the higher to lower state occurs, the radiation field gains energy from the fluid. The relative intensity of the two processes then depends upon the relative probability of finding the two initial states which is just the Boltzmann factor $\exp[-h\omega/K_BT]$.

The detailed balance factor in the infrared absorption case is different (equn. 6). The net power absorbed by the sample from the radiation field, which is what is observed, is hω (i.e. the photon energy) time the difference between the rate of stimulated absorption from some lower state to an upper state hω higher in energy and the rate of stimulated emission from the uppper to the lower. This difference is proportional to the difference between the probability of finding the two initial states or to $1-\exp(-h\omega/K_BT)$. Notice that the ratio of detailed balance factors between infrared absorption and light scattering is the $\omega[1-\exp(h\omega/K_BT)]$ factor which I referred to above. The similarity between the infrared and light scattering spectra when this factor is removed, which I mentioned then, reflects a similarity in the behaviour of the polarizability and dipole density correlation functions - it is to the analysis of these which I now turn.

§1.3 REDUCTION TO MOLECULAR PROPERTIES.

The polarization density in a small region of space surrounding the point $\underline{r}$ in the fluid may be written as a sum of contributions from all the molecules in the region i.e.

$$\Pi_a(\underline{r},t) = \sum_i \pi_a^i(t) \quad \delta(\underline{r}-\underline{r}^i(t))$$

$$+\sum_{i,j}{}' \pi_a^{ij}(t)\,\delta(\underline{r}-\underline{r}^i(t)) + \sum_{i,j,k}{}' \pi^{ijk}\ldots\ldots\ldots \qquad (7)$$

Where $\delta(\underline{r}-\underline{r}^i)$ is one if molecule i is within the small region of space around the point $\underline{r}$ and zero otherwise. The first term (in π_a^i) represents the contribution to the polarizability density at $\underline{r}$ from each molecule in the fluid with no other molecules present; I will call it the single molecule contribution. It arises from the polarization of molecule i by the incident electric field

$$\pi_a^i = \alpha_{ab}^i E_b^L$$

where E_b^L is the bth component of the (Laser) field and $\underline{\underline{\alpha}}^i$ is the molecular polarizability tensor. The remaining terms in the equation are "interaction-induced". They arise because the contribution of molecule i to the polarization density is affected by the presence of molecule j (π^{ij}), by the joint influence of molecules j and k (π^{ijk}) and so on. The time dependence of $\Pi(\underline{r},t)$, which is what determines the spectral shape, therefore arises in two ways:- from the translation of molecules into and out of the small region around $\underline{r}$ (i.e. from the time dependence of $\delta(\underline{r} - \underline{r}^i(t))$ and from the intrinsic time dependence of $\underline{\underline{\alpha}}^i$, $\underline{\pi}^{ij}$ etc.

The time dependence of a component of the isolated molecule polarizability (α^i_{ab}) may arise either from the reorientation of the molecule in space or because its conformation changes in time (i.e. the molecule vibrates). The time dependence due to rotation is illustrated for a cylindrical molecule, in figure 3. The polarizability ellipse represents the fact that the molecule is more polarizable along its long axis than perpendicular to it. Consequently when the molecule lies at some general angle (θ) with respect to an electric field $\underline{E}^L$ the induced dipole $\underline{\pi}^i$ is not parallel to the field. If the field were

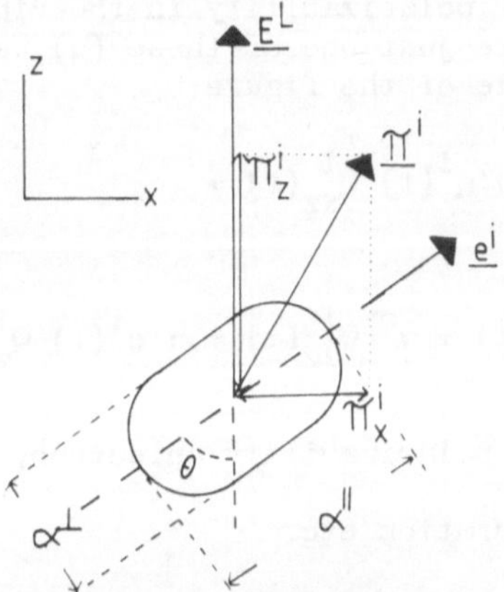

Figure 3 : Illustration of the induction of a dipole moment ($\underline{\pi}^i$) in a cylindrical molecule by an electric field.

oscillating, the induced dipole component parallel to the field (π^i_z) would give rise to an emitted field proportional to π^i_z with a parallel to E^L polarization (polarised scattering) whereas π^i_x would be responsible for depolarised scattering. If at some later time the molecule has reoriented to θ', the magnitudes of both π^i_z and π^i_x are altered. The change in orientation therefore alters the magnitudes of the emitted fields. The time dependent modulation of the emitted fields by the molecular reorientation thereby affects the spectrum of the scattered radiation (as in amplitude modulation in radio). For the illustrated case of the rotation in a plane the dependence of π_z and π_x on the molecular orientation is easily deduced:-

$$\pi_x = \alpha_{xz}E^L = (\alpha^{\parallel}\cos\theta\sin\theta - \alpha^{\perp}\cos\theta\sin\theta)E^L = \gamma E^L \frac{\sin 2\theta}{2} \qquad (8)$$

$$\pi_z = \alpha_{zz}E^L = (\alpha^{\|}\cos^2\theta + \alpha^{\perp}\sin^2\theta)E^L = (\bar{\alpha} + \gamma(\cos^2\theta - {}^1/_3))E^L \quad (9)$$

where γ $(= \alpha^{\|} - \alpha^{\perp})$ is the anisotropy and $\bar{\alpha}$ $(= 1/3\ (\alpha^{\|} + 2\alpha^{\perp}))$ the trace, or isotropic part, of the molecular polarizability. α_{zz} (responsible for the polarized scattering) contains a part, α, which is independent of the molecular orientation and a term in γ which does depend on θ. α_{xz} (responsible for the depolarized scattering) contains only an orientation dependent γ term; we shall see shortly that the time dependences of the γ contributions to α_{zz} and α_{xz} are the same.

The other source of time dependence of the polarizability of an isolated molecule is due to vibration, which may be described by an expansion of the molecular polarizability in the vibrational normal coordinates. If there were just one of these (q) we would have for the simple cylindrical molecule of the figure

$$\alpha^i_{xz}(t) = \gamma^o Q^i_{xz}(t) + \gamma' q^i(t)\ Q^i_{xz}(t) + \ldots. \quad (10)$$

and

$$\alpha^i_{zz}(t) = \bar{\alpha}^o + \bar{\alpha}' q^i(t) + \gamma^o\ Q^i_{zz}(t) + \gamma' q^i(t)\ Q^i_{zz}(t) \ldots\ldots. \quad (11)$$

where γ' is $\frac{\partial\gamma}{\partial q}$ and γ^o the polarizability anisotropy of the molecule in the mean molecular configuration etc.

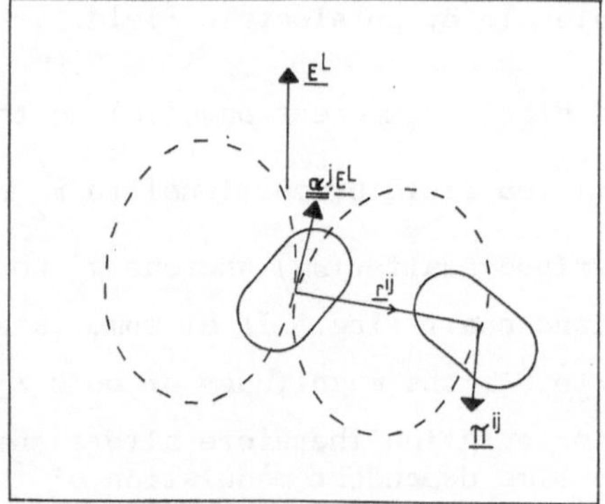

Figure 4 : Illustration of the DID mechanism; the dashed lines represent the lines of force of the field of the dipole induced in molecule j by the laser field.

Thus for the isolated molecule polarizability we are able to express the time dependence as explicit functions of the molecular coordinates. For an asymmetric top molecule with three vibrational normal coordinates, like water, these explicit functions are much more complex[9] than for the simple example above, but no new principles are involved in deriving them.

For the interaction-induced polarization it is not possible to make such a strong assertion until some specific mechanism is identified; general ideas about these terms may be found in refs 12 and 13. When the external field induces a dipole (and higher order multipoles) in one molecule, the field of the induced dipole in turn induces additional dipoles in neighbouring molecules. This additional source of polarization is one of the contributors to the total polarization described by the interaction-induced terms, it is a universal mechanism. Two-body ($\underline{\pi}^{ij}$), three-body ($\underline{\pi}^{ijk}$) and all higher order terms may arise as dipole-induced dipoles themselves induce additional dipoles. The dipole-induced dipoles (DID) for a pair of molecules in a field are illustrated in figure 4. The field due to a dipole ($\underline{\underline{\alpha}}^j.\underline{E}^L$) induced in a molecule j by the external field gives rise to a field given by $\underline{\underline{T}}(\underline{r}-\underline{r}^j).\underline{\underline{\alpha}}^j.\underline{E}^L$ at the point $\underline{r}$, where

$$T_{\alpha\beta}(\underline{r}) = (3\, r_\alpha r_\beta - r^2\delta_{\alpha\beta})/r^5 \tag{12}$$

The dipole induced in molecule i by this field is then

$$\underline{\pi}^{ij} = \underline{\underline{\alpha}}^i . \underline{\underline{T}}(\underline{r}^i - \underline{r}^j).\underline{\underline{\alpha}}^j.\underline{E}^L \tag{13a}$$

which may be written as a pair (DID) polarizability ($\underline{\pi}^{ij} = \underline{\underline{\alpha}}^{ij}.\underline{E}^L$):-

$$\underline{\underline{\alpha}}^{ij}(t) = \underline{\underline{\alpha}}^i(t).\underline{\underline{T}}\,(\underline{r}^i(t)-\underline{r}^j(t)).\underline{\underline{\alpha}}^j(t) \quad \text{(DID)} \tag{13b}$$

This shows that the time dependence of $\underline{\underline{\alpha}}^{ij}$ will arise from the relative translation of the molecules and also from their reorientation with respect to the intermolecular vector and vibration - see equns (10 & 11). Similar explicit functions may be written down for the higher-order terms.

The problem with the interaction-induced terms comes because other, more difficult to specify, mechanisms may also contribute. In particular the distortion of the charge density of one molecule by overlap interactions with its neighbours might be expected to be important. However, studies of fluids such as argon and CS_2 have shown that these terms are, in fact, much smaller than those of the DID type [12,13] and there seems no reason to suspect that water should be different in this respect. More worrying in this case are the possible contributions due to hydrogen-bond formation, about which little is known. Recently though, a number of studies[14,15] have suggested that the spectral features attributable to the interaction-induced polarizability may be understood without reference to an H-bond polarizability and this is the line I will take here. If such non-DID mechanisms are ignored than we are able to specify the functional dependence of the total polarizability on the molecular coordinates (as above) and thereby describe explicitly the source of its time dependence. Restricting the interaction-induced terms to those of DID type "solves" the <u>electrodynamic</u> aspect of the problem; we may now attempt to relate specific functions of the molecular coordinates to spectral features.

Before doing so it is appropriate to comment on the occurrence of similar problems in other types of experiment. The most closely related is infrared/dielectric spectroscopy in which the dipole density caused by the permanent charge distribution of the molecules is observed. In this case there are permanent dipole-induced dipoles (and those induced by other permanent multipoles) which are in many ways quite analogous to the dipole-induced dipole polarizabilities discussed above. The same issue, of whether dipoles attributable to the H-bond occur, must also be raised. In n.m.r. relaxation interaction-induced terms are rarely discussed. For the relaxation induced by magnetic interactions they should be negligible (such as dipole-dipole coupling in proton n.m.r.) but for the mechanisms involving electric fields (such as quadrupole relaxation) they should be significant. In this context they are usually manifest as a change in the quadrupole coupling constant with density; but in spherical systems, and notably ions, the intermolecular effects are the dominant source of relaxation.[16] Neutron scattering experiments are quite insensitive to interaction-induced terms.

§2.1 <u>The "single molecule" contributions to the spectra</u>.

To obtain an explicit equation for the spectrum we must insert our molecular expressions for $\underline{\underline{\Pi}}$ into the formal equation, 3. To actually deal with the real spectrum it is important to include <u>both</u> the single molecule <u>and</u> interaction-induced terms, as we shall see. However, it is convenient, for pedagogical reasons, to proceed by considering the two types of term separately and to deal with their interference later. In this section we shall consider only those dynamical events which can influence the spectrum via the single molecule polarizability.

From this limited viewpoint the depolarized spectrum is determined by (equns. 4 & 10) [for the time being we will stick to the simple cylindrical molecule with only one vibrational normal coordinate]

$$\langle \pi_x(\underline{K},t)\pi_x^*(\underline{K},0)\rangle = (E_z^L)^2 \langle \sum_i (\gamma^o Q_{xz}^i(t) + \gamma' q^i(t)Q_{xz}^i(t))$$

$$\times \sum_j (\gamma^o Q_{xz}^j(0) + \gamma' q^j(0)Q_{xz}^j(0)\exp\left[i\underline{K}.(\underline{r}^i(t)-\underline{r}^j(0))\right] \rangle \qquad (14)$$

This may be simplified, when multiplied out, as q^i is a variable which oscillates about zero, so that any average which contains an odd power of q^i must vanish. This leaves:-

$$\langle \pi_x(\underline{K},t)\pi_x^*(\underline{K},0)\rangle = (\gamma^o)^2 \langle \sum_{i,j} Q_{xz}^i(t)Q_{xz}^j(0)\ \exp[i\underline{K}.(\underline{r}^i(t)-\underline{r}^j(0))]\rangle$$

$$+ (\gamma')^2 \langle \sum_{i,j} Q_{xz}^i(t)Q_{xz}^j(0)q^i(t)q^j(0)\exp[i\underline{K}.(\underline{r}^i(t)-\underline{r}^j(0))]\rangle \qquad (15)$$

The correlation function which determines the polarized spectrum may be evaluated in a similar way. There is an additional simplifying feature here since the average of an odd power of a $Q_{\alpha\beta}^i$ vanishes[9] (try integrating Q_{zz}^i over all possible molecular orientations).
The result is:-

$$\langle \pi_z(\underline{K},t)\pi_z^*(\underline{K},0)\rangle = (\bar{\alpha}^o)^2 \langle \sum_{i,j} \exp\left[i\underline{K}.(\underline{r}^i(t)-\underline{r}^j(0))\right]\rangle$$

$$+ (\bar{\alpha}')^2 \langle \sum_{i,j} q^i(t)q^j(0)\exp\left[i\underline{K}.(\underline{r}^i(t)-\underline{r}^j(0))\right]\rangle$$

$$+ (\gamma^o)^2 \langle \sum_{i,j} Q_{zz}^i(t)Q_{zz}^j(0)\exp\left[i\underline{K}.(\underline{r}^i(t)-\underline{r}^j(0))\right]\rangle$$

$$+ (\gamma')^2 \langle \sum_{i,j} Q_{zz}^i(t)Q_{zz}^j(0)q^i(t)q^j(0)\exp\left[i\underline{K}.(\underline{r}^i(t)-\underline{r}^j(0))\right]\rangle \qquad (16)$$

The spectra therefore appear to involve a complex combination of molecular rotation, vibration and translation; in fact it is possible to go some way to separate the effects of the different types of motion.

Firstly we note that correlation functions which contain the vibrational coordinate q contain a factor which oscillates at a (very high) molecular vibrational frequency, ω_v, (i.e. $q^i(t)q^i(0) \sim \exp(i\omega_v t)$ so that these correlation functions contribute only to the high frequency spectra ($\omega > 1000\mathrm{cm}^{-1}$ in figure 2). The other correlation functions only contribute to the "low" frequency region ($\omega < 1000\mathrm{cm}^{-1}$ for water). Secondly, we note that a particular pair of molecules i and j, separated by $\underline{r}^{ij}$, may only make a contribution to an average involving $Q^i_{\alpha\beta}Q^j_{\alpha\beta}$ and/or q^iq^j, if the orientations and/or vibrational coordinates of the molecules are <u>correlated</u>; that is, if one molecule having a particular value for these coordinates systematically influences the values taken by the other. It is likely that molecules close to each other in the fluid are correlated, for these $\exp i\underline{K}.\underline{r}^{ij}$ is very close to unity. For distant molecules in normal fluids $Q^i_{\alpha\beta}Q^j_{\alpha\beta}$ vanishes on average (not true in liquid crystals). For all CFs with the appropriate Q or q factors then, we may replace the exponential factor by unity (see the footnote in reference 17).

When this simplification is made it is possible to show straightforwardly (see e.g. refs. 6 & 9) that the orientational CFs in the polarized and depolarized spectra are proportional to each other. We then have (refer to equns 15 & 16)

$$I^{VH}(\omega) = I^{ISO}(\omega) + {}^4\!/_3\, I^{ANIS}(\omega) \tag{17}$$

$$I^{VH}(\omega) = I^{ANIS}(\omega) \tag{18}$$

where, in the "low" frequency region for water

$$I^{ISO}(\omega)\ \alpha \int_0^\infty dt^{i\omega t}(\overline{\alpha}^o)^2 \left\langle \sum_{i,j}\exp\left[i\underline{K}\cdot(\underline{r}^i(t)-\underline{r}^j(0))\right]\right\rangle \quad \ldots.(\omega<1000\mathrm{cm}^{-1},\mathrm{SMO}) \tag{19}$$

$$I^{ANIS}(\omega)\ \alpha \int_0^\infty dte^{i\omega t}(\gamma^o)^2\left\langle \sum_{ij} Q^i_{xz}(t)Q^j_{xz}(0)\right\rangle \ldots.(\omega<1000\mathrm{cm}^{-1},\mathrm{SMO}) \tag{20}$$

Comparison of equations (17-20) shows that by recording and combining VV and VH spectra in the appropriate way we may separately determine CFs involving rotational and translational motion (but recall too that these are only the single molecule terms, hence SMO!). Similarly at high frequency a spectrum which involves only the vibrational phase appears:-

$$I^{ISO}(\omega)\ \alpha \int_0^\infty dt e^{i\omega t}(\overline{\alpha'})^2 \langle \sum_{ij} q^i(t)q^j(0) \rangle \ldots (\omega > 1000\text{cm}^{-1}, \text{SMO}) \qquad (21)$$

$$I^{ANIS}(\omega)\ \alpha \int_0^\infty dt e^{i\omega t}(\overline{\gamma'})^2 \langle \sum_{ij} Q^i_{xz}(t)Q^j_{xz}(0)q^i(t)q^j(0) \rangle$$

$$(\omega > 1000\text{cm}^{-1}; \text{SMO}) \qquad (22)$$

We will now consider what may be said about each of these correlation functions for water.

2.2 SMO Contributions to the Low Frequency Isotropic Spectrum

The shape of the isotropic Rayleigh spectrum is determined by the CF of $\alpha^o \sum_i \exp i\underline{K}.\underline{r}^i$ (equn. 19). This variable is the 'density wave', to which I referred in the introduction

$$\sum_i \exp(i\underline{K}\cdot\underline{r}^i) = \int d\underline{r} \exp(i\underline{K}\cdot\underline{r}) \sum_i \delta(\underline{r}-\underline{r}^i) = \int d\underline{r}\ \exp(i\underline{K}\cdot\underline{r})n(\underline{r}) \qquad (23)$$

where in the last step $\sum \delta(\underline{r}-\underline{r}^i)$ is identified as the density at $\underline{r}$, since it simply involves counting the number of molecules in a small region, around $\underline{r}$. Precisely the same CF (denoted $S(\underline{K},\omega)$ is seen in coherent neutron scattering but the important difference is in the size of $|\underline{K}|$, which determines the range over which the correlation in density fluctuations is sensed. In neutron scattering $|K|^{-1}$ is typically 1Å so that the atomic granularity of the fluid is very evident, in light scattering $|K|^{-1}$ might be 5000 Å so that the observed density correlations are transmitted via many molecules. On this large distance scale it is unnecessary, in a first approximation, to recognise the molecular nature of the fluid; the density is a conserved, hydrodynamic variable and its relaxation may be described by the equations of continuum hydrodynamics. This leads to the predicted spectrum (refs. 6, 9 & 10)

$$I^{ISO}(\omega) = S(K)\left\{(1-C_V/C_P)\frac{K^2D_T}{\omega^2 + (K^2D_T)^2} + \frac{C_V}{2C_P}\left[\frac{\Gamma K^2}{(\omega \pm V_S K)^2 + (\Gamma K^2)}\right]\right\} \qquad (24)$$

which consists of a central (ω=0) Lorentzian and two symmetrically displaced (at $\omega = \pm V_S K$) "Brillouin" lines. The central component arises from the coupling of the density fluctuations to temperature fluctuations and the width D_TK^2 reflects the relaxation time of the latter (D_T is the thermal diffusivity). The Brillouin lines reflect the coupling of the density fluctuations to sound waves, the shifts $\pm V_SK$ (V_S is the sound velocity) can be regarded as due to a Doppler shift of the light by sound waves of wavevector $\pm \underline{K}$. The widths of these lines, ΓK^2, is the sound wave absorption coefficient, it is primarily determined by the viscosity. The relative intensities of the central and Brillouin lines (the "Landau-Placzek ratio") is determined by the ratio of heat capacities C_P/C_V. $S(K)$ is the structure factor of the fluid; at the values of K of interest in light scattering it is K independent and given by the compressibility of the liquid.[9,10]

Nothing in what has been said above is peculiar to water, all liquids are alike as seen by hydrodynamics! A good example of the analysis of a Rayleigh-Brillouin spectrum for a "normal" fluid (CCl_4) is given in ref. 10 section 8.3. Even at this level though, water proves exceptional! Equation (24) shows that the central line vanishes if $C_P/C_V = 1$, but this is precisely what happens at 4^oC, the location of the density maximum. It was this prediction which was under examination in the remarkable recent work by Maisano et al[7], from which the inset in figure 2 was abstracted. The work is remarkable because perfectly dust-free water must be used, as even a trace of dust will give a spurious spike in the spectrum at zero frequency.

Detailed examination of R-B spectra shows that the simple hydrodynamic description is inadequate in two ways. The Brillouin frequency V_SK is typically of order 3GHz ($\sim 1/10cm^{-1}$) so that the fluid is probed on a timescale of order $5x10^{-10}s$. Although this is a long time compared to molecular collision times (say $10^{-13}s$) it is not long compared to all possible molecular relaxation times. In particular in water, especially in the supercooled region, slow structural relaxations may occur.[18] Simple hydrodynamics then fails as specific molecular events start to influence the spectral shape. This is accounted for phenomenologically by allowing the transport coefficients Γ and D_T to become frequency dependent.[9] The result is that the spectrum gains an additional central component (the "Mountain line"), the Brillouin lines broaden and become asymmetric and the Laudau-Placzek ratio is affected. All three effects are detected in the study of Maisano et al[7]; they provide potentially important information about supercooling.

Secondly, the detailed analysis shows evidence of interaction-induced effects. These do not strongly affect the spectral shape but do affect the distribution of intensity in the spectrum; in particular Maisano et al show that the predicted divergence of the Laudau-Placzek ratio is avoided because of interaction-induced effects.

2.3 The Low Frequency Anisotropic Spectrum.

The orientational CF (equn.20) which determines the anisotropic (or depolarized) low frequency spectrum ($\omega<1000$ cm^{-1} for water) in the single molecule polarizability picture is a "collective" CF[4], in that it involves not only the correlation of the orientation of one molecule with its own orientation at an earlier time (i=j) but also with the previous orientations of other molecules (i≠j). For the purposes of this discussion I wish only to consider the self orientational CFs, such as

$$C_2(t) = \langle Q^i_{xz}(t)\, Q^i_{xz}(0). \rangle \tag{25}$$

It can be shown (straightforwardly, e.g. refs. 9 and 19) for the model cylindrical molecule we have been using for illustration that

$$\langle Q^i_{xz}(t) Q^i_{xz} \rangle = \langle P_2(\underline{e}^i_z(t)\ \underline{e}^i_z(0)) \rangle \tag{26}$$

where $\underline{e}^i_z$ is a unit vector along the long axis of the molecule. For the rest of the article x,y and z will refer to molecular, rather than laboratory, axes. The self orientational CF is just the average value of the second order Legendre polynomial of the cosine of the angle swept out by this axis in time t (the subscript two on C_2 refers to the rank of the Legendre function).

Theoretical work and experimental studies on "normal" liquids suggest that we will not miss any new contributions to the spectra by considering only self terms. That is, if the self CF can be represented by some functional form then the corresponding collective CF can also be represented in that way but the functional may contain parameters which take different values. For example,[19,20] if the single particle CF decays exponentially ($C_2(t) = \exp(-t/\tau)$) then the collective CF will also decay exponentially - but with the relaxation time $\tau_{2,c}$.

$$\tau_{2,c} \cong g_2 \tau_2 \tag{27}$$

where g_2 is a parameter which measures the equal time orientational correlation between molecules.

The reason for making this simplification at the outset is because even the self CFs of water molecules are exceedingly complex. So far we have referred to a model cylindrical molecule, for which the polarizability tensor components in the laboratory frame are specified when only the long axis orientation (e^i_z) is given. Water is an asymmetric top, however; for it all that diagonal elements of the polarizability tensor in the molecular frame are unequal. To specify the polarizability in the laboratory frame the orientation of <u>all three</u> molecular axes ($\underline{e}^i_x$, $\underline{e}^i_y$ and $\underline{e}^i_z$, see figure 5) must be known. The depolarized light scattering spectrum is then expressed (ignoring collective effects) as a combination of orientational CFs for each of these molecular axes[21]

$$I^{ANIS}(\omega) \cong \int_0^\infty dt \exp(i\omega t)\left\{\gamma_x^2 C_2^x(t) + \gamma_y^2 C_2^y(t) + \gamma_z^2 C_2^z(t)\right\}$$

$$(\omega < 1000\text{cm}^{-2}, \text{SMO}) \qquad (28)$$

where the weighting factors are determined by the anisotropy in the polarizability tensor and C_2^x is $\langle P_2(\underline{e}_x^i(t)\cdot\underline{e}_x^i(0))\rangle$. Before considering the forms of these CFs, let us examine how related information occurs in other experiments.

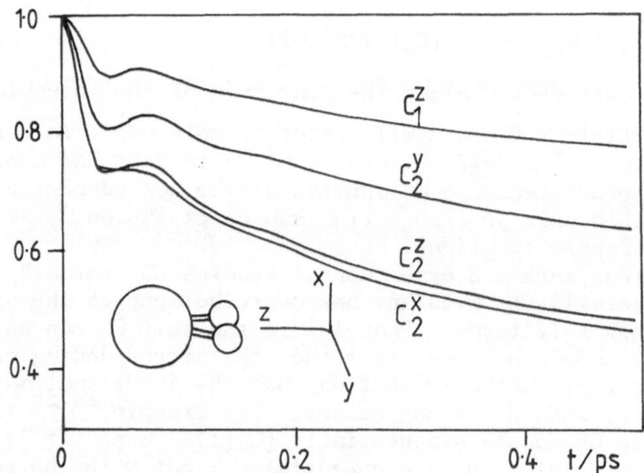

Figure 5 : The orientational correlation functions of unit vectors fixed along the molecular x,y and z directions (ref. 26).

In proton magnetic resonance of H_2O[22] the principal relaxation mechanism is intramolecular dipole-dipole coupling. The relaxation time is then determined by the reorientational correlation function of the interproton vector

$$T_1^{-1} = D \int_0^\infty dt\; C_2^y(t) \qquad \text{dipole-dipole} \qquad (29)$$

where D is a known constant. For a spin >1/2 nucleus, the quadrupole coupling provides another mechanism. For the deuteron in D_2O the reorientation of the O-D bond may be observed

$$T_1^{-1} = E \int_0^\infty dt \left\{ a\, C_2^z(t) + b\, C_2^y(t) \right\} \tag{30}$$

Note that in n.m.r. only the integral of a CF is found, not the spectrum. Neither of the n.m.r. mechanisms give the same combination of CFs as the light scattering; of course, this doesn't matter if the reorientation about all axes is the same. N.m.r. genuinely gives self correlation functions whereas light scattering is, in reality, collective. This, as we have seen, may mean that relaxation times obtained from the two experiments may differ - even for exponential decay, see equn (27).

In incoherent, inelastic neutron scattering[23] one may determine a correlation function involving the position of a single proton,

$$S_S(\underline{K},\omega)\ \alpha \int_0^\infty dt\ \exp(i\omega t) \left\langle \exp \underline{iK}\left[(\underline{r}_H^i(t) - r_H^i(0)) \right] \right\rangle \tag{31}$$

This may be rewritten by expanding the proton position about the centre of mass

$$\underline{r}_H^i(t) = \underline{r}^i(t) + H_z \underline{e}_z^i(t) + H\underline{e}_y^i(t) \tag{32}$$

so that

$$\exp(i\underline{K}\cdot\Delta\underline{r}_H^i(t)) = \exp\left[i\underline{K}\cdot\Delta\underline{r}^i(t) \right] \exp\left[i\ H_z \Delta(\underline{K}\cdot\underline{e}_z^i)(t) \right] \times \exp\left[i\ H_y \Delta(\underline{K}\cdot\underline{e}_y^i)(t) \right] \tag{33}$$

(where $\Delta(a(t))$ implies the change in a in the time t). This may be expresed in terms of orientational and translational CFs by assuming that the different motions are uncorrelated and formally expanding each exponential - but it is already clear that there is far too much information here for comparison with other experiments at this level of detail. I will refer to a more rough and ready scheme later.

The most closely related information comes from dielectric and far-infrared absorption measurements, though this does not mean that comparison with light scattering is straightforward. I-I effects occur in both spectra and a collective correlation function is observed in both cases, with consequences which are difficult to assess at a quantitive level.[24] At the single-molecule only, self CF level the dielectric/far-i.r. spectrum is given by the first-rank orientational CF of the dipole axis, i.e.

$$A(\omega) \cong \omega \tanh \left(\frac{h\omega}{2K_BT}\right) \int_0^\infty dte^{\,i\omega t}\, C_1^z(t) \qquad (34)$$

where

$$C_1^z(t) = \langle (\underline{e}_z^i(t)\cdot\underline{e}_z^i(0) \rangle \qquad (35)$$

In the Debye, rotational diffusion model the functions C_1^z and C_2^z are both predicted to be exponential and the decay times are such that

$$\tau_1^z = 3\,\tau_2^z \quad \text{(rotnl. diffn)} \qquad (36)$$

In general such a close relationship will not be found.

These considerations have been put forward to bring out the point that because of I-I effects, collective vs. self CFs and the asymmetric structure of the water molecule quantitative comparisons of the information obtained from different experiments is extremely difficult An alternative approach to the purely experimental is via computer simulation. Provided an experimental observable can be expressed in terms of CFs - no matter how complex - it may be calculated in a simulation (i.e. the electrodynamic problem must first be solved). A simulation may be refined so as to be able to reproduce several observations; it may then be interrogated to give more readily interpreted quantities than the observables themselves and an understanding of the interrelationships of different experimental results. Some examples of this approach have been discussed in a recent review:[25] the principles have only partially been followed in the simulation studies I will describe below.

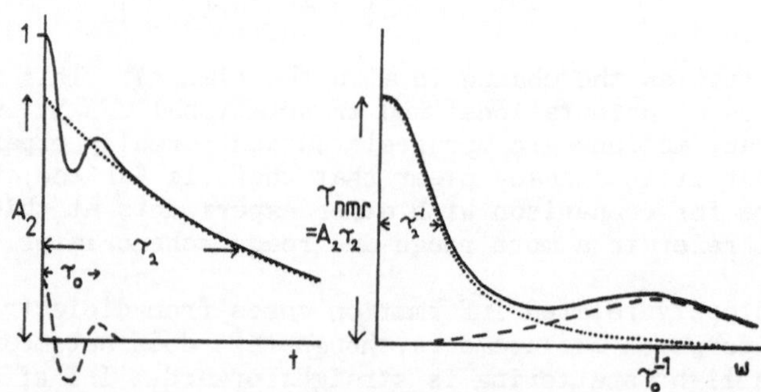

Figure 6 : Fourier transformation of a function of the form $A_2\exp(-t/\tau_2) + \varphi(t)$ representing the orientational correlation function shown in figure 5.

The orientational CFs of equn. (28) from a simulation of water (details in ref.26) are shown in figure 5. They indicate that the orientational motion of the different axes is not the same. The exponential relaxation at long times is typical of many fluids, the exponential may be characterised by the decay times τ_2^{α} (α = x,y,z). More remarkable is the short-time "glitch", which is only seen in the c.f.s of hydrogen-bonded liquids. It is the signature of the librational oscillation of the water molecules, which is found to have a characteristic amplitude of ~22°.[26]

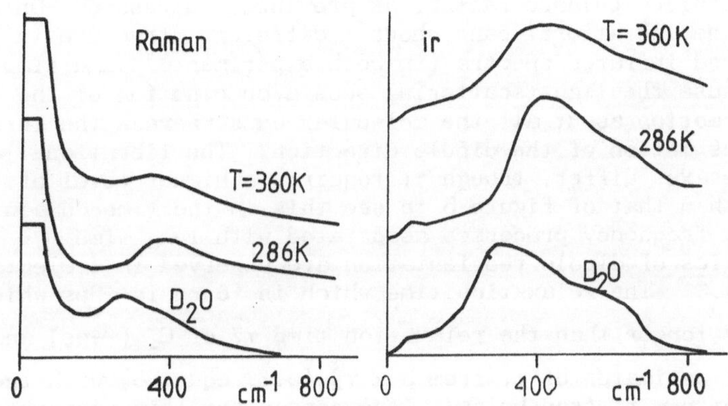

Figure 7 : Raman and infrared spectra calculated from simulations[26].

To see how CFs of this shape are associated with the spectral shape of figure 2, the process of Fourier transformation is illustrated in figure 6. The full function may be resolved into an exponential (decay time τ_2^{α}), which transforms into a Lorentzian centred at zero of width $(\tau_2^{\alpha})^{-1}$, and a rapidly decaying, oscillatory glitch-function, which transforms into a band shifted to the oscillation frequency with a width given by the glitch damping rate. The actual transforms of such simulated CFs (equn.27) bear some resemblance to the experimental spectra of figure 2. The high frequency glitch-structure may be associated with the librational[11] bands seen in the region 400-800cm^{-1}? of the real spectrum (fig.2). It exhibits the same isotope shift and characteristic changes with temperature as the observed librational band, though the characteristic frequency is too low - which reflects a deficiency of the simulation model. The zero frequency lorentzian is related to the low frequency structure seen in the depolarized Rayleigh spectrum inset in figure 2; its width is some combination of τ_2^x, τ_2^y and τ_2^z and therefore estimated to be ~(2.5ps)$^{-1}$ at 290K. Identifying such a feature in the observed spectrum is not straightforward, there is a lorentzian of roughly the right width but there are also other contributors to the spectrum not predicted in the SMO picture. Further, the observed relaxation time is collective and may differ from the relaxation of the self correlation functions. We shall return to this point below.

The relationship of the far-infrared absorption to the correlation function C_1^z (i.e. equn.34) may be investigated in a similar way. C_1^z is also shown in figure 4; it has the same shape as the C_2^α functions, in particular the glitch appears at the same time. Consequently the spectra of the two functions are very similar. The gross difference between the infrared and light scattering spectrum is due to the different detailed balance factor, as previously stressed. On closer inspection the librational band shows a different structure in light scattering and infrared spectra (in both experiment[1-3] and simulation). This is because the light scattering sees a combination of the librational motion about all the molecular axes whereas the infrared sees only the motion of the dipole direction. The librational motions of the three axes differ, though it requires a higher level of resolution than that of figure 5 to see this in the time domain.

The low frequency processes associated with long time characteristics of dipole reorientation are observed in dielectric spectroscopy.[27] The relaxation time which is found is ~9ps which is considerably longer than the relaxation time τ_1^z of C_1^z (~4ps) and experimental estimates of τ_1^z from $3 \times \tau_2^\alpha$ (c.f. equn.36) with the latter obtained from n.m.r. (see below). The reason for this apparent discrepancy is undoubtedly the collective nature of the dipole relaxation observed in dielectric relaxation, as in equn. 27 we expect[24]

$$\tau_{1,c}^z \cong g_1 \tau_1^z \tag{37}$$

The orientational correlation parameter may be estimated from the static dielectric constant[24] and a value of 3 seems not unreasonable for water.[27]

The relationship between the orientational CFs and the n.m.r. relaxation time is also illustrated in figure 6. Since T_1 is related to the integral (or transform at zero frequency) of the CF, it is sensitive only to the exponentially decaying part of C_2^y (equn.29). The time integral is only equal to τ_2^y if C_2^y is perfectly exponential; in water, the presence of the glitch means the integral is somewhat lower.[26]

As already indicated it is unlikely to be profitable to compare inelastic neutron spectra with other observables at the level of orientational correlation functions. A more illuminating, but less quantitative, scheme is suggested by the following. We note, integrating by parts, that (see also Dianoux's lecture[23])

$$\frac{\omega^2}{K^2} S_S(\underline{K},\omega) = \frac{1}{K^2}\int_0^\infty dt e^{i\omega t} \frac{\partial^2}{\partial t^2} \langle \exp\left[i\underline{K}\cdot(\underline{r}_H^i(t)-\underline{r}_H^i(0)\right]\rangle$$

$$= \int_0^\infty dt e^{i\omega t} \langle V_{H,K}^i(t)\, V_{H,K}^i(0) \exp\left[i\underline{K}\cdot(\underline{r}_H^i(t)-\underline{r}_H^i(0))\right]\rangle \qquad (38)$$

Where $V_{H,K}^i$ is the component of the proton velocity along $\underline{K}$ (the second equality involves switching the time derivative in the CF ref. 28). If the limit $K \to 0$ is taken (by performing a series of experiments and extrapolating to zero momentum transfer) the exponential term makes no contribution and

$$\lim_{K\to 0} \frac{\omega^2}{K^2} S_S(\underline{K},\omega) = \int_0^\infty dt e^{i\omega t} \langle V_{H,K}^i(t)\, V_{H,K}^i(0)\rangle. \qquad (39)$$

The proton velocity may be resolved into a centre of mass and an angular motion (cf. equn.32)

$$\underline{V}_H^i = \dot{\underline{r}}_H^i = \underline{V}^i(t) + \underline{\Omega}^i(t) \times (H_z\underline{e}_z^i(t) + H_y\underline{e}_y^i(t)) \qquad (40)$$

where $\underline{\Omega}^i$ is the angular velocity of molecule i and $\underline{V}^i$ its centre-of-mass velocity (i.e. $\dot{\underline{r}}^i$). So we see (schematically) that

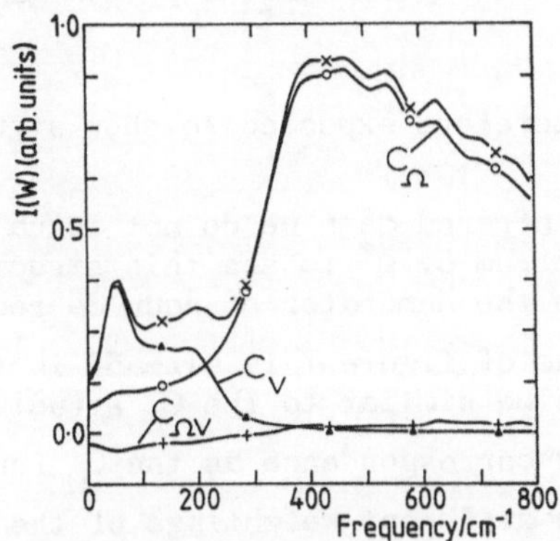

Figure 8 : The spectral density of the proton velocity and its resolution into spectra associated with translation and rotation.

$$\lim_{K\to 0} \frac{\omega^2}{K^2} S_S(\underline{K},\omega) = \int_0^\infty dt e^{i\omega t} \left\{ a\, C_V(t) + bC_\Omega(t) + cC_{\Omega\, V}(t) \right\} \quad (41)$$

Where C_V is the velocity CF, the C_Ω a CF involving the angular velocity and angular position ($\underline{e}^i_\alpha$), $C_{\Omega\, V}$ the cross term between the two types of velocity and a, b and c are coefficients determined by the molecular geometry. The spectrum of the proton velocity from the simulation is shown in figure 8, its shape is confirmed by the limited experimental data which is available.[29] There is a broad, intense band with much subsidiary structure from about 300cm^{-1} (~35meV) upwards and weaker features at 60cm^{-1} (8meV) and 200cm^{-1} (25meV). The figure also shows the resolution of the full spectrum into the three component spectra of equation (41). It can be seen that the broad high frequency band is associated with rotational motion through C_Ω, and the 60 and 200 cm^{-1} peaks with translational motion through C_V; the cross term is (fortunately) negligible.

The information in the orientational correlation functions (and therefore in the light scattering and far infrared spectra) can be represented in the same way. Consider, for example, the spectrum of C_1^z (equn.35), other orientational functions give similar results:-

$$\omega^2 \int_0^\infty dt e^{i\omega t} C_1^z(t) = \int_0^\infty dt e^{i\omega t} \left\langle \frac{\partial}{\partial t} \underline{e}^i_z(t) . \frac{\partial}{\partial t} \underline{e}^i_z(0) \right\rangle$$

$$= \int_0^\infty dt e^{i\omega t} \left\langle \underline{\Omega}^i(t) \times \underline{e}^i_z(t) \cdot \underline{\Omega}^i(0) \times \underline{e}^i_z(0) \right\rangle \quad (42)$$

This reduced spectrum is therefore expected to show structure characteristic of C_Ω.

To deal with the far-infrared data we do not actually have to go so far as multiply the spectrum by ω^2 to see this structure. The $\omega(1-\exp(-h\omega/K_BT))$ factor in the numerator of equn 34 reduces to ω^2 as $T \to \infty$, so that the spectrum of figure 7 is already in the appropriate form. It is indeed seen to be similar to the C_Ω structure in figure 8. We cannot expect a detailed correspondence as the C_Ω functions in equations 41 and 42 contain different weightings of the angular velocity components along each molecular axis.

Light scattering spectra have to be explicitly reduced (multiplied by ω^2) to bring them to the appropriate form. Some workers[30] prefer to use $\omega(1-\exp(-h\omega/K_BT))$ as the reducing factor for the (specious) reason that if the liquid were a harmonic solid (!) this would be the Bose population factor for phonons of frequency ω - this generates the so-called $R(\omega)$ representation. The $R(\omega)$ spectrum from the simulated C_2^αs is very similar to the far infrared spectrum of figure 7 and I have not shown it separately. Instead in figure 9 I have shown experimental $R(\omega)$ spectra from reference 30; the focus of attention here is on peaks at ~60cm^{-1} and 200cm^{-1} <u>which are not present</u> in the simulated functions (but are also observed in experimental far infrared spectra. We know (from neutron studies c.f. figure 8) that the origin of these features is translational, yet the single molecule polarizability picture presents no mechanism whereby translation may influence depolarized light scattering spectra. We return to this in the next section; for the present we simply note how readily the reduced representation has enabled us to compare these spectra.

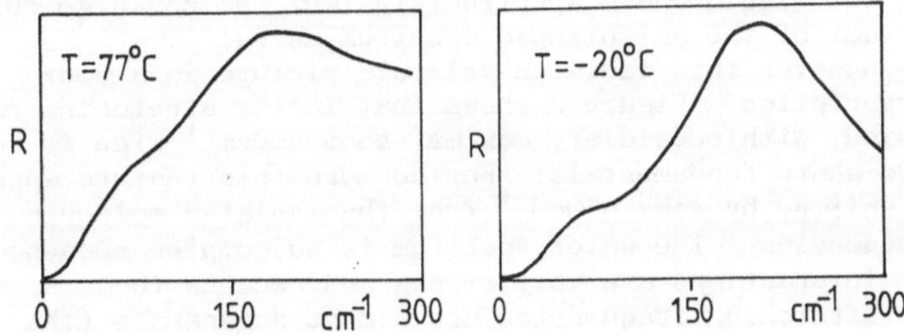

Figure 9 : Experimental $R(\omega)$ spectra for H_2O (ref 30).

§2.4 <u>The High Frequency Region</u>.

The appropriate correlation functions for a discussion of the high frequency region (ω>1000cm^{-1}) are of the kind appearing in equations 21 and 22 (within the single molecule polarizability picture). In fact I shall only deal with the isotropic spectrum, which depends simply on the vibrational phase; this will already present many problems without the additional orientation dependent factor present in equation (22).

Equation 21 was obtained under the simplifying assumption that there was only one vibrational degree of freedom. In fact water has three and equation 21 must be generalised accordingly

$$I^{ISO}(\omega) \; \alpha \; \mathrm{Re}\int_0^\infty dt \; e^{i\omega t} \sum_{a,b} \varpi^a \varpi^b \langle \sum_{ij} q_a^i(t) q_b^j(0) \rangle \tag{43}$$

where q_a^i is an appropriate coordinate for describing the molecular deformation and ϖ^a the associated derivative of the isotropic polarizability. For the isolated water molecule the appropriate coordinates are normal coordinates: the symmetric stretch q_1 at frequency ω_1 = 3832cm^{-1}, the asymmetric stretch q_2 at ω_2 = 3943cm^{-1} and the bend q_3 at ω_3 = 1648cm^{-1}. For non-interacting molecules i and j are uncorrelated and the normal coordinates on a given molecule are also independent so that

$$I^{ISO}(\omega) \propto \mathrm{Re}\int_0^\infty dt\, e^{i\omega t} N\sum (\alpha^a)^2 \langle q_a^i(t) q_a^i(0) \rangle$$

$$\propto \mathrm{Re}\int_0^\infty dt e^{i\omega t} \sum_a (\alpha^a)^2 \langle (q_a^i)^2 \rangle \cos(\omega_a t) \qquad (44)$$

which is just three sharp lines at the isolated molecule frequencies. For most "normal" liquids this would also be a good first-order description of the liquid phase spectrum; in fact, we could go further and make sense out of the anisotropic spectrum as well.

For water however this isolated molecule picture is a poor first-order description. Figure 2 shows that in the stretching region only a broad band, with subsidiary maxima, some $400cm^{-1}$ wide is seen, rather than two sharp fundamentals. Furthermore this feature appears at $\sim 3400cm^{-1}$, shifted by some $450cm^{-1}$ from the isolated molecule stretching frequencies. The water spectrum is so complex because the intermolecular interactions are very strong and because the intramolecular stretching frequencies are almost degenerate (the situation is compounded by the fact that the overtone of the bend, at $2\omega_3$, is also nearly degenerate with the two stretches). Not only are these three resonances of a single molecule strongly coupled to each other by the intermolecular interactions (and also by intramolecular effects such as coriolis coupling) but they are also coupled to the resonances of neighbouring molecules by the hydrogen bonding interactions ("resonance transfer"). In these circumstances there is no hope of simplifying equation 43 by neglecting the coupling between different modes ($a \neq b$) or different molecules ($i \neq j$). We note that precisely the same problems will occur in interpreting infrared or inelastic neutron scattering data on the stretching region.

Happily, a means exist whereby the complexities caused by inter and intramolecular coupling may be avoided. In the HDO[31] molecule the O-D stretching vibration lies at $2824cm^{-1}$ (isolated molecule) and the O-H stretch at $3890cm^{-1}$. Consequently the O-D stretch of a single HDO molecule in H_2O has no nearby (inter or intramolecular) resonances with which it may be coupled by the intermolecular interactions. The O-H stretch of a single HDO in D_2O has the same property. Therefore by studying dilute solutions of HDO in H_2O or D_2O the spectrum of an isolated O-H bond caused by the intermolecular interactions in water may be isolated.

Such spectra are sketched in figure 10. They are still extremely broad and strongly shifted from the isolated molecule resonance positions. The relative breadth and shift of the two bands may be understood through the fact that the proton makes larger amplitude oscillations than the deuteron, so that the OH bond is most strongly influenced by interactions with other molecules. The shape of the lines is quite complex, and appears as the superposition of two peaks.

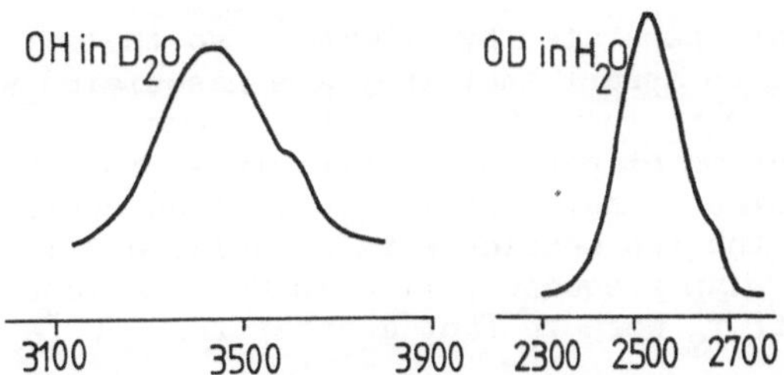

Figure 10 : Spectra of isolated O-H and O-D stretching motion[1].

The relative intensity of the peaks changes with temperature, though the total intensity of the band is temperature independent (at least for T>20°C). It is found that the lineshape may be described as

$$I(\omega)=(1-K(T))A(\omega) + K(T)B(\omega) \tag{45}$$

where A and B are temperature independent subspectra.[1,32] This form is consistent with the occurrence of an isosbestic frequency; that is, a frequency at which the spectra at many different temperatures pass through a point. An isosbestic frequency is also seen in the spectra of pure water.[32] The natural interpretation of equation 45 is that water molecules may be in one of two states, with characteristic spectra A and B and that K(T) is fixed by an equilibrium between these two states. The higher frequency band is due to "free" or "open" water, with a smaller degree of hydrogen bonding. It becomes most intense as the temperature is raised.

Much of the literature on the spectroscopy (and other properties) of water is dominated by a discussion of this and more complex, multi species schemes. There are two principal objections to the introduction of such ideas as a basis for a general understanding of the properties of water. The "species" tend to be identified by the value of some particular type of observation; for example in spectroscopy - by a high or low value for a characteristic frequency. Yet several structures might yield the same value for this property but totally different values for some other observable - so that the interpretation of one experiment would yield nothing of value for another. Secondly, each class of observation is associated with a particular timescale. Whether or not a "species" is identified in an experiment depends upon the relationship between the timescale of the experiment and that of the structural relaxation which causes the species to interconvert. In spectroscopy if two species are associated with characteristic frequencies which differ by $\Delta\omega$ then they will only be identifiable as distinct peaks if they interconvert on a timescale (τ_S) such that

$$\Delta\omega_V > \tau_S^{-1} \tag{46}$$

From the vibrational spectrum of O-H in D_2O we see that the peaks of the A and B subspectra are separated by ~200cm^{-1}, so that it is consistent with the data to assert that they are associated with two species with lifetimes longer than 10^{-13}s. This seems a reasonable lower bound for the lifetime of particular structures in a hydrogen bonding fluid. On the other hand, this interpretation could not be confirmed by n.m.r.; if the two species were associated with even an enormous difference in Larmor frequency, like 10MHz, they could only be seen as distinct peaks if τ_S were in fact greater than 10^{-8}s.

The most direct way to justify a multispecies model for liquid water would be to identify features observed in the atom-atom distribution function, obtained by direct neutron and X-ray diffraction measurements, with distinct intermolecular structures. Then to show that these structural entities could be identified with the species whose presence is inferred from more indirect observations, such as the interpretation of the vibrational spectrum. As I have already stressed, computer simulations have a lot to offer in bringing together results from such complex observations. The structure of water has been extensively examined in simulations using potential models in which the intramolecular distances are held rigid. These have shown that a hydrogen bond can be identified using an energetic criterion[33] (i.e. a hydrogen bond is considered formed between molecules A and B if the potential energy of their interaction is less than a critical parameter). It has also been shown that the number of hydrogen bonds, identified in this way,[33] accords well with the coordination number under the first peak of the O-H distribution function. That is, the formed H-bond has a distinct structural signature. However, there is no distinct feature associated with the broken hydrogen bond. The H-bond may assume a range of geometries, the range of the O-H-O angle being of order 60°. The distribution narrows and the peak moves towards 180° as the temperature is lowered, consistent with a more ice-like structure for the water.[33]

The rigid model simulations cannot help with understanding the vibrational spectrum. For this a flexible model must be used so that the spectrum may be obtained, either from the correlation functions of the internal coordinates of the molecule, as in equn.(43), or by some other means. (Note that ideally the correlation functions must be calculated quantum mechanically - a formidable task). Since work of this kind will be discussed by Bopp[5], I shall simply summarise a few of the findings of a paper by Reimers and Watts[34] - which should be regarded as a brave but incomplete attempt to calculate the spectrum.

In the most interesting calculations carried out by these authors, in their so-called local mode analysis, the inter- and intra- molecular couplings are greatly simplified so that it becomes most appropriate to regard the calculated spectra as a result of superimposing the spectra of three independent resonances, in particular, with no contribution from intermolecular resonance transfer. Nevertheless, the calculated spectra have many properties in common with those observed. In particular, the occurrence of the large shift from the gas-phase resonance position, the large width of the band, the two-peak structure and the isosbestic frequency are all reproduced. These findings point to the anharmonicity of the intramolecular potential as the dominant factor in determining the lineshape, with a lesser role for intermolecular couplings than is commonly supposed. Reimers and Watts[34] went on to examine whether the two peaks (corresponding to the A and B subspectra of the experimental data) could be associated with distinct intermolecular structures seen in the simulation. Their conclusion on this point was negative - the pair interaction energy, O-H-O bond angle, O•••H bondlength all varied systematically and continuously with frequency and the two peaks could not be associated with distinct structures. The findings support the continuous random network ideas of the water structure rather than simplified ideas based on a small number of hydrogen bonded species.

§3. Interaction-Induced Effects.

So far I have discussed only those contributions to the spectra which arise from the first term in equation 7, in this section I will consider the consequences of the existence of interaction-induced (I-I) contributions to the polarizability density. These consquences can be resolved into two broad camps:- the I-I polarizability may lead to new spectral features due to motions which are not spectrally active in the single molecule polarizability picture; secondly, the I-I polarizability may alter the intensities of spectral features whose basic characteristics are attributable to the single molecule polarizability - this may cause a redistribution of intensity across the spectrum and thereby alter its shape. A good deal is known about these effects in "normal" fluids and they have been discussed in review articles.[12,13] An example of a "new feature" is the Raman scattering associated with u-symmetry vibrations of a molecule which possesses an inversion centre.[12] Alterations of band intensities have long been discussed in terms of "local field factors"; more recently a molecular picture, in which the I-I effects are explicitly recognised, has been found quantitatively useful.[35]

§3.2 New (I-I) Features in Water.

As I have already remarked, the single molecule polarizability picture has a number of shortcomings in dealing with the light scattering spectrum in the low frequency (< 1000cm^{-1}) régime; the absence of 60cm^{-1} and 200cm^{-1} translational features has been noted as well as extra contributions in the depolarized Rayleigh spectrum. Here I will examine the types of spectral contribution which arise from the simple pair DID polarizability (equn. 13). To understand the role of this term it is instructive to begin by noting that the anisotropy in the molecular polarizability tensor of water is very small (typically 10% of the isotropy, ref. 31 gives α_{xx} = 1.162 α_{yy} = 1.069 and α_{zz} = 1.279 Å^3). Consequently, to a good approximation, we replace the $\underline{\underline{\alpha}}^{i}$ factors in equn. 13 by the simple scalar mean polarizability, i.e.

$$\underline{\underline{\alpha}}^{ij}(t) = (\bar{\alpha})^2 \underline{\underline{T}} (\underline{r}^i(t) - \underline{r}^j(t)) \tag{47}$$

This term now simply depends upon the relative position of molecules (not their orientation). It is now easy to see that $\underline{\underline{\alpha}}^{ij}$, in this approximation, is traceless and thus contributes only to the anisotropic light scattering spectrum (the neglected anisotropy in the molecular polarizability can lead to isotropic scattering). The importance of the DID scattering in water is caused, in part, by the smallness of the single molecule contribution to the anisotropic scattering. If as noted above the molecular anisotropy is roughly 1/10 of α which itself is ~1.1 $Å^3$, we see that the DID polarizability due to a neighbouring pair in water (separation ~2.8 Å) is comparable to the molecular anisotropy, i.e.

$$\alpha_{zz} \sim \frac{1}{10}\bar{\alpha}(= \frac{1.1}{10}) \sim (\bar{\alpha})^2 T_{zz} \sim 2\frac{(1.1)^2}{(2.8)^3} \tag{48}$$

This consideration suggests that the DID mechanism might be responsible for a large part of the total intensity of the anisotropic spectrum.

In a fluid of spherical molecules there is no molecular anisotropy and the anisotropic light scattering spectrum is determined solely by the correlation function

$$\langle \sum_{ij} T_{\alpha\beta} (\underline{r}^{ij}(t)) \sum_{k,l} T_{\alpha\beta}(\underline{r}^{kl}(0)) \rangle \tag{49}$$

which involves the relative translation of molecules. The spectrum is reasonably well understood.[12,13] For liquid rare gases, for example, the spectrum is basically exponential $\exp(-\omega/\omega_0)$ and broad. ω_0 is expected to scale roughly as $(K_B T/m\sigma^2)^{1/2}$, where σ is the interatomic separation and m the mass and is about $25 cm^{-1}$ for liquid argon. At low frequencies there is an additional Lorentzian whose width is much more state dependent (like the diffusion coeffi ient) and which is about $7cm^{-1}$ wide for argon. The low frequency feature reflects a cooperative relaxation of the local structure of the fluid whereas the higher frequency exponential is associated with the elementary translational motions of the atoms within their "cage".

In taking these ideas over to water, we must bear in mind a number of complicating factors. Firstly, we must allow for the simultaneous presence of the molecular and I-I effects. Although, as argued above, the DID integrated intensity might be larger than the single molecule contribution, it is likely to be distrbuted over a broad spectrum and cannot be assumed to dominate in every region. The full correlation function which must be considered is

$$< \sum_i (\gamma\, Q^i_{\alpha\beta}(t) + \alpha^2 \sum_j T(\underline{r}^{ij}(t)) \sum_k (\gamma\, Q^k_{\alpha\beta}(0) + \alpha^2 \sum_l T(\underline{r}^{kl}(0)) > \qquad (50)$$

The autocorrelation function of the orientation variables ($Q_{\alpha\beta}$) we discussed in 2.3; as we saw there it will contribute a low frequency lorentzian whose width (determined by the molecular reorientation time) is likely to be comparable to that of the low frequency feature anticipated from the translational term (equn. (49)). Thus, separating the two features is always likely prove difficult. Secondly, the full CF contains an interference or cross term between the $\underline{\underline{Q}}$ and $\underline{\underline{T}}$ variables; the characteristics of this term cannot be deduced readily (see next section). Finally, as we have already noted, the translational motion of the water molecules is likely to be significantly different from that of argon, due to the effect of the H-bonding interactions. Therefore, even the translational correlation function above, equn. (49), might exhibit novel behaviour. In particular, we may expect the low frequency structure due to the cooperative structural relaxation to differ in character (especially in the supercooled region) and also the pronounced oscillatory nature of the translational motion apparent in the velocity CFs to show up in the high frequency wings.

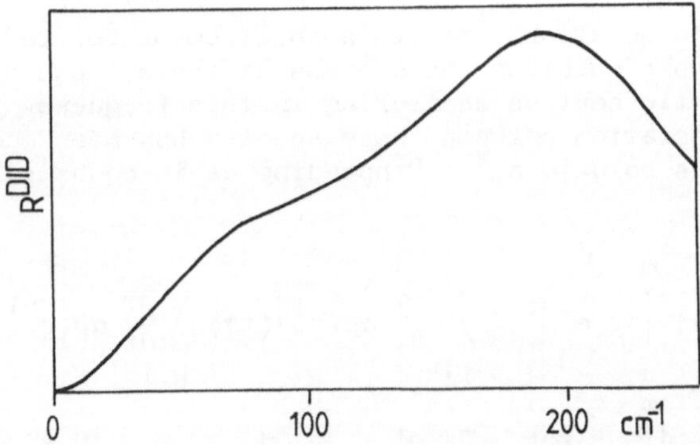

Figure 11 : Simulated R(ω) spectrum from the DID mechanism[14].

This last point has been examined in simulation studies of the translational CF[14] (equn. 49). The other issues could also be addressed by simulating the full CF (equn. 50), but this has not yet been done. The spectrum is found to be basically exponential as in argon, with a decay rate (ω_0) roughly that expected from scaling the inert gas data, as described above. There are, however, significant differences from argon as anticipated. The low frequency spectrum is relatively intense, showing that the character of the structural relaxation differs, and very high frequency translational shoulders are apparent. These may be made more obvious by displaying the simulated spectrum in the R(ω) representation, as is done in figure 11. It may be seen that the R(ω) DID spectrum exhibits peaks at $60cm^{-1}$ and $200cm^{-1}$, as were present in the experimental spectrum (figure 9) but absent from the spectrum calculated on the single molecule polarizability picture.

From the above findings it may be asserted that the DID mechanism generates a spectrum with the right kind of shape and intensity to account for the light scattering spectrum of water between about $20cm^{-1}$ and the onset of the librational region at about $450cm^{-1}$. Furthermore, this mechanism makes an important contribution in the low frequency region which cannot yet be fully characterised. The importance of the DID mechanism is consistent with the high depolarization ratio (~0.72) of the water spectrum between 5 and $500cm^{-1}$. It is difficult (for me) to see why an interaction-induced polarizability which originates in hydrogen-bond-induced electrodynamic effects should generate a depolarized spectrum. Very recently, the small isotropic spectrum from $20\text{-}1000cm^{-1}$ has been studied[15] very carefully by M.A. Ricci and co-workers. It is here that any electrodynamic effect of H-bonding should be most obvious. She has compared the isotropic spectrum of H_2O with that of H_2S (which is considered not to form H-bonds) and shown that the two spectra may be scaled (using the scaling factor discussed above) so as to coincide. The spectrum of H_2S itself may be quantitatively understood on the basis of DID and extended DID models. It would appear from this work that the assumption that the H-bond has little influence on the electrodynamics may be well-founded.

The appearance of the $60cm^{-1}$ and $200cm^{-1}$ peaks in the R(ω) spectrum prompts the question - can this spectrum be related to that of the velocity CF, whose spectrum shows peaks at these frequencies and hence to the inelastic neutron scattering in this frequency range (c.f. equn. 41). An association between these spectra has been made in concentrated aqueous solutions.[23] Proceeding as in equns (42) et seq. we see that

$$R^{DID}(\omega) \;\alpha\; (\bar{\alpha})^4 \int_0^\infty dt\, e^{i\omega t} \left\langle {\sum_{ij}}' \frac{\partial T}{\partial t}\, \alpha\beta(\underline{r}^{ij}(t)) \sum_{k,l} \frac{\partial T}{\partial t}\, \alpha\beta(\underline{r}^{kl}(0)) \right\rangle$$

$$\alpha\; (\bar{\alpha})^4 \int_0^\infty dt\, e^{i\omega t} \left\langle \sum_{ij} (\underline{v}^i(t) - \underline{v}^j(t)).\, \nabla T_{\alpha\beta}(\underline{r}^{ij}(t)) \right.$$

$$\left. \times \sum_{kl} (v^k(0) - \underline{v}^l(0)).\, \nabla T_{\alpha\beta}(\underline{r}^{kl}(0)) \right\rangle \qquad (51)$$

can thus be seen that $R^{DID}(\omega)$ shows the relative velocity of pairs of particles which must be near neighbours (otherwise the ∇T(rij) factor, which is proportional to $(r^{ij})^{-4}$ would be small). To pass from here to the velocity CF we must make drastic assumptions.[14] We must assume that the particle velocities are uncorrelated from the configurational variables, that these configurational variables relax much more slowly than the velocities and that the velocities of different particles are uncorrelated with each other. In my view this makes it extremely unlikely that the light scattering and neutron scattering can be usefully compared at a quantitative level. What we do learn, qualitatively, is that oscillations exhibited by the relative velocity of near neighbours occur at the same frequencies as seen in the single particle motion - which gives a stimulating insight into the nature of interparticle motions.

As we noted in section 1.3 a similar permanent dipole-induced dipole mechanism can contribute to the dielectric and far infrared spectrum of water. If we repeat the order of magnitude estimates though we find that the induced dipole ($\alpha T\mu$) for a pair of near neighbours is only likely to be about one tenth ($\sim\frac{2\bar{\alpha}}{r3}$) of the single molecule permanent dipole. We might therefore expect the I-I effects to be less all-persuasive in the dielectric case and this does seem to be the case. In particular the low frequency dielectric régime reflects molecular reorientation (alone). However, the detailed balance factor in the far infrared absorption coefficient renders high frequency phenomena more prominent and easy to observe than in the light scattering spectrum. The translational $60cm^{-1}$ and $200cm^{-1}$ bands are observed,[37] their presence may be accounted for by a similar analysis to that given above.[14]

§3.3. Redistribution of I-I Intensity.

So far we have examined the spectral features which arise from the autocorrelation functions of the single molecule and interaction -induced polarizabilities; in general, however, there is also an interference or cross term (as in equn. (50)). The presence of this greatly complicates the attribution of a feature to the SMO or I-I category. The influence of the cross term on the shapes of light-scattering and dielectric spectra of "normal" liquids has been extensively investigated.[38] The general conclusion from this work is that in the cross CF features observed in both the SMO and I-I auto-CFs are seen with no new distinctive structure. The effect on the spectrum is thus to redistribute intensity in the spectrum and thereby to alter its appearance from that predicted by superimposing the spectra of the two auto-CFs alone.

Under a particular condition - in which a <u>timescale separation</u> between the relaxation of the single molecule and interaction-induced auto-CFs holds - it is possible to go beyond these general considerations and regain a quantitative description of the spectra.[14] The idea may be illustrated by considering the contribution to the polarization of a given molecule (Π), which includes the isolated molecule (Π^{IM}) and I-I terms (Π^{II})

$$\Pi = \Pi^{IM} + \Pi^{II} \tag{52}$$

The spectrum is related to the CF $\langle \Pi(t)\Pi \rangle$ and, so far, we have been looking at the two ACFs $\langle \Pi^{IM}(t)\Pi^{IM} \rangle$ and $\langle \Pi^{II}(t)\Pi^{II} \rangle$. However, a more useful representation to understand the full correlation function might be:-

$$\Pi(t) = (1 + f)\Pi^{IM}(t) + \delta\Pi^{II}(t) \tag{53}$$

where

$$f = \langle \Pi^{IM}\Pi^{II} \rangle / \langle \Pi^{IM}\Pi^{IM} \rangle \tag{54}$$

and

$$\delta\Pi^{II}(t) = \Pi^{II}(t) - f\,\Pi^{IM}(t). \tag{55}$$

f is the "projection" of the I-I term onto the molecular one.[39] The value of Π^{IM} is fixed by the configuration of the chosen molecule (say, by its orientation) Π^{II}, on the other hand, is determined by the instantaneous positions of the neighbours of that molecule with respect to it. However, since some relative positions are favoured, if we average Π^{II} over all neighbour configurations for a fixed configuration of the chosen molecule we get a non-zero answer (see e.g. ref. 14, page 447) - this is the quantity $f\Pi^{IM}$. Thus in equation (53) this average II term has been lumped in with the single-molecule contribution and simultaneously removed from $\delta\Pi^{II}$. This redistribution will be useful if when the orientation (say) of the central molecule changes (which occurs on the characteristic timescale of the relaxation of Π^{IM}) the neighbours of the chosen molecule <u>rapidly</u> return to their mean position about it. The mean II term then will adiabatically follow the molecular reorientation and be described by $f\Pi^{IM}$, i.e. it will behave as an additional contribution to the molecular polarizability. The removal of $f\Pi^{IM}$ from $\delta\Pi^{II}$ simultaneously will mean that any slow motion relating to the molecular reorientation which is present in Π^{II} will be removed from $\delta\Pi^{II}$, i.e. the auto-correlation function of the latter will only represent the rapid structural relaxation about the central molecule. The scenario is described by saying that there is a timescale separation between the single molecule and structural relaxations.

It can be shown that if the timescale separation does indeed hold then

$$\langle \Pi(t)\Pi \rangle \sim (1 + f)^2 \langle \Pi^{IM}(t)\Pi^{IM} \rangle + \langle \delta\Pi^{II}(t)\delta\Pi^{II} \rangle \tag{56}$$

i.e. the cross-term between $\delta\Pi^{II}$ and Π^{IM} may be neglected. The spectrum then consists of a superposition of the single molecule only spectrum - but with the intensity altered by the $(1+f)^2$ factor - and the spectrum of structural relaxation. The $(1+f)^2$ factor is a "local-field" factor or a contribution to an effective molecular polarizability. This type of description has been shown to be useful in some cases.[40]

One region of the water spectrum to which the timescale separation scenario should certainly apply is to the very low frequency isotropic Rayleigh-Brillouin spectrum, which was discussed in section 2.2. This is confined to frequencies less than 0.2cm^{-1}, corresponding to times longer than 10^{-11} secs, whereas we expect structural relaxation to be at least an order of magnitude faster than this (though in the supercooled region this may cease to be the case). Rather than describe the spectrum as the isotropic isolated molecule plus I-I terms, as in section 2, i.e.

$$\Pi = \sum_i (\bar{\alpha}^o + \sum_j \bar{\alpha}^{ij}(t))\ \delta(\underline{r}-\underline{r}^i(t))E^L(\underline{r}) \tag{57}$$

the above considerations suggest

$$\Pi(\underline{r},t) = \sum \bar{\alpha}^o(1 + f)\ \delta(\underline{r}-\underline{r}^i(t))E^L(\underline{r}) + \sum \delta\alpha^{ij}(t)\ \delta(\underline{r}-\underline{r}^i(t))E^L(\underline{r}) \tag{58}$$

Since the relaxation of the second term (structural) is rapid it will only contribute a flat background to the spectrum at the low frequencies of interest. On the simple isolated molecule picture of section 2.2 (i.e. neglecting f) the relaxation of the first term was shown to be just that of the density fluctuations of the fluid. However, the presence of f may now cause an additional dependence on the temperature fluctuations since the correlation function which determines f reflects the intermolecular structure of the fluid, which will be altered by a change in temperature. The hydrodynamic prediction discussed in section 2.2 may be generalised to allow for the effect of the temperature fluctuations.[7] No new spectral features arise but intensity is shifted between the central Rayleigh and Brillouin lines. This is sufficient to cause the divergence of the Landau-Placzek ratio to be avoided, as has been carefully documented by Maisano et al.[7]

I have discussed this example in some detail because it is related to recent observations and illustrates the principles involved. However such redistributions of intensity may be expected to influence lineshapes in other spectral regions. In particular, Reimers and Watts[34] in their study of the vibrational spectra draw attention to large effects of this kind.

§4. Concluding Remarks.

With regard to the title of this book what I have said above should be regarded as merely an introduction! - to the subject of light scattering in solutions. However, I have recently written a review article on spectroscopic studies of ionic solutions, at least in so far as computer simulations are relevant to these studies. I will content myself then with a reference to this work[41] but draw attention to the fact that a number of articles in a similar vein have been appeared, by Heinzinger, Bopp and co-workers,[42] since it was first prepared. Furthermore my article does not consider the important subject of vibrational spectroscopy in aqueous solutions.

REFERENCES

1. F. Franks (ed.), '*Water - A Comprehensive Treatise*' **Vol. 1** (Plenum, N.Y. 1972) - see article by G.E. Walrafen, pg. 151.

2. B.E. Conway, *Ionic Hydration in Chemistry and Biophysics*, (Elsevier, N.Y. 1981).

3. D. Eisenberg and W. Kauzman, *The Structure and Properties of Water* (Oxford U.P., 1969).

4. D. Kivelson and P.A. Madden, *Ann. Rev. Phys. Chem.* **31**, 523 (1980).

5. P. Bopp - in this volume.

6. B. Berne and R. Pecora, *Dynamic Light Scattering* (Wiley, N.Y. 1976).

7. G. Maisano et al. *Mol. Phys.* 57, 1083 (1986).

8. C. Montrose et al. *J. Chem. Phys.* 60, 5025 (1974).

9. C.H. Wang, *Spectroscopy of Condensed Media*, (Academic Press, N.Y. 1985).

10. W.H. Flygare, *Molecular Structure and Dynamics* (Prentice-Hall, N.J., 1978).

11. R.G. Gordon, *Adv. Magn. Reson.* **3**, 1 (1968).

12. A.J. Barnes, et al (eds) *Molecular Liquids* (Reidel, Dordrecht 1984) - see article by P.A. Madden on pg 431.

13. G. Birnbaum (ed) *Phenomena Induced by Intermolecular Interactions* (Plenum, N.Y., 1985).

14. P.A. Madden and R.W. Impey, *Chem. Phys. Lett*, **123**, 502 (1986).

15. A. de Santis et al 'Raman Spectra of Water in the Translational Region' *Chem. Phys. Lett.* - to be published.

16. H. Versmold, *Mol. Phys.* 57, 201 (1986).

17. In some fluids long-range correlations between molecular orientations are induced *via* coupling to shear modes[9]. In such a case, the K dependence of the correlation functions cannot be ignored and the VH and HH spectra are not equivalent.

18. C.A. Angell, *Ann. Rev. Phys. Chem*, **34**, 583 (1983).

19. Ref. 12 - see article by W.A. Steele, pg. 111.

20. T. Keyes and D. Kivelson, *J. Chem. Phys.*, **56**, 1057 (1972); see also ref. 4.

21. Equation 28 is only approximate; coupling between the reorientation of different molecular axes may occur, see R.M. Lynden-Bell in ref. 12.

22. Ref. 12 - see article by H. Versmold on pg. 309; see also Dianoux - this volume.

23. F. Volino and A.J. Dianoux, Chapter 2 in *Organic Liquids* eds. A.D. Buckingham et al (Wiley, N.Y. 1978); see also Dianoux - this volume.

24. P.A. Madden and D. Kivelson, *Adv. Chem. Phys.*, **56**, 467 (1984).

25. P.A. Madden, 'Simulation of Properties of Spectroscopic Interest' in *Enrico Fermi Summer School in Phys. No.* 97 eds. G. Ciccotti and W. Hoover (North-Holland, 1986).

26. R.W. Impey et al *Mol. Phys.* **46**, 513 (1982).

27. Reference 1 - see article by J.B. Hasted on pg. 255.

28. J.-P. Hansen - this volume.

29. Reference 1 - see article by D.I. Page on pg. 333.

30. S. Krishnamurthy et al *J. Chem. Phys.* **79**, 5863 (1983).

31. Reference 1 - see article by C.W. Kern and M. Karplus on pg. 21.

32. G. D'Arrigo et al *J. Chem. Phys.* **75**, 4264 (1981).

33. W.L. Jorgensen and W.L. Madura, *Mol.Phys.* **56**, 1381 (1985).

34. J.R. Reimers and R.O. Watts, *Chem. Phys.* **91**, 201 (1984).

35. Ref. 13 - see article by B. Ladanyi.

36. A. de Lorenzi et al. *Phys. Rev.* **A33**, 3900 (1986).

37. J.B. Hasted et al *Chem. Phys. Lett.*, **118**, 622 (1985).

38. Ref.13 - see article by P.A. Madden.

39. T. Keyes et al *J. Chem. Phys.* 55, 4096 (1971).

40. P.A. Madden and D. Tildesley, *Mol. Phys.*, 55, 969 (1985).

41. P.A. Madden '*Dynamics of Coordinated Water*' in Proceedings of N.Y. Acad. Sci. eds D. Beveridge and W.L. Jorgensen - to be published.

42. Gy. I. Szasz and K. Heinzinger, *J. Chem. Phys.*, 79, 3467 (1983).

MOLECULAR DYNAMICS SIMULATIONS OF AQUEOUS IONIC SOLUTIONS

Philippe Bopp
Institut für physikalische Chemie
Technische Hochschule Darmstadt
Petersenstrasse 20
D-6100 Darmstadt , FRG

ABSTRACT. The general principles of Molecular Dynamics (MD) computer simulations are briefly sketched. Models used in simulations of aqueous ionic solutions are reviewed and static and dynamic results obtained from such simulations are discussed and compared with experimental ones.

1. INTRODUCTION

Simulation studies have contributed a great deal to the progress in the understanding of liquid systems in the past decade. Contrary to the other states of the matter, where the typical interparticle interactions are either strong (solids), weak (gases) or altogether negligible (ideal gas), the liquid state has resisted rigorous theoretical treatment for a long time. The intermediate strength of the interactions and the resulting short to medium range order render the application of the approximations necessary for a traditional theoretical treatment very difficult. Due to the additional long range electric interactions the situation is particularly complicated in liquids containing ions in finite concentrations.

With the emergence of powerful computers in the late 60's and in the 70's a vast range of problems connected with real liquids became tractable with simulation methods, which had been developed about 10 years before (1,2) and tested with simple systems. The field has been expanding at a tremendous speed since then and applications have emerged in almost every field of physics, chemistry, the biosciences, the engineering sciences and many other fields. Because of the underlying approach, which is in philosophy often close to the experimental approach, simulations have also sometimes been called 'computer experiments'.

It is the purpose of this lecture to give a brief overview of the applications of the Molecular Dynamics (MD) computer simulation method to the study of the structural and dynamic properties of aqueous ionic solutions. Of course time and space do not allow a halfways complete review or discussion of the developments in all the many aspects of this field. It is also impossible to cite all the relevant literature to a satisfactory extent. I have attempted to present recent results; the reader interested in the genesis of the field or in more detailed discussions should turn to the cited papers and use them as guides to the aspects neglected here.

M.-C. Bellissent-Funel and G. W. Neilson (eds.), The Physics and Chemistry of Aqueous Ionic Solutions, 217–243.

In the first section of this lecture the MD method itself and the basic underlying assumptions will be briefly reviewed. We shall then discuss some of the 'models' used to represent the water molecule and the ions in such simulations. In the third chapter we shall study the structural properties with special attention given to the phenomenon of ionic hydration. This information will be compared with experimental evidence, mostly from X-ray and neutron scattering experiments. Remarks about the distribution of the interaction energies in the liquid will lead to a brief discussion of quantum effects and in the last part of this section a few results of MD simulations of anisotropic systems will be briefly described.

The subsequent chapters will focus on the dynamic properties. The translational, librational, and in the case of simulations with so called 'flexible' models also the vibrational motions of the ions and molecules will be discussed in terms of autocorrelation functions and their Fourier transforms (power spectra). Single ion values for the self-diffusion coefficients, intramolecular frequency shifts due to specific solute-solvent interactions can for instance be determined from the simulation and compared with the ones obtained experimentally.

2. THE MOLECULAR DYNAMICS COMPUTER SIMULATION METHOD

The aim of an MD simulation is to compute macroscopic properties of a chemical system assuming essentially that the microscopic interaction potentials are known. More specifically the following assumptions are made:

- The atoms or molecules constituting the chemical system are represented (modelled) by point masses or rigid bodies subject to classical mechanics.
- All interaction potentials between these masses and/or rigid bodies are known.
- The classical trajectories of the point masses and rigid bodies under the influence of their mutual forces and torques are representative of the motions of the atoms and molecules in the chemical system.
- Statistical averages taken over the ensemble of phase space points of the trajectories over a sufficient length of time reproduce the structural and dynamic properties of the system.

For the simulation of a continuous (or bulk) liquid a few more assumption have to be made:

1. The liquid is represented by a finite number, usually a few hundred, point masses or rigid bodies located in a regular cell, usually a cube. The cell and all the particles replicate themselves infinitely in all directions. This construction avoids surface effects which would otherwise dominate the behaviour of such a small sample (vide infra). The walls of the cell are transparent and particles can move freely between the cell and its periodic replica, thus keeping the density in the cell constant. Particles in the replica cells are often called 'mirror particles', and as they are replica of particles in the cell, their motions are the same and do not have to be computed. This construction is called 'periodic boundary conditions ' and is illustrated for a simple two dimensional case in FIG.1. Once the number of particles that one wants to use in a simulation is known, the volume of the cell is determined by the desired density of the system. The dependency of the simulated results on the number of particles used and thus on the cell size has been of great concern and has been studied; a few examples will be shown here.

2. The 'minimum distance convention' is applied. This is a prescription how the interaction potentials (and thus the interparticle forces) exerted by all particles j in the system on a given particle i are to be computed: The smallest of all distances between i and either j or any mirror particle of j, called here j', is used to compute the potential

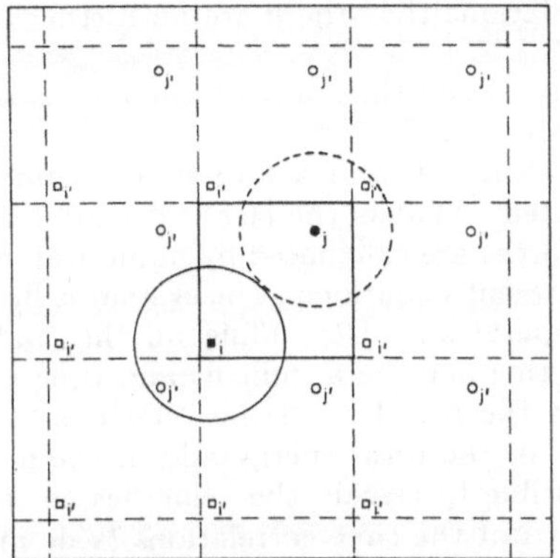

Figure 1. Periodic boundary conditions: The central box holding two particles i and j is surrounded by image boxes holding mirror particles i' and j'. The spheres indicate the 'minimum distance convention' applied to evaluate the interaction between i and j: it is seen that either the interaction between i and j' or between j and i' may be used.

energy and the forces. Together with the periodic boundary conditions this procedure avoids surface effects and allows (in principle) the simulation of an infinitely extended system with a finite number of particles.

3. The interaction potentials between the molecules are usually assumed to be pair-wise additive. The potential energy of a particle i with respect to all other particles j is thus written:

$$V_i = \sum_j V_{ij}(r_{ij}) \tag{1}$$

where the sum is (in principle) over all particles j or mirror particles j'. This assumption is not a fundamental one, but dropping it and using eg. three-body-forces leads to computer time requirements about one order of magnitude higher than for present simulations. Some attempts have nevertheless been made recently in this direction, mostly for simple systems (3), but also for water (4-7). On the other hand, three body terms can easily be used in the intermolecular potentials of flexible molecules (8). The increase of computer time is minimal in this case.

4. If electric charges or dipoles occur in the system studied, like the point charges of the water (vide infra) and the ions , the long range Coulomb forces have to be dealt with. Their correct treatment in the periodic system of the infinitely repeating cells requires special procedures like the Ewald summation (9) or the 'reaction field' method (10), which cannot be treated in the context of this lecture. In certain cases the so called 'shifted force potential method' (11), which is much less computer time consuming, may be applied. A comparison of these and other summation methods applied to MD simulations of molten salts and aqueous systems as well as estimations of errors due to the truncation of interactions at finite distances in MD and MC simulations, have been published (12,13). The influence of the periodic boundary conditions on the calculated properties has also been considered (14,15).

The input to an MD simulation program is thus: The number of particles in the cell, their masses and moments of inertia, the desired density and, most importantly, their interaction potentials. (If the 'reaction field' method is used to treat the long range interactions the dielectric constant which determines the reaction field is also needed).

It is one of the main properties of such simulations that there are no further adjustable parameters.

Under the assumptions made, the results depend thus only on the microscopic interactions between the atoms and molecules in the system.

Once this set of information has been obtained, and a few more technical details have been settled, the simulation can be started. That is the classical trajectories of the particles under the influence of their mutual forces are calculated by numerical integration of Newton's equation or any other set of equivalent equations. The system is first allowed to equilibrate and to reach the desired temperature. Depending on the configuration chosen to start from and on the model used this may be a difficult and time consuming procedure involving many manipulations like the rescaling of the velocities to dissipate energy. During the simulation the constancy of the total energy and of the momenta is monitored. At this point it is no longer possible to rescale the velocities to control the temperature of the system as this would interrupt the time correlations (vide infra).

The locations, velocities, forces and other relevant informations about the particles are stored on some data storage device at regular time intervals. Depending on the types of interaction between the particles and on their mathematical representation the simulation may be a very computer time consuming affair. A considerable programming effort is usually required to ensure reasonable machine performances and to master the management of the large amouts of data generated in such a run. A few so called 'canned programs' have recently been made available for the simulation of certain classes of simple problems (16).

The aim is now to calculate as many macroscopic physical properties as possible from the simulated trajectories. One usually distinguishes between the so called 'static' properties and the so called 'dynamic' ones. The static properties can be calculated without knowledge of the time evolution of the system, ie. just by taking averages over all available configurations. They may thus also be obtained from the so called Monte Carlo (MC) simulations, and we will compare later on results of MC simulations with ones from MD simulations. Typical examples of static properties are the various pair distribution functions characterizing the structure of a system, or thermodynamic functions like the internal energy. The determination of dynamic properties is only possible from MD simulations as it does require the knowledge of the evolution of the system in time. Typical examples are here the self-diffusion coefficients or characteristic reorientation times for certain motions of a molecule in the solution.

The next step is the comparison of the simulated results with experimental ones. This comparison can be made at two levels.

1. 'Experimental' curves, eg. scattering functions, can be computed from the simulation and compared directly with the measured ones (usually after the experimental corrections have been made). This comparison indicates whether the simulated system, especially the choice of interaction potentials, is a reasonable representation of the chemical system with respect to the property studied. Examples of this type of comparison will be given in the present lecture.

2. Often assumptions have to be made to allow a microscopic interpretation of experimental results. In many cases, one is for instance interested in so called 'single ion' values, that is the effect of a single ionic species, on a certain physical property. But only 'salt' values, (i.e. cationic and anionic effects together) are usually accessible experimentally. On the other hand, single ion values can be obtained quite easily from simulations. Their comparisson with experimental ones provides a test of consistency on the assumptions made in the interpretation of the experiment and a check on the values themselves as well. A deeper insight into the microscopic nature of the measured phenomena may thus be gained.

The comparison of simulations results of the same chemical system using different models is also very instructive. As we shall see in the following discussions, different models sometimes lead to somewhat different results. Up to now no model has been able to describe satisfactorily all physical-chemical properties of a liquid as complex as liquid water. Whether the development of such a model will eventually be feasible or not remains to be seen, and whether such a hypothetical model would still be simple enough to be of any use is another question.

Consequently at the present time most simulations of water and aqueous systems have been used as what might be called an 'analytical' tool as described above. With increasing confidence in the ability of a certain model to describe a certain class of properties (eg. structural properties, dielectric properties, vibrational properties etc.) adequately one may then move on to use the simulations as a predictive tool and study systems which are difficult or impossible to study experimentally (eg. liquids under high pressure, biological systems, interfaces).

3. MODELS

In the following the term 'model' will be used to describe the set of (Born-Oppenheimer) interaction potentials associated with a chemical species or, more generally, the term 'model for a certain molecule' will be used to designate all parameters associated with the point mass or rigid body describing the molecule, that is for instance the moments of inertia, the locations of the force centres in the molecular frame, the types of the force centres, eg. partial charges, etc..

As already mentioned, the results of an MD simulation depend critically on the choice of interaction potentials. To describe an aqueous solution of a 1,1 salt, at least 6 microscopic interaction potentials are required: solvent-solvent, solvent-cation, solvent-anion, cation-cation, cation-anion and anion-anion if the solvent molecules are counted as one unity. This number increases eg. to 10 if the water molecule is considered as consisting of two distinct atomic species. In the case of rigid molecules each molecule-molecule interaction may consist of many site-site central force potentials. More complicated functions like Gaussian overlap potentials (17,18) have also been used to model the interactions in molecular liquids.

Interaction potentials can be obtained in many ways, eg. from quantum mechanical calculations, from experiments or from simple ansätze according to chemical intuition. We note in passing that as in most cases we have to restrict ourselves to pair potentials, high accuracy ab initio potentials may not necessarily be the best ones to approximate a liquid in a simulation. The potentials used here need to be effective ones, ie. incorporating in an approximate way many body interactions. This may be the reason for the succes of empirical effective potentials based on chemical intuition in MD and MC simulations.

A number of models for liquid water has been developed in the past 15 years. The conceptionally simplest models consider the water molecule as a rigid unit, among those are the ST2 model (19), an effective potential and probably the most widely used one, the MCY model (20), derived from ab inito calculations, the TIP4P model (21), and many others. In the second type of models the water molecule is considered as consisting of 3 distinct atoms. The Central Force (CF) models (22), the BJH model (8), the RWK2M model (23), the models by Toukan and Rahman (24), Lie and Clementi (25) and Dang and Pettitt (26), which were developed from existing rigid models, belong to this class. These flexible models allow to study the deformations of the molecules in the liquid due to their neighbourhood, an effect which is not negligible in hydrogen bonded systems. As a consequence of these deformations, on the other hand, approximate procedures become necessary to study the rotational and vibrational motions separately (27) (vide infra). Also due to the fast intramolecular motions a shorter timestep is usually required in simulations using a fexible model. As examples of the two classes, we shall look at the ST2 model and the BJH model.

The ST2 model describes the water molecule as a Lennard-Jones (LJ) sphere with empirically adjusted parameters. Its centre is identified with the location of the oxygen. The charge distribution is modelled by 4 partial charges in tetrahedral directions around the LJ centre, two positive ones (0.23e , e: elementary charge) at 1 Å distance representing the hydrogens and two negative ones (-0.23e) at 0.8 Å from the centre representing the eletron orbitals. The tetrahedrally arranged point charges render possible the formation of 'hydrogen bonds' in the right directions, a switching function prevents unreasonably large interactions between the charges. The ST2 model, which is sketched in FIG.2, is thus a 5 centre model. One of its main advantages is that combination rules (28) may be used to combine it with other models, eg. for ions.

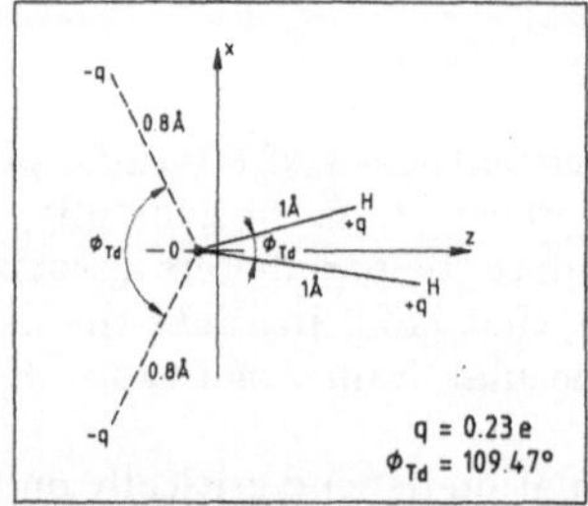

Figure 2. The ST2 model for water. A LJ sphere is centred on the oxygen and 4 point charges are arranged tetrahedrally at 0.8 and 1 Å distance . For the discussion in section 8.3 the lone pairs are supposed to be in the x-z plane (sketched) and the hydrogens in the y-z plane.

The BJH model was developed from the last version of the CF potentials. The aim was to obtain a model which would yield vibrational motions in better agreement with experiment (29). It consists of an intermolecular and an intramolecular part. The intermolecular part is very similar to the CF potential in the intermolecular region. Point charges are located on the hydrogens and on the oxygen, which are also the centres of empirical, mostly repulsive potentials. The intramolecular part was adapted from a potential obtained from vibrational spectroscopy in the gas phase (30). It is formulated in the internal coordinates (31) stretch and bend and thus contains three body interactions. A somewhat different approach was proposed by Demontis et al. (32). They chose to correct the unsatisfactory behaviour of the CF potential by introducing empirical intramolecular stretch-stretch coupling terms in the CF potential. The CF and BJH models are 3 centre models, and as the empirical parts of the intermolecular potentials are not written as expressions for which combination rules are available, their combination with other potentials is a little more involved than in the ST2 case.

Two more advanced model for the water molecule should also be mentioned here, the 'polarization model' by Stillinger and David (33) and the models with three body and higher order terms by Clementi et al. (5). Both models require very large amounts of computer time, and no large scale simulations of ionic systems have been reported with these models. Comparative studies of various water models with respect to various properties have for instance been made by Morse et al. (34) and by Neumann (35).

The interactions of the ions with each other and with the water molecules have been represented in many different ways . One approach is to describe the ions as loaded LJ spheres . The LJ parameters may be taken from isoelectronic noble gases, this works quite well for cations, or fitted from Pauling radii (36). The combination of such potentials with each other and with the ST2 water model is possible by using Kong's combination rule (28). Another possibility is to use ab initio potentials and attempt to fit them to simple analytical functions. This fit is by no means a trivial procedure and Bounds and Bounds (37) have developed procedures to determine more easily ion-water force and potential functions from ab-initio results in connection with the TIP4P water model. One may also try to fit the ab initio potential surface by using spline functions but because of the necessity to describe the potentials up to quite high interaction energies this may require very large quantities of fast storage on the computer.

The reader is referred to the literature for a more detailed discussion of the problems and difficulties related with the determination and representation of potentials suitable for simulations. The potentials used in the simulations discussed in the following sections are briefly described in TABLE I together with details of the simulations and literature references.

TABLE I . Selected simulations of aqueous ionic solutions: potentials, numbers of particles, total simulation times and temperatures.

	water	ion-water	ion-ion	$n^+ : n^- : n_w$		t [ps]	T[K]	Ref
A MD	ST2	charged LJ	charged LJ	8:8:200	LiI	10	305	52
					$NaClO_4$	5.9	294	40
					NH_4Cl	3.5	301	38
					CsF	6.5	307	53
B MC	MCY	ab initio	-	1:0:215	Li^+, Na^+	-	298	54
				0:1:215	F^-, Cl^-	-	298	55
C MD	MCY	ab initio	-	1:0:64	Na^+	30	282	56
				0:1:64	Cl^-	30	287	
				0:1:125	Cl^-	30	287	
D MD	TIP4P	ab initio	-	1:0:64	Na^+	25-30	287	57
				0:1:64	Cl^-	25-30	279	37
E MD	CF	ab initio%	*	8:8:200	$NaCl$	1.4	290	58
	CF	ab initio%	*	4:8:200	$MgCl_2$	3.3	307	59
	BJH	ab initio%	*	4:8:200	$CaCl_2$	10.0	300	60
	BJH	ab initio%	*	8:8:200	$NaCl$	5.0	300	51
	BJH	ab initio%	*	36:36:144	$LiCl$	1.0	314	61

% fitted with simple ion-oxygen and ion-hydrogen central force potentials.
* Ion-ion interactions:
-Charged LJ in the NaCl simulation with CF water and in the LiCl simulation,
-Ab-initio in the $MgCl_2$ and $CaCl_2$ simulations,
-Charged LJ for the $Na^+ - Na^+$ and $Na^+ - Cl^-$ potentials and ab-initio for the $Cl^- - Cl^-$ potential in the NaCl simulations with BJH water.

4. STRUCTURAL PROPERTIES

The discussion in this section will focus on the hydration of simple monoatomic ions. Simulations of systems containing more complicated ions eg. NH_4^+ (38,39), ClO_4^- (40) or the dimethyl-phosphate anion (41) have also been reported in the literature. Quite complex biological sytems have for instance been simulated by Skerra (42) or Clementi et al. (43). New and very interesting studies of the behaviour of water and aqueous solutions near interfaces have also been reported recently (44-50), we shall come back very briefly to these results at the end of the present section.

4.1 Radial Distribution Functions

The first properties usually derived from a simulation are the various radial distribution functions (RDF) $g_{xy}(r)$ or the closely related radial pair correlation functions $h_{xy}(r) = g_{xy}(r) - 1$, where $r = \mid \vec{r}_x - \vec{r}_y \mid$ is the distance between particles x and y. The RDF may be interpreted as the probability to find a particle y at a radial distance r from a central particle x, normalized to the average probability given by the number density ρ_O. The number of particles y within a sphere of radius r from the central particle x may thus be obtained by

$$n_{xy}(r) = 4\pi\rho_O \int_o^r r'^2 g_{xy}(r') dr' \tag{2}$$

The function $n_{xy}(r)$ is often termed the 'running average', if y designates the water molecules, its value at a certain distance from a central ion x (eg. at the first minimum of $g_{xy}(r)$) may be used as a convenient definition of the hydration number of x.

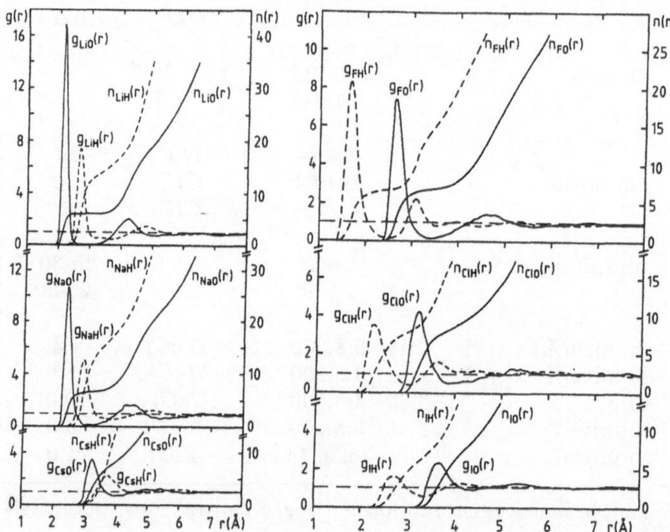

Figure 3. Ion-oxygen (full) and ion-hydrogen (dashed) RDFs and running integration numbers from MD simulations of 2.2m solutions of LiI (52),$NaClO_4$ (40),NH_4Cl (38) and CsF (53) with the ST2 model at room temperature, see TABLE I.

FIG.3 shows ion-oxygen and ion-hydrogen RDFs from various simulations of 2.2m solutions with the ST2 water model , type A in TABLE I. The first peaks in these functions are representative of the first hydration shell of the ions; roughly speaking one might say that the higher and narrower the peaks, the better defined or 'stronger' the hydration. With increasing size of the ion the hydration shell becomes less and less pronounced, as expected. The hight of the first peak decreases and the first minimum gets filled up. Accordingly the plateau in n(r) disappears and the end of the first hydration shell becomes less well defined. The existence of a second hydration shell with about 12 water molecules around the Li^+ ion is well established and has been confirmed by diffraction studies (100), in the case of Na^+ and F^- the formation of a second shell is indicated in FIG.3. FIG.4 shows a comparison of Na^+-oxygen and Cl^--oxygen RDFs obtained from simulations with various models which are described in TABLE I. Note that the temperatures and ionic concentrations are not identical in these simulations. There is general good agreement between the different simulations with respect to the first hydration layer of the sodium ion, independently of temperature or concentration. This peak has also been found to be very little pressure dependent (51). A more or less strongly pronounced second layer is found for this ion in all studies , but as could be expected its height and position are quite sensitive to the circumstances of the simulation, and also to the models employed.

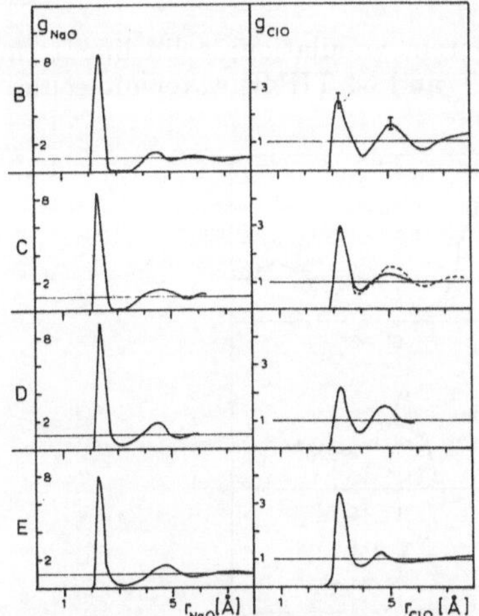

Figure 4. Sodium-oxygen and chloride-oxygen RDFs from various simulations. The letters refer to the descriptions in TABLE I.

Larger discrepancies are found in the case of the Cl^- ion . The positions of the first peak agree within a few tenths of an Å, but the peak height is found to vary by about 30 % between the various simulations. A counterion dependence of the hydration of this ion has been found in several instances (60,62). The structure of the hydration layer of Cl^- is also known to be more sensitive to pressure than eg. the hydration of the Na^+ ion (51,57). Consequently the agreement between the hydration numbers obtained for this ion from the simulations by integrating the ion-oxygen RDF up to the first minimum is not very good.

As an example of the hydration of a divalent ion, FIG.5 shows the ion-oxygen RDF for the Ca^{++} ion from two different simulations (57,60). The sharp first peak and the pronounced second hydration shell are immediately appearant, there is a discrepancy (0.15 Å) between the positions of the first maxima. Comparisons with experimental results (63) lead Bounds (57) to ascribe this to a systematic error in the ion-water potential. The unexpectedly high hydration numbers (9.2 (60) and 9.3 (57)) agree well with each other and with neutron scattering results (63) at low to medium high concentrations. Because of its special biological interest the structure and the symmetry of the hydration shell of Ca^{++} have recently been studied in detail (64).

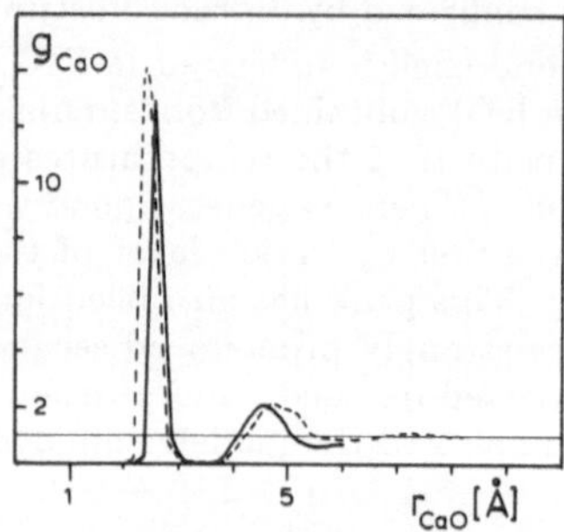

Figure 5. Calcium-oxygen RDFs from two different simulation: solid line 1.1m $CaCl_2$ solution with BJH water, dashed line: one Ca^{++} and 64 TIP4P water molecules.

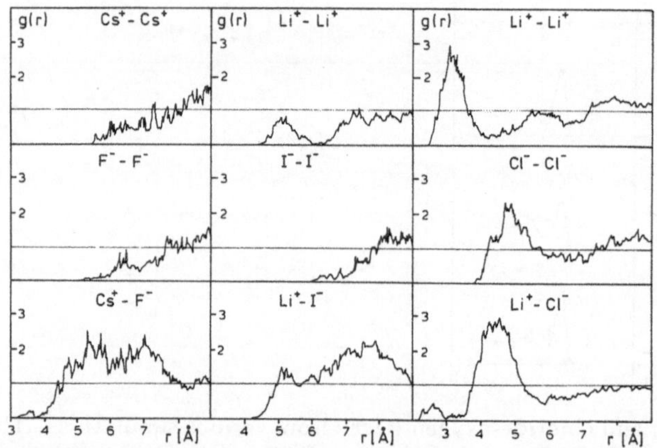

Figure 6. Ion-ion RDFs for a 2.2m CsF solution, a 2.2m LiI solution (both with the ST2 model) and a 13.9m LiCl solution (BJH water).

Ion-ion pair distribution functions are of course of particular interest in chemistry and several attempts have been made to determine them experimentally (65,66). Unfortunately it is quite difficult to obtain reliable results for these functions from simulations of solutions of low to medium concentration of the type described here. The difficulties arise from the small number of solute molecules in the simulated system. Very long runs are thus neccessary to obtain a sufficient number of averages (109). As an example of what can be achieved FIG.6 shows the ion-ion RDFs for two 2.2 molal solutions (8 cations,

8 anions and 200 water molecules) studied with the ST2 model (52,53) and for a solution of much higher concentration (13.9m) studied with the BJH model (61).

4.2 Comparison with Experimental Results

As it has already been stressed, the only way to check the reliability of the potentials employed in a simulation is the comparison of the results with data resulting unambiguously from experiments. As far as the structure of a solution is concerned, a comparison with results of scattering experiments may be carried out both on the level of the structure function or total RDF, or on the level of the partial RDFs like the ones discussed in the preceding section. Both methods have their advantages and disadvantages. As an example of the latter case FIG.7 shows the comparison of the neutron weighted Cl^-water RDF from an MD simulation of a $MgCl_2$ solution (59) (cf. E in TABLE I) and from several neutron diffraction studies with isotopic substitution (67,68). The experimental curves have been rescaled for concentration but the influence of counterions and concentration on the RDFs themselves is not known. The differences between these curves can be usefully compared with the ones in FIG.4 between simulations with different models and under different ambient conditions.

As an example of a comparison at the scattering function level FIG.8 shows the simulated and measured X-ray structure function for a very highly concentrated LiCl solution ($LiCl \cdot 4H_2O$) (61). This simulation was again carried out with technique E of TABLE I and contained 36 ion pairs and 144 water molecules. The agreement is rather good, discrepancies are found in the higher k region where the diffraction data might contain relatively large uncertainties due to low scattering intensities. From this comparison and similar ones with neutron scattering data one can conclude that the model employed in the simulation is a reasonable representation of the system studied, at least with respect to the structure. This could not be expected in the first place as for instance the water model used was developed to describe bulk water, and the typical water-water arrangements and thus their average interactions are quite different here from pure water or low concentration solutions.

The Cl^--Cl^- RDF in a 14.9m LiCl solution in D_2O has recently been determined from a neutron second order difference experiment (66). This function shows an extremely marked and well separated first peak at 3.75 A , the integral over which is roughly two, indicating a high degree of association between the chlorine ions. The peak amplitude is about 1.6, compared to typical values of about two (69) for molten salts, and the RDF drops to zero at 4.8 A. The second peak at 6.38 A with an amplitude of about 1.8 is even more pronounced than in molten salts. This very strong structure in the concentrated solution is quite surprising, even taking into account the difference in temperature between the present system and a molten salt (700-1000 K). On the other hand virtually no anion-anion association is seen in the simulated RDF at the somewhat lower concentration of 13.9m (FIG.6), this discrepancy cannot be resolved at the present time.

4.3 Orientations

The orientation of the water molecules in the hydration shell of the ions could in principle be determined from the locations of the first maxima in the ion-oxygen and ion-hydrogen RDFs (cf. FIG.3). This is the only way in which this information can be deduced from diffraction studies (63), and because of the widths of the peaks it is difficult to obtain accurate values by this method (68). On the other hand it is quite straightforward to calculate the average orientations and even the distributions of orientations directly from a simulated ensemble. An interesting consistency check between experiment and theory

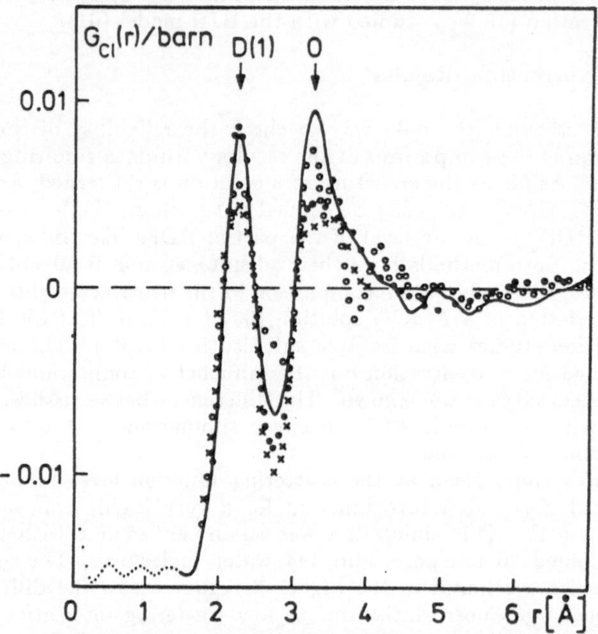

Figure 7. Comparison of the weighted Cl^--water radial distribution function from an MD simulation of a 1.1m $MgCl_2$ solution (solid line) with results from neutron diffraction studies of a 5.32m NaCl (ooo) , a 3m $NiCl_2$ (xxx) and a 9.95m LiCl (...) solution (68).

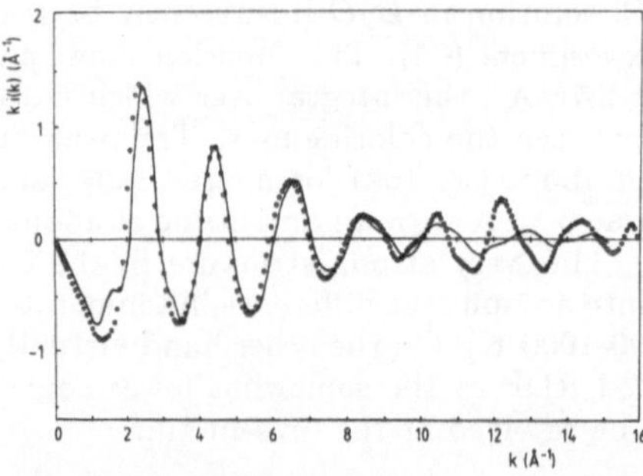

Figure 8. X-ray structure function for a 13.9m LiCl solution from an MD simulation (solid line) and from experiment.

becomes thus possible. FIG.9 shows the probability distribution of $cos(\theta)$, where θ is the angle between the dipole moment of the water molecule and the vector connecting the oxygen and the ion, as illustrated in the insertion, for various monovalent ions from the simulations A. Only water molecules in the first hydration shells of the ions are counted. It is seen that in all cases the maximum probability of this angle corresponds to the formation of a 'hydrogen bond' between anions and water. A negative point charge, corresponding

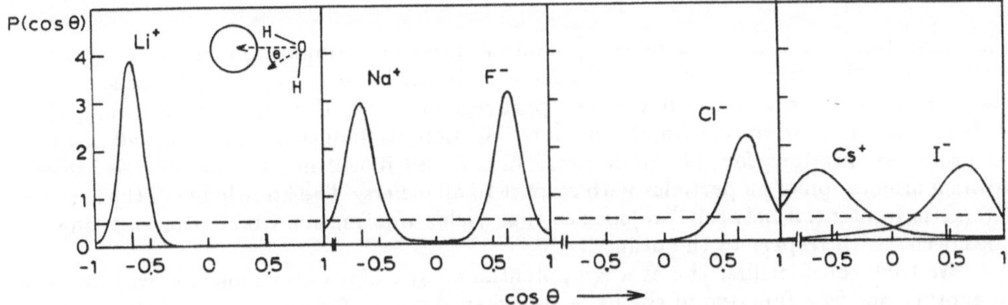

Figure 9. Distribution of $cos(\theta)$ for the water molecules in the first hydration shells of various alkali and halide ions for the same simulation as in FIG.4. The dashed lines indicate uniform distributions.

to the lone pair orbital of the water molecule in the ST2 model, is oriented toward the cations. The distributions have a sizable width which increases with increasing ion size.

The orientation of the water molecules is an instance where the results of simulations with different models show a qualitative disagreement. This is demonstrated in FIG.10 where the distributions of $cos(\theta)$ are shown for the water molecules in the first hydration shells of Na^+ and Cl^- from simulations of 2.2 m solutions with the ST2 model (A) and the CF/BJH models (E). In the case of the Cl^- ion both simulations agree in showing a preference for the formation of a linear hydrogen bond. For Na^+ the simulation with the three-centre models results in a preferential trigonal orientation of the water molecules while in the ST2 case a preference exists for a 'lone pair orbital' directed toward the cation. The origin of this discrepancy is easily traced to the different charge distributions assumed in the two water models. On the other hand, the averages over the two distribution for Na^+, which are also indicated in FIG.10, fall within a range of about 130 to 140 degrees, in agreement with (57) and slightly higher than in (56) (the definition of the tilt angle is different in these two papers). As mentioned above this value can also be inferred from neutron scattering experiments. A value of 135 degrees is for instance reported for Li^+ (70). At this point the discrepancy between the two qualitatively different distributions can thus not be resolved.

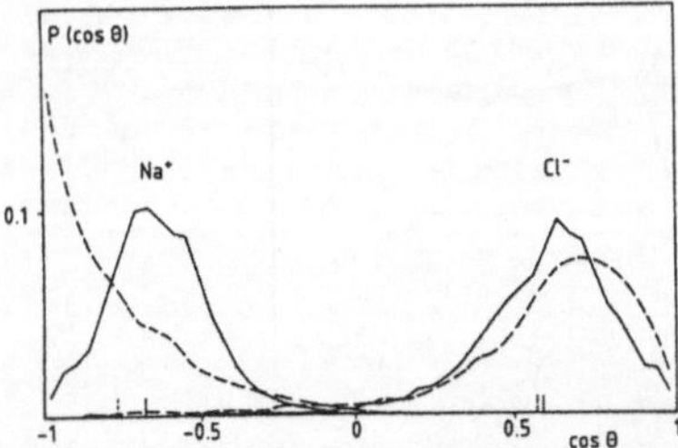

Figure 10. Distribution of $cos(\theta)$ for the water molecules in the first hydration shells of the Na^+ and Cl^- ions from simulations of 2.2m solutions with the ST2 model (solid line) and with the CF model (dashed line). The average orientations are indicated on the abscissa.

5. ENERGETICS

The importance of the choice of the potentials used in a simulation has been stressed many times. The potential curves as they are shown in plots (71) usually refer to the lowest possible energy between the two particles involved, thus usually assuming that their mutual orientation is optimal. Of course the potential must be known for all possible reciprocal orientations for the simulation. In a liquid it will not be possible to achieve optimal orientation of all particles with respect to all others. The knowledge of the average interaction potential between two species thus yields information about the 'ordering' of one species with respect to the other.

We shall consider first the average potential energy of a water molecule with respect to another one as a function of the oxygen-oxygen distance for pure water, water under a pressure of 20 kbar (72) and water in a very highly concentrated LiCl solution (61). All these simulation were carried out with model E of TABLE I. FIG.11 shows the two pure water cases. The total interaction energy has been decomposed in the Coulombic and non-Coulombic components. It is interesting to note that with this model a small region where the average interaction potential is positive appears around an O-O separation of 3.5 Å in the pure liquid. Simulation with the ST2 model display this feature only in the presence of a salt. The value of the average potential energy at the minimum becomes less negative with increasing density. This is mainly due to the change in the non-Coulombic part and leads to values for the change of the internal energy upon isothermal compression, in disagreement with experiment (72). The same is true for the MCY model (73) while the ST2 model seems to lead to better agreement (74,75). On the other hand, the increase of the repulsive interactions leads to a more realistic presssure dependence of the diffusion for the BJH model than for the ST2 model .

The partition between Coulombic energy, modelling the hydrogen bonds, and the other potential terms, representing the van der Waals interactions between the molecules, which is also shown in FIG.11, is also in broad agreement with results from infrared spectroscopy. From the empirical Badger-Bauer (77) rule, which correlates the shift of the infrared absorption frequencies with the hydrogen-bond energies it is concluded that about one third of the total water-water interaction energy should be ascribed to van der Waals type interactions (76).

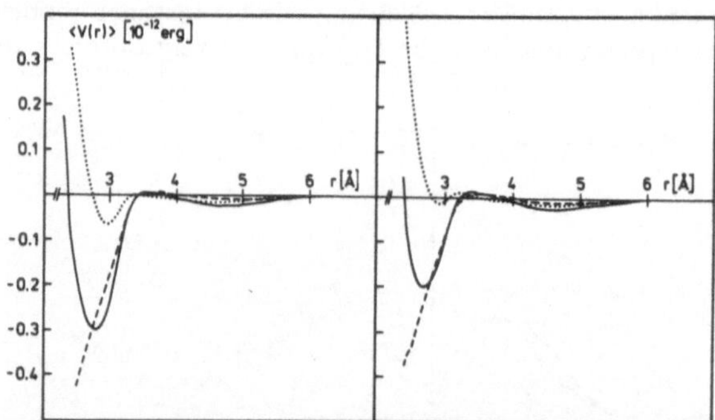

Figure 11. Average potential energy between two water molecules in pure water. Left: under ambient condition, right: at a density corresponding to an external pressure of 22 kbar. Solid line: total potential energy, dashed line: Coulomb part from the partial charges, dotted line: non-Coulomb contributions.

The effect of ions on the average water-water potential is demonstrated very drastically in FIG.12. At the very high salt concentration used in this simulation (61) all water molecules are coordinated to one (or more) ions. The water-water interaction energies become completely positive while the ion-water energies approach quite closely the curve for optimal orientation. For lower salt concentrations the curves are intermediate between the two extreme examples given here. The order of the water molecules with respect to an ion is always quite close to optimal in the immediate vicinity of the ion (but see FIG.9 and FIG.10 and the large width of the distributions) and decreases with increasing distance from the ion, depending on the ion charge and size. For the 2.2m salt solutions studied the water-water curves remain qualitatively similar to the pure water ones, small differences appear first in the depth of the average potential and in the region around 3.5 to 4 Å (78).

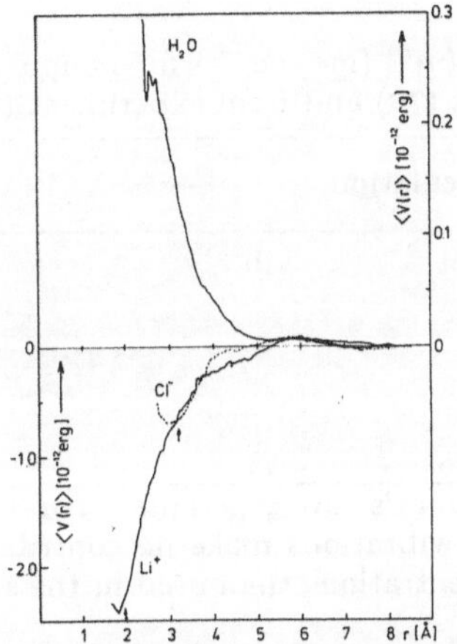

Figure 12. Average potential energy between a central particle and a water molecule as a function of the interparticle separation for a 13.9m LiCl solution at room temperature. Top: water-water; bottom: ion-water. The arrows indicate the positions of the peak maxima in the corresponding RDFs. Note that the scale for the water-water case is magnified by a factor of ten compared to the ion-water cases.

6. A FEW QUANTUM EFFECTS

While microscopic phenomena are of course governed by quantum mechanics, MD simulations utilize classical mechanics to describe them. It seems reasonable to expect that classical mechanics describes adequately phenomena the typical quantum energy of which is small compared to the thermal energy k_BT. An interesting test has been made in this context for a simulation of solutions of Argon and Helium and other inert solvents in water using the CF model (79). In spite of the fact that these solutions do not fall exactly within the framework of the present lecture, it might be worthwhile to give here a brief account of this work as it can easily be extended to ionic solutions.

The isotope effect on the Henry constant, ie. the difference between the Henry's law constant say for the dissolution of two different Argon isotopes in water, can be regarded as the isotope effect on the equilibrium constant k_1/k_2 of the solute atoms between the aqueous phase and the gas phase. In deriving the following equation it has (like usually) been assumed that the gas phase is ideal and that the translations in the gas phase behave classically. Within the framework of the so called 'first quantum approximation', which is applicable to systems in the high temperature regime, the isotope effect on the Henry's law constant for a monoatomic solute can be written as:

$$ln(\frac{k_1}{k_2}) = \frac{\hbar^2}{24(k_BT)^3} < F^2 > (\frac{1}{m_1} - \frac{1}{m_2}) \tag{3}$$

m_1 and m_2 are the masses of the two solute isotopes, the other symbols have their usual meanings. The interesting quantity is $< F^2 >$, the classically evaluated mean of the sum of the squares of the three Cartesian components of the force acting on a solute particle in the solution, a quantity easily obtainable from an MD simulation.

TABLE II . Effective 'force constants' $< F^2 > /k_BT$ (mdyne/Å) for nonpolar species in aqueous solutions at 295 K from MD simulations (79) and from experiment (81,82).

Species	Experiment *	MD calculation		
		trans	rot	vib
Argon	-	0.36-0.40	-	-
Helium	0.12	0.25	-	-
Oxygen	0.23	-	-	-
Hydrogen	0.23	-	-	-
Nitrogen	0.35	.44	.05	.18

* From isotope effect measurements on Henry's law constants. For the diatomic molecules it is assumed here that rotations and vibrations make no contributions. The experiments also correspond to much lower concentrations than used in the simulations.

It is now possible to compare the quantities $< F^2 >$ or $< F^2 > /k_BT$ which are formally equivalent to the sums of the harmonic force constants in the usual theory of isotope effects (80) with the ones obtained from experimental measurements of the isotope effect on the Henry constant (81,82). A few results have been collected in TABLE II. The order of magnitude and the relative order for the different solutes seem to be reproduced satisfactorily.

7. SOME STRUCTURAL RESULTS FOR ANISOTROPIC SYSTEMS

Two examples of MD simulations of non-homogenous ionic solutions shall be briefly sketched, one of more electrochemical interest and one of biological interest. The first one (83) deals with the behaviour of an aqueous 2.2m LiI solution in the vicinity of the [100] surface of Platinum. The water model employed is the ST2 model and the ions are modelled like in the analogue simulations of bulk solutions. The crystal-water and crystal-ion potentials consist of LJ and Coulombic contributions where the crystal is treated as an instantenously polarizable medium with infinite dielectric constant. One of the main points of interest here is the surface induced order in the liquid as the interatomic spacing of 2.77 Å of the surface atoms is very similar to the average O-O distance in liquid water. FIG.13 shows the probability distribution of finding a water molecule as a function of the Cartesian coordinates of its distance from to a given central water molecule in a layer of 2.5 Å depth above the crystal surface. A similarly strong surface induced order in the first

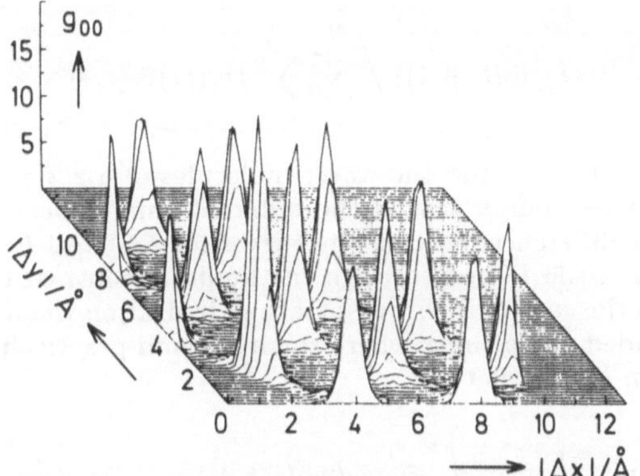

Figure 13. Probability of finding another water molecule as a function of the x and y distance coordinate from a central water molecule within a layer of 2.5 Å depth from a Platinum surface (85)

liquid layer is also found in a simulation of pure (BJH) water near the same surface, but with a more realistic surface-water potential (84,85).

MD simulations are also particularely suited for the study of certain biological problems. The diffusion of ions with and without water through transmembrane channels has been studied repeatedly both by MC (86) and by MD (42,43,87,88) simulations. The structure of water around a Na^+ ion in a transmembrane channel has eg. been investigated in some detail (42). The channel was modelled using Urry's gramicidin A atomic coordinates (89). The TIP4P water model was used together with the ion-water potentials developed by Bounds (37,57). Ions and water interacted with the channel through LJ and Coulomb potentials. Here again a very strong order induced by the channel has been found for the water. The diffusion has also been studied and the motions of the ion and the water molecules in the channel were also found to be highly correlated.

8. DYNAMIC PROPERTIES

8.1 Translational Diffusion and Hindered Translations

The dynamical behaviour of large systems like the liquids studied here is most conveniently studied by calculating the correlation functions of the quantities of interest. Detailed studies have been made for pure water using eg. the MCY model (90) or more recent models (25). In this brief discussion we shall restrict ourselves to the autocorrelation functions (acf) of the translational and angular velocities of single particles and to a few other autocorrelation functions which have the same general structure. The velocity autocorrelation function (vacf) is eg. defined as:

$$\hat{C}_{vv}(t) = \frac{1}{N_i N_j} \sum_{j}^{N_j} \sum_{i}^{N_i} \langle \vec{v}_i(t_j) \vec{v}_i(t_j + t) \rangle \tag{4}$$

where $\vec{v}_i(t_j)$ is the velocity of particle i at time t_j, N_i denotes the number of particles and N_j the number of time origins. This function is often normalized to 1 at its origin (normalized vacf, nvacf):

$$C_{vv}(t) = \sum_j^{N_j} \sum_i^{N_i} (\vec{v}_i(t_j)\vec{v}_i(t_j + t)) \Big/ \sum_j^{N_j} \sum_i^{N_i} (\vec{v}_i(t_j)\vec{v}_i(t_j)) \tag{5}$$

As an example the nvacf for the ions and the water molecules of a 2.2m LiI solution are shown in FIG.14a. For a better comparison of the different time dependencies they are drawn only up to 0.5 ps. The difference between the curves for Li^+ and I^- reflect the rather slow translational motion of the I^- ion through the liquid in contrast to the more oscillatory motions of the Li^+ in the cage of its firmly attached hydration water molecules. The integral over the vacf, extended up to times where the integrand practically vanishes, is the diffusion coefficient (Green-Kubo relation):

$$D = \frac{1}{3} \lim_{t \to \infty} \int_0^t \hat{C}_{vv}(t')dt' \tag{6}$$

These times are obviously different for the three functions plotted in FIG.14a. As the underlying simulation has a finite length there is obviously an inverse correlation between the length of the vacf and the number of time origins entering the averaging process in Eq.4 . The statistical quality of the diffusion coefficients reported in TABLE III is thus not the same for all species (independently of the numbers of particles, which are also different). Experimental values are also given (91-93). The experimental error can be estimated to be about 10 % so that there is good agreement within the limits of error of both methods.

This agreement encourages one to attempt a finer analysis of the diffusion phenomena in the solution. Quantities which can be inferred from experiment only with great difficulty can be calculated from the simulation and new experiments can then be attempted to confirm or infirm the predictions of the simulation. In order to study the single ion contribution to the difference in the self-diffusion coefficients between pure water and the water in a solution the vacf has been determined separately for three water subsystems: water in the first hydration layers of the cations and anions, and water not in any first hydration layer, called here bulk water. The three vacf's for the three subsystems in the above mentioned LiI solution are shown in FIG.14b and the resulting self-diffusion coefficients are given in TABLE III. The following conclusions (94) can be drawn : 1) In all three water subsystems the translatory diffusional motion is reduced relative to pure water under the same conditions. 2) The difference in the self diffusion coefficients between bulk water and pure water contributes more than half to the change which occurs in going from pure water to the 2.2m solution. 3) The self-diffusion coefficients of the Li^+-hydration water and the ion itself are smaller by more than a factor of two than that of bulk water. 4) The self diffusion coefficient of I^- is much smaller than that of its hydration water, which is approximately equal to the one for bulk water.

The difference in the self-diffusion coefficients between bulk water and pure water is easily explained: As can be seen from FIG.3 about 12 water molecules form a second hydration shell around the Li^+ ion, but are counted here as bulk water. If they were counted as hydration water, one would roughly have $8 * (6 + 12) = 144$ molecules in the Li^+-hydration water subsystem , about $8 * 8 = 64$ in the Cl^--hydration water subsystem (hydration shells overlap), and thus virtually none remaining on the average in the bulk subsystem. At the concentration studied it is thus doubtful to try to construct models to explain experimental results by assuming that bulk water has the same diffusional properties as pure water.

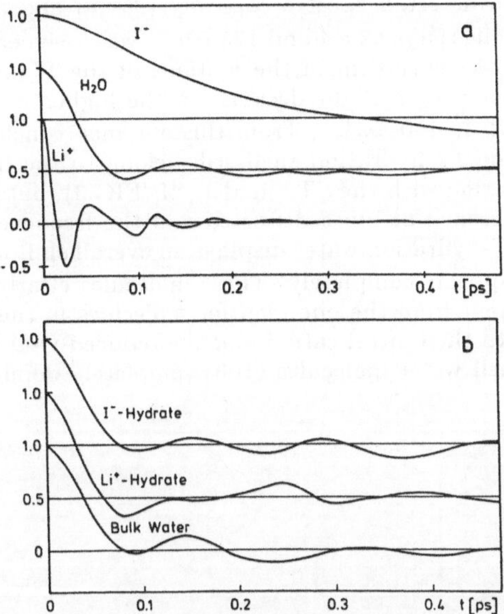

Figure 14.Normalized velocity autocorrelation functions (nvacf's) from an MD simulation of a 2.2m LiI solution with the ST2 model (108). a) water molecules, lithium ions and iodide ions b) separately for 'bulk' water and the hydration water of the ions.

TABLE III. Self-diffusion coefficients of the ions and of the water from an MD simulation of a 2.2m LiI solution at 305 K and from experimental results (91) converted to the same temperatur (103) in units of $10^{-5} cm^2/s$. For comparison the value for ST2 water under the same circumstances is 3.4±0.15 , the experimental value for pure water is 2.73.

	Li^+	I^-	water	bulk water	Li^+ hydration water	I^- hydration water
exp.	1.0	1.47	2.35			
MD	0.7±0.3	1.4±0.15	2.48±0.06	2.85±0.08	1.33±0.1	2.67±0.1

Further insight into the dynamical behaviour of the liquid can be gained by studying the power spectra of the nvacf's. As the nvacf's are even functions of time the power spectra may be defined as:

$$\tilde{C}(\omega) = \int C_{vv}(t) cos(\omega t) dt \tag{7}$$

($\tilde{C}(\omega)$ in arbitrary units), where here again the integral is carried out up to sufficiently high values of t. In certain instances it may be necessary to apply special procedures to avoid spurious oscillations in the transform functions.

As an example FIG.15 shows the power spectra of the nvacf's for the three water subsystems of the 2.2m LiI solution. The shape of curve for bulk water is very similar to the curve obtained for pure water under the same conditions (shown here). The peak at about 50 cm^{-1} has also been detected spectroscopically (95-98) and has been assigned to O-O-O bending motions while the peak around 175 cm^{-1} was assigned to O-O stretching type motions (99). The power spectrum of the motions of the I^--hydration water resembles the one for bulk water with a slight decrease in the higher frequencies and a slight increase in the range of 75 to 150 cm^{-1}. From this one may conclude that dynamically the water molecules in the I^--hydration shell still belong to the hydrogen bond network (especially in this simulation with the ST2 model , cf. FIG.3), but that the hydrogen bonds may be somewhat loosened or more distorted and the frequencies thus somewhat lower. The curve for the Li^+-hydration water displays an overall shift to higher frequencies, the peak at 50 cm^{-1} disappears completely. These molecules clearly show a dynamical behaviour completely different from the one seen for molecules in the general H-bond network. The strong binding to the central cation and the reduced O-O distance between first and second hydration shell water molecules (100) completely dominate the motions.

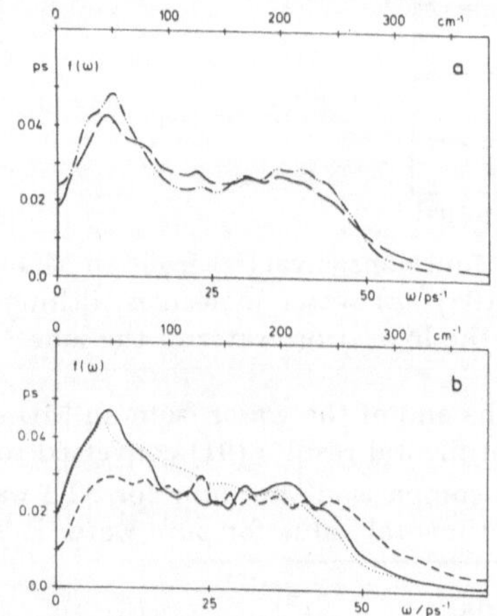

Figure 15. Power spectra of the water nvacf's from FIG.14 (108). a: total water (dashed) and pure water (dash-dotted) for comparison; b: see FIG.16

8.2 Residence Times

The residence time of water molecules in the hydration spheres of the ions can also be determined by a correlation function technique (56,58). Vectors containing informations about the subsystem to which each water molecule belongs are constructed at regular time intervals in the simulation. These vectors are then correlated in very much the same manner as is done for the velocities in equation 4. Of course these correlations have to be extended much further in time than eg. the vacf's and are thus statistically less reliable. TABLE IV gives results from simulations and a few experimental results (101). Not unexpectedly the residence times obtained with the ST2 model seem to be in general somewhat larger than the ones obtained with other models. In view of the large scattering between the experimental results the simulated values are in good agreement with each other and except in one case are always within the experimental error margins. As expected the residence times are found to increase with decreasing size of cations and anions. The residence time of a water molecule in the first hydration layer of the Cl^- ion is generally found to be quite similar to the residence time of a water molecule in the neighbourhood of another water molecule (56). For the divalent ions studied it is virtually impossible to determine residence times from presently feasible simulations.

TABLE IV. Residence times of the water molecules in the hydration shells of various ions (102) from MD simulations of 2.2m solutions (cf. TABLE I, A and E), from simulations at infinite dilution and from experiments (101), in ps.

Ion	MD	experiment		Ion	MD	experiment	
Li^+	85	8 - 40		F^-	12	6 - 60	
	33		(56) at 278 K		20.3		(56) at 278 K
	6		(56) at 368 K	Cl^-	5	3 - 10	
Na^+	15	5 - 30			4.5		(56) at 287 K
	9.9		(56) at 282 K		3.8		(58) at 290 K
K^+	4.8	3 - 10	(56) at 274 K	Br^-	-	2 - 7	
Rb^+	-	4		I^-	2.5	1.3 - 4	
NH_4^+	9	-		ClO_4^-	2.5	-	
Cs^+	5	3 - 4		H_2O	2.5	2.5 - 10	

8.3 Librations and Molecular Reorientations

In simulations carried out with rigid molecules the acf's of the angular velocities can easily be obtained . As an example FIG.16 shows the power spectra of the rotational motions around the three principal axes of the water molecule for the three water subsystems (108) of the 2.2m LiI solution described above. Again the spectra for the hydration water of the I^- ion are fairly similar to the ones for bulk water while the spectra for the hydration water of the Li^+ ion are markedly shifted to higher frequencies. The librations around the y-axis (the axis in the plane of the water molecule and perpendicular to the dipole moment direction) are most strongly affected by the ion as a strong preference exists for a 'lone pair orientation' (cf. FIG.9) of the water molecules in the first hydration shell.

Detailed studies of the rotational correlations of the water molecules have also been carried out for this system (103). The results can be compared to results of proton relaxation measurements. The discussion is quite involved and cannot be repeated here, its general conclusions are along the line of the ones drawn from the discussion of the self-diffusion coefficients. Here again the usefulness of simulations for the microscopic interpretation of macroscopic properties is demonstrated.

8.4 Vibrations

Several approaches have been developed to study the intramolecular vibrational motions of the water molecules in liquid water. One possibility is to use the forces acting on the different force centres of a rigid model molecule to estimate the shifts in vibrational frequencies (40). Another method uses expansions of the potential energy around instantaneous configurations of the system to determine normal or local mode frequencies and to calculate absorption intensities (23). On the other hand if a flexible model is used, ie. a model where relative motions between the oxygen and the two hydrogens forming a

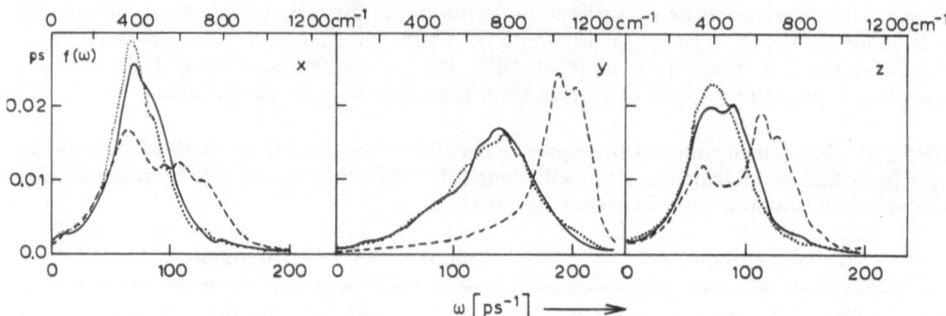

Figure 16. Power spectra of the librational motions around the three principal axes (see FIG.2) from an MD simulation of a 2.2m LiI solution with the ST2 model calculated separately for bulk water and the hydration water of the ions (108). Solid lines: bulk water, dotted line: hydration water of I^-, dashed line: hydration water of Li^+.

molecule are permitted, one may turn directly to the analysis of the actual motions occuring in the system. Of course motions like the vibrational motions of the water molecule, ie. with frequencies in the range of several thousand cm^{-1} , are not in the high temperature regime in the sense described above. The assumption is thus that reasonable effective potentials can be constructed (and this is certainly true in harmonic approximation) which will reproduce the vibrational frequencies and their shifts due to external interactions in (classical) MD simulations. Several models have been constructed recently with this aim in mind (8,32,24-26).

In a typical power spectrum of the nvacf of the hydrogens from a simulation with flexible water three regions are immediately evident. Firstly a peak below 1000 cm^{-1}, ascribed to translations and rotations, similar to the ones discussed in detail for the LiI solution above. Secondly a peak around 1700 cm^{-1} which ought to be the bending motion of the molecule and thirdly a broad peak centred around 3500 cm^{-1} which must be correlated with the O-H stretching motions. Such functions can of course also be obtained for the three water subsystems defined above, and FIG.17 shows the result of this decomposition for the frequency range between 3000 and 4000 cm^{-1} from a simulation of a 1.1m $CaCl_2$ solution with the BJH model. The curves have been normalized to the same area to show clearly the single ion effects, the total effect is of course proportional to the number of molecules in each subsytem. The peak maxima are shifted with respect to pure water by about 15 to -20 cm^{-1} for bulk water and the hydration water of Cl^-, respectively, and by a surprisingly high -300 cm^{-1} for the hydration water of Ca^{++} (110). This result is in contradiction to the generally accepted view that the main effect on the IR absorption frequencies is due to the anions (104). Since then new experimental studies (40,105,106) have been conducted with the aim to determine the single ion effect of various cations on the O-H vibrational frequencies and surprisingly large red shifts have been found for water molecules in the vicinity of certain cations like Mg^{++}, but not for Ca^{++} (106). The discussion of this matter is still open.

We shall now try to characterize in more detail the oscillatory motions of the water molecules in an aqueous solution. As already discussed, due to the deformations of the molecules in the liquid, a rigorous separation of vibrations and rotations is not feasible. FIG.18 shows the bands obtained from an approximate method (27) for the three normal mode type motions in the bulk water subsystems of the above mentioned 1.1m $CaCl_2$

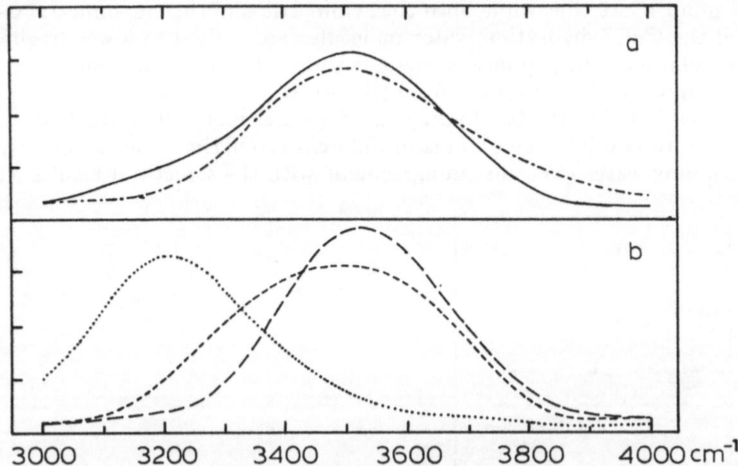

Figure 17. Normalized power spectra of the vacf's of the hydrogens in the range of the O-H stretching frequencies in arbitrary units from an MD simulations of pure water and of a 1.1m $CaCl_2$ solutions with the BJH model. a) Dash-dotted line: pure water, solid line: 1.1m $CaCl_2$ solution. b) long dashs: bulk water, short dashs: hydration water of the Cl^- ions, dots: hydration water of the Ca^{++} ions.

solution, and FIG.19 shows the positions of the peak maxima of the bands obtained for the two stretching modes in the three subsystems together with the peak positions from the vacf's of the hydrogens. The error bars for the hydration water subsystems are larger than for the bulk water because of the smaller number of molecules in the averaging process, the gas phase values obtained under the same circumstances with the water model used are also indicated. This method also allows to determine approximate lifetimes for the vibrational motions.

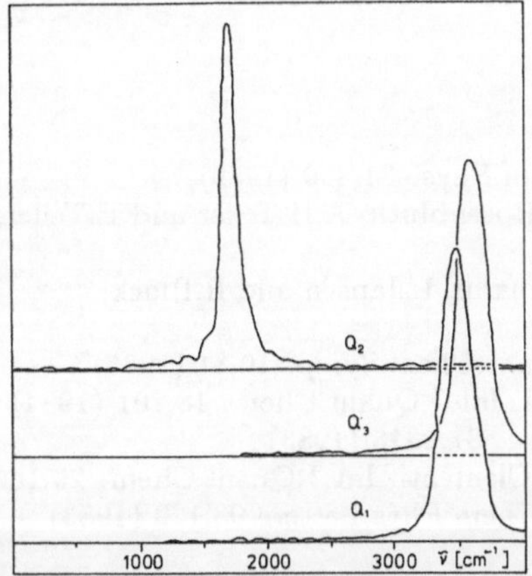

Figure 18. Power spectra from autocorrelation functions approximating the three normal modes of water for the bulk water of a 1.1m $CaCl_2$ solution

It is seen in FIG.19 that while both the symmetric and the assymmetric O-H stretching modes of the Ca^{++}-hydration water molecules are shifted to lower frequencies compared to the bulk, only the symmetric mode is shifted for the hydration water of the Cl^- ion. This is in agreement with results from IR spectroscopy of water-base complexes (107) and is due to the fact that the two hydrogens of a water molecule in the hydration shell of an anion belong to two hydrogen bonds of different strengths: One to the anion and one to the surrounding water molecules,in agreement with the structural results ,cf. FIG.3.

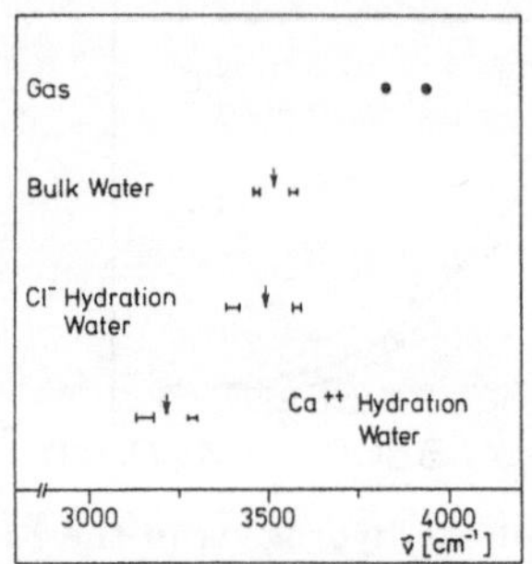

Figure 19. Positions of the peak maxima of the power spectra from acf's approximating the symmetric and antisymmetric stretching modes of a water molecule for the three water subsystems in a 1.1m $CaCl_2$ solution . The arrows indicate the peak positions from the nvacf's. The values obtained under the same circumstances for a gas phase water molecule with the same model are also given for comparison.

9. CONCLUSIONS

Molecular Dynamics simulations of aqueous ionic solutions have been able to reproduce many properties of these systems and have lead to a more detailed and consistent microscopic picture. Remaining discrepancies, indicating a lack of either theoretical understanding or experimental precision, provide the challenge for future work.

10. REFERENCES

(1) B.J.Alder and T.E.Wainwright, J.Chem.Phys. 31,459 (1959)
(2) N.Metropolis, A.W.Rosenbluth, M.N.Rosenbluth, A.H.Teller and E.Teller, J.Chem.Phys. 21,1087 (1953)
(3) E.E.Polymeropoulos, P.Bopp, J.Brickmann, L.Jansen and R.Block, Phys.Rev.A 31,3565 (1985)
(4) E.Clementi and G.Corongiu, Int.J.Quant.Chem.Symp. 10,31 (1983)
(5) J.Detrich, G.Corongiu and E.Clementi, Int.J.Quant.Chem. 18,701 (1984)
(6) P.Habitz and E.Clementi, J.Chem.Phys. .87,2815 (1983)
(7) P.Habitz, P.Bagus, P.Siegbahn and E.Clementi, Int.J.Quant.Chem. 23,1803 (1983)
(8) P.Bopp, G.Jancsó and K.Heinzinger, Chem.Phys.Letters 98,129 (1983)
(9) P.P.Ewald, Ann.Physik 64,253 (1921)
(10) J.A.Barker and R.O.Watts, Chem.Phys.Lettres 3,144 (1969)
(11) W.B.Street, D.J.Tildesley and G.Saville, ACS Symposion Series, 86,144 (1978)
(12) D.J.Adams, E.M.Adams and G.J.Hills, Mol.Phys. 38,387 (1979);

D.J.Adams, J.Chem.Phys. 78,2585 (1983)
(13) M.Mezei, Chem.Phys.Letters, 74,105 (1980)
(14) L.R.Pratt and S.W.Haan, J.Chem.Phys. 74,1864 (1981)
(15) D.Levesque, this volume
(16) A list of available programs and other helpful informations about activities in the field of simulations are found in an informal newslettre 'Information Quarterly for Computer Simulations of Condensed Phases' (CCP5 Newslettre) edited by W.Smith and M.Leslie, Science and Engineering Research Council, Daresbury Laboratory, Daresbury, Warrington WA4 4AD, Great Britain
(17) B.J.Berne and P.Pechukas, J.Chem.Phys. 56,4213 (1972)
(18) J.G.Gay and B.J.Berne, J.Chem.Phys. 74,3316 (1981)
(19) F.H.Stillinger and A.Rahman, J.Chem.Phys. 60,1545 (1974)
(20) O.Matsuoka,E.Clementi,M.Yoshimine, J.Chem.Phys. 64,1351 (1976)
(21) W.L.Jorgensen, J.Chandrasekhar, J.D.Madura, R.W.Impey and M.L.Klein, J.Chem.Phys. 79,926 (1983)
(22) H.L.Lemberg and F.H.Stillinger, J.Chem.Phys. 62,1677 (1975)
F.H.Stillinger and A.Rahman, J.Chem.Phys. 68,666 (1978)
(23) J.R.Reimers and R.O.Watts, Chem.Phys. 91,201 (1984);
R.O.Watts, Chem.Phys. 26,367 (1977)
(24) K.Toukan and A.Rahman, Phys.Rev.B 31,2643 (1985)
(25) G.C.Lie and E.Clementi, Phys.Rev.A, 33,2679 (1986)
(26) L.X.Dang and B.M.Pettitt, submitted
(27) P.Bopp, Chem.Phys., 106,205 (1986)
(28) C.L.Kong J.Phys.Chem. 59,2464 (1973)
(29) G.Jancsó and P.Bopp, Z.Naturforsch. 38a,206 (1983)
(30) G.D.Carney, L.A.Curtiss and S.R.Langhoff, J.Mol.Spectry. 61,371 (1976)
(31) E.B.Wilson, J.C.Decius and P.C.Cross, Molecular Vibrations, McGraw Hill, New York 1955
(32) P.Demontis, G.B.Suffritti, E.S.Fois and A.Gamba, J.Mol.Struct. THEOCHEM 21,201 (1985) ; Chem.Phys.Letters 127,456 (1986)
(33) F.H.Stillinger and C.W.David, J.Chem.Phys. 69,1473 (1978)
(34) M.D.Morse and S.A.Rice, J.Chem.Phys. 76,650 (1982)
(35) M.Neumann, Mol.Phys. 57,97 (1986) and work in progress
(36) L.Pauling, The Nature of the Chemical Bond, Cornell University Press, Ithaka 1948
(37) D.G.Bounds and P.J.Bounds, Mol.Phys. 50,25 (1983); ibid 50,1125 (1983)
(38) Gy.I.Szász and K.Heinzinger, Z.Naturforsch. 34a,840 (1979)
(39) H.J.Böhm and I.R.McDonald, J.Chem.Soc., Faraday Trans.2 80,887 (1984)
(40) G.Heinje, Thesis, Philipps Universität Marburg, 1986
(41) G.Alagona, C.Ghio and P.Kollman, JACS 108,185 (1986)
(42) A.Skerra and J.Brickmann, work in progress
(43) K.S.Kim, H.L.Nguyen, P.K.Swaminathan and E.Clementi, J.Chem.Phys. 89,2870 (1985)
(44) N.Anastasiou, D.Fincham and K.Singer, J.Chem.Soc., Faraday Trans.2 79,1639 (1983)
(45) G.Barabino, C.Gavotti and M.Marchesi, Chem.Phys.Letters 104,478 (1984)
(46) N.I.Christou, J.S.Whitehouse,D.Nichols and N.G.Parsonage, Faraday Symp.Chem.Soc. 16,139 (1981)
(47) B.Joenson, Chem.Phys.Letters 82,520 (1981)
(48) C.Y.Lee, J.A.McCammon and P.J.Rossky, J.Chem.Phys. 80,4448 (1984)
(49) R.Sonnenschein and K.Heinzinger, Chem.Phys.Letters 102,550 (1983)
(50) E.Spohr and K.Heinzinger, Chem.Phys.Lettres 123,218 (1986)
(51) G.Jancsó, K.Heinzinger and P.Bopp, Z.Naturforsch. 40a,1235 (1985)

(52) Gy.I.Szász, K.Heinzinger and W.O.Riede, Z.Naturforsch. 36a,1067 (1981)
(53) Gy.I.Szász and K.Heinzinger, Z.Naturforsch. 38a,214 (1983)
(54) M.Mezei and D.L.Beveridge, J.Chem.Phys. 74,6902 (1981)
(55) D.L.Beveridge, M.Mezei, P.K.Mehrota, F.T.Marchese, G.R.Shanker, T.Vasu and S.Swaminathan, in: Molecular-Based Studies of Fluids, (J.M. Haile and G.A.Mansoori, Eds.) The American Chemical Society, Washington 1983
(56) R.W.Impey, P.A.Madden and I.R.McDonald, J.Phys.Chem. 87,5071 (1983)
(57) D.G.Bounds, Mol.Phys. 54,1335 (1985)
(58) P.Bopp, W.Dietz and K.Heinzinger, Z.Naturforsch. 34a,1424 (1979)
(59) W.Dietz, W.O.Riede and K.Heinzinger, Z.Naturforsch. 37a,1038 (1982)
(60) M.M.Probst, T.Radnai, K.Heinzinger, P.Bopp and B.M.Rode, J.Phys.Chem. 89,753 (1985)
(61) P.Bopp, I.Okada, H.Ohtaki and K.Heinzinger, Z.Naturforsch. 40a,116 (1985)
(62) M.F.Mills,J.R.Reimers and R.O Watts, Mol.Phys. 57,777 (1986)
(63) N.A.Hewish, G.W.Neilson and J.E.Enderby, Nature,Lond., 297,138 (1982); G.Neilson and J.E.Enderby, Proc.Royal Soc.A 353,390 (1983)
(64) G.Pálinkás and K.Heinzinger, Chem.Phys.Letters, 251,126 (1986)
(65) A.Corrias, A.Musinu and G.Pinna, Chem.Phys.Letters 120,295 (1985)
(66) A.P.Copestake,G.W.Neilson and J.E.Enderby, J.Phys. C 18,4211 (1985)
(67) S.Cummings, J.E.Enderby, G.W.Neilson, J.R.Newsome, R.A.Howe, W.S.Howells and A.K.Soper, Nature 287,714 (1980)
(68) Gy.I.Szász, W.Dietz, K.Heinzinger, G.Pálinkás and T.Radnai, Chem.Phys.Letters 92,388 (1982)
(69) F.Lantelme and P.Turq, J.Chem.Phys. 77,3177 (1982)
(70) J.R.Newsome, G.W.Neilson and J.E.Enderby, J.Phys. C 13,L923 (1980)
(71) K.Heinzinger, Pure and Appl.Chem. 57,1031 (1985)
(72) G.Jancsó, P.Bopp and K.Heinzinger, Chem.Phys. 85,377 (1984)
(73) R.W.Impey, M.L.Klein and I.R.McDonald, J.Chem.Phys. 74,647 (1981)
(74) C.Pangali, M.Rao and B.J.Berne, Mol.Phys. 40,661 (1980)
(75) F.H.Stillinger and A.Rahman, J.Chem.Phys. 61,4973 (1974)
(76) H.Kleeberg and W.A.P.Luck, private communication
(77) R.M.Badger and S.H.Bauer, J.Chem.Phys. 5,839 (1937)
(78) K.Heinzinger and P.C.Vogel, Z.Naturforsch. 31a,463 (1976)
(79) L.X.Dang, Thesis, University of California, Irvine, 1985
(80) M.Wolfsberg and M.J.Stern, Pure Appl.Chem. 8,225 (1964)
(81) B.B.Benson and C.E.Klotz, J.Chem.Phys. 38,890 (1963); B.B.Benson, D.Kraus and M.A.Peterson, J.Solution Chem. 8,655 (1979); B.B.Benson and D.Kraus, J.Solution Chem. 9,895 (1980)
(82) J.Mucitelli and W.Y.Wen, J.Solution Chem. 7,257 (1978)
(83) E.Spohr and K.Heinzinger, J.Chem.Phys. 84,2304 (1986)
(84) S.Holloway and K.H.Bennemann, Surface Science 101,327 (1980)
(85) E.Spohr, Thesis, Johannes Gutenberg Universität Mainz , 1986
(86) S.L.Fornili, D.P.Vercauteren and E.Clementi, J.Biomol.Struct.Dyn. 1,1281 (1984)
(87) U.Kappas, W.Fischer, E.E.Polymeropoulos and J.Brickmann, J.theor.Biol. 112,459 (1985)
(88) D.H.J.Mckay, P.H.Berens, K.R.Wilson and A.T.Hagler, Biophys.J. 46,229 (1984)
(89) D.W.Urry, Proc.Natl.Acad.Sci. 69,1610 (1972)
(90) R.W.Impey, P.A.Madden and I.R.McDonald , Mol.Phys. 46,513 (1982)
(91) L.Endom, H.G.Hertz, B.Thuel and M.D.Zeidler, Ber.Bunsen.Phys.Chem. 71,1008 (1967)
(92) M.I.Emelyanov, Zhurn.Struct.Khim. 6,295 (1965)

(93) D.W.McCall and D.C.Douglas, J.Phys.Chem. 69,2001 (1965)
(94) K.Heinzinger, P.Bopp and G.Jancsó, Acta Chimica Hungarica, in press
(95) J.O.Burgman, J.Sciensci and K.Skold, Phys.Rev. 170,808 (1968)
(96) G.E.Walrafen, J.Chem.Phys. 40,3249 (1964)
(97) B.Curnutte and A.Williams, in: Structure of Water and Aqueous Solutions, edited by W.A.P. Luck, Verlag Chemie, Weinheim 1974
(98) A.J.Dianoux, this volume
(99) M.G.Sceats and S.A.Rice, J.Chem.Phys. 72,3236 (1980)
(100) T.Radnai, G.Pálinkás, Gy.I.Szász and K.Heinzinger, Z.Naturforsch. 36a,1076 (1981)
(101) G.Jancsó, P.Bopp and K.Heinzinger, Hungarian Academy of Sciences, Report No. KFKI-1977-101 (1977)
(102) P.Bopp, unpublished results
(103) Gy.I.Szász, K.Heinzinger and W.O.Riede, Ber.Bunsen.Phys.Chem. 85,1056 (1981)
(104) W.A.P.Luck in Structure of Water and Aqueous Solutions, edited by W.A.P. Luck, Verlag Chemie, Weinheim 1974
(105) H.Kleeberg, G.Heinje and W.A.P Luck, J.Phys.Chem, in press
(106) O.Kristiansson, A.Eriksson and J.Lindgren, Acta.Chem.Scand. Ser.A 38,609 (1984) and work in progress
(107) D.Schiöberg and W.A.P.Luck, J.Chem.Soc. Faraday Trans. 75,762 (1979)
(108) Gy.I.Szász and K.Heinzinger, J.Chem.Phys. 79,3467 (1983)
(109) J.M.Caillol, D.Levesque and J.J.Weis, preprint ORSAY 86/23
(110) M.M.Probst,P.Bopp, K.Heinzinger and B.M.Rode, Chem.Phys.Letters 106,317 (1984)

STRUCTURAL MODELS OF THE ELECTRODE-ELECTROLYTE INTERFACE

M.P. Tosi, P. Ballone and G. Pastore
International Centre for Theoretical Physics
P.O. Box 586
34100 Trieste
Italy

ABSTRACT. Models of theoretical interest for the charged electrode/electrolyte interface are briefly reviewed. Attention is given to tests of approximate theories for inhomogeneous charged liquids against simulation data and to microscopic treatments of the electron distribution in a metallic electrode.

1. INTRODUCTION

The charged interface between a blocking metallic electrode and an electrolyte solution is an important example of an electrical double layer (1,2). The complexity of the physico-chemical structure of such a real interface has motivated the study of various models. These, though crude, can shed light on important aspects of the problem.

Schematically, one would first like to achieve control of the theory for two separate problems, regarding the effect of external charges on (i) a semi-infinite metal of ions and electrons and (ii) a semi-infinite classical liquid of ions in a solvent. Further schematization leads us to the semi-infinite jellium model, both in quantum mechanics (for metallic electrons) and in classical statistical mechanics (for ions), and to the restricted primitive model for an electrolyte solution or a molten salt facing a charged hard wall. We shall briefly review progress in the study of these models and some developments towards more complex models for the interface. Reference is also made to recent reviews by Carnie and Torrie (3) and by Blum (4).

2. DENSITY FUNCTIONAL THEORY AND THE JELLIUM MODEL

Density functional theory (DFT) allows treatment of both classical and quantal systems (5,6) and could be used in principle to treat quite real-

M.-C. Bellissent-Funel and G. W. Neilson (eds.), The Physics and Chemistry of Aqueous Ionic Solutions, 245–253.

istic models of the metal-electrolyte interface. It rests on a theorem by Hohenberg, Kohn and Mermin, stating that the Helmholtz free energy F of an inhomogeneous system in external one-body potentials is a functional of the particle density profiles. Writing F as the sum of (a) the ideal gas term, (b) the 'excess' term due to the interactions between the particles and (c) the energy of interactions with the external potentials, one sees upon minimization that the evaluation of the equilibrium density profiles is equivalent to treating an ideal gas immersed in effective one-body potentials. At the lowest level of approximation for an inhomogeneous fluid, these can be determined through an expansion in density gradients from properties of the bulk homogeneous liquid, which are its excess free energy density and linear density response functions at the local values of the particle densities.

An important distinction arises between classical and quantal fluids in the handling of the ideal-gas term (a): this can be written explicitly in terms of the density profiles for a classical fluid, whereas an effective one-body Schrödinger equation is set up for a degenerate electron system. This approach, originally proposed by Kohn and Sham, allows an exact account of the quantal single-particle kinetic energy. A less trustworthy alternative is offered by a low-order gradient expansion for the quantal kinetic energy, which is available from the theory of the inhomogeneous Fermi gas. For classical fluids, on the other hand, DFT is closely related to the cluster expansion and alternative theoretical approaches can, of course, be followed (see section 3 below).

2.1. Degenerate Jellium

Early work by Bardeen (7) proposed a model for a degenerate electron fluid on a semi-infinite neutralizing background, for which a number of exact analytic results are known (7-10). The model considers only exchange and confines the electrons by a hard wall, located by electrical neutrality at distance $\xi = 3\pi/(8k_F)$ outside the edge of the background. The capacitance of the Bardeen model at the point of zero charge (pzc) has been evaluated by Newns (11), with the result that the induced electronic charge by a weak uniform electric field has its centre of mass at distance $d \simeq k_{TF}^{-1} + \pi/(4k_F)$ from the wall. Here, k_F and k_{TF} are the Fermi wavenumber and the Thomas-Fermi screening wavenumber. The result of Newns should be compared with the well-known result of the Gouy-Chapman theory for a classical ionic fluid, in which the classical Debye screening length determines the interfacial capacitance at the pzc. At metallic densities d in the Bardeen model lies slightly inside the edge of the background.

DFT calculations of surface properties of simple crystalline metals were initiated by Lang and Kohn (12), who used the jellium model and also considered effects of metal ions on a lattice as represented by appropriate pseudopotentials. A review has been given recently by Lang (13). In particular, Lang and Kohn (14) have evaluated the electron redistribution induced in jellium by a weak uniform field, finding that its centre of mass lies at distance $d' \simeq 1.2 \div 1.6$ a.u. outside the edge of the background, in the metallic density range. The same distance determines the effective location of the surface for the purpose of evaluating the image potential of a point charge well outside the surface. The location relative to the last lattice plane of the crystal is assessed by considering the latter to be at one-half of an interplanar spacing inside the background edge (thus neglecting surface relaxation of the lattice).

It has been recognized for some item that the spillover of the metallic electrons can have an effect on the capacitance when jellium is in contact with other media (15). Various authors (16-19) have evaluated the inner layer capacitance of models in which jellium is superposed in succession by (a) a gap containing only the metal electron cloud, (b) a dielectric layer penetrated by the electron spillover and (c) a diffuse ionic atmosphere. Thus, Feldman _et al._ (18) show in such a model that the shape of the capacitance curve is strongly influenced by the difference between the location of the centre of mass of the excess charge q on the electrode and the width of the gap, the latter being allowed to depend on q. Halley _et al._ (19) report Lang-Kohn-type calculations for various metal electrodes, including pseudopotential corrections to the jellium model in order to describe dependences on the metal crystal plane.

With regard to liquid metal surfaces, the ionic and electronic density profiles ought to be treated on an equal footing. Such an approach has been examined for the free surface of the liquid alkalis from Na to Cs by Evans and Hasegawa (20), using variational monotonic profiles. Approximate contact can be achieved with data on surface tension and surface entropy. On the other hand, computer simulation of liquid Na clusters in a pseudoatom model by D'Evelyn and Rice (21) has drawn attention to the possibility of local ordering at the liquid-vapour interface, arising primarily from density-dependent electronic contributions to metallic cohesion. Oscillatory atomic profiles are indicated by a comparison of similar results for Cs with X-ray reflectivity data (22).

2.2. Classical Jellium

The classical plasma model deserves mention in the present context for being exactly soluble in special cases (23) and because of the availability of simulation data (24), which provide a test for classical theories of inhomogeneous liquid structure on what is a prototype for classical Coulomb liquids. In fact, two classes of problems have been considered (25-32), namely (a) the behaviour of the plasma near a hard wall and (b) the equilibrium of two plasmas at different densities, these being idealizations of the situation which obtains in the presence of a blocking electrode and of a reversible electrode, respectively.

Some brief comments on the results can be made as follows. Firstly, oscillations arise in the density profile with increasing coupling strength, at $\Gamma = \beta e^2/a \simeq 2$. Secondly, considerable care is needed for a precise evaluation of the profile, both in the simulation (finiteness of the simulation box) and in the theory (bridge diagram contributions). Thirdly, the potential drop from an uncharged hard wall can be satisfactorily assessed from rough profiles which embody the correct asymptotic behaviour (as determined by bulk properties via the gradient expansion) and the contact value of the density (as determined by the 'contact theorem' expressing mechanical equilibrium between the interface and the bulk).

3. RESTRICTED PRIMITIVE MODEL AND CLASSICAL STRUCTURE THEORIES

The restricted primitive model (RPM) for the diffuse layer considers a fluid of charged hard spheres facing a charged hard wall in a dielectric continuum. A decisive step forward in the study of this model was taken by Torrie and Valleau (33,34) by Monte Carlo simulation. The basic simulation cell in their method is a rectangular prism which is part of an infinite slab confined between two parallel walls. The rest of the slab consists of replicas of the central cell, having a charge distribution which is the average charge distribution of the central cell as measured over all preceding configurations. Sampling of configuration space is carried out at constant chemical potential corresponding to a given bulk density and temperature, thus allowing an unambiguous identification of the bulk fluid state that would be in equilibrium with the charged interface, notwithstanding the finite size of the simulation sample.

The original work of Torrie and Valleau (33) refers to a symmetric 1:1 electrolyte at several concentrations, with special attention to concentrations of 0.01 M, 0.10 M and 1.0 M. The data concern potential profiles and density profiles for counterions and coions at several values of the surface charge density over a wide range. Data on other systems (2:1, 2:2 and asymmetric electrolytes, image charge effects) have subsequently become available (34). We have recently examined in more detail the symmetric 1:1 system at 1.0 M by the same Monte Carlo method (35), assessing the capacitance curve of the RPM in this case both by numerical differentiation of potential vs charge data and by a fluctuation-theory formula which relates the capacitance to the mean square fluctuation of the dipole moment of the system (36).

The data produced by the simulation work display the energence of interesting structural features in the model with increasing charge density, depending on the ionic coupling strength. Thus, in the 1:1 electrolyte at 1.0 M, a layering of counterions accompanied by expulsion of coions is found to emerge next to the wall. This restructuring of the interface leads to large values of the potential drop and to structure in the capacitance curve. In 2:2 electrolytes, on the other hand, the counterions are seen to be drawn at large charge densities into a monolayer against the wall, so that charge inversion and oscillations in the potential profile occur.

We must refer the reader to the original papers and to the review of Carnie and Torrie (3) for further discussion of the Monte Carlo data and comparisons with a number of approximate theories of the diffuse double layer. We shall here only summarize some main points and indicate recent developments in the theory which have been stimulated by these data.

It turns out that the Gouy-Chapman theory of the diffuse double layer, simply modified by using the ionic radius as the distance of closest approach of the ions to the wall, does quite well over a significant range of concentration and charge density. Corrections to such a mean field approximation have been systematically developed, on the other hand, by treating the process of charging of an ion in the modified Poisson-Boltzmann (MPB) approach stemming from the Kirkwood hierarchy. In particular, the MPB5 approximation developed by Outhwaite and Bhuiyan (37) for the evaluation of fluctuation potential and excluded volume effects gives a very good overall account of the data, for what concerns also the inclusion of image charges, over à rather wide range of concentration and charge density. Use of the Kirkwood hierarchy equation for the density profiles in combination with an hypernetted-chain (HNC) evaluation of inhomogeneous pair distribution functions has been developed by Kjellander and Marcelja (38).

Quite delicate is also the adaptation to the present problem of theories which are commonly and quite successfully used in the evaluation of bulk liquid structure, stemming from the Bogolubov-Born-Green-Yvon hierarchy (BBGY) and from the cluster expansion. The BBGY relates the evaluation of density profiles to the inhomogeneous two-body densities, for which a closure approximation has to be introduced. The appropriate 'contact theorem' relating the species averaged contact density to the bulk pressure and to the surface charge (39) is exactly satisfied in principle, but great care has to be taken in ensuring local electroneutrality (40) and in avoiding the appearance of negative values of the pair functions near the wall. An appropriate ansatz for the two-body densities has been evaluated by Caccamo et al. (41). Their approach yields good general agreement with the computer results on density profiles and potential drops for high electrolyte densities and surface charges, reproducing in particular the qualitative features of counterion layering in 1:1 systems and of charge inversion in 2:2 systems.

In the absence of image charge effects the cluster expansion (or equivalently DFT) relates instead the evaluation of density profiles to the inhomogeneous two-body direct correlation functions. Replacement of these functions by their homogeneous bulk values is commonly referred to in the present context as the HNC closure, from the analogy between the form taken by the equilibrium equations for the density profiles (i.e. the wall-ion correlation functions) and that of the HNC integral equations for bulk liquid structure. Forstmann et al. (42) in the study of the RPM for the double layer have combined the HNC closure with an evaluation of the two-body functions which accounts for the local unbalance of the two ionic species near the charged hard wall by appeal to the theory of a charged binary liquid immersed in a uniform neutralizing background. They report satisfactory results for various types of electrolytes.

The formally exact functional expansion of Lebowitz and Percus (43), on the other hand, corrects the HNC closure through terms involving three-body and higher direct correlation functions for the bulk liquid. These terms in the expansion correspond to bridge diagram contributions in modified-HNC (MHNC) theories of bulk liquid structure. Considerable experience has been gained in recent years in the estimation of bridge functions in bulk-structure calculations (44), and we have recently evaluated the three-body wall-ion bridge functions in a superposition approximation and their effect on density profiles and potential drop for the 1:1 system at 1.0 M (35).

In summary, it appears that there is a substantial measure of agreement on the RPM double layer between 'experiment' and theory in a variety of approaches. A good theoretical description of the double layer in the model requires quite a considerable degree of sophistication, to correct for deficiencies in simple structural theories which are not so strikingly evident in calculations of the bulk liquid structure, at least for monatomic liquids. It is particularly satisfactory, we feel, that a qualitatively correct interfacial structure emerges in our MHNC calculations from a systematic development of the basic theory. Ironically, however, the numerical effort needed to implement such a theory is not vastly different from that required for simulation runs by the method of Torrie and Valleau.

4. ION-DIPOLE AND JELLIUM-ION-DIPOLE MODELS

Attention has also been given in the theoretical literature to effects due to the discrete nature of the solvent, through the study of model mixtures of ions and dipoles. A review of this work can be found in the article of Blum (4). In concluding this brief review, we shall instead mention the work of Badiali et al. (45) and of Schmickler and Henderson (46), in which a model of hard-sphere ions and dipoles is interfaced with a jellium model for a metallic electrode. These calculations refer to the region near the potential of zero charge. The two models differ in some details and we shall focus for definiteness on the work of Schmickler and Henderson.

The edge of the jellium background is taken by these authors as the location of a hard wall against penetration of the ions and dipoles into the metal. The mixture of ionic and point-dipole hard spheres is treated in the MSA, an orientating effect by the jellium field being allowed even at the pzc, and a variational monotonic profile is taken for the electron density. The electrons experience electrostatic interactions with ions and dipoles as well as short-range repulsions from solvent molecules, described by an effective potential barrier V of height 3 eV for the mercury-water interface.

There is a pleasing measure of agreement between the results of this simple model of the metal/electrolyte solution interface and experimental data on the capacitance of laboratory systems at the pzc. The comparisons with experiment that Schmickler and Henderson discuss concern (a) the concentration dependence of the double layer capacitance and the temperature dependence of the inner layer capacitance for the mercury-water interface; (b) the dependence of the inner layer capacitance on the electrode material, which is described by taking the barrier height V as following the variation of the jellium work function with electron density, and (c) the dependence on the solvent, which is due to changes of the barrier height, the solvent dielectric constant and the size of the solvent molecules.

ACKNOWLEDGMENTS

Our work on electrode/electrolyte systems has been sponsored by the Ministero della Pubblica Istruzione and by the Consiglio Nazionale delle Ricerche of Italy. We have greatly benefited from discussions and collaboration with Dr. P.J. Grout and Prof. N.H. March.

REFERENCES

1. J.O'M. Bockris, B.E. Conway and E. Yeager, Comprehensive Treatise of Electrochemistry - The Double Layer (Plenum, New York 1980).
2. R. Parsons, this volume.
3. S.L. Carnie and G.M. Torrie, Adv. Chem. Phys. 56, 141 (1984).
4. L. Blum, in press.
5. S. Lundqvist and N.H. March, Theory of the Inhomogeneous Electron Gas (PLenum, New York 1983).
6. R. Evans, Adv. Phys. 28, 143 (1979); G. Rickayzen, in Amorphous Solids and the Liquid State (ed. N.H. March, R.A. Street and M.P. Tosi; Plenum, New York 1985).
7. J. Bardeen, Phys. Rev. 49, 653 (1936).
8. R. Stratton, Phil. Mag. 44, 1236 (1953).
9. I.D. Moore and N.H. March, Ann. Phys. 97, 136 (1976).
10. L. Miglio, M.P. Tosi and N.H. March, Surf. Sci. 111, 119 (1981).
11. D.M. Newns, Phys. Rev. B1, 3304 (1970).
12. N.D. Lang and W. Kohn, Phys. Rev. B1, 4555 (1970).
13. N.D. Lang, in ref. 5.
14. N.D. Lang and W. Kohn, Phys. Rev. B7, 3541 (1973).
15. A.A. Kornyshev, W. Schmickler and M.A. Vorotyntsev, Phys. Rev. B25, 5244 (1982).
16. J.P. Badiali, J. Goodisman and M. Rosinberg, J. Electroanal. Chem. 143, 73 (1983).
17. W. Schmickler, J. Electroanal. Chem. 150, 19 (1983).
18. V.I. Feldman, A.A. Kornyshev and M.B. Partenskii, Solid State Commun. 53, 157 (1985).
19. J.W. Halley, B. Johnson, D. Price and M. Schwalm, Phys. Rev. B31, 7695 (1985).
20. R. Evans and M. Hasegawa, J. Phys. C14, 5225 (1981); M. Hasegawa and M. Watabe, J. Non-Cryst. Solids 61/62, 707 (1984).
21. M.P.D'Evelyn and S.A. Rice, Phys. Rev. Lett. 47, 1844 (1981).
22. D.S. Sluis, M.P. D'Evelyn and S.A. Rice, J. Chem. Phys. 78, 1611 (1983).
23. B. Jancovici, Phys. Rev. Lett. 46, 386 (1981).
24. J.P. Badiali, M.L. Rosinberg, D. Levesque and J.J. Weis, J. Phys. C16, 2183 (1983).
25. P. Ballone, G. Senatore and M.P. Tosi, Lett. N. Cim. 31, 619 (1981).
26. H. Totsuji, J. Chem. Phys. 75, 871 (1981).
27. J.P. Badiali and M.L. Rosinberg, J. Chem. Phys. 76, 3264 (1982).
28. M.L. Rosinberg, J.P. Badiali and J. Goodisman, J. Phys. C16, 4487 (1983).
29. P. Ballone, G. Senatore and M.P. Tosi, Physica 119A, 356 (1983).
30. P. Ballone, G. Pastore and M.P. Tosi, Physica 128A, 631 (1984).

31. M. Hasegawa and M. Watabe, J. Phys. C18, 2081 (1985).
32. M. Hasegawa and M. Watabe, J. Stat . Phys., in press.
33. G.M. Torrie and J.P. Valleau, J. Chem. Phys. 73, 5807 (1980); see also W. van Megen and I. Snook, J. Chem. Phys. 73, 4656 (1980).
34. G.M. Torrie, J.P. Valleau and G.N. Patey, J. Chem. Phys. 76, 4615 (1982); J.P. Valleau and G.M. Torrie, J. Chem. Phys. 76, 4623 (1982); G.M. Torrie and J.P. Valleau, J. Phys. Chem. 86, 3251 (1982); G.M. Torrie and J.P. Valleau, J. Chem. Phys. 81, 6291 (1984); G.M. Torrie, J.P. Valleau and C.W. Outhwaite, J. Chem. Phys. 81, 6296 (1984).
35. P. Ballone, G. Pastore and M.P. Tosi, J. Chem. Phys. (in press).
36. L. Blum, D. Henderson, J.L. Lebowitz, C. Gruber and P.A. Martin, J. Chem. Phys. 75, 5974 (1981).
37. C.W. Outhwaite and L.B. Bhuiyan, J. Chem. Soc. Faraday Trans. II 79, 707 (1983).
38. R. Kjellander and S. Marcelja, J. Chem. Phys. 82, 2122 (1985).
39. D. Henderson, L. Blum and J.L. Lebowitz, J. Electroanal. Chem. 102, 315 (1979); L. Blum and D. Henderson, J. Chem. Phys. 74, 1902 (1981).
40. R.W. Pastor and J. Goodisman, J. Chem. Phys. 68, 3654 (1978).
41. C. Caccamo, G. Pizzimenti and L. Blum, J. Chem. Phys. 84, 3327 (1986).
42. P. Nielaba and F. Forstmann, Chem. Phys. Lett. 117, 46 (1985); F. Forstmann et al., in the course of publication.
43. J.L. Lebowitz and J.K. Percus, J. Math. Phys. 4, 116 (1963).
44. See e.g. Y. Rosenfeld and N.W. Ashcroft, Phys. Rev. A20, 1208 (1979); R. Bacquet and P.J. Rossky, J. Chem. Phys. 79, 1419 (1982); H. Iyetomi and S.I. Ichimaru, Phys. Rev. A27, 1241 (1983); P. Ballone, G. Pastore and M.P. Tosi, J. Chem. Phys. 81, 3174 (1984).
45. J.P. Badiali, M.L. Rosinberg, F. Vericat and L. Blum, J. Electroanal. Chem. 158, 253 (1983).
46. W. Schmickler and D. Henderson, J. Chem. Phys. 80, 3381 (1984).

ELECTRODE/ELECTROLYTE INTERFACES : EXPERIMENTAL RESULTS

Roger Parsons
Department of Chemistry
The University
Southampton S09 5NH, U.K.

1 - ELECTRIFIED INTERFACES

ABSTRACT. Single ionic properties are not usually considered to be measurable but, by analogy with the thermionic work function of an electron, a similar property may be defined for ions. The experiment by which this can be measured will be discussed as well as the way this quantity may be used to relate physical and chemical energy scales. Consideration of the work function of different faces of a single crystal leads to further enquiry into the components of this quantity and thence to some information about surface structure. Various types of evidence for the structure of liquid surfaces will be considered for pure and mixed solvents in the light of simple molecular models. The quantitative difference in the electrostatic capacity of a free surface and of an interface leads to qualitative differences in the behaviour of solvent molecules, but a useful comparison is possible by considering interfaces with no net charge separation i.e. at the potential of zero charge (pzc). The measurement of this parameter and its relation to the electronic work function will be discussed and the information that this gives about solvent structure at the metal/electrolyte interface. Further informatioon can be obtained from classical electrochemical experiments and modern spectroscopic techniques together with theoretical models.

1. INTRODUCTION

The aim of this lecture is to describe methods for dealing with the equilibrium of a charged particle between two phases, and how this is related to the structure of the interfacial region, or interphase [1]. A clear understanding of these problems leads on to a solution of the difficulties associated with such concepts as single electrode potentials, single ionic activity coefficients, Gibbs energies of ionic transfer between solvents, ionic solvation energy etc. At the same time this leads to some information about the structure of the interphase.

M.-C. Bellissent-Funel and G. W. Neilson (eds.), The Physics and Chemistry of Aqueous Ionic Solutions, 255–290.

2. EQUILIBRIUM OF CHARGED SPECIES AND MEASUREMENT OF ENERGIES

It is helpful to begin from a familiar example: the electronic work function of a metal. This quantity Φ is the energy required to extract an electron from an electrically uncharged piece of the metal; the electron comes from the Fermi level and is transferred to a point in field-free space far from the metal. Equilibrium of electrons between two different metals in contact is achieved when the energies of their Fermi levels are equal. In order to reach this condition, electrons flow from the metal of lower work function, M, to the metal of higher work function M_2. This redistribution of charge at the interface produces a potential difference between the two pieces of metal, known as the contact potential. The equilibrium condition is equality of the electrochemical potential of the electrons in the two metals:

$$\tilde{\mu}_e^1 = \tilde{\mu}_e^2 \qquad (1)$$

The electrochemical potential of a charged particle in a conducting phase depends on the state of the electrical charge of the phase as expressed by its outer potential ψ which is the potential due to the presence of free charge on the phase. Thus $\tilde{\mu}_i^\alpha$ may be expressed as [2]

$$\tilde{\mu}_i^\alpha = \alpha_i^\alpha + z_i e \psi^\alpha \qquad (2)$$

where z_i is the charge number of the particle, i is the unit charge and α_i^α is the electrochemical potential of i in the uncharged phase, α, known as the 'real potential'. It is evident that

$$\alpha_e^i = - \Phi^i \qquad (3)$$

or that the real potential is the energy change accompanying the <u>insertion</u> of a charged particle into an uncharged phase.

From equations (1), (2) and (3) it follows that the contact potential is given by:

$$\Delta\psi = \psi^1 - \psi^2 = (\alpha_e^2 - \alpha_e^1)/e = (\Phi^1 - \Phi^2)/e \qquad (4)$$

Direct measurement of this quantity is difficult and it is more usual to measure the 'compensation' potential in the arrangement shown in Fig. 1 which is due to Kelvin. The applied potential ΔE is adjusted so that the two pieces of metal making up the parallel plate condenser are uncharged. This is tested by verifying that no current flows round the circuit when the distance between the plates is altered (often a vibrating condenser is used). It may then be shown that

$$\Delta E = (\alpha_e^2 - \alpha_e^1)/e \qquad (5)$$

or in fact that the compensation potential is equal to the contact potential (provided that the value of α is independent of the state of charge; this is quite accurately satisfied).

Essentially similar concepts can be used to discuss the behaviour of ions in electrolytes, the equilibrium of an ionic species, i, between two solvents being subject to a condition analogous to that of equation (1)

$$\tilde{\mu}_i^1 = \tilde{\mu}_i^2 \tag{6}$$

and a work function or real potential α_i being a measurable quantity. The latter is in fact a measurable single solvation energy. Differences of the ionic real potential may be determined by the analogue of the Kelvin experiment described above. However, for liquid electrolytes the uncharged condition is more conveniently determined by a method devised by Kenrick [3], and shown schematically in Fig. 2. A jet of one electrolyte flows down the axis of a cylinder, down the surface of which the second electrolyte flows. If the surfaces are charged, current must flow in the external surface to maintain the charge on the expanding surface. Thus the absence of current is the condition of an uncharged surface. The potential applied then gives the difference of real potentials as in equation (5). Individual real potentials may be obtained by replacing one of the electrolytes by a metal whose electronic work function is known. This was first achieved successfully by Randles [4] who used a mercury jet in the Kenrick apparatus. Improved accuracy was obtained, using a redesigned apparatus, by McTigue [5] and an alternative method has been used by Gomer and Tryson [6] to avoid the effect of possible contamination of the metal surface.

The real potential measured in this way is the Gibbs energy of a single charged species in a given phase with respect to its Gibbs energy in field-free space. It is concentration dependent for ions in an electrolytic solution and this concentration dependence may be used to investigate the activity coefficients of the ionic species concerned as well as to determine a standard value of the real potential at a standard concentration. Once a value of this standard real potential has been determined for one ionic species in a given solvent, values for other species may be calculated from the ordinary Gibbs energies of solvation of salts since these are the sums of the ionic contributions

$$\mu_{salt} = \nu_+\mu_+ + \nu_-\mu_- = \nu_+\alpha_+ + \nu_-\alpha_- \tag{7}$$

Other thermodynamic functions for single ionic species can be obtained from the Gibbs energy by standard methods e.g. entropy, enthalpy, etc.

Much discussion has centred around the problem of single (or 'absolute') electrode potentials. Much of this has little practical use, but the quantity which is useful is that which enables electron energy levels in a metal or semiconductor to be related to the electron levels of a redox species in solution. This is often called the

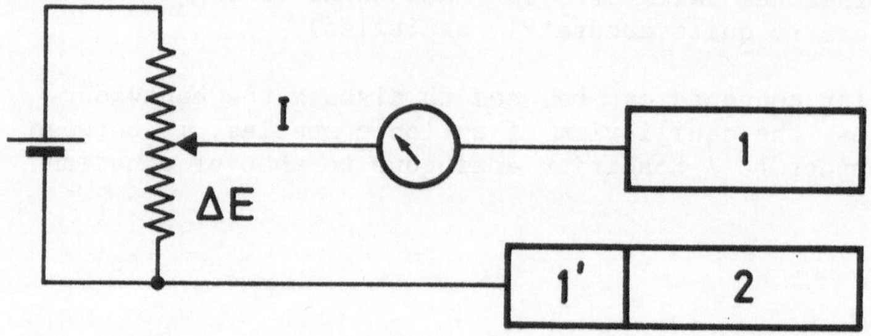

Figure 1

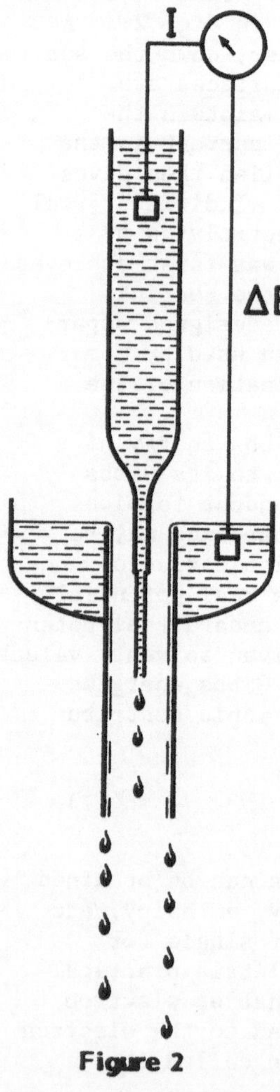

Figure 2

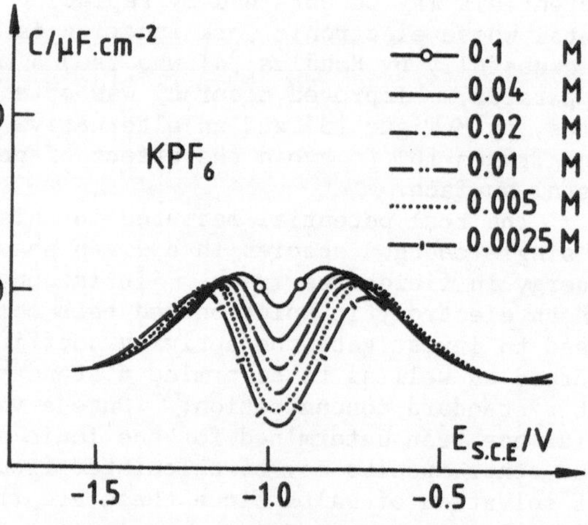

Figure 3

'absolute potential of the hydrogen electrode' but in fact it is the fictitious real potential of an electron in equilibrium with the hydrogen couple in the given solvent i.e.

$$\alpha_e^* = \tfrac{1}{2}\mu_{H_2}^g - \alpha_{H+}^S \qquad (8)$$

where the standard chemical potential of gaseous hydrogen (μ_{H2}^g) and the standard real potential of the solvated proton are referred to field-free space. The asterisk indicates the fictitious nature of α_e^* which refers to the non-existent free electrons in the solution [7].

3. THE SURFACE DIPOLAR LAYER

The ideas outlined in the previous section are sufficient for the description of real experiments involving the transfer of charges between phases. However, it is interesting to investigate the nature of the real potential further. A direct indication that it includes a surface contribution as well as one from the interaction of the charged particle with the bulk of the phase comes from experiments with the field emission microscope. The figures obtained from these show at once that the work function (or α) depends on the structure of the crystal faces from which electrons are emitted (it also shows that an adsorbed layer greatly affects the value of α). This can be explained by the existence of a surface dipole. Electrons tend to overshoot the surface defined by the positive atomic cores. Thus there is a dipolar layer with its negative pole pointing outwards [8]. The magnitude of the surface dipole depends on the packing of the atomic cores and the magnitude of the work function increases with the density of the atomic packing [9]. Thus, it is possible to split the real potential into a bulk contribution, μ, and a surface contribution χ, known respectively as the chemical potential and the surface potential:

$$\alpha_i = \mu_i + z_i e\chi \qquad (9)$$

μ_i is clearly independent of the surface structure. The division described by equation (9) cannot be made experimentally but can be made using models such as those proposed for metals by Bardeen [10] and by Kohn and Lange [11].

Similar considerations apply to the real potentials of ions in solution. The surface potential then arises from the orientation of surface solvent molecules in the simplest case, although especially in more concentrated solutions ionic double layers may occur at the surface. Various indirect methods for estimating the magnitude of the solvent surface potential have been devised. Randles and Schiffrin [12] measured the temperature coefficient of the real potential of ions in water and used what are probably good estimates of the entropy of single ionic species to calculate the temperature coefficient of the surface potential. The value of -0.43 mV K^{-1} was then used with a simple Boltzmann model for the orientation and the assumption that χ

must vanish at the critical point, to estimate that χ for water at room temperature is about +80 mV. This would indicate a small degree of orientation with the proton of the water molecule pointing towards the bulk of the solution. In general, estimates agree that the degree of orientation is very small [13] - complete orientation of a monolayer of water molecules would lead to $\chi \simeq 10V$. The reason for the small degree of orientation appears to lie in the strong dipole interactions between the water molecules. Simple calculations of dipole-dipole forces between molecules in a monolayer lead to the conclusion that the energy missing from this is a minimum when the dipoles are oriented parallel to the plane of the monolayer [14]. Interaction with bulk solvent seems unlikely to modify this conclusion. A model using a water molecule embedded in a continuum with a permittivity gradient [15] suggests that the quadrupole moment is responsible for the slight preferential orientation with the protons towards the bulk. Ellipsometric measurements indicate that the surface region in which properties differ from those in the bulk is essentially a monolayer [16].

Changes in surface potential may be found more directly in some mixed solvent systems [17]. For example in ethanol + water, the real potential changes quite sharply at low ethanol contents in the region where surface tension measurements show that ethanol is strongly adsorbed at the surface. Once the surface layer is entirely composed of ethanol (above a bulk mole fraction of about 0.1) the real potential changes much more slowly. This suggests that in the first region the changes in α are primarily due to changes in χ while in the second they are primarily due to changes in μ. This leads to an estimated difference of χ of about 350 mV between water and ethanol.

4. THE INTERFACIAL REGION BETWEEN TWO CONDENSED PHASES

Here the modifications to the above conclusions when two condensed phases are brought into contact are considered. One major source of different behaviour is the large difference of electrostatic capacity associated with a free surface and an interphase. The capacity of a metal sphere 1 mm radius in vacuum is 1.1×10^{-13}F while a sphere of gold of the same size in aqueous electrolyte will have a capacity of about 2.5×10^{-2}F. This enormous difference means that very large potentials are required to build up charges on the sphere in vacuum, while quite large charges can be obtained in solution with modest potentials. This is why surface dipole orientations are rather independent of applied potential in vacuum but can vary greatly in solution. The essential reason for the high capacity in solution is that the counter charge is made up of ions which can approach the metal surface within a molecular diameter whereas in vacuum the counter charge is at a macroscopic distance from the sphere (assumed infinite distance).

The closest comparison between the free surface and the interface may be made when each is uncharged. It is then interesting to enquire whether the electron overlap at the metal surface and the dipolar orientation at the solution surface are substantially perturbed when the two are brought into contact. This may be investigated by comparing the potential at which the charge on the metal (in solution) is zero - the potential of zero charge, pzc with the electronic work function of the metal.

The pzc of an electrode may be obtained from experiments in which the capacity of the electrode is measured. Under conditions where no electrode reaction can occur (no faradaic process), the electrode behaves electrically as a capacitor and its capacity can be measured by conventional methods such as an alternating current bridge or a transfer function analyser. As its capacity is strongly dependent on the applied (d.c.) potential, a rather small alternating signal must be used ($\sim$ 5 mV). A typical set of results is shown in Fig. 3 where a silver electrode in an electrolyte containing no surface active species is shown. Such results may be interpreted in terms of a simple model in which the ions interact in a simple electrostatic way with the electrode charge σ on unit area and are treated as point charges in a dielectric continuum except that they approach the metal surface to a distance of closest approach determined by the size of the real solvated ions. Gouy [18] and Chapman [19] developed this model along the lines used later by Debye and Hückel for their theory of interionic attraction. If the potential at a distance x from the electrode surface is ϕ, then the local concentration of ions of species i was expressed by the Boltzmann relation

$$n_i^x = n_i^b \exp(-z_i e\phi/kT) \tag{10}$$

where n_i^x is the concentration at x and n_i^b that in the bulk of the solution. The local charge density can be expressed in terms of these local concentrations

$$\rho = \sum_i z_i e n_i \tag{11}$$

and these equations can be substituted in the Poisson equation in its 1 dimensional form:

$$\nabla^2\phi = -\rho/\varepsilon = d^2\phi/dx^2 \tag{12}$$

to obtain the Poisson-Boltmann equation:

$$d^2\phi/dx^2 = -\varepsilon^{-1} \sum_i z_i e n_i^b \exp(-z_i e\phi/kT) \tag{13}$$

This is the basic equation for the potential in the region of the double layer occupied by the ions - the diffuse part of the double layer. The potential profile may be obtained by integrating it twice, but it is more useful at present to integrate it once and use the Gauss equation to obtain a relation between the surface charge density on the metal σ and the potential at the distance of closest approach (the outer Helmholtz plane oHp) ϕ_2.For a binary symmetrical electrode where $z = z_+ = -z_-$ this results in

$$\sigma = (2kT\varepsilon/ze\ L_D)\ \sinh(ze\phi_2/2kT) \tag{14}$$

where L_D is the Debye length.

$$L_D = (\varepsilon kT/2z^2 d^2 n^b)^{\frac{1}{2}} \tag{15}$$

The potential of the metal with respect to the solution ϕ^M can be expressed as

$$\phi^M = \phi^i + \phi_2 \tag{16}$$

where ϕ^i is the potential drop across the inner layer, i.e. the potential of the metal with respect to that of the outer Helmholtz plane and ϕ_2 is the potential of the latter with respect to the solution. From (16) it follows immediately that

$$(d\phi^M/d\sigma) = (d\phi^i/d\sigma) + (d\phi_2/d\sigma) \tag{17}$$

or

$$C^{-1} = (C^i)^{-1} + (C^d)^{-1} \tag{18}$$

since each of the terms in (17) is a differential capacity. C^d is the differential capacity of the diffuse layer which can be calculated from (14) directly:

$$C^d = (\varepsilon/L_D)\cosh(ze\phi_2/2kT) \tag{19}$$

This capacity is effectively in series with the inner layer capacity C^i. At the pzc it follows from eq. (14) that ϕ_2 is zero and from (19) that C^d has a minimum value ε/L_D. In aqueous solution at 25°C ε/L_D is equal to 224 $\sqrt{c}$ μF cm^{-2} where c is the concentration of a uni-uni valent electrolyte in mol l^{-1}. Thus in a 1 mol l^{-1} C^d is always large in comparison with the experimental capacity which is consequently almost equal to the inner layer capacity C^i. As the solution is diluted the contribution of the diffuse layer capacity becomes noticeable close to the pzc and a marked minimum in the experimental curve in the more dilute solutions gives a good indication of the pzc. This is in fact the most reliable method for the determination of this parameter on solid electrodes. For liquid electrodes confirmation of this method is given from measurements of the interfacial tension and use of Lippmann's equation [2] which identifies the pzc with the maximum of the interfacial tension/potential curve (the electrocapillary curve).

A simple thermodynamic analysis [2] shows that the electronic work function and the pzc (E_z) of a given metal are related by:

$$E_z = \Phi - \chi^M + g_{dipole}^{M/S} + \text{const} \tag{20}$$

where χ^M is the surface potential of the metal, $g_{dipole}^{M/S}$ is the total dipole potential across the metal/solution interface and the constant is independent of the nature of the metal but dependent only on the reference electrode used. Since $g_{dipole}^{M/S}$ includes contributions from the metal dipolar layer (electron overshoot) and from the solvent orientation, it is clear that a plot of Φ against E_2 will have unit slope if the electron overshoot is unperturbed by the presence of the solvent and if the solvent dipoles in the surface layer remain oriented parallel to the plane of the interface, or if any perturbation of the electron overshoot is cancelled out by an opposing orientation of the solvent. The simple correlation

$$E_z = \Phi + \text{const} \tag{21}$$

was proposed many years ago by Frumkin [20] and is approximately borne out by modern data [21]. Although the details of this relationship are still open to discussion, largely because there is no direct evidence for the separate contributions of χ^M and the components of $g_{dipole}^{M/S}$, there is some agreement that the perturbation of the metal dipole is small and the solvent orientation remains almost parallel to the interface. The latter is supported by a simple calculation of the orientation of dipoles in a monolayer adjacent to an imaging plane [14] as well as by experimental observations that the adsorption of the highly polar zwitterionic molecule glycine at the mercury/water interface causes no potential shift of the electrode held at constant charge (e.g. at the pzc). The only reasonable interpretation of the latter is that highly dipolar molecules are strongly oriented parallel to the plane of the interface.

Further evidence for the structure of water (and of other solvents) at the interface with a metal has been obtained from the behaviour of the inner layer capacity C^i. This has been obtained from the experimental capacity using equation (18) together with C^d calculated from equation (19). The outer Helmholtz potential is obtained from equation (14) using the value of σ obtained from experiment as

$$\sigma = \int_{E_z}^{E} C \, dE \tag{22}$$

The values of C^i obtained are found to be independent of concentration, provided that the electrolyte is not surface active, but they are strongly dependent on σ. This behaviour is attributed to the polarization of water in the region between the metal surface and the oHp. That is C^i is simulated as a parallel plate condenser with a monolayer of orientable water as dielectric. Contributions to C^i may also arise from change in electron overshoot with σ and from water further from the metal surface but these seem likely to be less important. Early models of the water derive from the approximation due to Watts-Tobin

[22] that water may take up two positions only with dipole components perpendicular to the plane of the interface pointing to or away from the metal. These components were taken to be a fairly large fraction of the dipole moment of water. This model and its many successors give a good account of the C^i/σ relation with the assumption of a number of parameters but they do not take account of the argument that the dipoles must be essentially parallel to the interface at the pzc.

Recently, a more rigorously based statistical mechanical approach has been used in particular by Henderson et al. [23] who represented the ions and the solvent by hard spheres carrying respectively a charge and a dipole. Using the hypernetted chain theory and the mean spherical approximation, they found that the reciprocal capacity of the pzc could be expressed as the sum of two items: one arising from the ions and is equivalent to $(C^d)^{-1}$ while the other arises from the solvent, but from the solvent in the whole double layer. The magnitudes of the two terms are in good agreement with experiment without need for arbitrary assumption of values of parameters. However, at present the extension to non-zero values of σ is not feasible. The simplicity of the physical model does not permit an explanation of some of the more subtle experimental observations. For example C^i has been found to depend significantly on the isotopic composition of the solvent, particularly close to the pzc [26] being as much as 5% lower in D_2O than in H_2O. The only model which incorporates an explanation for such an effect has been developed by Guidelli [28]. This is a descendent of the Watts-Tobin type of treatment but allows a much greater number of solvent molecule orientations which are hydrogen bonded and their equilibria are discussed according to a quasi-chemical model. The presence of hydrogen bonding leads to the account of the isotope effect and although the parameters are adjusted they are physically reasonable.

Even more subtle effects remain beyond the reach of current theories. Solvent interaction seems likely to be responsible for the remarkable differences in the adsorption behaviour of the isomeric compounds mannitol and sorbitol [26] which differ in the configuration at one carbon centre.

More direct experimental examination of water structure at interfaces is becoming possible. For example water monolayers can be studied on metal surfaces by electron spectroscopy. An EELS study of water on Pt(111) [27] suggested that a monolayer showed a hexagonal ice-like structure. The direct relevance of this to an electrode in solution is however doubtful because the monolayer volatilized well below room temperature and the orientation of the water molecules would produce a large surface potential. In situ experiments using Raman or infrared spectroscopy seem more likely to yield relevant data. These have been shown to be feasible in the last few years. Fleischmann et al [28] showed that it was possible to obtain Raman signals from pyridine adsorbed on silver electrodes. It was shown later by Jeanmaire and Van Duyne [29] and by Albrecht and Creighton [30] that these resulted from an anomalous enhancement which has come to be known as the Surface Enhanced Raman Effect. More recently the vibrational bands of water in the double layer have been observed by SERS [31] although only in the presence of specifically adsorbed anions. There is a clear potential dependence of the O-H bonding and stretching modes but at present it is not clear how relevant these results are to water in the double layer unperturbed by ionic adsorption.

A second important technique for in situ study of the electrode/electrolyte interface is reflection spectroscopy using infrared. That external reflection could be used in these experiments was demonstrated by Bewick at al [32] and a variety of techniques were developed [33]. The vibrational spectrum of water was detected in some of the early experiments [32,34] and interpreted in terms of the presence of dimeric water, but later experiments suggest that the situation may be more complex with water present in various degrees of polymerization as well as in the monomeric form [35]. It is likely that evidence from this powerful method will accumulate in the next few years and provide specific data for models of this region.

References

[1] see R. Parsons and S. Trasatti, J. Electroanal. Chem., 205 (1986) 359. for a discussion of nomenclature.

[2] see for example R. Parsons in 'Modern Aspects of Electrochemistry' Vol.1. Ed. J. O'M Bockris and B.E. Conway, Butterworths, London 1954, p.103.

[3] F.B. Kenrick, Z. physikal. Chem., 19 (1896) 625

[4] J.E.B. Randles, Trans. Faraday Soc. 52 (1956) 1573.

[5] P. McTigue and J. Farrell, J. Electroanal. Chem., 139 (1982) 37.

[6] R. Gomer and G. Tryson, J. Chem. Physics, 66 (1977) 4413

[7] see R,. Parsons in 'Standard Potentials in Aqueous Solution', Ed. A.J. Bard, R. Parsons and J. Jordan, Dekker, New York, 1985, Chap. 2.

[8] C. Herring and M.H. Nicholls, Rev. Mol. Physics, 21 (1949) 185.

[9] R. Smoluchowski, Phys. Rev. 60 (1941) 661.

[10] J. Bardeen, Phys. Rev., 49 (1963) 653

[11] N.P. Phang and W. Kohn, Phys. Rev., B, 1 4555; B, 3 (1971) 1215

[12] J.E.B. Randles and D.J. Schiffrin, J. Electroanal. Chem., 10 (1965) 480.

[13] J.E.B. Randles, Phys. Chem. Liquids, 7 (1977) 107.

[14] R. Parsons and R.M. Reeves, J. Electroanal. Chem., 123 (1981) 141.

[15] F.H. Stillinger and A. Ben-Naim, J. Chem. Physics, 47 (1967) 4431.

[16] J.W. McBain, R.C. Bacon and H.D. Bruce, J. Chem. Physics, 7 (1939) 818.

[17] R. Parsons and B.T. Rubin, J.C.S. Faraday Transactions, 70 (1974) 1636.

[18] G. Gouy, J. Phys. Chim., 9 (1910) 457.

[19] D.L. Chapman, Phil. Mag, 25 (1913) 475.

[20] A.N. Frumkin, J. Colloid Sci., 1 (1946) 290.

[21] S. Trasatti, J. Electroanal. Chem., 33 (1971) 351.

[22] R.J. Watts-Tobin, Phil. Mag. 6 (1961) 133.

[23] D. Henderson, L. Blum and M. Lozada-Cassou, J. Electroanal. Chem., 150 (1983) 291.

[24] R. Parsons, R.M. Reeves and P.N. Taylor, J. Electroanal. Chem. 50 (1974) 149.
[25] R. Guidelli, J. Electroanal. Chem., 197 (1986) 77.
[26] R. Peat and S. Shannon, J. Electroanal. Chem., 159 (1983) 229.
[27] H. Ibach and S. Lehwalt, Surface Sci., 91 (1980) 187.
[28] M. Fleischmann, P. Hendra and A.J. McQuillan, Chem. Phys. Lett. 26 (1974) 163.
[29] D.L. Jeanmaire and R.P. Van Duyne, J. Electroanal. Chem. 84 (1977) 1.
[30] M.G. Albrecht and J.A. Creighton, J. Amer. Chem. Soc., 99 (1977) 5215.
[31] M. Fleischmann, P. Graves, I. Hill, A. Oliver and J. Robinson, J. Electroanal. Chem., 150 (1983) 33.
[32] A. Bewick, K. Kunimatsu and B.S. Pons, Electrochim. Acta, 25 (1980) 465.
[33] A. Bewick and S. Pons, 'Adv. in Infrared and Raman Spectroscopy', Eds. R.J.H. Clark and R.E. Hester, Wiley Heyden, Chichester, 12 (1985) 1.
[34] A. Bewick and K. Kunimatsu, Surface Sci., 101 (1981) 131.
[35] K. Kunimatsu, Personal Communication.

2 - ADSORPTION ON SOLID METAL SURFACES

ABSTRACT. The importance of making measurements using clean and well-defined surfaces is stressed and the techniques by which this may be done will be outlined. The classical thermodynamic route for the study of adsorption is summarized. The significance of capacity/potential and current/potential curves (cyclic voltammograms) is discussed. The work function/pzc relation shows that the pzc will depend on crystal orientation and the experimental evidence that this is so will be presented. The immediate consequence is that analysis of results for polycrystalline surfaces becomes difficult if not impossible. This is illustrated for a series of systems of increasing complexity:
(i) Interface of 'ideal' structure with no 'specific adsorption'. The model consists of a 'space charge' of ions, a solvent monolayer and a more or less simple surface charge on the metal. Modern developments of this model will be discussed to assess how accurate the simple model may be; (ii) Interface with specific ionic adsorption. An explanation of the resultant modification of the capacity curves will be given with consideration of the possible occurrence of chemisorption with partial charge transfer; (iii) Interface as in (i) but with adsorption of non-ionic species. The consequent modification of the capacity will be discussed in terms of a simple molecular model; (iv) interface with ionic adsorption and non-ionic adsorption. The problems of analysing this complex but common situation is considered; (v) Adsorption with complete charge transfer will be discussed and the mobility of electrode surfaces exemplified.

1. INTRODUCTION

A considerable body of information about the behaviour of metal/electrolyte interfaces has been built up using mercury electrodes. These have the advantage of being readily renewed and so measurements can be made on fresh, clean surfaces. They are also uniform and structureless. However most practical electrodes are, of course, solid and it is important to know what differences in behaviour may occur with solid metals.

The principal problems in obtaining reliable results with solid metal electrodes are their cleanliness and their structure. Rather small amounts of impurities (10^{-9} - 10^{-10}mol will cover a cm^2) may modify their properties substantially and can accumulate at the surface during the time of an experiment. Surface structures prepared under controlled conditions may not be maintained when the electrode is put in contact with the electrolyte. Satisfactory solutions to these problems have been developed only in the past few years. In the previous three decades electrochemists used empirical procedures [1] such as repetitive cycling of the electrode potential using a triangular wave form until a steady state cyclic voltammogram was achieved (the cyclic voltammogram is a diagram of current flowing as a function of the potential which is varied linearly with time). Internal evidence such as charge balance between oxidation and reduction of surface species was used in favour of the cleanliness of the electrode and reproducibility could be obtained. However, the actual state of the surface was unknown and there were controversies, such as that over the question as to the extent of penetration of oxygen into a Pt electrode, which could not be settled by these methods.

2. THE PREPARATION OF CLEAN, WELL-DEFINED ELECTRODES

Three methods have been developed in the past decade for the study of electrochemistry on electrodes whose surfaces are well-defined.

The earliest and still the most satisfactory in that control of the electrode is the most rigorous was pioneered by Hubbard [2]. The electrochemical cell is contained in a compartment attached directly to the UHV system in which surface characterization by Low Energy Electron Diffraction, Auger Spectroscopy, etc. can be carried out. After preparation of the surface of the crystal and characterization it is transferred to the adjoining compartment where the pressure can be raised by the admission of an inert gas. The electrolytic cell is then introduced and the electrode put in contact with the electrolyte. Two configurations have been used: an electrode in the form of a prism with all faces cut to give the same surface structure is immersed completely in the electrolyte, or a single face is put in contact with a thin layer of electrolyte supported on the counter electrode. The latter has the advantage of using a very small quantity of electrolyte, thus reducing the risk of contamination from this source. In favourable cases, the reverse process of emersing the crystal from the electrolyte and returning it to the UHV chamber can be achieved successfully and the surface state monitored. The operations at the higher pressure are particularly susceptible to contamination and great care is necessary in the purification of the inert gas and the electrolyte as well as in their manipulation in the system.

The second method is a variation on the first in that the transfer from UHV to electrochemistry is achieved using a chamber detachable from the UHV system [3]. This is then attached to a glove box and filled with the high purity inert gas circulating in it. The electrode can then be transferred to a conventional electrochemical cell within the glove box. The electrochemical experiments can be carried out within the glove box, or the cell can even be removed from it. The major problem here is the maintenance of a large volume of gas in a satisfactory state of purity, but it has been demonstrated that quite satisfactory transfers in both directions are achievable.

In a third method [4,5] no direct transfer is attempted, but parallel experiments are done on a pair of identically treated electrodes. One is used in the electrochemical experiment while the other is put into the UHV analysis chamber. Such a procedure of course depends on the possibility of preparing two identical electrodes. The high sensitivity of the electrochemical experiments can be used to verify this. An important aspect of the practice of this third method was the emphasis on the use of the complete electrochemical biography of the electrode. The current was monitored from the first moment of contact with the electrolyte, which was made at a potential controlled by a potentiostat. Thus oxidation or reduction processes amounting to ∿ 1% of a monolayer or more can be observed and used to follow any changes in the state of the electrode. This technique is not exclusive to the third method but can, in principle, be used with both the others.

3. IN SITU VERIFICATION OF THE STATE OF AN ELECTRODE SURFACE

Even with a perfect system of transfer between a UHV analysis chamber and an electrochemical cell, there must remain the possibility that a change in state of the electrode surface could occur in transfer which is reversed on reverse transfer. Hence methods of study of the electrode <u>in situ</u> are of great interest. Surface composition can be studied <u>in situ</u> by the spectroscopic methods mentioned in the previous article but these have not yet developed to the stage where a complete surface analysis is possible. The electrochemical techniques described in the previous section remain the most convincing evidence that the surface is free from contamination, although the cleanliness of the electrode transferred back into the UHV chamber is strong confirmation of this.

It is now known that surfaces of metals are remarkably mobile even at small fractions of the melting point of the bulk metal (see below). Hence the possibility of a reversible structural change is perhaps more difficult to exclude than a reversible composition change. For this reason an <u>in situ</u> probe of surface structure is much to be desired. While the possibility of X-ray [6] and neutron [7] diffraction has been demonstrated, they have not yet reached the stage of development sufficient for the characterization of the surface layer of a single crystal electrode. At present, the most direct evidence comes from modulated UV/visible spectroscopy. Here a beam of light is reflected from the metal/electrolyte interface while the electrode potential is modulated for example by a 40 mV sine wave signal. The signal from the photomultiplier receiving the reflected beam is amplified using a phase-sensitive detector (lock-in amplifier) with the modulating signal as a reference. The change in reflectivity due to the potential modulation is thus recorded. Since the change of electrical potential can penetrate into the metal only to a distance of the order of the Thomas-Fermi screening length, the modulation in the reflectivity must be due to changes within the electrical double layer including the first layer of metal atoms. The spectra obtained with single crystal electrodes are found to depend markedly on the orientation of the single crystal [8]. Normal incidence spectra are azimuthally anisotropic for orientations whose surface symmetry is lower than binary while surfaces of high symmetry like the (111) and (100) show isotropic reflectivity [9]. This is the most direct structural information obtained so far in situ but it does not distinguish surface structures having similar symmetry. The metal contribution to electro-reflectance spectra has been interpreted in terms of surface states [10,11] and it is conceivable that surface structure might be determinable in more detail this way.

4. THE ELECTRLCAL DOUBLE LAYER AT SOLID METAL ELECTRODES

Information about the double layer at solid electrodes is most conveniently obtained by the measurement of their differential capacity. Despite some residual problems with the frequency dispersion and the lack of a completely capacitative behaviour [12], it has proved possible to obtain satisfactory data for a number of sp metals [13,14]. Of the relatively high melting metals, silver appears to provide the most reliable results and recent experiments [15] have suggested that the least strongly adsorbed anions on this metal are BF_4^- and PF_6^- (unlike Hg where the F^- ion seems to be least adsorbed). Hence the simplest behaviour will be observed in solutions of a salt like KPF_6.

The correlation between the work function and the pzc mentioned in the previous article together with the known orientation dependence of the work function (as seen in the field emission microscope) leads immediately to the idea that the pzc depends on the crystal orientation [16]. This is indeed found experimentally and the denser planes are found to have higher (more positive) potentials of zero charge. Differences of several hundred mV have been found e.g. Ag(111):E_z = -0.69V; Ag(110): E_z = -0.98V (SCE). These values have been determined from the minimum capacity in dilute solutions as described in the previous article. The fact that the pzc depends on the nature of the crystal plane exposed to the electrolyte leads immediately to an idea of the difficulty of dealing with a polycrystalline electrode, which can be considered as made up of small patches of the various possible orientations. An electrode maintained at a given potential will then have patches which carry different charges. For example, on a polycrystalline Ag electrode held at 0.8V (SCE), the (111) patches would carry a negative charge while the (110) patches would carry a positive charge. This example shows how the double layer structure becomes complex and necessarily three-dimensional for a polycrystalline electrode, even in the simplest case when there is no specific adsorption. Valette and Hamelin [17] discussed this problem in detail and were able to produce a capacity curve approximating to that of a polycrystalline electrode by combining the data for the three low index planes in the proportion that these were observed on the etched polycrystal. Clearly this is a simplification and this example shows the impossibility of disentangling the true behaviour of homogeneous parts of an electrode from the average results obtained using a polycrystal, even with this simplest double layer structure. Even the concept of a pzc of a polycrystal is difficult. Experimentally it is found that the minimum in the capacity curve in dilute solutions is close to that of the least dense plane (110) which has the most negative pzc. As Valette and Hamelin showed, this is simply because the contributions of the various patches on the surface are in parallel and their capacities sum. The low capacities at the more positive potentials are masked by the high capacities due to other parts of the electrode which carry positive charges at these potentials. The model of a poly crystal under these simple conditions was discussed further by Grigoryev [18], Bagotskaya et al [19] and Vorotyntsev [20] who considered the effect of the patch size.

5. ADSORPTION OF UNCHARGED MOLECULES

It is evident from the above discussion that understanding of adsorption at solid electrodes can come only from studies of single crystals, the data from polycrystals being virtually impossible to analyse, except perhaps for low melting metals whose pzc depends very little on crystal orientation. Early measurements of the capacity of Ag(111) and Ag(100) in solutions containing n-butanol were presented by Vitanov and Popov [21]. These show the behaviour that studies of such adsorption on Hg have indicated as characteristic of neutral molecule adsorption. The capacity is lowered in the region of the pzc where adsorption is strongest. This region is bounded in both directions by a sharp rise of the capacity to a peak well above that of the base solution. At the extremities the capacity returns to that of the base solution indicating that the neutral species is not adsorbed at the strongly positively or negatively charged surface. On Hg these capacity peaks are strongly frequency dependent as a result of the slowness of the adsorption-desorption process which is controlled by diffusion of the molecule to and from the interface and perhaps also by the adsorption process itself. The results obtained by Vitanov and Popov [21] indicate that these peaks are not fully developed at the ac frequency used for the measurement, suggesting that the adsorption process is much slower than similar processes on Hg.

Similar results were found for the adsorption of diethyl ether on Au single crystals [22] although the measurement of electroreflectance showed that the adsorbed film is transparent and does not perturb the spectrum of the gold. This suggests that weak physical adsorption is occurring. Direct measurement of the capacity does not lead to equilibrium values although the behaviour of each low index surface is clearly characterized. Thermodynamic analysis to obtain adsorbed amounts was based on direct measurement of charge [23] which could be extended to longer times. This showed that the parameters of the adsorption isotherm depend on the geometry of the surface as well as the way these depend on the surface charge. Thus even in the case of physical adsorption the adsorption is specific to the crystal face; this is most likely to be due to the interaction of water with the gold surface, since adsorption of ether necessarily involves the displacement of solvent.

More complex behaviour is observed with molecules that are more strongly adsorbed such as pyridine [24] or when there is anionic adsorption as well as neutral molecule adsorption as in the example of pyridine adsorbed on Ag in chloride solutions [25].

6. SPECIFIC IONIC ADSORPTION

The simple model of the ionic part of the interphase described in the previous article is valid only for a limited number of electrolytes. In a larger number of cases it is quantitatively applicable for electrodes bearing a negative charge when the cation in solution is a simple inorganic species like an alkali metal ion. For the majority of anions, substantial deviations from the simple model occur at positively charged electrodes. Adsorption of the anion is much greater than the Gouy-Chapman theory would predict, so much so that the cation is also positively adsorbed. Also in contrast to the simple electrostatic theory, the adsorption is strongly dependent on the nature of the ion and not just on its charge. The term 'specific adsorption' is therefore used to denote this; such a nomenclature avoids the implication of a particular mechanism and is used operationally by determining the deviation of the adsorption from that predicted at the given electrode charge from Gouy-Chapman theory.

Detailed analysis of the specific adsorption has been carried out on rather few solid metals, for example Cl^- on Ag(100) [26] using capacitance measurements and Br^- on Au single crystals using chronocoulometric measurements [27]. As in the adsorption of neutral molecules there are specific differences between the different crystal faces, but some general features may also be noted. The amount of anion specifically adsorbed from a solution of given concentration increases approximately linearly with the charge on the metal up to a more or less well-defined limit which corresponds to that expected for a monolayer. This suggests that the anions are bonded to the metal by short range forces which may well be thought of as chemical bonds. Although this relation between amount adsorbed and electrode charge is rather featureless, the specific adsorption produces marked effects on the capacity curve which can be summarized as a broad peak at the more negative potentials where a very small degree of adsorption is present ($\sim$ 5% of the monolayer), a second broad peak at more positive potentials where the monolayer is about half completed, and a sharp peak at the most positive potentials where the monolayer is almost complete. These features are separated to an extent which depends on the crystal face and the particular metal and anion.

The first peak can be understood by considering the modification to equation (17) of the previous article due to the presence of a specifically adsorbed monolayer carrying a charge density σ^i (here the adsorbed ions are assumed to carry their full charge i.e. the charge they have in solution). Equation (17) must then be written

$$d\phi^M/d\sigma = (d\phi^i/d\sigma) + (d\phi_2/d\sigma^d)(d\sigma^d/d\sigma) \tag{1}$$

where σ^d is the charge on unit area of the diffuse layer and

$$\sigma^{d} = -(\sigma + \sigma^{i}) \tag{2}$$

It follows then that

$$C^{-1} = (C^{i})^{-1} + (C^{d})^{-1}\,[1 + (d\sigma^{i}/d\sigma)] \tag{3}$$

At low amounts adsorbed, the coefficient $d\sigma^{i}/d\sigma$ starts from zero and, as $|\sigma^{i}|$ increases it decreases through -1 and then limits at a value of about -1.4. In the region when its value is about -1 it is clear from equation (3) that the effect of the diffuse layer is eliminated and the observed capacity is closely equal to the inner layer capacity. When this situation occurs close to the pzc for Au or Ag electrodes a peak related to the reorientation of water molecules is observed as described in the previous article. The second peak arises because the inner layer capacity is dependent on the slope $d\sigma^{i}/d\sigma$. The σ^{i}/σ curve has the form of a rather flat sigmoid with a maximum slope in the region of the half completed monolayer. The third, sharp peak has been ascribed to a reconstruction of the monolayer accompanied by a substantial transfer of charge from the anions to the electrode, but clear evidence for this has not yet been obtained.

This interpretation may be applied in a qualitative way to the systematic variations in the capacity curve observed when higher index, stepped, surfaces are used [28] but no detailed analysis has been been made. However, it must be noted that all the measurements discussed so far in this section were made in mixed electrolytes of constant ionic strength where the salt of the anion in question was mixed with a fluoride of the same cation. As mentioned above, recent studies by Valette [15] have shown that fluoride ion is appreciably adsorbed on Ag electrodes and that the PF_6^- is a better choice for a non-adsorbed ion. Studies of the adsorption of Br^- and Cl^- on Ag(100) have shown significant differences [29] although these do not appear to invalidate the general interpretation given above. A more serious objection may be the assumption that anions retain their full charge up to relatively large fractions of the monolayer. The shift of pzc under conditions where the diffuse layer charge is zero may be estimated from these experiments; it is found to be linear with the amount adsorbed with a slope comparable to that found for the change in work function in the adsorption of the same halogen atom from the gas phase. This slope is much less than would be expected for ionic adsorption e.g. it corresponds to a dipole of unit charges separated by about 10 pm. If this is a consequence of partial transfer of the charge from the anion, it occurs at the lowest densities of adsorption.

7. CHEMISORPTION WITH CHARGE TRANSFER AND THE MOBILITY OF METAL SURFACES

In the case of platinum electrodes, large capacities are observed at potentials extending in the positive direction from the reversible hydrogen potential. These are generally accepted to arise from the adsorption of protons from the solution with electron transfer to form adsorbed H atoms. The form of these capacity curves, which is most conveniently studied using cyclic voltammetry, is a sensitive indicator of the surface structure of the electrode and some results will be summarized here.

{111} surface [30]

A crystal annealed at high temperature, quenched with pure water and brought into contact with 0.5M H_2SO_4 at a potential of 0.6V RHE (with respect to a reversible hydrogen electrode in the same solution) shows a reduction transient which can be identified with oxygen adsorbed from the atmosphere. Cycling in the region 0.05 to 0.6V RHE shows a broad region of current well above that expected for the double layer charging and stretching from 0.05 to nearly 0.5V RHE. The oxidation current on the positive sweep is virtually identical at each point to the reduction current on the negative sweep and this symmetry is maintained to sweep speeds of up to 50V s^{-1}. It therefore corresponds to a fast process. The charge corresponding to this region is 255 μC cm^{-2} which will include a double layer contribution which is probably quite small. The charge calculated for an ideal {111} plane assuming one electron transferred per Pt atom is 243 μC cm^{-2}. Extension of the potential sweep to 1.2V (RHE) showed no other features above what might be expected from the double layer charging. This should be contrasted with the behaviour of polycrystalline Pt where the currents attributed to hydrogen adsorption do not extend above 0.3V and those attributed to oxygen adsorption begin already at 0.8V. This unusual voltammogram is stable for many cycles between 0.05 and 1.2V (RHE) (denoted here type C). However, further extension of the potential sweep to 1.5V leads in the first cycle to an oxidation charge of 550 μC cm^{-2} and an equal reduction charge in a peak centred at 0.72V RHE having a form very like the reduction peak of adsorbed oxide on polycrystalline Pt. At the same time the form of the current at potentials negative of 0.5V is substantially modified although the total charge involved is unchanged. The modification can be described by saying that the most positive third of the charge disappears and is replaced by a highly peaked region with the peak potential at 0.12V. The voltammogram is now stable for many cycles and has almost exactly the characteristics of that published by Hubbard et al [31] for this orientation (type H voltammogram).

The anion present in the solution has a marked effect on the form of a type C voltammogram [32]. If 1M perchloric acid is used in place of 0.5M sulphuric, the region of current between 0.05 and 0.5V splits into two unequal parts. The larger corresponding to ⅔ of the charge and at the more negative potentials is unaffected by the change of acid while the more positive ⅓ moves to more positive potentials becoming located between 0.6 and 0.82V (RHE). This voltammogram has the characteristics described above when the upper limit of the potential sweep is 0.9V i.e. the anodic and cathodic currents are equal at each potential up to 50V s^{-1} and the voltammogram is stable. In addition variation of the acid concentration over two orders of magnitude leaves the position of both regions unchanged on the RHE scale i.e. they both move following the hydrogen electrode. Addition of progressive amounts of sulphuric acid causes the more positive ⅓ to move progressively towards the more negative ⅔.

Extension of the upper limit of potential cycling to 1.15V yields a sharp anodic peak at 1.06V with a charge corresponding to 130 μC cm^{-2}. At the same time the cathodic peak at 0.8V is enlarged by the same amount of charge. Nevertheless the voltammogram remains stable and the change to a type H is achieved only by an extension to higher potentials and larger charges. Recently a careful LEED study by Wagner and Ross [38] has indicated that the change from voltammograms of type C to type H is associated with the disappearance of long range order and the formation of a disordered stepped surface. This confirms earlier suggestions by Clavilier that a structural change was involved and is consistent with Motoo's observations [34] of the effect of oxygen on the annealing process.

Disagreements still remain about the interpretation of the voltammogram. It seems to be generally accepted that oxidation currents above 0.8V (RHE) are associated with the deposition of adsorbed oxygen species. It then seems necessary to accept that in the presence of sulphuric acid (HSO_4^- ions) this deposition is strongly inhibited up to 1.2V. It seems unlikely that this is due to contamination since the oxidation up to 1.5V yields a very precise charge balance with the associated reduction peak, so it must be concluded that oxide species adsorb weakly on this type of surface. Once they are forced to adsorb and a monolayer (550 μC cm^{-2}) is formed and removed the surface structure changes and the oxygen species adsorb in the normal way. This surface reconstruction is progressive; if less than a monolayer is adsorbed it is less complete. On the other hand in the absence of adsorbed anions ($HClO_4$) there is a threshold of oxygen adsorption (130 μC cm^{-2}) before reconstruction is initiated. This may be correlated with the observation in UHV that adsorption of oxygen up to a 2 x 2 overlayer does not cause reconstruction of a (1 x 1) Pt(111) surface [35]. Under electrochemical conditions this behaviour seems to be maintained in the presence of water and weakly adsorbed anions but considerably modified by the more strongly adsorbed anion.

{100}

The behaviour of surfaces with the {100} orientation is not less interesting although less spectacular in that no such unusual hydrogen adsorption states are observed. The voltammograms show usually two peaks in the hydrogen region at +0.27 and +0.37V (RHE). The total charge in these peaks if integration is continued up to +0.15V, where there is a minimum in the current, is 205 ± 5 μC cm^{-2}. This is close to that calculated for an ideal (1 x 1) Pt(100) plane which is 209 μC cm^{-2}. However, the relative height of the two peaks depends on the pretreatment before electrolyte contact as well as in the electrochemical cell [36]. A series of experiments in which the electrode was cooled in different atmospheres and quenched with pure water in equilibrium with different atmospheres at different temperatures showed that the electrode adsorbed different amounts of oxygen before contact with the electrolyte. This could be demonstrated by contacting at a potential of +0.9V where no transient current was observed. On the first negative sweep a cathodic peak at 0.785V could be observed which was attributed to the reduction of thermally adsorbed oxygen. The largest peak obtained corresponded to a charge of 210 μC cm^{-2} which could be related to the 2 x 2 overlayer [37]. This could be reduced to a very small amount by cooling the electrode in hydrogen and quenching in water saturated in hydrogen. In the first case the peak at 0.37V was largely suppressed at the expense of the peak at 0.27V on the negative sweep whereas in the second case the peak at 0.37V was much enhanced and that at 0.27V much diminished. Similar effects could be produced by electrochemical treatment in the hydrogen or oxygen regions. Thus the electrode in the first case above showed a slightly different peak distribution on the following positive sweep as the peak at 0.37V began to reappear, presumably as a result of the adsorption of hydrogen to saturation. Cycling between 0.08V and 0.55V (i.e. without oxygen adsorption) gradually produced a peak distribution like that observed in the second case above. The process could be reversed by cycling into the oxygen region. It seems that adsorbed oxygen stabilizes a structure in which the more weakly hydrogen adsorbing sites predominate again related to the absence of long-range order. The fact that these changes are reversible under electrochemical conditions is particularly interesting. It contrasts with the behaviour of the {111} plane where the reverse process from voltammograms of type H to those of type C has only been achieved by high temperature annealing.

{110}

Electrodes of this orientation prepared by high temperature annealing show apparently simple behaviour [36] with a sharp peak at 0.12V which with its shoulders corresponds to 220 μC cm^{-2}. The oxygen adsorption begins close to 0.8V and has a structure somewhat different shape in detail from that of polycrystalline electrodes. Subsequent cycles show only subtle changes in the peak structure compared with those of the other low index planes. In fact, the agreement with other published results [38] [39] for this orientation is good indicating that this plane is less sensitive to the details of its preparation.

Experiments with high index, stepped, surfaces have confirmed [40] these interpretations and have enabled the behaviour of polycrystalline electrodes to be understood. Such experiments show that simple electrochemical techniques can provide accurate and useful information on the surface properties of metals. They also show how remarkably mobile, even Pt surfaces can be at temperatures far below the melting point. This could be an important factor in the widespread catalytic activity of Pt.

REFERENCES

[1] See for example: R. Parsons, Surface Science 101 (1980) 316.
[2] R.M. Ishikawa and A.T. Hubbard, J. Electroanal. Chem., 69 (1976) 317.

[3] J.P. Bellier, J. Lecoeur and A. Rousseau, J. Electroanal. Chem. 200 (1986) 55.
[4] J. Clavilier and J.P. Chauvineau, J. Electroanal. Chem., 97 (1979) 109/
[5] J. Clavilier and J.P. Chauvineau, J. Electroanal. Chem., 100 (1979) 461.
[6] M. Fleischmann, P. Graves, I. Hill, A. Oliver and J. Robinson, J. Electroanal. Chem., 150 (1983) 33.
[7] G. Bomchil and C. Rickel, J. Electroanal. Chem., 101 (1979) 133.
[8] T.E. Furtak and D.W. Lynch, Phys Ref. Letters, 35 (1975) 960.
[9] T.E. Furtak and D.W. Lynch, J. Electroanal. Chem., 79 (1977) 1.
[10] K.M. Ho, C-L. Fu, S.H. Liu, D.M. Kolb and G. Piazza, J. Electroanal. Chem., 150 (1983) 235.
[11] S.H. Liu, C. Hinnen, C. Nguyen Van Huong, N.R. Tacconi and K.M. Ho, J. Electroanal. Chem., 176 (1984) 325.
[12] G.J. Brug, A.L.G. Van den Eeden, M. Sluyters-Rehbach and J.M. Sluyters, J. Electroanal. Chem., 176 (1984) 273.
[13] A. Hamelin, T. Vitanov, E. Sevastyanov and A. Popov, J. Electroanal. Chem., 145 (1983) 225.
[14] A. Hamelin, 'Modern Aspects of Electrochemistry', Ed. B.E. Conway, R.E. White and J. O'M Bockris, 16 (1985) 1.
[15] G. Valette, J. Electroanal. Chem., 122 (1981) 285, 138 (1982) 37.
[16] R. Parsons, Surface Science, 4 (1964)
[17] G. Valette and A. Hamelin, J. Electroanal. Chem., 45 (1973) 301.
[18] N.B. Grigoryev, Dokl. Akad. Nauk. S.S.R. 229 (1976) 647.
[19] I.A. Bagotskaya, B.B. Damaskin and M.D. Levi, J. Electroanal. Chem., 115 (1980) 189.
[20] M.A. Vorotyntsev, J. Electroanal. Chem., 123 (1981) 379.
[21] T. Vitanov and A. Popov, Trans, S.A.E.S.T. 10 (1975) 5.
[22] C. Nguyen Van Huong, C. Hinnen, J.P. Dalbera and R. Parsons, J. Electroanal. Chem., 125 (1981) 177.
[23] J. Lipkowski, C. Nguyen Van Huong, C. Hinnen, R. Parsons and J. Chevalet, J. Electroanal. Chem., 143 (1983) 375.
[24] A. Hamelin and G. Valette, Compt. Rend. Acad. Sci. (Paris), Ser C 267 (1968), 127, 211.
[25] M. Fleischmann, J. Robinson and R. Waser, J. Electroanal. Chem., 117 (1981) 297.
[26] G. Valette, A. Hamelin and R. Parsons, Zeit Physikal Chem., 113 (1978) 71.
[27] C. Nguyen Van Huong, C. Hinnen and A. Rousseau, J. Electroanal. Chem., 151 (1983) 149.
[28] A. Hamelin and J.P. Bellier, Surface Sci., 78 (1978) 159.
[29] G. Valette and R. Parsons, J. Electroanal. Chem., 191 (1985) 377.
[30] J. Clavilier, R. Faure, G. Guinet and R. Durand, J. Electroanal. Chem., 107 (1980) 205.
[31] A.T. Hubbard, R.P. Ishikawa and J. Katekara, J. Electroanal. Chem. 86 (1978) 271.
[32] J. Clavilier, J. Electroanal. Chem., 107 (1980) 211.
[33] F.T. Wagner and P.N. Ross, J. Electroanal. Chem., 150 (1983) 141.
[34] S. Motoo and N. Furuya, J. Electroanal. Chem., 172 (1984) 339.
[35] J. Gland, Surface Sci., 93 (1980) 487.

[36] J. Clavilier, D. Armand and B.L. Wu, J. Electroanal. Chem., 135 (1982) 159.
[37] B. Lang, P. Legare and G. Maire, Surface Sci., 47 (1975) 89.
[38] K. Yamamoto, D. Kolb, R. Kotz and G. Lempfuhl, J. Electroanal. Chem., 96 (1979) 233.
[39] P.N. Ross, J. Electrochem. Soc., 126 (1979) 67.

3 - ELECTROCATALYSIS

ABSTRACT. The specific features of electrode reactions will be summarized and the origin of the potential-dependent rate constant discussed. Limiting types of electrode reaction will be defined in terms of degree of interaction with the interface and the further discussion will be limited to electrocatalytic reaction. The broad characteristics of an idealized electrocatalytic reaction will be derived and shown to lead to a 'volcano' type of relation. This will be compared with experimental results and its deficiencies as a model discussed particularly as concerned structural effects for example on hydrogen adsorption. The experimental results for the oxidation of small organic molecules will be considered together with their implications for models of electrocatalytic reactions. The effects of surface geometry and of underpotential deposition will be described and current explanations of these discussed in terms of modern spectroscopy for the study of adsorbed species. Correlations with catalytic studies in UHV may or may not be relevant to reactions under electrochemical conditions. How soon can electrocatalysts be designed?

1. INTRODUCTION

The term 'catalysis' is used to describe the acceleration of a reaction by an agent which itself remains unchanged. The reaction may be homogeneous or heterogeneous depending on whether the catalyst is a species soluble in the reaction medium or is a separate phase. The extension of this term to electrode reactions involves some difficulty of definition because electrode reactions are inevitably heterogeneous. Nevertheless two basic classes of electrocatalysis can be distinguished. The analogue of homogeneous catalysis may be considered to be a reaction which is catalysed by a species in solution which is generated in an electrode reaction [1]. The analogue of heterogeneous catalysis in contrast is a reaction where the electrode provides a favourable path by way of an adsorption site. Here the second type - heterogeneous electrocatalysis will be discussed.

2. KINETICS OF ELECTRODE REACTIONS (2)

Electrode reactions by definition occur with a transfer of charge between two phases. Often this is associated with the change of charge carrier e.g. from electrons in a metal to ions in an electrolyte. The interfacial nature of the reaction means that reacting species must be transported to and from the interface. Although this process often plays a significant rôle in determining the rate of the overall process, it will be ignored here, where the discussion is concerned only with the rate of the interfacial reaction itself.

A formal description of a single step electrode reaction is

$$R_{(S)} \longrightarrow O_{(S)} + e^-_{(M)} \tag{1}$$

where R is a reduced form of a species and O the corresponding oxidized form, the subscript S indicating that both are present in solution; $e^-_{(M)}$ is an electron in the metal. The charges on R and O are not indicated but it is evident that they differ by one electronic charge. If the rate of the electrode reaction is uniform over the electrode surface, it is convenient to define a rate on unit area $\vec{v}$ in the forward (anodic) direction of equation (1) and $\overleftarrow{v}$ for the reverse (cathodic) direction. These rates are dependent on the reactant concentration in the normal way for a first order chemical reaction

$$\vec{v} = \vec{k}\,[R] \tag{2}$$

$$\overleftarrow{v} = \overleftarrow{k}\,[O] \tag{3}$$

where the square brackets indicate concentrations close to the electrode surface, and the k's are rate constants. Since the rates have dimensions of amount of substance $(\text{area})^{-1}$ $(\text{time})^{-1}$ and the concentrations amount of substance $(\text{volume})^{-1}$ it follows that the rate constants have dimensions of length $(\text{time})^{-1}$, i.e. a linear velocity.

The particular feature of rate constants of electrode reactions is that they depend on the potential applied to the electrode. This is an inevitable result of the fact that the reaction involves transfer of charge across the interface. This means that the Gibbs energy change accompanying reaction (1) depends on the electrode potential and hence the ratio of forward to reverse rates must also depend on potential. The potential dependence of the rate constant is usually expressed in the form

$$\vec{k} = k_s \exp[(1-\alpha)e\eta/kT], \tag{4}$$

$$\overleftarrow{k} = k_s \exp[-\alpha e\eta/kT], \tag{5}$$

where k_s is the standard rate constant, i.e. that at the equilibrium potential of the electrode, η is the overpotential, i.e. the deviation of the electrode potential from its equilibrium value (E_e)

$$\eta = E - E_e. \tag{6}$$

α is a proper fraction known as the transfer coefficient. The rate of an electrode reaction is conveniently measured as a current because each elementary act of reaction (1) is accompanied by the transfer of unit charge across the interface. The current density is then

$$j = e(\vec{v} - \overleftarrow{v}) \tag{7}$$

At the equilibrium potential the two terms on the right-hand side of equation (7) are equal and each is known as the exchange current density, j_o

$$j_o = e\vec{v} = e\overleftarrow{v} = ek_s[R] = ek_s[O]$$

These equations can be used for more complex reactions which involve several steps, using the usual principle of chemical kinetics. In the simplest cases it is necessary only to replace e in equations (4), (5), (7) and (8) by ne, where n is the number of unit charges transferred in a unit electrode reaction, and the concentration factors in (2), (3) and (8) by others which may be of higher order.

3. NON-ELECTROCATALYTIC REACTIONS

Electrode reactions which occur without heterogeneous electrocatalysis are those for which the interaction of the reactants with the electrode material is so weak that it is sufficient only to allow the transfer of electrons between reactant and electrode [3]. The electron transfer is then closely analogous to the homogeneous electron transfer known as 'outer sphere' [4]. Models for this type of reaction were developed particularly by Hush [5] and by Marcus [6] and later by Levich et al [7] and Dogonadze et al [8]. In these the importance of the interaction of the ionic charge with its surroundings is emphasized. If the fluctuations of the environment are represented as harmonic oscillators, the vertical (Franck-Condon) transfer of the electron in equation (1) from R in its equilibrium state will result in the formation of O in an environment of energy λ greater than its equilibrium value. This route for the reaction would lead to a rate many order of magnitude slower than that observed, because of the low probability of the high energy electron transfer. A more favourable route can be found when electron transfer occurs during a fluctuation of the ion's environment. The calculated rate constant then takes the form:

$$k_s = Z \exp(-\lambda/4kT) \tag{9}$$

where Z may be regarded as the probability that the ion is close to the electrode surface where it can react. The parameters of equation (9) depend essentially on the nature of the reacting ion, with no dependence on the nature of the electrode material (although there is a dependence of Z on the density of electronic states in the metal at the Fermi level see [9]). This is due to the absence of strong interaction between ion and electrode assumed in the model and it is the characteristic of non-electrocatalytic reactions. Simple electron transfer reactions are found to follow this type of behaviour and some examples are given in Table I. Here it may be seen that the rate constant for a given reaction varies over a range of about an order of magnitude as the electrode material is changed.

Table I. Rates of 'simple electron transfer reactions at 25°C.

Redox system	Electrode	Solvent	Electrolyte	k_s/cm s^{-1}	α_c
MnO_4^{2-}/ MnO_4^-	Pt	H_2O	1M KOH	$1.2x10^{-2}$	0.30
"	Au	"	"	$8x10^{-3}$	0.35
"	Pd	"	"	$6.7x10^{-3}$	0.30
"	C	"	"	$2.1x10^{-2}$	0.25
"	Cu	"	"	$2.6x10^{-2}$	0.25
Q/Q^-	Pt	DMF	0.3M Et_4NClO_4	0.12	0.5
	Au	"	"	0.18	0.5
	Hg	"	"	1.3	0.5
	Pt	AN	"	0.31	0.5
$Fe(CN)_6^{4-}$/ $Fe(CN)_6^{3-}$	Pt	H_2O	1M KNO_3	$6.6x10^{-2}$	0.40
"	C	"	"	$6x10^{-3}$	0.50

4. HETEROGENEOUS ELECTROCATALYTIC REACTIONS

In contrast to the reactions discussed in the previous section, these electrocatalytic reactions involve strong interaction with the electrode surface, usually in the formation of one or more adsorbed species in close analogy to the heterogeneous catalytic reactions occurring in the gas phase. A relatively simple example of this type of reaction is the hydrogen evolution reaction which may be written

$$2S\text{-}H^+_{(S)} + 2e^-_{(M)} \longrightarrow H_2(g) + 2S_{(S)} \qquad (10)$$

where $S\text{-}H^+$ represents a proton attached to some other species in solution - normally H_2O in acid solution and OH^- in alkaline solution. The reaction is written here as a cathodic reaction, because this is what is normally studied, although the convention that anodic reactions are positive is retained.

Reaction (10) is of considerable practical importance as it occurs in many industrial electrolytic processes as a main reaction or as a side reaction. The reverse reaction occurs in most practical fuel cells in use at present. It is often a component reaction in corroding systems. Hence it is of importance to understand this reaction and how to modify its rate.

Reaction (10) is usually assumed to occur via three elementary reaction steps:

$$S\text{-}H^+_{(S)} + e^-_{(M)} \longrightarrow H_{(ads)} + S_{(S)} \qquad \text{(A)}$$

$$2H_{(ads)} \longrightarrow H_2(g) \qquad \text{(B)}$$

$$S\text{-}H^+_{(S)} + H_{(ads)} + e^-_{(M)} \longrightarrow H_2(g) \qquad \text{(C)}$$

The kinetics of these reactions are written in the same way as described above except that allowance must be made for the fact that the intermediate species $H_{(ads)}$ forms a chemisorbed monolayer. Thus for reaction A the rates are expressed for a uniform surface as

$$\vec{v}_A = \vec{k}_A\,[H^+]\,(1-\theta), \qquad (11)$$

$$\overleftarrow{v}_A = \overleftarrow{k}_A\,\theta, \qquad (12)$$

where θ is the fraction of the monolayer occupied by H_{ads}. Thus the factor $(1-\theta)$ expresses the fraction of the surface free of H_{ads} and available for the deposition of H_{ads}. Similar rate equations may be written for B and C.

At equilibrium (11) and (12) may be written (cf 8)

$$j_{O,A} = e\vec{k}_{A,e}[H^+]\,(1-\theta_e) = \overleftarrow{k}_{A,e}\,\theta_e, \qquad (13)$$

where the subscript e represents the equilibrium value of k or θ. Hence

$$\frac{\theta_e}{1-\theta_e} = \frac{\vec{k}_{A,e}}{\overleftarrow{k}_{A,e}}\,[H^+], \qquad (14)$$

which is a Langmuir isotherm expressing the adsorption process described by equation A. The adsorption constant

$\vec{k}_{A,e}/\overleftarrow{k}_{A,e}$ is related to the standard Gibbs energy of this reaction by the usual thermodynamic relation

$$\vec{k}_{A,e}/\overleftarrow{k}_{A,e} = \exp(-\Delta G/kT). \tag{15}$$

If (14) is solved for θ_e or $(1-\theta_e)$ and the result substituted into (13), the exchange current density can be expressed as

$$j_{0,A} = e\vec{k}_{A,e}[H^+]/\{1 + \vec{k}_{A,e}[H^+]/\overleftarrow{k}_{A,e}\} . \tag{16}$$

It is evident now that this exchange current density depends strongly on the strength of adsorption of the atomic H as expressed by the adsorption constant, equation (15). In particular, for weakly adsorbing surfaces, the adsorption constant is small and (16) reduces to

$$j_{0,A} = e\vec{k}_{A,e}\,[H^+] \tag{17}$$

while for strongly adsorbing surfaces, the adsorption constant is large and (16) reduces to

$$j_{0,A} = e\overleftarrow{k}_{A,e} \tag{18}$$

It is qualitatively evident that the rate of adsorption will increase with increase of the strength of adsorption while the rate of desorption will decrease. Hence it follows from (17) and (18) that the exchange current of reaction A will increase with the strength of adsorption initially (while the latter is relatively weak) and then decrease (when the latter is relatively strong). This is an expression of the principle enunciated by Sabatier [10] that for efficient catalysis the catalyst must adsorb the intermediate neither too strongly nor too weakly.

Further analysis [10] of the rate constants of reaction A leads to the conclusion that a plot of ln $j_{0,A}$ against $\Delta G/kT$ will have a slope of $-(1-\alpha)$ when ΔG is large and of α when ΔG is large and negative, the maximum in the curve occurring at $\Delta G/kT = 0$. This plot is of the same type as that designated a 'volcano' curve by Balandin [12] in the field of heterogeneous catalysis. It is in fact found experimentally for a large number of such reactions.

A similar type of analysis with analogous conclusions can be carried out for reactions B and C, the principal difference being that the slopes of the branches of the volcano curve for reaction B are twice as great because this reaction involves two adsorbed hydrogen atoms. Hence the predictions of the simple model: of a volcano curve, and wide variations of the rate as a function of the nature of the electrode material are independent of the nature of the detailed mechanism of the electrode reaction. This is confirmed by experiment as shown in fig.1 where it is also evident that the rate of this reaction can vary over several orders of magnitude in contrast to the rate of the reactions

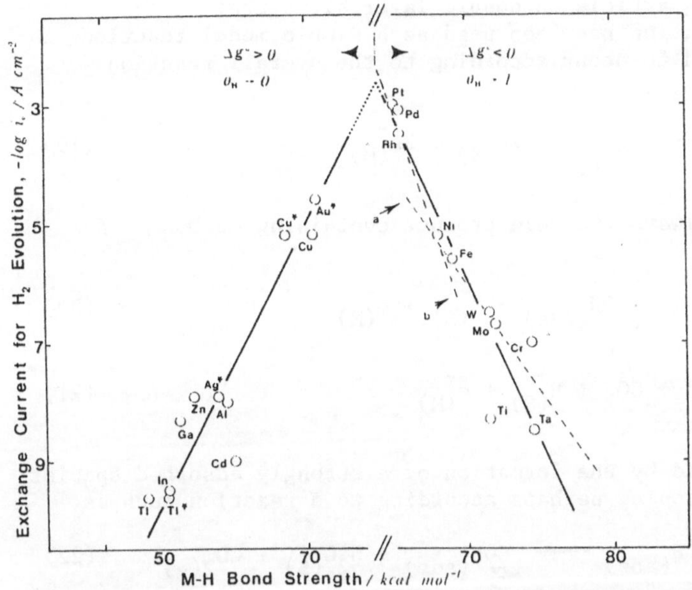

Exchange currents for electrolytic hydrogen evolution *vs.* strength of metal–hydrogen bond derived from heat of hydride formation[124] in the case of *sp* metals, and from heat of adsorption from gas phase[123] in the case of transition metals. Starred values refer to spectroscopic dissociation heat[18]. Value of adsorption heat for W from Eley and Norton[151]. Δg^0 is standard free energy of hydrogen adsorption. θ_H is surface coverage with atomic hydrogen. Arrows indicate theoretical slopes[110] for (a) ion + atom, (b) combination reaction.

shown in Table I.

The model described here is very simple. It can be made more realistic by considering the behaviour of a heterogeneous surface or of interaction between the adsorbed species. Only the region of the maximum of the volcano curve is affected by this improvement; it is made flatter, but otherwise there is no qualitative modification. Extension to reactions where there is more than one adsorbed intermediate is substantially more difficult because of lack of information about mixed adsorption and the way it depends on the nature of the adsorbent. The subsequent section is devoted to an account of simple organic oxidations where there is more than one intermediate.

5. OXIDATION OF METHANOL AND FORMIC ACID ON NOBLE METAL ELECTRODES

A considerable amount of work has been done on the electro-oxidation of small organic molecules because of the interest of methanol as a fuel for fuel cells. However, the reaction remains poorly understood partly because of its intrinsic complexity, involving two or three adsorbed intermediate states and partly because much of the work has been done using polycrystalline electrodes. For a structure sensitive reaction this leads to difficulties in the interpretation similar to those mentioned in the previous article on double layer structure.

Formic acid oxidation has been used as a simple model reaction [13]. It is believed to occur according to the overall reaction

$$HCOOH_{(S)} \longrightarrow CO_{2_{(g)}} + 2H^+_{(S)} + 2e^-_{(M)} \tag{19}$$

although CO_2 is not always the sole product containing carbon. A possible scheme is

$$HCOOH_{(S)} \longrightarrow \underset{x}{C}OOH_{(ads)} + H^+_{(S)} + e^-_{(M)} \tag{20}$$

$$COOH_{(ads)} \longrightarrow CO_2 + H^+_{(S)} + e^-_{(M)} \tag{21}$$

but this is complicated by the formation of a strongly adsorbed species which causes self-poisoning perhaps according to a reaction such as

$$2COOH_{(ads)} + H_{(ads)} \longrightarrow \underset{xxx}{C}OH_{(ads)} + H_2O_{(S)} + CO_{2(g)} \tag{22}$$

Because of the self-poisoning of the reaction, the rate is strongly time dependent and steady state measurements are difficult to interpret. Cyclic voltammetry has been used in preference partly because it defines the time scale of the experiment and partly because it enables surface species to be detected under controlled conditions.

The voltammogram for formic acid oxidation on Pt (Fig. 2) shows

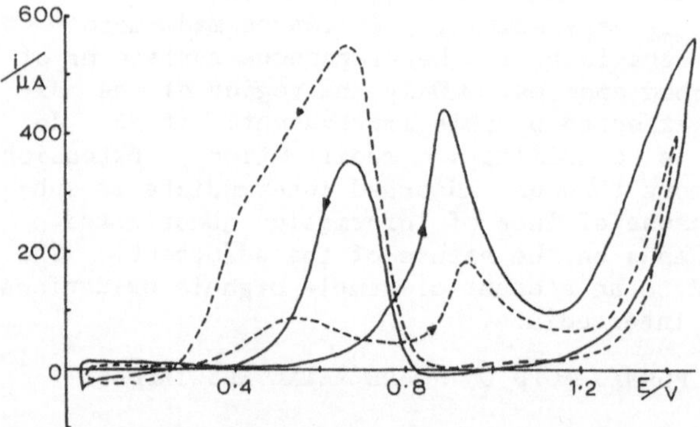

Fig. 2. Cyclic voltammograms of a smooth Pt electrode in 0.5 *M* H_2SO_4 containing: (-----) 1.0 *M* HCOOH, (——) ~1.0 *M* CH_3OH. Sweep speed 140 mV s^{-1}.

that hydrogen adsorption which normally appears in the region 0 to 0.3V (RHE) is strongly suppressed. This is due to the more strongly adsorbed species derived from formic acid, as the potential is swept to more positive values. There is a weak peak at ∿ 0.55V (RHE) (peak 1) and a somewhat stronger one at ∿ 0.95V (RHE) (peak 2) in the region where an oxygen monolayer begins to form on Pt. A larger current appears at 1.3V which is due to oxidation of formic acid on oxidized Pt. On the reverse sweep there is no current in the region of peak 2, nor is there a cathodic current due to the removal of the oxygen monolayer. The latter seems to be due to fact that it is masked by a large oxidation of formic acid on the newly liberated bare metal sites. The poisoning intermediate is then reformed as the potential passes through the more negative region. Thus the peak 1 is attributed to oxidation of formic acid on a poisoned surface while peak 2 is due largely to the oxidation of the poisoning intermediate (see [13] for more detailed evidence). The voltammogram for methanol shows similar poisoning characteristics except that peak 1 does not appear on the positive sweep.

The possibility of understanding this behaviour is much increased by the use of single crystal electrodes as well as the techniques described in the previous article. These showed, in fact, remarkable differences of behaviour of the three low index planes on the first sweep [14]. To avoid modification of the surface structure as far as possible the sweep was limited to an upper potential limit before oxygen adsorption began. The simplest behaviour is that of the (111) plane (Fig. 3) where a single broad peak extends from 0.2 to 0.8V (RHE) and is almost the same on the positive and negative sweeps. However, the

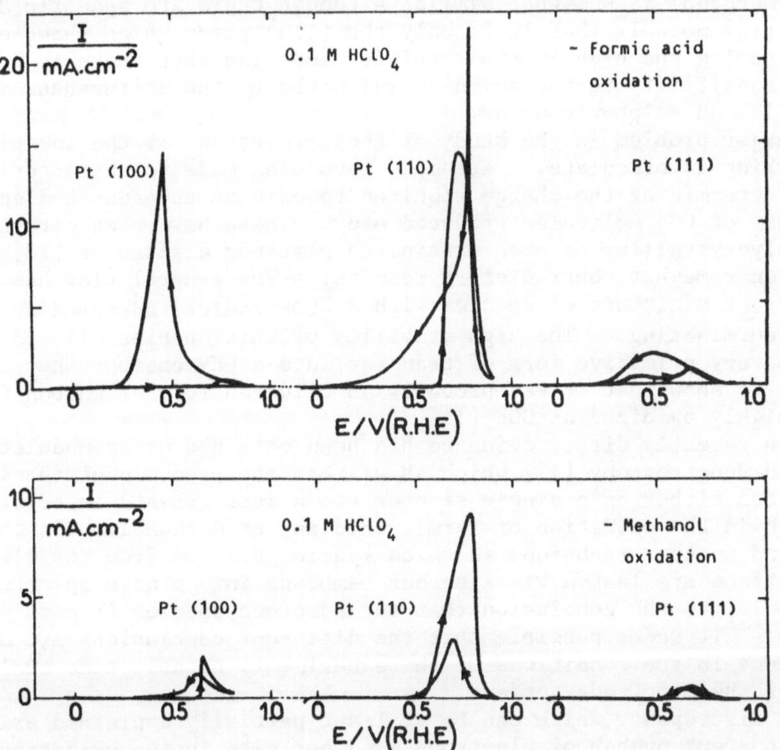

peak current is rather small being about 1.5 mA cm^{-2}. This indicates a rather low catalytic activity for this plane, which has the advantage that it does not lead to the formation of the blocking intermediate. This is indicated by the close similarity of the positive and negative sweeps in the first cycle as well as the fact that continued cycling causes very little change in the voltammogram.

The other two low index planes show much greater activity. The (100) plane is completely blocked in the positive sweep, no current appearing until 0.8V (RHE). Reversal of the sweep at 0.85V however leads to a large anodic peak on the negative sweep whose maximum is nearly 10 times that on the (111) plane. Evidently the blocking intermediate is formed rapidly as the electrode contacts the solution. The high activity of this plane results also in the formation of blocking species which cause a progressive decrease in the peak current on continued cycling. The behaviour of the (110) surface is somewhat similar except that the surface blocking is removed quite sharply on the first positive sweep at 0.66V to produce a high peak at 0.72V. After reversal of the sweep the anodic peak is lower and broader. Again continued cycling causes a progressive decay of the currents. As a result of this progressive poisoning of the (100) and (110) planes and the absence of this on the (111), the latter is the best operating catalyst for this reaction in the long term.

The behaviour of the low index planes of Pt in the electrooxidation of methanol is somewhat similar although there are specific differences [15] notably that it is only the (110) plane which shows extensive poisoning and high initial activity and also that there is a greater sensitivity to the anion as indicated by the differences in perchloric and sulphuric acids.

A major problem in the study of these reactions is the identity of the blocking intermediate. Attempts have been made to characterize this by determining the charge required to oxidize the adsorbed species, the number of CO_2 molecules produced etc. These have been carried out using polycrystalline or even platinized platinum electrodes [13] and have given somewhat contradictory results. The general view has been in favour of a mixture of species with a $\begin{smallmatrix}x\\x\\x\end{smallmatrix}$COH radical adsorbed on three sites predominating. The high stability of this species allowed the use of a very primitive form of transfer into a UHV chamber where XPS was used to show that it was probably an oxidized form of carbon though not as highly oxidized as COH [16].

More recently direct evidence has been obtained using modulated infra-red spectroscopy [17] which shows that the predominant species is CO adsorbed either on a single site or (to a less extent) in bridged form both in the oxidation of formic acid and of methanol. On the other hand another technique in which species derived from the electrode surface are leaked via a porous membrane into a mass spectrometer [18] has led to the conclusion that the adsorbed species is more probably COH. It seems possible that the different conclusions are due to differences in the conditions of the experiment, notably time-scale and nature of the electrode surface.

One discrepancy which can be at least partially explained arises in the apparent number of electrons used per site in the oxidation of this strongly bound intermediate. A predominance of single site adsorbed CO would require an oxidation charge corresponding to 2 electrons per site. This is larger than the charge observed in most previous work but the discrepancy may arise from two causes: the ill-defined nature of the electrodes used or the different time scales of the experiment. Hence it is of great interest to carry out experiments on well defined surfaces at time scales comparable to the usual electrochemical experiments.

This problem has been solved quite recently in the laboratories at Meudon by Clavilier and Sun [19]. Electrodes prepared as described above are brought in contact with the electrolyte containing the reacting species. They are then transferred to a cell containing the base electrolyte. This transfer is done with the electrode protected by a drop of the electrolyte containing the fuel. After contact with the pure electrolyte under potentiostatic control, time is allowed for the unadsorbed fuel to diffuse away.

A negative sweep is then used to investigate the fraction of the surface blocked by adsorbed species and the latter is oxidized on the subsequent positive sweep. The form of the oxidation peak is independent of the excursion into the hydrogen region. The area of the peak corresponds closely to a charge of 2 electrons per site in agreement with the proposed single bonded CO. Experiments with CO adsorbed directly yield somewhat similar oxidation peaks although the detailed form depends on the coverage. This demonstrates that CO evidently exists not only under the long term conditions of the early IR experiments. It is notable that the voltammograms on the single crystal have sharp peaks with well-defined fine structure, in great contrast to the broad unstructured peaks obtained in the dipping experiments with high area Pt electrodes [13].

The nature of the intermediate in these reactions remains incompletely understood, as does the detailed kinetics of these "simple" oxidation reactions. Without this understanding, it is difficult to reach a generalisation like that achieved with the volcano curve for the simpler type of reaction and so to be in a position to predict catalytic behaviour. The use of single crystal electrodes as well as of modern surface science techniques opens up a promising route to the solution of these problems and hence to the possibility of the rational design of specific and selective electrocatalysts.

REFERENCES

[1] C.P. Andrieux, J.M. Dumas-Bouchat and J.M. Saveant, J. Electroanal. Chem., **87** (1978) 39.

[2] W.J. Albery, 'Electrode Kinetics', Oxford, 1975.

[3] R. Parsons, Surface Science, **2** (1964) 418.

[4] W.L. Reynolds and R.W. Lumry, 'Mechanism and Electron Transfer, Ronald Press, New York, 1966.

[5] N.S. Hush, J. Chem. Physics, **28** (1958) 962.

[6] R.A. Marcus, J. Chem. Physics, **24** (1956) 966.

[7] V.G. Levich, Adv. Electrochem. Electrochem. Eng. 4, Eds P. Delahay and C.W. Tobias, Wiley, New York 1966, p.249.
[8] R.R. Dogonadze, J. Ulstrup and Yu I. Kharkhats, J. Electroanal. Chem., 39 (1972) 47.
[9] M.V. Vojnoviċ and D.B. Sepa, J. Chem. Physics, 51 (1969) 5344.
[10] P. Sabatier, Le catalyse en chimie organique Berange, Paris, 1920.
[11] R. Parsons, Trans Faraday Soc., 54 (1958) 1053.
[12] A. Balandin, 'Advances in Catalysis' ed. D.D. Eley, H. Pines and P.B. Weisz, Academic Press, New York, 19 (1969) 1.
[13] A. Capon and R. Parsons, J. Electroanal. Chem., 45 (1973) 205.
[14] J. Clavilier, R. Parsons, R. Durand, C. Lamy and J.M. Leger, J. Electroanal. Chem., 124 (1981) 321.
[15] J. Clavilier, C. Lamy and Leger, J. Electroanal. Chem., 125 (1981) 249.
[16] G.C. Allen, P.M. Tucker, A. Capon and R. Parsons, J. Electroanal. Chem., 50 (1974) 335.
[17] B. Beden, A. Bewick, K. Kunimatsu and C. Lamy, J. Electroanal. Chem., 142 (1982) 205.
[18] J. Willson and J. Heitbaum, J. Electroanal. Chem., 185 (1985) 181.
[19] J. Clavilier and Sun Shi-gang, J. Electroanal. Chem., 194 (1986) 471.

BASIC THEORY OF POLYELECTROLYTES

J-M.Victor
Laboratoire de physique theorique des liquides
Universite P.& M.Curie, Tour 16
4. pl. Jussieu
75252 Paris Cedex 05
France

ABSTRACT. We attempt to give a theoretical overview of the main features of polyelectrolytes. We first discuss the one-chain problem: the condensation of counterions is analyzed from Manning's argument and developped by means of the Poisson-Boltzmann equation; after having defined the persistence length of a neutral polymer, we present the derivation of the electrostatic persistence length of a polyion as given by Odijk. We then discuss the scaling laws appearing in the many-chains problem: the standard scaling theory of neutral polymers is applied to salted solutions of polyelectrolytes; finally the salt-free solutions are studied using conjectural scaling laws and an hypothetic phase diagram is derived thereabout.

1. PRESENTATION

1.1. What is it?

We consider here polyelectrolytes as **charged linear polymers in aqueous solution**. They are then different from colloidal particles, whose shape is rather compact. Extensions to branched polymers are possible but not treated here. The solution is composed of: **polymers**, **counterions**(c-i) and **salt**.

Examples: polyacids(DNA), polybases(PSSNa, PSSTMA), proteins above or below the isoelectrical point. These examples give an idea of the **huge biological importance** of polyelectrolytes.

1.2 Why is it a distinct area?

Polyelectrolytes are by no way a mere superposition of electrolytes and polymers properties. New and rather unexpected behaviours are observed:

(i) Whereas polymers exhibit only excluded-volume effects, the long-ranged coulombic interactions, which are present in polyelectrolytes, give rise to new critical exponents.

(ii) The main difference with electrolytes is that one kind of ions(i.e. c-i) are stuck together along a chain: **the collective**

M.-C. Bellissent-Funel and G. W. Neilson (eds.), The Physics and Chemistry of Aqueous Ionic Solutions, 291–310.

contribution of the charged monomers causes a strong field in the vicinity of the chain, even at very low dilution.

2. COUNTERIONS CONDENSATION

2.1. Manning's argument[1]

Consider an infinitely thin rod, whose linear density is $\lambda = -e/A$; there may be some salt added and the valence of the counterions is $Z>0$. Let call r_n the distance of the nth counterion from the rod and look at a configuration where the 1st counterion is at $r_1<r_0<<\chi^{-1}$ whereas any other counterion is farther than χ^{-1}.

For $r<r_0$ the screening is negligible so that the potential energy of the 1st counterion is merely:

$$E_p(r_1) = \frac{2Ze^2}{4\pi\varepsilon A} \mathrm{Log}(r_1) \tag{1}$$

For $j\geq 2$ the screened potential is

$$E_p(r_j) \approx K_0(\chi r_j) \tag{2}$$

$$\text{with } \chi^2 = 4\pi Q\left(\sum_a n_a Z_a^2\right) \quad \text{and} \quad Q = \frac{\beta e^2}{4\pi\varepsilon} \tag{3}$$

(the sum runs over each ionic species a)

2.1.1. Onsager's treatement. The partition function of the system is

$$\mathbb{Z}_N = \int dr_1 \ldots \int dr_N \exp\left[-\beta \sum_{j=1}^{N} E_p(r_j)\right]$$

$$> \int r_1 dr_1 \exp[-\beta E_p(r_1)] \left[\int_{r_0}^{\infty} r dr \exp(-\beta E_p(r))\right]^{N-1}$$

$$= f(r_0) \int_0^{r_0} r_1^{\,1-2Z\xi} dr_1 \tag{4}$$

$$\text{with } \xi = \frac{Q}{A} \equiv \frac{\text{Bjerrum length}}{\text{charge separation}} \tag{5}$$

If $2Z\xi\geq 2$, the partition function $\mathbb{Z}_N$ is diverging, so that for $\xi\geq 1/Z$ the system presents a **phase transition.**

At $t=25°C$ $\varepsilon_r=78.5$ ($\varepsilon=\varepsilon_0\varepsilon_r$) hence $Q\approx 7\mathring{A}$

2.1.2. Application of this treatment to DNA: two consecutive pairs of phosphates are separated by A=1.7A, then $\xi=4.2$ so that any ion, regardless to its valence, is bound to condense.

Criticisms: (i) ε should be modified by the very condensation, even at low dilution.

(ii) real polyelectrolytes have a finite diameter; Onsager's treatment does not apply anymore and we may wonder whether an actual phase transition occurs, and of which kind.

2.2 Results of the Poisson-Boltzmann equation

2.2.1. Analytic solution for a pure polyelectrolyte solution[2,3].
Consider an infinite, uniformly charged cylinder:
Let a be the radius of the infinite cylinder and R>a the radius of a coaxial cylinder containing the solution of counterions. The counterions are supposed to be dimensionless and to bear valence 1.

The Poisson-Boltzmann(P.B.) equations are:

$$n(r) = n^* \exp(-\tilde{\Phi}) \tag{6}$$

$$\tilde{\Phi} = e\Phi/kT \tag{7}$$

$$\Delta\Phi + ne/\varepsilon = 0 \tag{8}$$

which give

$$\Delta\Phi + 4\pi Q n^* \exp(-\tilde{\Phi}) = 0 \tag{9}$$

The boundary conditions are:

$$\text{(i) at } r=a : \frac{\partial\tilde{\Phi}}{\partial n}(a) = -4\pi Q\sigma \quad \text{(charge density } \sigma = -\pi a A/2) \tag{10}$$

$$\text{(ii) at } r=R : \frac{\partial\tilde{\Phi}}{\partial n}(R) = 0 \tag{11}$$

Putting:

$$\begin{cases} \chi^2 = 8\pi Q n(R) \\ u = \mathrm{Log}(\chi r) \\ f(u) = \tilde{\Phi}(r) \end{cases} \tag{12}$$

we obtain an equation for f(u):

$$f'' = \frac{1}{2} e^{2u+f} \tag{13}$$

with

$$f'(a) = 2\xi \quad \text{and} \quad f'(R) = 0 \tag{14}$$

giving a solution for f of the following form:

$$e^{f} = \left\{2|\lambda|/\chi r\cos[\lambda \mathrm{Log}(r/R_m)]\right\}^2 \tag{15}$$

with

$$\begin{cases} \lambda^2 = (\chi R/2)^2 - 1 \\ \lambda \mathrm{tg}[\lambda \mathrm{Log}(r/R_m)] = 1 \end{cases} \tag{16}$$

2.2.2. Relationship with condensation. Let R(F) be the radius of the cylinder containing the fraction F of the counterions;if $\xi>1$, a critical behaviour of R(F) occurs at $F=F_o=1-1/\xi$:

$$F<F_o : R(F) \longrightarrow a*\exp\{F/[(\xi-1)(F_o-1)]\} \text{ when } \chi \longrightarrow 0 \tag{17}$$

$$F>F_o : R(F) \longrightarrow \infty \text{ as } \chi^{-1} \tag{18}$$

$$F=F_o : R(F_o) \sim \chi^{-1/2} \ (R(F_o) = R_m) \tag{19}$$

Note that F_o is exactly the fraction of condensed counterions in the Manning's picture.

Example: DNA

$$\xi = 4.2 \text{ and } a = 10\text{Å} \tag{20}$$

$$F_o = .76 \text{ so that } R(F=50\%) \approx 18\text{Å} \tag{21}$$

Hence half of all counterions are located in a sheath of thickness 8Å.

For very low bulk concentrations ($<10^{-4}$M), the polyion is still surrounded by **molar concentrations**: the Manning's condensation happens to be a **'biological trick '** to trapp ions along the DNA chain, where they are biologically useful.

2.3. Corrections to P.B. equation

The P.B. equation has to be corrected so as to take into account correlations between c-i (electrostatic and hard-core effects): These correlations give rise to oscillations[4] in the radial distribution of the c-i (observed in Monte-Carlo calculations). Discrepancies between P.B. and Monte-Carlo results diminish with increasing c-i diameter[4].

2.4. Practical attitude for the following

From now on, we will use Manning's requirements[1]:

(i) If $\xi\leq1$ there are no condensed c-i.

(ii) If $\xi>1$ a fraction $1-1/\xi$ condenses on the polyion so that the initial electric linear density along the chain goes from -e/A to -e/Q (i.e. ξ goes to 1); the uncondensed c-i are treated through the Debye-Hückel equation with:

$$\chi^2 = 4\pi Q\xi^{-1}nZ^2 \quad \text{(for a salt-free solution)} \tag{22}$$

$n\xi^{-1}$ is the concentration of the uncondensed c-i.

3. PARAMETERS OF A NEUTRAL POLYMER

Notations:

N ≡ number of monomers per chain

A ≡ monomer length

L ≡ total length of a chain

3.1. Persistence length

3.1.1. Effects of local stiffness. Consider the following chain structure:

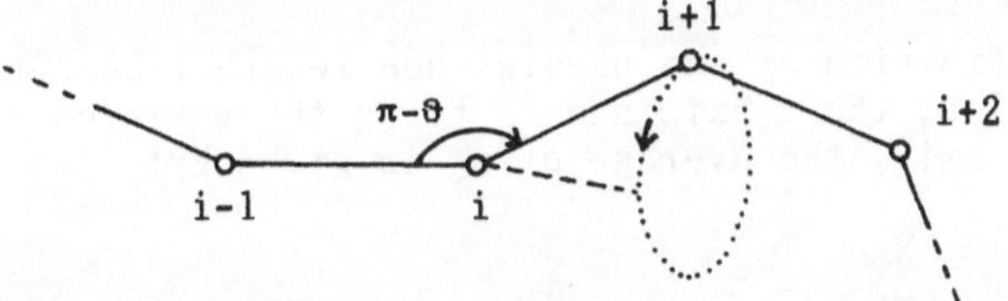

Figure 1. Structure of a model polymer chain: any monomer freely rotates around its neighbours with a fixed angle θ.

The local structure imposes short-range correlations. Let u_i be the unit vector between monomers of indices i and i+1:

$$\langle u_i . u_j \rangle = (\cos\theta)^{|i-j|} \tag{23}$$

The Flory radius R_F defined through

$$\langle (r_N - r_1)^2 \rangle = A^2 \sum_{1 \le i \le j \le N} \sum \langle u_i . u_j \rangle = R_F^2 \tag{24}$$

is equal to

$$R_F^2 = \frac{1+\cos\theta}{1-\cos\theta} NA^2 \tag{25}$$

More general models (with restricted angular rotation) give

$$R_F^2 = \sigma N A^2 \tag{26}$$

with σ of order 1

3.1.2. Equivalent Kuhn chain. Is it possible to build a gaussian chain (i.e. a chain for which $\langle u_i . u_j \rangle = \delta_{ij}$) having the same length L and the same Flory radius R_F as the original chain?

Such a gaussian chain should have N_K monomers of length A_K satisfying the following equations:

$$L = NA = N_K A_K \tag{27}$$

$$R_F^2 = \sigma N A^2 = N_K A_K^2 \tag{28}$$

It then follows

$$A_K = \sigma A \tag{29}$$

$$N_K = N \sigma^{-1} \tag{30}$$

These are the parameters of the so-called **Kuhn chain**, thus showing that **the local stiffness can be eliminated by a finite renormalization.**

3.1.3. First definition of the persistence length (discrete case). u_1 being aligned along the z-axis and z_N being the position of the last monomer on this axis, the average of z_N is given by:

$$\langle z_N \rangle = A \sum_{j=1}^{N} u_1 . u_j \longrightarrow \frac{A}{1-\cos\theta} \quad \text{as } N \longrightarrow \infty \tag{31}$$

This limit is called the **persistence length** of the chain and will be noted L_p.

3.1.4. Continuous model (Kratky & Porod[5]). Suppose that A and θ are now going to 0 so that L=NA and L_p=A/(1-cosθ) remain fixed.

Now

$$1-\cos\theta \sim \theta^2/2 \tag{32}$$

Hence

$$(\cos\theta)^i \sim \exp(-iA/L_p) \tag{33}$$

We can define a curvilinear abscissa s(m) along the chain in such a way that monomer 1 is fixed at s(1)=0 and monomer N at s(N)=L. s being the abscissa of any monomer, u(s) defines the unit vector tangent to the curve at the point s.

Monomer i is now fixed by its abscissa $s(i)=iA$ and u_i is replaced by $u(s)$.Then

$$\langle u(s).u(t)\rangle \sim \exp(-|s-t|/L_p) \tag{34}$$

The angular correlations show an exponential decay whose characteristic length is L_p. This is a second definition of L_p, relevant to the continuous case.

We can still compute the average position $z(L)$ of one extremity, the other one being fixed at the origin with $u(0)$ along the z-axis:

$$\langle z(L)\rangle = L_p[1-\exp(-L/L_p)] \tag{35}$$

And we find that $\langle z(L)\rangle \longrightarrow L_p$ as $L \longrightarrow \infty$.

As for the Flory radius we find

$$R_F^2 = 2L_pL\{1-L_p/L+(L_p/L)\exp(-L/L_p)\} \tag{36}$$

And then

$$R_F^2 \sim 2L_pL \tag{37}$$

Hence the equivalent Kuhn chain defined in the limit $L \gg L_p$ has the parameters:

$$A_K = 2L_p \tag{38}$$

$$N_K = \frac{L}{2L_p} \tag{39}$$

This continuous chain model is usually called the wormlike chain. According to the value of L_p/L the wormlike chain exhibits 3 different behaviours:

$L \gg L_p$: gaussian
$L \sim L_p$: flexible
$L \ll L_p$: rod-like

3.1.5. Relation between L_p and the elastic energy of curvature[6]. Still consider the wormlike chain and let ρ be the curvature vector:

$$\rho = \frac{du}{ds} \tag{40}$$

where $u(s)$ is the unit tangent vector to the chain at the point s.

The free energy of the chain is a functional of the curvature:

$$F[\rho]-F_o = \alpha/2\int_0^L \rho^2 ds \tag{41}$$

($\alpha \equiv$ elastic constant, $F_o \equiv$ free energy of the rod)
We prove now that

$$\alpha = L_p kT \tag{42}$$

($k \equiv$ Boltzmann constant)

Suppose that the chain is embedded in a plane (2-dimensional case) with a constant curvature ρ. The angle between $u(0)$ and $u(s)$ is $\vartheta(s)=\rho s$; ϑ has a maximum at $s=L$ and $\vartheta_m=\rho L$. The elastic free energy is then

$$\Delta F[\rho] = \frac{\alpha}{2}\rho^2 L = \frac{\alpha\vartheta_m^2}{2L} \tag{43}$$

Hence

$$\langle\vartheta_m^2\rangle \sim \frac{\int_0^\infty \vartheta^2 \exp(-\frac{\alpha\vartheta^2}{2L})\, d\vartheta}{\int_0^\infty \exp(-\frac{\alpha\vartheta^2}{2L})\, d\vartheta} = \frac{L}{\alpha} kT \tag{44}$$

In the 3-dimensional space there are 2 degrees of freedom for the curvature ρ and then

$$\langle\vartheta_m^2\rangle \sim 2 \frac{L}{\alpha} kT \tag{45}$$

Now

$$\langle z(L)\rangle = \int_0^L \langle\cos\vartheta(s)\rangle ds \tag{46}$$

and

$$\langle\cos\vartheta(s)\rangle \sim 1 - \frac{\langle\vartheta^2(s)\rangle}{2} \sim 1 - \frac{kTs}{\alpha} \sim \exp(-\frac{kTs}{\alpha}) \tag{47}$$

Hence

$$\langle z(L)\rangle = \frac{\alpha}{kT}[1-\exp(-\frac{kT}{\alpha}L)] \tag{48}$$

which is to be compared to (35), giving (42).

3.2. Excluded-volume parameter

3.2.1. Introduction of the excluded-volume. In the wormlike chain model the potential energy for the excluded-volume condition is defined as follows: any two points on the chain, noted 1 and 2, located at r_1 and r_2 interact through a potential energy

$$w(r_1,r_2) = \beta_e kT\delta(r_1-r_2) \tag{49}$$

βe is the excluded-volume. Obviously, excluded-volume effects appear only for $|s(r_1)-s(r_2)|>L_p$.

3.2.2. Influence of the excluded-volume on the Flory radius. If there is no excluded-volume(i.e. $\beta_e=0$) the Flory radius is given by (36), but when $\beta_e \neq 0$ the chain is not allowed to intersect itself anymore: the polymer is now a self-avoiding random walk (SAW) whose Flory radius is given by

$$R_F \sim L^\nu \text{ with } \nu \approx 3/5. \tag{50}$$

Information about the prefactor of L^ν is given by the 2-parameters model

$$R_F^2 = \alpha^2(Z)\langle R_F^2\rangle_\theta \tag{51}$$

θ is the so-called θ-temperature at which the excluded-volume repulsion is exactly cancelled by the Van der Waals attraction: the behaviour of the chain becomes gaussian, which allows to define the equivalent Kuhn-chain by

$$\begin{aligned} L &= N_K A_K \\ \langle R_F^2\rangle_\theta &= N_K A_K^2 \end{aligned} \tag{52}$$

Z is the excluded-volume parameter defined by

$$Z = (3/2\pi)^{3/2} A_K^{-7/2} \beta_e L^{1/2} \tag{53}$$

and $\alpha(Z)$ is given by the Yamakawa-Tanaka equation

$$\alpha^2(Z) = .541+.459(1+6.04*Z)^{2/5} \tag{54}$$

As $L \longrightarrow \infty$, $\alpha^2(Z) \sim Z^{2/5}$ so that $R_F \sim L^\nu$ with $\nu=3/5$.

4. EQUIVALENT PARAMETERS FOR A CHARGED POLYMER

4.1. Odijk's calculation of the electrostatic persistence length[7]

The free energy of a wormlike charged chain is given by a sum of elastic and electrostatic contributions:

$$\frac{\Delta F}{kT} = \frac{1}{2} L_p \int_0^L \rho^2 ds + \frac{\Delta F_{el}}{kT} \tag{55}$$

$$\frac{\Delta F_{el}}{kT} = \frac{Q}{\xi^2} \sum_{1 \le i < j \le N} \sum \left[\frac{\exp(-\chi r_{ij})}{r_{ij}} - \frac{\exp[-\chi(j-i)A]}{(j-i)A} \right] \tag{56}$$

Suppose now that there exists some length λ over which the polyelectrolyte is stiff ($\lambda \sim L_p$) and much bigger than the screening length ($\lambda >> \chi^{-1}$). Then the short-range interactions (short **along** the chain, i.e. persistence effects) and the long-range ones (**along** the chain, i.e. excluded-volume effects) are well separated.

Setting ΔF_s(resp.ΔF_ℓ) for the short-(resp.long-)range part of the free energy, we obtain

$$\frac{\Delta F_{el}}{kT} = \frac{\Delta F_s}{kT} + \frac{\Delta F_\ell}{kT} \tag{57}$$

with

$$\frac{\Delta F_s}{kT} = \frac{Q}{\xi^2} \sum_{|i-j| < \lambda} \sum \left[\frac{\exp(-\chi r_{ij})}{r_{ij}} - \frac{\exp[-\chi(j-i)A]}{(j-i)A} \right] \tag{58}$$

which gives in the continuous limit

$$\frac{\Delta F_s}{kT} = \frac{1}{2} L_e \int_0^L \rho^2 ds \tag{59}$$

with

$$L_e = \frac{Q}{4\chi^2 A^2 \xi^2} \tag{60}$$

which is the so-called **Odijk's formula.**

This shows the existence of a total persistence length $L_t = L_p + L_e$

The validity of this result is submitted to the condition $L_t >> \chi^{-1}$.

Now

$$L_p > \frac{A^2 \xi^2}{Q} \Rightarrow L_t > \chi^{-1} \tag{61}$$

and L_p usually verifies this condition.

Example: DNA

A more exact calculation of the persistence length has been made by Le Bret[8] on a torus, using the Poisson-Boltzmann equation. His result is non-analytic, but recovers Odijk's formula when χa and ξ go to 0.

Table 1. Comparison of the theoretical predictions for L_e as given by Odijk's formula(O) and Le Bret (LB) vs experimental results (Exp) for L_T.

C_M	L_e(LB) Å	L_e(O) Å	L_T(Exp) Å
10^{-4}	2300	3250	
10^{-3}	489	325	
5.10^{-3}	222	65	440
10^{-2}	168	32.5	
2.10^{-2}	137	16.3	360
10^{-1}	93	3.3	325
.6	55	.5	265
1	42	.3	254

L_t-L_e(LB) is nearly constant between 10^{-2}M and 1 M, and around 230 Å. It then follows that the bare persistence length (i.e. of the uncharged DNA) is about 230 Å.

4.2. Electrostatic excluded-volume[7]

ΔF_1 is then treated as a perturbation: the excluded-volume effects occur now between renormalized monomers whose length is $2L_t$ (equivalent Kuhn segment) and section $\sim 2\chi^{-1}$, giving an excluded-volume

$$\beta_e \sim 8\pi L_t^2 \chi^{-1} \tag{62}$$

An exact calculation gives logarithmic corrections to this result[9]:

$$\beta_e = 2\pi L_t^2 \chi^{-1}[\mathrm{Log}(4\pi)+\gamma-1/2-\mathrm{Log}(\chi Q)] \tag{63}$$

($\gamma \equiv$ Euler's constant)

4.3. Conclusions of the one-chain problem

4.3.1. Counterions condensation occurs whenever A, the interspace between two neighbouring charges, is less than Q, the Bjerrum length. This is rather general for a completely charged polymer (i.e. when each monomer is charged) because Q∿7Å at the room temperature.

Q>A fixes the strong-coupling regime, for which ξ=Q/A.

4.3.2. There exists a persistence length, varying with the ionic strength. The dependence in the ionic strength is given through Odijk's formula

$$L_t = l_p + \frac{Q}{4\chi^2 A^2 \xi^2} \tag{64}$$

giving in the strong-coupling regime

$$L_t = L_p + \frac{1}{4Q\chi^2} \tag{65}$$

The function $L_t(\chi^{-1})-\chi^{-1}$ has a positive minimum in $\chi^{-1}=2Q$ when $L_p>Q$, which means that the validity condition $L_t>\chi^{-1}$ is equivalent to $L_p>Q$, a condition usually met in practice. Moreover, in weak screening, for which $\chi^{-1}\gg Q$, the χ^{-2} dependency of L_t gives $L_t\gg\chi^{-1}$.

Remarks:

(i) When the polyion length L is comparable to χ^{-1}, Odijk's formula is more involved:

$$L_e = 1/12\ QN^2 h(\chi L) \tag{66}$$

with

$$h(y) = e^{-y}(y^{-1}+5y^{-2}+8y^{-3})+3y^{-2}-8y^{-3} \tag{67}$$

which gives the usual result when $\chi L\gg 1$.

Hence Odijk's formula is valid whenever $L\gg\chi^{-1}\gg Q$.

(ii) Near the rod limit $L\sim L_t$, and for $\chi^{-1}\gg Q$, $L_t\gg\chi^{-1}$, so that $L\gg\chi^{-1}$ and Odijk's formula is valid.

5. SCALING LAWS FOR SALTED SOLUTIONS OF POLYELECTROLYTES

5.1. Scaling laws for neutral polymers[10]

5.1.1.The problem is to find intensive properties of the system which are independent of the contour length,in the limit of $L \longrightarrow \infty$.

5.1.2. Correlation length in a polymer solution. Let c^* be the concentration of monomers at which chains begin to overlap. Then

$$c^* \sim \frac{L/A}{R_F^3} \tag{68}$$

For $c>c^*$ there exists a correlation length, noted ξ and defined as follows:

(i) $r<\xi$: only intra-chain interactions are relevant, inter-chains ones being screened.

(ii) $r>\xi$: all interactions are screened.

Interactions are then confined inside a mesh of dimension ξ; polymers behave as gaussian chains whose renormalized segments are parts of the original chain with Flory radius ξ. Such segments are called blobs and can be viewed in the following way:

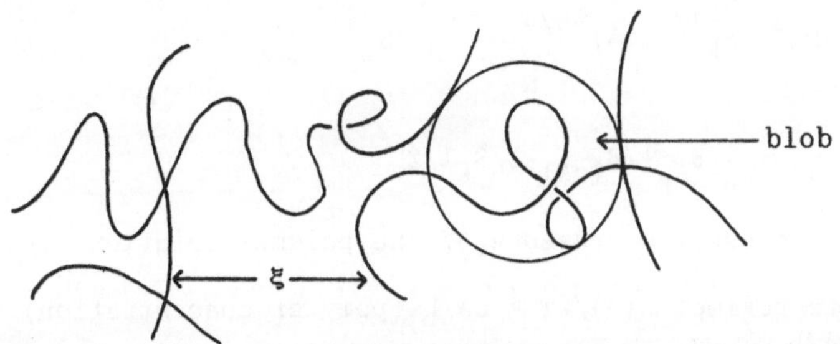

Figure 2. illustration of the blob concept. The solution of polymers is sketched as a lattice whose mesh is as large as one correlation length.

5.1.3. Calculation of the correlation length. The following scaling procedure is standard:

(i) at c^*: $\xi \sim R_F$

(ii) at $c^* \ll c \ll 1/A^3$(semi-dilute regime): ξ scales like a power of c. Then

$$\xi \sim R_F(c/c^*)^x \tag{69}$$

(iii) ξ must become independent of L as $L \longrightarrow \infty$.

Using (53) we find

$$Z \sim \beta_e A_K^{-7/2} L^{1/2} \tag{70}$$

and with (51) and (54)

$$R_F \sim \beta_e^{1/5} A_K^{-1/5} L^{3/5} \tag{71}$$

Hence

$$x = -3/4 \tag{72}$$

and

$$\xi \sim \beta_e^{-1/4} A_K^{1/4} (cA)^{-3/4} \tag{73}$$

Moreover c^* is given by:

$$c^* \sim (L/A)^{-4/5} (\beta_e/A^3)^{-3/5} (A_K/A)^{3/5} A^{-3} \tag{74}$$

5.1.4. We can then compute the Flory radius beyond c^*: Each polymer behaves now as a gaussian chain of blobs, each blob having a Flory radius of length ξ. Hence

$$R_F(c) \sim \{L/(gA_K)\}^{1/2}\xi \quad (75)$$

where g is the number of Kuhn segments per blob. Now ξ is the Flory radius of a chain of length gA_K, giving, by comparison with (71),

$$g \sim \beta_e^{-3/4} A_K^{-1/4} (cA)^{-5/4} \quad (76)$$

and finally

$$R_F(c) \sim \beta_e^{1/8} A_K^{-1/8} (cA)^{-1/8} L^{1/2} \quad (77)$$

5.1.5. As for the osmotic pressure of the polymer solution, the scaling procedure is:

(i) dilute regime: $\pi(c)/kT \approx cA/L$ (polymer concentration) so that $\pi(c^*)/kT \sim c^*A/L$

(ii) semi-dilute regime:

$$\pi(c)/kT \sim \pi(c^*)/kT(c/c^*)^y \quad (78)$$

(iii) $\pi(c)$ must be independent of L. Hence

$$y = 9/4 \quad (79)$$

and

$$\pi(c)/kT \sim 1/\xi^3 \quad (80)$$

with ξ given by (73).

5.2. Modelization of a salted solution.

We suppose here that the solution is composed of: polyions, counterions and an excess of a 1-1 electrolyte(salt). The salt concentration is noted n and is supposed to be so much greater than the c-i concentration that this can be neglected. The screening is then given by

$$\chi^2 = 8\pi Q n \quad (81)$$

The polyions are supposed to be wormlike-chains with a persistence length L_t given by (64). The equivalent Kuhn segments have a length

$$A_K = 2L_T \quad (82)$$

and the excluded-volume is

$$\beta_e \sim L_T^2 \chi^{-1} \quad (83)$$

The excluded-volume parameter given by (53) and (83) is

$$Z \sim L_T^{-3/2} \chi^{-1} L^{1/2} \quad (84)$$

and the Flory radius is

$$R_F \sim L_T^{1/5}\chi^{-1/5}L^{3/5} \tag{85}$$

showing the χ-dependence of the prefactor of $L^{3/5}$.

Conditions of validity:

(i) The excess of salt (strong screening regime) makes $L_T \ll L$, thus justifying (85).

(ii) As for L_T, we have to reexamine Odijk's formula . As a matter of fact, the contribution of other chains to the persistence length of one given chain is no longer negligible as soon as the correlation length ξ becomes smaller than L_e. De Gennes & al.[11] computed such a contribution and found

$$L_T \sim (cA)^{-1} \tag{86}$$

It thus appears that intra-chain and inter-chains contributions to the persistence length are of the same order, and it is then reasonnable to keep

$$L_e \sim (16\pi QAc)^{-1} \tag{87}$$

Now $L_e \gg L_p$ whenever $c \ll (QAL_p)^{-1}$ which is generally verified, unless L_p is very large (as for DNA). Hence

$$L_T \sim (16\pi QAc)^{-1} \tag{88}$$

Remark: Odijk's formula as well as de Gennes' result for L_e are submitted to the condition $L \gg \chi^{-1} \gg Q$; but if $\chi^{-1} \gg Q$ we know (cf 4.3.2.) that $L_T \gg \chi^{-1}$ and because $L_T \ll L$ we have $L \gg \chi^{-1}$. It is then sufficient that $\chi^{-1} \gg Q$, which is the rule.

5.3. Application of the scaling concepts to polyions

5.3.1. Polyions begin to overlap at c^*, given by (74), (82) and (83), i.e.

$$c^* \sim (L/A)^{-4/5}(L_T/A)^{-3/5}(\chi A)^{3/5}A^{-3} \tag{89}$$

or, if $L_e \gg L_p$

$$c^* \sim (L/A)^{-4/5}(\chi A)^{-9/5}(\chi Q)^{3/5}\chi^3 \tag{90}$$

5.3.2. Correlation length: (73) and (83) give

$$\xi \sim L_T^{-1/4}\chi^{1/4}(cA)^{1/4} \tag{91}$$

Moreover, if $L_e \gg L_p$, $L_T \sim (4Q\chi^2)^{-1}$ so that

$$\xi \sim Q^{1/4}\chi^{3/4}(cA)^{-3/4} \tag{92}$$

5.3.3. Flory radius beyond c*: (76) and (83) give the number of Kuhn segments per blob

$$g \sim (L_T/A)^{-7/4}(\chi A)^{3/4}(cA)^{-5/4} \tag{93}$$

and

$$R_F(c) \sim L_T^{1/8}\chi^{-1/8}(cA)^{-1/8}L^{1/2} \tag{94}$$

If $L_e >> L_p$

$$g \sim (Q/A)^{5/4}(\chi Q)^{1/2}(c\chi^{-3})^{-5/4} \tag{95}$$

$$R_F(c) \sim Q^{-1/8}\chi^{-3/8}(cA)^{-1/8}L^{1/2} \tag{96}$$

Condition of validity: $g>>1$, i.e.

$$c << (Q/A)(Q\chi)^{2/5}\chi^3 \tag{97}$$

5.3.4. The osmotic pressure is still given by (80) with ξ from (88) or (89) according to the value of L_e/L_p.

6. SALT-FREE SOLUTIONS

6.1. New problems

The main difference with the salted case is that, here, the distribution **and** the conformation of the polyions depend ,in a **self-consitent way**, on the monomer concentration c.

Moreover the screening now depends **only** on counterions and then on c. The scaling is thereabout more intricate. For simplicity we assume a strong coupling, i.e. Q>A, and we shall suppose that the **screening is due only to uncondensed ions**. Hence

$$\chi^2 = 4\pi Ac \tag{98}$$

The aim of this study is to **build the phase diagram** of the polyelectrolyte solution as a function of the monomer concentration c.

6.2. Behaviour in the dilute regime

6.2.1. Conformation of the polyions: If the solution is so dilute that the polyions are not yet entangled, we can apply results from the one-chain problem. Provided that $L>>\chi^{-1}$ and $L_e>>L_p$, the persistence length of a polyion is given through Odijk's formula

$$L_T \approx (16\pi QAc)^{-1} \tag{99}$$

Polyions have a rod-like behaviour as long as $L_T > L$, i.e. for

$$c < c_b \sim (16\pi QAL)^{-1} \tag{100}$$

Now $L_p \ll L_e$ as soon as $c \ll c_p = (16\pi QAL_p)^{-1}$. And $c_b \ll c_p$ whenever $L > L_p$. The case where $L_p > L$ goes back to solutions of rigid rods (whatever the concentration may be). An example is given by the Tobacco Mosaic Virus (TMV)[12]. Because we are interested here in scaling laws we only consider the case of very long polyions. Therefore L_p is to be neglected in the dilute regime.

On another part, $L \gg \chi^{-1}$ is equivalent to $c \gg (4\pi AL^2)^{-1}$, which is obviously verified by c_b.

As a conclusion we can say that:

(i) For $c < c_b$ polyions are rigid rods.

(ii) For $c_p \gg c > c_b$ polyions are flexible with a persistence length given by (99).

6.2.2. Distribution of the rods ($c < c_b$). The following discussion is rather conjectural. The problem is to describe the pair distribution function in a solution of rods: we just examine here which order is likely to appear in such a system.

Entanglement sets on at

$$c^* \sim (6/\pi)A^{-1}L^{-2} \tag{101}$$

and $c_b > c^*$ whenever $L/Q > 100$, i.e. for a number of monomers $N > 100$ (remember that Q and A are of the same order). Hence **entanglement generally sets on before bending.**

When $c \ll c^*$ the solution is a gas of rods. The average volume occupied by a rod is

$$v = (cA/L)^{-1} = (4\pi/3)a^3$$

The mean distance between neighbouring rods is then

$$a = (\chi^2/3L)^{-1/3}$$

We go on by comparing a and χ^{-1}. Because $\chi a = (3\chi L)^{1/3}$ we get as a result that $\chi^{-1} > a$ whenever $\chi^{-1} > 3L$, i.e. for $c < c^*/200$. In that case the OCP becomes a reasonable approximation for the system. After sphericalization (i.e. by averaging over the angular dependency of the rods), an effective coupling constant Γ can be evaluated as usual

$$\Gamma = \frac{(L/Q)^2 e^2}{akT} = L^2/(Qa)$$

We know from computer simulations[13] that the OCP undergoes a transition from a fluid to a crystalline(BCC) phase when $\Gamma > 178$. We may then expect some crystalline ordering in a salt-free solution of polyelectrolytes whenever

$$3L < \chi^{-1} < (L/250Q)^{3/2}L \tag{102}$$

which imposes $L/Q > 500$. Hence a polyion lattice is likely to exist at very low concentrations; however any experiment attempting to exhibit such a lattice is bound to fail, because of the very smallness of the relevant concentrations (the maximum packing fraction at which this is to occur is of order 10^{-8}!).

When $\chi^{-1} < a$ the rods behave like hard cylinders of length L and radius χ^{-1}. A criterium due to Onsager[13] then predicts a nematic order when $L/(2\chi^{-1}) > 8$, i.e. when

$$c > 10c^* \tag{103}$$

The self-consistency imposes $c_b > 10c^*$, i.e. $L/Q > 1000$.

6.3. Behaviour above the bending concentration

We attempt here to understand the behaviour of the polyion solution for $c>c_b$. The polyions are now rod-like only over distances of order L_T, each segment of length L_T being surrounded by a sheath of radius χ^{-1} and then occupying a volume $v \sim \pi L_T \chi^{-2}$. Using (98) we get

$$cv \sim L_T/A \tag{104}$$

which means that the segments are nearly at close-packing: rather surprisingly it turns out that the radius of the sheath (i.e. the Debye length) is exactly the average distance between two neighbouring polyions, whatever the concentration may be.

We show now that χ^{-1} is precisely the correlation length ξ. The standard scaling procedure runs as follows:

(i) at c_b: the preceding discussion shows that the only relevant length is χ^{-1} because $L_T=L$ at this concentration.

(ii) for $c>c_b$:

$$\xi(c) \sim \xi(c_b)(c/c_b)^y \tag{105}$$

(iii) $\xi(c)$ must be independent of L

Hence

$$\xi(c) \sim (Ac)^{-1/2} \sim \chi^{-1} \tag{106}$$

This scaling law remains valid as long as $\xi \ll L_T$, i.e. for

$$16\pi Q(AC)^{1/2} \ll 1 \tag{107}$$

that is to say for

$$\chi Q \ll 0.1 \tag{108}$$

Since $\xi \ll L_T$, the polyions behave like a gaussian chain whose Kuhn segments have a length $2L_T$; the Flory radius is then given by

$$R_F(c) \sim [L/(8\pi QAc)]^{1/2} \tag{109}$$

6.4. Behaviour in the semi-dilute regime

As the concentration of monomers grows up, the screening becomes stronger and stronger, so that we expect to recover the behaviour of a salted solution. Quantitatively this is submitted to the condition (97) where χ is now given by (98). Hence for

$$\chi Q \gg 0.1 \tag{110}$$

we recover the same situation as in the salted case, i.e. a quasi-neutral solution of polymers with a renormalized persistence length L_T and a excluded-volume $\beta_e \sim L_T^2 \chi^{-1}$. We can then use formula (92) and (98) to find

$$\xi(c) \sim Q^{1/4}(cA)^{-3/8} \tag{111}$$

The Flory radius of a polyion is given by

$$R_F(c) \sim Q^{-1/8}(Ac)^{-5/16}L^{1/2} \tag{112}$$

6.5. Transition from the dilute to the semi-dilute regime.

The dilute regime is characterized by $\xi(c) \ll L_T$ whereas in the semi-dilute regime $\xi(c) \gg L_T$. In view of (108) and (110) we then expect a transition at $\chi Q \sim 0.1$, i.e. for a monomer concentration

$$c^{**} \sim 10^{-3}/(AQ^2) \tag{113}$$

Now for $c < c^{**}$, the potential in the vicinity (i.e. at a distance $d \ll L_T$) of a polyion is nothing but the potential of an infinite rod with the same linear density (i.e. $-e/Q$). Two interacting polyions are separated by less than the correlation length ξ, and because $\xi \sim \chi^{-1} \ll L_T$, the interaction potential in the solution may well be approximated by a logaritmic potential. An effective coupling constant Γ is then built by dividing the potential energy of a Kuhn segment (length $2L_T$) by its kinetic energy, which is precisely of order kT. Hence

$$\Gamma \sim 4QL_T/A^2 \tag{114}$$

The polyelectrolyte solution is then well approximated by a two-dimensional OCP, with the above coupling constant. Thereabout we get a striking answer to the question of the behaviour around c^{**}: computer simulations[14] have shown a phase transition from a liquid to a solid state (hexagonal lattice) at $\Gamma \sim 140$. We may therefore expect some order-disorder transition in the polyelectrolyte solution when the effective coupling constant Γ as given in (114) is about 140, i.e. for

$\chi A \sim 0.1$: this is exactly the result given in (113) (remember $Q \sim A$).

Moreover, when $c_b < c < c^{**}$, Γ is greater than 140 and we then expect that the polyions are locally parallel (over a cylindrical volume whose length and radius are respectively of the order of L_T and a few ξ).

6.6. Conjectural phase diagram

The above discussion can be sum up in the following phase diagram (the temperature being fixed at t=25°C):

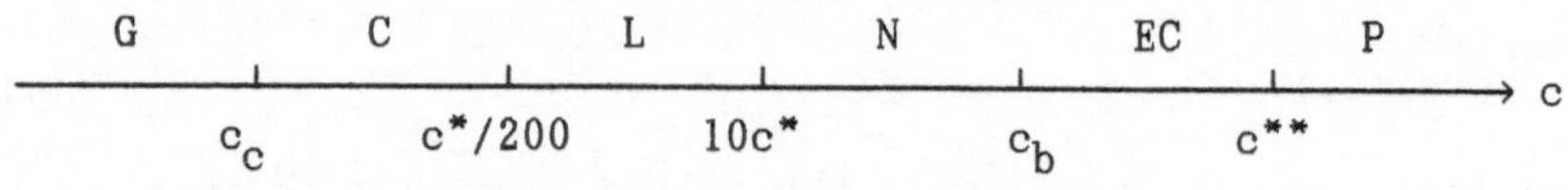

figure 3. Phase diagram (at fixed temperature) of a solution of pure polyelectrolyte. Six phases are predicted:

G: gas of rods } point-like behaviour of rods
C: crystal of rods }
L: liquid of rods
N: nematic order
EC: entangled crystallites
P: polymer-like structure

REFERENCES

1. G.S. Manning, J.Chem.Phys. 51, 924 (1969).
2. R.M. Fuoss, A. Katchalsky & S. Lifson, Proc.Nat.Acad.Sci.Wash 37, 579 (1951).
3. M. Le Bret & B.H. Zimm, preprint.
4. P. Mills, C.F. Anderson & M.T. Record, J.Phys.Chem. 89, 3984 (1985).
5. O. Kratky & G. Porod, Rec.Trav.Chim. 68, 1106 (1949).
6. L.D. Landau & E.M. Lifshitz, Statistical Physics, 3rd Ed. Part 1, §127: Fluctuations in the curvature of long molecules, Pergammon Press, 1980.
7. Th. Odijk & A.C. Houwaart, J.Polym.Sci.:Polym.Phys.Ed. 16, 627 (1978).
8. M. Fixman & J. Skolnick, Macromolecules 11, 863 (1978)
9. M. Le Bret, J.Chem.Phys. 76, 6243 (1982).
10. P-G. de Gennes, Scaling concepts in polymer physics, Cornell University Press, 1979.
11. J. Hayter, G. Janninck, F. Brochard-Wyart & P-G. de Gennes, J.Phys.Lett. 41, L-451 (1980).
12. W.L. Slattery, G.D. Doolen & H.E. De Witt, Phys.Rev. A 26, 2255 (1982).
13. L. Onsager, Ann.N.Y.Acad.Sci. 51, 627 (1949)
14. J.M. Caillol, D. Levesque, J.J. Weiss & J.P. Hansen, J.Stat.Phys. 28, 325 (1982).

EXPERIMENTS ON POLYELECTROLYTE SOLUTIONS

G. Jannink
*Laboratoire Léon Brillouin**
CEN-Saclay
91191 Gif-sur-Yvette Cedex France

**Laboratoire Commun CEA-CNRS*

Abstract

Polyelectrolyte solutions have been examined by neutron scattering experiments in order to determine separately the polyion interference term and the polyion form factor. The results are used to test hypothesis on the strong persistence length decrease with concentration, predicted by the theory of Odijk. The concentration diffusion coefficient is calculated from the quasi-elastic scattering experiment : the data reveal the decrease of this coefficient with the wave vector, typical of ionic solutions. More characteristic is the mobility wave vector dependance, related to the existence of the persistence length.

These experiments complement earlier results obtained from osmotic pressure measurements and viscosimetry.

It is confirmed that polyelectrolytes made of flexible polymers have a structure whose disorder increases with concentration, in a very broad concentration range.

M.-C. Bellissent-Funel and G. W. Neilson (eds.), The Physics and Chemistry of Aqueous Ionic Solutions, 311–336.

1. POLYELECTROLYTES

There is a large class of macromolecules which has the property to dissociate into disymmetric ions when dispersed in water. The larger ion is called the polyion ; its molecular weight, M_1, is of the same order of magnitude as the molecular weight of the macromolecule ; the number Z_1 of ionizable sites is greater than unity. It determines the "surface" charge Z_1e. The smaller ion is called the counterion, with molecular weight $M_2 \ll M_1$; it has usually a single ionizable site ($Z_2 = 1$).

As an example, consider the polysalt derived from sulfonated polystyrene

$$(-CH_2 \quad - \quad \underset{\displaystyle C_6H_4SO_3Na}{\overset{}{\underset{|}{CH}}} \quad -)_N$$

where N is the degree of polymerization. The number Z_1 of ionizable sites is here equal to N.

In case of full ionization, the polysalt is transformed into the polyion

$$(-CH_2 \quad - \quad \underset{\displaystyle C_6H_4SO_3^-}{\underset{|}{CH}} \quad -)_N$$

and if all the $Z_1 = N$ counterions N_a^+ are free, the "effective" charge $\bar{Z}_1$ of the polyion equals the "surface" charge $Z_1 (=N)$. In reality, the effective charge $\bar{Z}_1$ is not exactly equal to Z_1. It is a quantity to be measured and we shall see that Z_1 is smaller, but of the order of $Z_1 = N$.

In this example, the polyion is made of linear flexible polymers. Other polyions are formed with rigid spherical macromolecules. For instance, sphere with a diameter $\Phi \simeq 80$ nm can be obtained by emulsion polymerisation of styrene (using $K_2S_2O_8$ as catalyser and sodium dodecyl sulfate as emulsifier[1]). During the synthesis, acidic groups are covalently bound to the surface[2]. They are the ionizable sites. The surface charge is typically $Z_1 e = 10^3 \times e$ and the counterions are H^+(or Na^+) molecules.

Linear flexible and spherical rigid particles represent two extreme cases for carriers of multiple charges. In between, we find several types of macromolecules, such as branched polymers, etc.

In the process of synthesis, it is possible to control the number Z_1 of ionizable sites per polyion. Let $y = Z_1/N$ be the fraction of ionizable sites per available site. This parameter can be varied from 0 to 1. In the case of sulfonated polystyrene, the result is for instance obtained by adding a controled amount of NaOH : the parameter y is then equal to the fraction of NaOH molecules per ionizable site.

For the study of polyelectrolytes in aqueous solutions, the basic parameters are :

- the concentration $\mathbb{C}$ of polyions
- the average molecular weight ratio M_1/M_2 (we assume that the molecular weight distribution is sharply peaked around M_1)
- the fraction y of ionizable sites per available site.

The polyion concentration $\mathbb{C}$ is directly calculated from the concentration by weight, ρ, of the polyelectrolyte prior to dissociation

$$\mathbb{C} = \frac{\rho \mathcal{A}_v}{M_1 + M_2}$$

where $\mathcal{A}_v$ is the Avogadro number.

We may already point out a characteristic feature of polyelectrolyte solutions, namely the non uniform distribution of the charges belonging to the polyions. Such a distribution becomes uniform if the bonds holding the monomers together are distroyed, but because of the existence of such bonds, there are accumulations of electric charges in the solution. Consider as an example the case of an aqueous polyelectrolyte solution, with a concentration by weight $\rho = 10$ g/ℓ. Supposing $Z_1 = N$ and a monomer molecular weight equal to 200, the average concentration of ionizable sites is $C = 3 \times 10^{-2}$ nm^{-3}. However, in the volume occupied by the polyion the (local) concentration of ionizable sites is much larger. Typically $C(\vec{r}) = 1$ nm^{-3}, for $\vec{r}$ belonging to the molecular volume of the polyion. This accumulation of charges at the site of the macromolecule is the cause of specific behaviours which are observed in aqueous polyelectrolyte solutions. For instance, the osmotic pressure against pure water is extremely high, but there are many other unusual experimental results. Physical properties of polyelectrolytes are known to be very sensitive to chemical impurities, when the solutions are made in pure water. Even though ionic impurities are carefully washed out, it is often difficult to obtain totally reproducible experimental results. On the contrary, experiments are performed without such difficulties when a simple salt is added to the polyelectrolyte solution. A way to determine experimental data on polyelectrolyte solutions in pure water, is to extrapolate at zero added salt, results obtained with finite quantities of added salt.

Systematic experimental investigations of aqueous polyelectrolyte solutions began [2] around 1950. Experiments aim at a better knowledge of both atomic and dielectric structures of these systems. We are interested here in the experimental determination of the following quantities :

- effective charge of polyions
- polyion and counterion structure functions
- polyion mobility
- hydration
- screening length.

2. PHASE DIAGRAM OF POLYELECTROLYTE SOLUTIONS

Latex charged spheres dispersed in water are known to form different states of order according to their charge and concentration, to the temperature of the solution and to the concentration of added simple salt. In very dilute aqueous solutions, the spheres are said to be in the liquid phase (within a liquid) : their diffraction pattern is similar to the one observed in pure liquid at higher wave vector transfer. As the concentration of the spheres increases, a first order transition occurs : the spheres form a periodic lattice, whose rigidity modulus increases with concentration. A criterion is known for the existence of the cristalline structure (derived from Lindemann rule[3])

$$\frac{(\bar{Z}_1 e)^2 \beta}{\epsilon d} \geqslant 155 \qquad (2.1)$$

where $\bar{Z}_1$ is the effective charge per macromolecule and d the average nearest neighbour distance. Relation (2.1) is based on the comparison between the potential energy increase corresponding to a small relative displacement of the spheres, and thermal energy β^{-1} It reasonably interpolates experimental observations on latex spheres. For instance, at a concentration $\mathbb{C} = 2.56 \times 10^{-9}\ nm^{-3}$ of spheres bearing 3×10^{3} ionizable sites, the ordered phase in pure water is observed for temperatures below 39°C. Above this temperature the structure is disordered, liquid like. (There are however exceptions to this transition : melting of crystalline structures by density *increase* is encountered in Wigner crystals).

When formula (2.1) is applied to linear flexible polyelectrolytes, it does not fit the observations. For instance there is no clear indication of the existence of an ordered state ; only recent results of light scattering experiment suggest an ordered structure at very low concentration, and for $Z_1 \gg 1$. However, an increase in concentration does not stabilize this structure as suggested by equation (2.1). On the contrary it seems to introduce a greater disorder. This point was discussed by de Gennes et al[4,5] and by Odijk[6] (see also lecture notes by J.M. Victor). As written in equation (2.1), the criterion for order applies to rigid molecules. Linear polyions are rigid rods only in the limit of zero concentration. At finite concentration, they melt into smaller subunits. Odijk[6] pointed out the rapid decrease of the persistence length with concentration ; this decrease is faster than the decrease of the screening length K^{-1} and obviously of the distance d between nearest neighbour polyions. Therefore the overall polyion structure tends to be less ordered when concentration increases. The precise evaluation of the polyion flexibility and its relation to the counterion binding is the object of the experiments.

3. BASIC EXPERIMENTAL RESULTS

3.1. Determination of an effective polyion charge

We assumed that all ionizable sites give rise to a charge, called the surface charge. The number of such charges per polyion is $Z_1 e$. Because of the strong interaction with the counterions in the vicinity of the polyion, the effective charge $\bar{Z}_1 e$ is smaller than the surface charge. The effective charge is directly related to the conductivity σ. In the Nerst Einstein approximation, the conductivity is written as (see also lecture note by J.P. Hansen)

$$\sigma_{NE} = e \sum_{\mathcal{A}=1}^{2} \mathbb{C}_{\mathcal{A}} \mu_{\mathcal{A}} Z_A \tag{3.1}$$

where $\mathbb{C}_{\mathcal{A}}$ is the molecular concentration (polyion, $\mathcal{A}=1$; counterion, $\mathcal{A}=2$), $\mu_{\mathcal{A}}$ is the mobility and $Z_{\mathcal{A}}$ the surface charge (number of available ionizable sites per molecule). The measured conductivity, σ, is however found to be smaller than σ_{NE}

$$\sigma = \Phi_{\sigma} \sigma_{NE} \tag{3.2}$$

where $\Phi_{\sigma} < 1$. This reduction factor Φ_{σ} accounts for the strong interaction between counterion and polyion in the vicinity of the polyion.

Of great interest is the observed decrease of σ measured as a function of the fraction y of ionizable sites per available sites on the polyion. The result[7] (see fig. 1) can be summarized by the two limiting numbers

$$\begin{aligned} \Phi_{\sigma}(y) &= 1 \quad , \quad y \to 0 \\ &\propto 0.4 \; , \quad y \to 1 \; , \quad Z_1 \gg 1 \end{aligned}$$

The accumulation of charges on the polyion ($y \to 1$) is seen to have considerable influence on the solution conductivity. Exact calculation of σ is in principle feasable[8]. It requires a knowledge of the counterion distribution and mobility.

2.2. Determination of an effective concentration of counterions

Let $M = M_1 + M_2$ be the molecular weight of the polyelectrolyte prior to dissociation, and $\rho = \rho_1 + \rho_2$ its concentration by weight. For $\rho \geqslant 10^{-2}$ g/ℓ, the reduced osmotic pressure $\pi M/RT\rho$ (fig. 2) of the aqueous polyelectrolyte solution against pure water, is seen[9] to be nearly independent of concentration, as in the case of a dilute solution without interaction. We may write the ideal osmotic pressure as

$$\pi_{id}\, \beta = \mathbb{C}_1 + \mathbb{C}_2 = C\left(1 + \frac{1}{N_1}\right) \tag{3.3}$$

The measured osmotic pressure, π , is however smaller than π_{id}

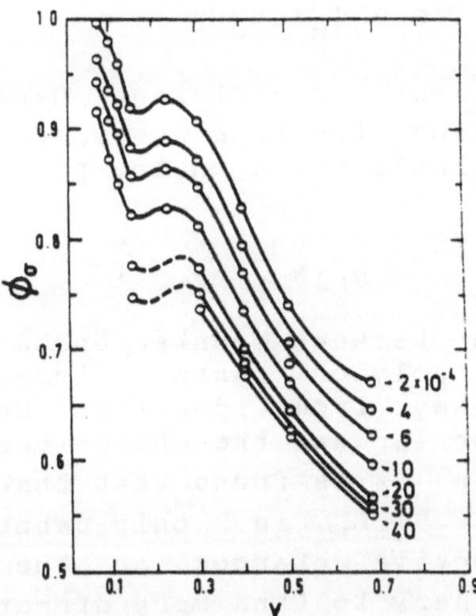

Figure 1. Reduction coefficients for the ionic solution conductivity, as a function of the fraction y of ionizable sites per available site on the polyion. This coefficient is obtained[7] from the comparison between the measured conductivity and the conductivity calculated in the Nerst Einstein approximation.
The different curves correspond to concentration of added salt.
The polyelectrolyte is PMA.

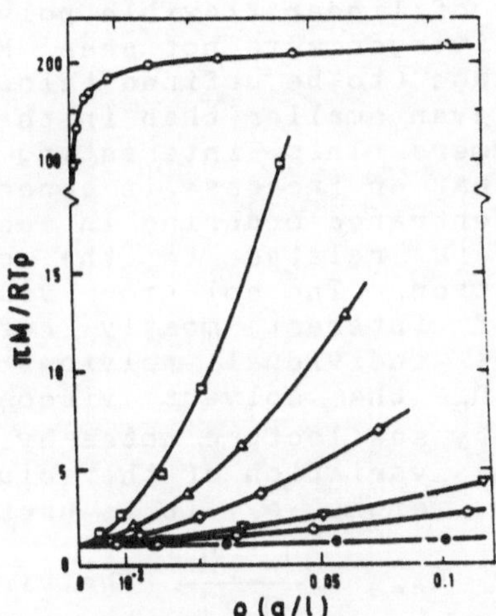

Figure 2. Reduced viscosity $\pi M/RT\rho$ as a function of concentration by weight ρ , The existence of a plateau is compatible with an ideal behaviour. Its height indicates on the other hand, strong deviation from ideality. Sodium sulfonated polystyrene M_w = 220 000, Ref. 9.
The curves in the lower part of the figure correspond to polyelectrolyte solutions with added simple salt.

$$\pi = \Phi_\pi \pi_{id} \quad (3.4)$$

where $\Phi_\pi \leqslant 1$ is the osmotic coefficient. Here again, Φ_π is found to be a decreasing function of y, the fraction of ionizable sites to available sites on the polyion

$$\Phi_\pi(y) = 1 \;, \quad y \to 2$$
$$\simeq 0,16 \;, \quad y \to 1$$

The strong interaction between counterions and polyions in the vicinity of the polyion again drives the osmotic pressure behaviour away from ideality. Certainly, the osmotic reduction factor Φ_π, and the charge reduction factor Φ_σ are related. However, it is found that the concentration of free counterions, $\Phi_\pi C$, is only about half the concentration of effective charges on the polyion, $\Phi_\sigma C$. Therefore, it is necessary to find more direct experimental evidence for the counterion-polyion correlation.

3.3. Reduced viscosity ; variation with concentrations

For polyelectrolytes made of rigid spheres and high charge, the viscosity shows a characteristic divergence when concentration increases towards critical concentration of ordered cristalline structure. On the contrary, for polyelectrolytes made of linear flexible polymers (with the same charge), such a divergence is not seen. Moreover, there is a concentration range (to be defined below) in which the viscosity increase is even smaller than in the corresponding case of neutral polymers. This interesting observation is the first evidence that an increase in concentration tends to reduce rather than entrance ordering in such systems, and that this reduction is related to the collapse of the polyion rod configuration. The solution viscosity, in the concentration range of interest, mostly reflects momentum transfers due to the individual polyions. Let η be the solution viscosity, η_0 the solvent viscosity. (for the definition of viscosity see lecture notes by J.P. Hansen). In order to study the variation of the solution viscosity with concentration by weight, ρ, it is useful to consider the reduced viscosity $\eta_{red} = \dfrac{\eta - \eta_0}{\eta_0 \rho}$. The variation of η_{red} versus ρ is given[10] in figure 3. The data are well fitted by the formula

$$\eta_{red} = \frac{A}{1 + B\,\rho^{1/2}} \quad (3.5)$$

in the concentration range in which η_{red} is a decreasing function of ρ. If, in this range, we neglect polyion-polyion interactions, due for instance to entanglements, we can write the reduced viscosity directly in terms[11] of the polyion hydrodynamic radius of gyration, R_h :

$$\eta_{red} = R_h^2 \mathcal{A}_v \pi a_h / m \tag{3.6}$$

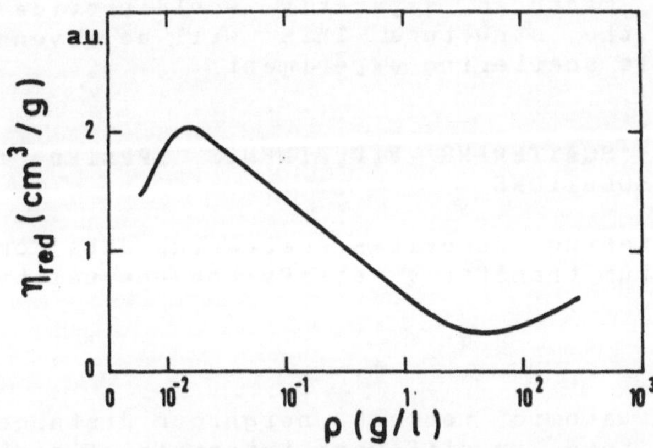

Figure 3. Schematic representation of the reduced viscosity versus concentration by weight (from C. Wolff[10]).
The part of the curve discussed in this section is the decreasing part. At higher concentration, the viscosity increases as a result of polyion interaction effects, such as entanglements.

where a_h, m are respectively the *monomer* hydrodynamic radius and molecular weight. In this hypothesis, the data indicate that the hydrodynamic radius R_h^2 decreases with concentration ρ, as $\rho^{-1/2}$. However this rate of decrease will be found too slow, if polyion-polyion interactions are accounted for in the calculation of the reduced viscosity. [Note that (3.5) differs from the usual expansion formula

$$\eta_{red} = [\eta] + k_H [\eta]^2 \rho \qquad (3.7)$$

where $[\eta] = \lim_{\rho \to 0} \eta_{red}$ and where k_H is the Huggins coefficient. Such an expansion is not suited to reflect rapid variation of the coil shape with concentration]. An experiment in which both *intra* and *intermolecular* correlations are measured separatly would provide better understanding of the structure. This will be given by the neutron small angle scattering experiment.

4. SMALL ANGLE SCATTERING EXPERIMENTS APPLIED TO POLYELECTROLYTE SOLUTIONS

Small angle scattering describes scattering data for which the elastic momentum transfer q satisfy the inequality :

$$q \ll \frac{1}{a} \qquad (4.1)$$

where a is the atomic nearest neighbour distance. This inequality holds true in different intervals of scattering angles, according to the value of λ/a, where λ is the incoming radiation wavelength. If $\lambda/a \lesssim 1$ (neutron scattering), the data of interest are measured at small scattering angles. On the contrary, for $\lambda/a \gtrsim 1$ (light scattering) these data are measured over a large interval of scattering angles.

4.1. Small angle scattering versus diffraction experiments. Hydration effects.

In the range $q \ll 1/a$, radiation scattering is insensitive to overall density fluctuations : the aqueous polyelectrolyte solution can be considered as incompressible. Let $\varphi_{\mathcal{A}}(r)$ be the local volume fraction of species $\mathcal{A}$ ($\mathcal{A}$ = 0, solvent ; $\mathcal{A}$ = 1, polyion, $\mathcal{A}$ = 2, counterion). Incompressibility implies :

$$\sum_{\mathcal{A}=0}^{2} \varphi_{\mathcal{A}}(r) \equiv 1 \tag{4.2}$$

and therefore, one of the unknown volume fraction fluctuation, say $\varphi_0(r)$, can be expressed as a function of the others. This is the basic assumption used to interpret small angle scattering experiment. The scattering cross section per unit volume for a 3 components system is

$$\Xi(q) = \sum_{\mathcal{A}=0}^{2} \sum_{\mathcal{B}=0}^{2} b_{\mathcal{A}}\, b_{\mathcal{B}}\, C_{\mathcal{A}}\, C_{\mathcal{B}}\, \mathbb{H}_{\mathcal{A}\mathcal{B}}(\vec{q}) \tag{4.3}$$

where $b_{\mathcal{A}}$ are the (monomer) scattering length, $C_{\mathcal{A}}$ the (monomer) concentration and

$$\mathbb{H}_{\mathcal{A}\mathcal{B}}(\vec{q}) = \int d^3r\, e^{i\vec{q}.\vec{r}} \left(\frac{\langle C_{\mathcal{A}}(0)\ C_{\mathcal{B}}(r) \rangle}{C_{\mathcal{A}}\, C_{\mathcal{B}}} - 1 \right) \tag{4.4}$$

the structure functions. The relation between volume fraction and concentration is

$$\varphi_{\mathcal{A}}(r) = \frac{C_{\mathcal{A}}(r)}{v_{\mathcal{A}}}$$

where $v_{\mathcal{A}}$ is the partial volume. Assumption (4.2) and the equality $C_1 = C_2$ lead to the simplified equation :

$$\Xi(q) = C^2 \sum_{\mathcal{A}=1}^{2} \sum_{\mathcal{A}=1}^{2} \bar{b}_{\mathcal{A}}\, \bar{b}_{\mathcal{B}}\, \mathbb{H}_{\mathcal{A}\mathcal{B}}(\vec{q}) \tag{4.5}$$

where

$$\bar{b}_{\mathcal{A}} = b_{\mathcal{A}} - b_0 \frac{v_0}{v_{\mathcal{A}}} \tag{4.6}$$

is the contrast length of species $\mathcal{A}$ against pure water. Equation (4.5) is our basic formula.

In the diffraction range

$$q \geqslant \frac{1}{a}$$

the discrete nature of the atomic distribution in space becomes conspicuous. Equation (4.2) does not hold true because of local fluctuations in density. Equation (4.5) is then incomplete ; the full equation (4.3) must be used to interpret the diffraction experiment.

The difference between small angle scattering and diffraction techniques appears clearly in the study of hydration. This phenomenon describes the binding of water molecules to the ions, in this vicinity. It is essential for preservation of ionic dissociation.

In the diffraction pattern, hydration is revealed by diffraction peaks, especially those attributed to the $\mathbb{H}_{01}(\vec{q})$ and $\mathbb{H}_{02}(\vec{q})$ partial structure function (see lecture notes by G.W. Neilson and J.E. Enderby). In the small angle scattering range (4-1), hydration also affects the scattering cross section, but in a different manner. The scattering intensity is modified by the fact that the volume fraction of water molecules bound to the ions is larger than the volume fraction of bulk water. As a consequence, an exact description of the aqueous solutions uses 4 different components ($\mathcal{A}$ = 0, 1, 2, 3), where $\mathcal{A}$=0 represents bulk water and $\mathcal{A}$=3, bound water. The contrast length (4.6) is now defined with respect to bulk water : a contrast between bound and bulk water exists because of molecular volume differences. The small angle scattering cross section of the solution is written as in (4.5), but with a summation extending from $\mathcal{A}$ = 1 to $\mathcal{A}$ = 3.

Plestil[12,13] and coworkers have evaluated hydration of linear polyions, using neutron and X-rays small angle scattering data on identical aqueous solutions of linear polyelectrolytes. A typical result of the small angle scattering experiment is given in figure 4, which displays the characteristics maximum scattered intensity at finite wave vector transfer. Let R_G be the polyion radius of gyration. Assuming that in the asymptotic range $qR_G >> 1$ and $q \geqslant q_{Max}$, the main contribution to the cross section (4.5) is given by the polyion form function and the adsorbed water and counterion molecules, the cross-section (4.4) is written

$$\Xi(q) \simeq (b^{-})^{2} \; C \; N \; H_{app}(\vec{q}) \qquad (4.7)$$

where $H_{app}(\vec{q})$ is the apparent form function of the polyion plus water and counterion complex, and b^{-} an average contrast length. In the interval of reciprocal space of interest, the apparent and true polyion form function differs by a factor

$$\exp\left(-\frac{q^2}{2}R_\perp^2\right) \qquad (4.8)$$

where $R_\perp^2$ is the *lateral* apparent radius accounting for the adsorbed water and counterion molecules. A general expression[14] for this radius is

$$(R^\perp_{app})^2 = \frac{1}{N^\perp}\sum_{\mathcal{A}=1}\sum_{\mathcal{B}=1}\frac{\bar{b}_\mathcal{A}\,\bar{b}_\mathcal{B}}{(\bar{b})^2}N^\perp_\mathcal{A}N^\perp_\mathcal{B}(R^\perp_{\mathcal{A}\mathcal{B}})^2 \qquad (4.9)$$

The unknown quantities are the number of molecules $N^\perp_\mathcal{A}$ ($\mathcal{A}$=2, 3) belonging to the clouds, and the effective radii $R^\perp_{\mathcal{A}\mathcal{B}}$. A first result of Plestil and coworkers gives an excess of 1.1 water molecules in a cylindrical shell formed by one monomer of the polyion and a radius of 0.7 nm. This group also finds a remarkable nonlinear increase of the hydration effect, as the parameter y ratio of ionizable sites to available site) is varied from 0 to 1 (see figure 5). Precise evaluation of hydration and (counterion) condensation effects in (4.7) is however a difficult exercise. More rigorous results are expected from this interesting investigation.

4.2. Partial structure functions and the charge distribution

In this section and the following, we assume equation (4.5)

$$\Xi(q) = C^2\sum_{\mathcal{A}=1}^{2}\sum_{\mathcal{B}=1}^{2}\bar{b}_\mathcal{A}\,\bar{b}_\mathcal{B}\,\mathbb{H}_{\mathcal{A}\mathcal{B}}(\vec{q}) \qquad (4.10)$$

to be the adequat expression of the small angle scattering cross section per unit volume for the aqueous polyelectrolyte solution[15] ($\mathcal{A}$=1, polyion ; $\mathcal{A}$=2, counterion). The contrast length $\bar{b}_\mathcal{A}$ with respect to bulk water, should then account for hydration effects :

$$\bar{b}_\mathcal{A} = b^*_\mathcal{A} - b_o\frac{v_o}{v^*_\mathcal{A}} \qquad (4.11)$$

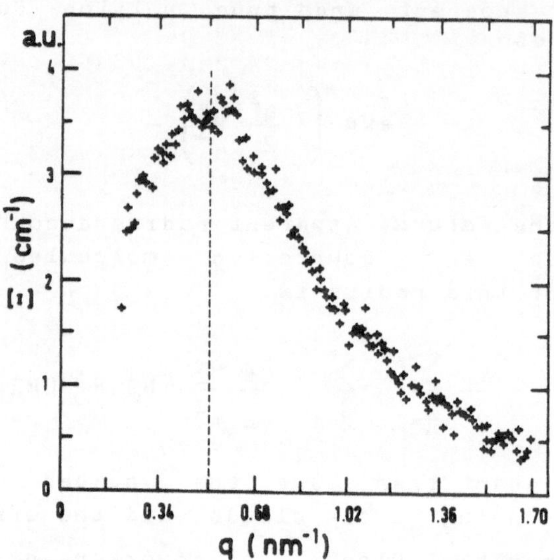

Figure 4. Small angle scattering cross section of a sodium sulfonated polystyrene aqueous solution, measured with X-rays. Concentration by weight $\rho = 30$ g/ℓ
$M_w = 300\,000$ (C. Williams, Lure, Orsay).

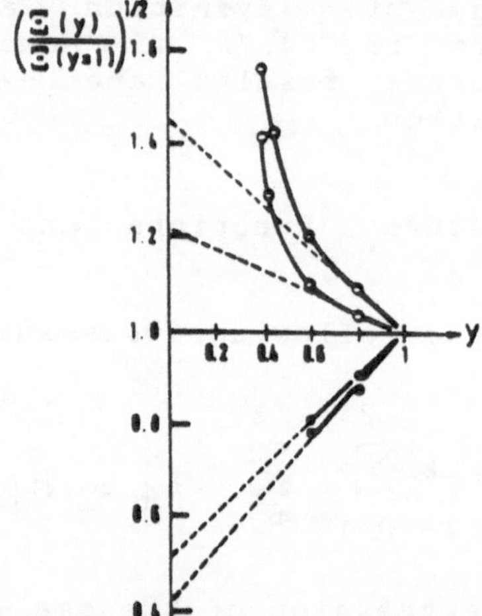

Figure 5. Variation of scattered intensities with the charge ratio y, at fixed $q(>.q_m)$ PMA aqueous solution. $M_w = 15\,000$.
upper part : neutron scattering
lower part : X-ray scattering
The hydration of the polyion is derived from the data derived respectively from X-ray and neutron scattering (Plestil et al, refs. 12 et 13).

where $b^*_{\mathcal{A}}$, $v^*_{\mathcal{A}}$ are respectively collision length and partial volume averaged over monomer (species $\mathcal{A}$) and surrounding hydration shell.

The function displayed in figure 4 is according to (4.10), a combination of 3 partial structure function : polyion-polyion (11) polyion-counterion (12 and 21) and counterion-counterion (22). These functions are weighted by collision lengths $b_{\mathcal{A}}$, which depend on the radiation used for the experiment.

In order to determine the partial structure functions $\mathbb{H}_{\mathcal{AB}}(\vec{q})$ from the small angle scattering experiment, the cross section $\Xi(q)$ of several samples is measured. These samples differ only with respect to their contrast length. These lengths can be modified by use of isotopic substitution, either on the polyion (sulfonated polystyrene)

$$-(-CH_2 - \underset{\underset{C_6H_4SO_3^-}{|}}{CH} -)_N \rightarrow (-CD_2 - \underset{\underset{C_6D_4SO_3^-}{|}}{CD} -)_N$$

or on the counterion

$$N^+(CH_3)_4 \rightarrow N^+(CD_3)_4$$

Results of neutron scattering data [16] on the 4 different combinations of deuterated and non deuterated species are shown in figure 6. The partial structure functions are derived from the data, by adequat linear recombinations. To achieve this result, exact knowledge of the contrast length $b_{\mathcal{A}}$ is required. The result corresponding to the table 3 in reference 16 are displayed on figure 7. Notice the similarity of polyion and counterion structure functions, indicating a strong correlation.

The normalisation of the polyion partial structure function is given by the approximate relation

$$c\mathbb{H}_{11}(\vec{0}) \simeq (\Phi_\pi + \frac{1}{N})^{-1} \simeq \Phi_\pi^{-1} \qquad (4.12)$$

where Φ_π is the osmotic reduction coefficient.

Equation (4.12) identifies the polyion structure function at zero angle to the osmotic compressibility. (This type of relation is exact for a solute made of a single component). The normalization of structure functions derived from experimental observations, is important for the determination of the charge structure function

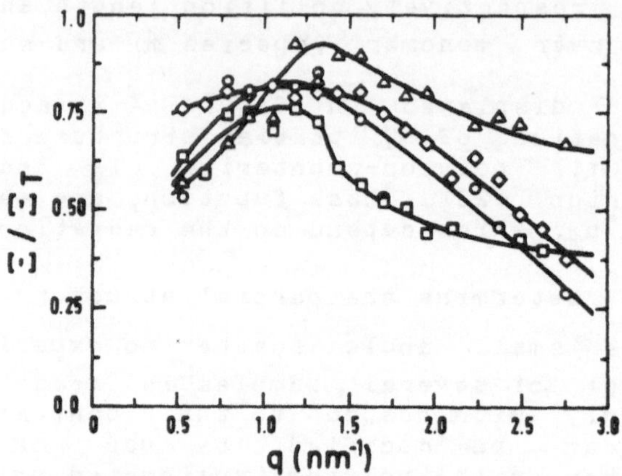

Figure 6. Neutron small angle scattering cross section ratios $\Xi(q)/\Xi_T(q)$, where Ξ_T is the total cross section including spin incoherence. Result obtained with polarized neutron beams. Tetramethylamonium sulfonate polystyrene, M_w = 300 000
Concentration by weight ρ = 160 g/ℓ.
+ deuterated polyion, non deuterated counterion.
o deuterated polyion, deuterated counterion
Δ non deuterated polyion, deuterated counterion
x non deuterated polyion, non deuterated counterion (ref.16).

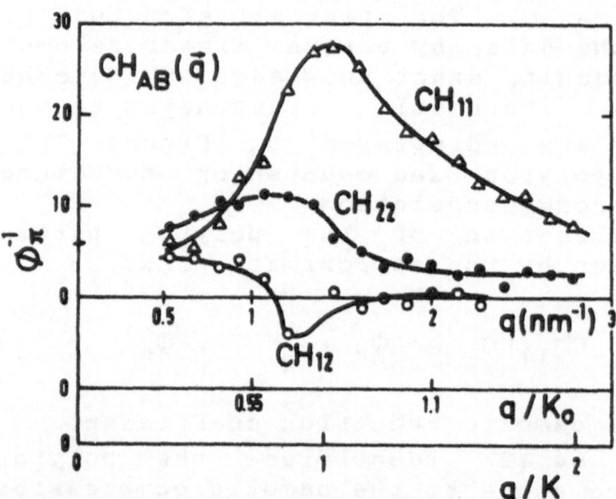

Figure 7. Partial structure functions derived from the results displayed in fig.6. The normalization of the functions is taken from condition (4.12), with Φ_π = 0.16.

$$S_{ZZ}(q) = C \sum_{\mathcal{A}=1}^{2} \sum_{\mathcal{B}=1}^{2} (-1)^{\mathcal{A}} (-1)^{\mathcal{B}} \mathbb{H}_{\mathcal{A}\mathcal{B}}(\vec{q}) \quad (4.13)$$

(see figure 8). Sum rules to which this function obeys, are (see lecture notes by H.L. Friedman)

$$S_{ZZ}(0) = 0$$

$$S_{ZZ}(q) = q^2/\bar{K}^{-2} \quad , \quad q^2 \ll \bar{K}^{-2} \quad (4.14)$$

where $\bar{K}$ is here an apparent inverse screening length. This length takes into account both ionic Coulomb interactions *and* intramolecular correlations. The respective contributions have not yet been calculated.

For polyelectrolyte solutions (in contradiction to simple electrolytes) the inverse screening length is usually defined by the relation

$$K_o^2 = \frac{4\pi\beta e^2 C}{\epsilon} \quad (4.15)$$

or[6]

$$K_{oo}^2 = K_o^2 \, \Phi_\pi \quad (4.16)$$

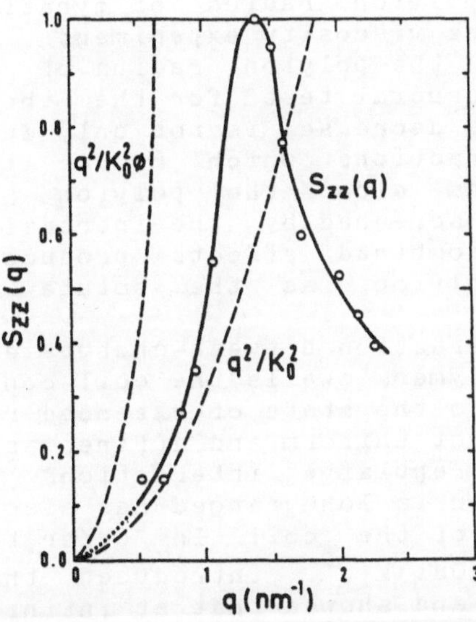

Figure 8. Charge structure function as defined by equation (4.13), and calculated functions K_o^2/q_o^2, $\Phi_\pi K_o^2/q_o^2$.

Only Coulomb interactions are considered. Equation

(4.15) and (4.16) also indicate that the polyions are not supposed to contribute to the screening. equation (4.16) relates the Coulomb effective interaction to the "free" counterions. In principle, such equations are only valid at very large dilutions. When the polyelectrolyte concentration increases, the polyions become flexible and the smaller polyion rigid units are able to screen the interaction. Therefore, (4.16) must be revised.

For all these reasons, we expect $\bar{K}$ to be different from K_{oo} (for instance greater). For ρ = 160 g/ℓ, the calculated value of K_{oo} is 1 nm^{-1}, and the experimental data are consistent with $\bar{K} \simeq 10^{-1}$ nm^{-1} as shown in fig. 8. Thus the sum rule (4.14) cannot be interpreted here exclusively in terms of the dielectric structure (determination of the dielectric constant, respective contributions of counterion and polyions to the screening).

4.3. Study of the polyion structure function. Decomposition into inter- and intramolecular correlations.

In the past decade, one of the main purposes of the neutron small angle scattering experiment has been to determine the flexible polyion form function, at different polyelectrolyte concentrations above the polyion overlap concentration. The result of such an experiment gives in principle a better estimation of the polyion radius of gyration, at finite concentration, than the viscosity experiment.

The decrease of the polyion radius of gyration with concentration is a crucial test for the theory of linear polyelectrolyte. This decrease is not only due to enhanced polyion-polyion interactions which reduce the effect of repulsive interactions along the polyion ; the Coulomb interaction is also screened by the increasing number of counterions. These combined effects produce a stronger collapse of the polyion, as the solute concentration increases.

The Coulomb interaction between charges distributed on a flexible linear polymer, swells the coil considerably and eventually brings it to the state of extended rod.

The swelling is not uniform and affine, as it is in the case of short range repulsive interactions. The fact that the Coulomb interaction is long ranged has also an effect on the local stiffness of the coil. In order to account for this correlation, Odijk [6] introduced the "electric" persistence length L_p and showed that at infinite dilution

$$L_p = 1/4QK^2 \tag{4.17}$$

where $Q = e^2\beta/\epsilon$ is the Bjerrum length and K is the inverse screening length. Later, de Gennes [5] indicated that this relation holds true in the range of concentration where polyions strongly overlap.

The neutron scattering experiment provides the data necessary to determine the polyion form function, at any concentration. The experiment is carried along the following steps:

- the (pure) water is replaced by a heavy water light water mixture. The fraction α^* of heavy water is such that $b_2^-(\alpha^*) = 0$. As a result the cross section depends only on the polyions

$$\Xi(q) = C^2 \, (b_1^-(\alpha^*))^2 \, \mathbb{H}_{11}(\vec{q}) \qquad (4.18)$$

- the polyion structure function is naturally decomposed in intra-and intermolecular contributions

$$\mathbb{H}_{11}(\vec{q}) = \frac{N}{C} H_1(\vec{q}) + \mathbb{H}_{11}^{II}(\vec{q}) \qquad (4.19)$$

where $H_1(\vec{q})$ is the polyion form factor and $\mathbb{H}_{11}^{II}(\vec{q})$ the intermolecular structure function.

Introducing a fraction X of deuterated polyions, the total monomer concentration is

$$C = CX + C(1-X) \qquad (4.20)$$

The monomer contrast lengths are respectively $b_{1D}^-(\alpha^*)$ and $b_{1H}^-(\alpha^*)$. The cross section is now

$$\Xi(q) = C\left\{X(b_{1D}^-(\alpha^*))^2 + (1-X)(b_{1H}(\alpha^*)^2\right\} NH_1(\vec{q})$$

$$+ C^2 \left\{X\, b_{1D}^-(\alpha^*) + (1-X)\, b_{1H}(\alpha^*)\right\}^2 \mathbb{H}_{11}^{II}(\vec{q}) \qquad (4.21)$$

Intramolecular contributions vary linearly with X, whereas intermolecular contributions vary quadratically. Measuring the cross section at different values of X but at constant concentration C, we can determine both the polyion form function $H_1(\vec{q})$ and the polyion intermolecular structure function $\mathbb{H}_{11}^{II}(\vec{q})$. The decomposition of $\mathbb{H}_{11}(\vec{q})$ into $H_1(\vec{q})$ and

$\mathbb{H}_{11}^{II}(\vec{q})$ from $\Xi(q;C,X)$, is in fact very sensitive to the value given to the contrast length $\bar{b}_{1D}(\alpha^*)$ and $\bar{b}_{1H}(\alpha^*)$. In particular the result differs if hydration effects are included or not. Figure 9 shows the result obtained by Nierlich et al[17]. Values used by these authors are :

$$b_{1H} = 12.07 \times 10^{-5}\ \text{nm} \ ; \ v_1 = 0.18\ \text{nm}^3$$

It is of interest to note that the form function differs from the total structure functions, even at higher values of the momentum transfer : the contribution of intermolecular correlations to $\mathbb{H}_{11}(\vec{q})$ is not negligible in the range $qR_G \gg 1$. From the analysis of the polyion form function, one gets :

- in the "Guinier" range ($qR_G \ll 1$) the squared radius of gyration, R_G^2.
- in the asymptotic range ($qR_G \gg 1$), the persistence length L_p. The derivation of this result is not trivial : the model used for the polyion configuration is the Brownian coil with persistence length. For this model $R_G^2 = LL_p$, where L is the extended polyion length. Values of R_G^2 determined experimentally from neutron scattering data at several concentrations by weight ρ, are displayed on figure 10.

The data indicate a decrease of R_G^2 with ρ :

$$R_G^2 \propto 1/\rho \qquad (4.22)$$

which is faster than the one derived from viscosity data ($R_h^2 \propto \rho^{-1/2}$, see section 2). This is expected, since intramolecular correlations are not the only contributions to viscosity. An interesting consequence of (4.22) is given by the behaviour of the overlap ratio R_G^2/d^2, where d is the average nearest neighbour distance between polyions. According to (4.22) this ratio decreases with concentration as:

$$\frac{R_G^2}{d^2} \propto \rho^{-1/3} \qquad (4.23)$$

This result reflects the complex mechanism for the polyion swelling reduction ; the screening of the Coulomb repulsive interaction is not primarely due to the polyions. The main

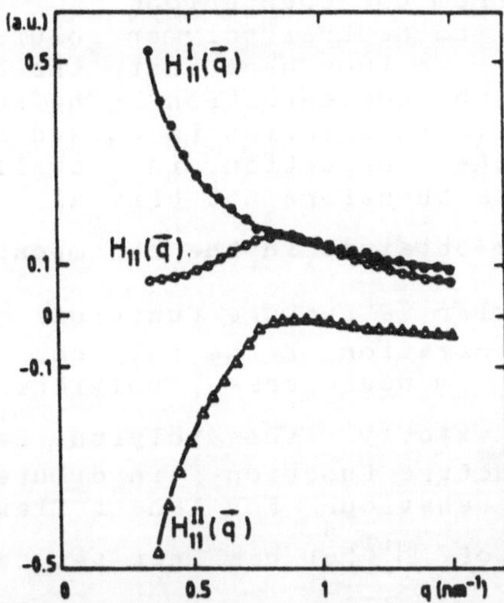

Figure 9. Representation of the polyion structure function $\mathbb{H}_{11}(\vec{q})$ as a sum of the polyion form function $\frac{N}{C} H_1(\vec{q}) = \mathbb{H}^{I}_{11}$ and the polyion intermolecular structure function $\mathbb{H}^{II}_{11}(\vec{q})$ Results of neutron small angle scattering experiments. M_w = 80 000 ; ρ = 20 g/ℓ. (reference 18).

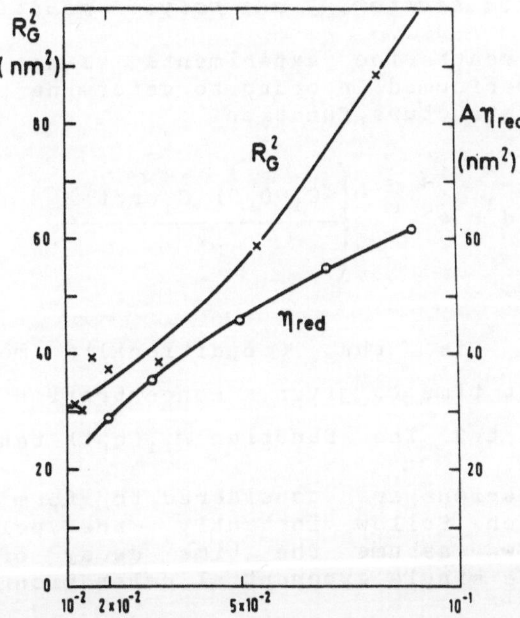

Figure 10. Variation of the polyion radius of gyration with concentration (M_{1w} = 40 000).
Variation of the reduced viscosity with concentration (M_{1w} = 41 000).

contribution comes from the counterions.

If we compare to neutral polymer solutions, we find a different behaviour. We know here that, the overlap between coils increases with concentration. The screening of the short range repulsive interaction is caused by the polymers themselves, and the reduction in swelling is weak. Observation (4.23) is therefore not trivial.

The second term obtained in the decomposition of $\mathbb{H}_{11}(\vec{q})$, is the intermolecular structure function, $\mathbb{H}^{II}_{11}(\vec{q})$, related to the mutual organization of the polyions. In the case of rigid, spherical monodisperse polyions, the ratio $C\mathbb{H}^{II}_{11}(\vec{q})/H_1(q)$ is exactly the polyion center of mass intermolecular structure function : in dilute solutions, it has a liquid like behaviour. For linear flexible polyions, the interpretation of $\mathbb{H}^{II}_{11}(\vec{q})$ has not yet been worked out. Experimental data for $\mathbb{H}^{II}_{11}(\vec{q})$ indicate deviations from the Debye-Heïckel model, which are attributed to local order. Such qualitative observations are confirmed by electric birefringence measurements[17]. Models and calculations are needed to make proper use of these experimental results.

4.4. *Quasielastic scattering and polyion mobility*

Quasielastic scattering experiments using neutrons and photons, were performed in order to determine the time decay of the polyion structure function

$$\mathbb{H}_{11}(\vec{q},t) = \int d^3r\, e^{\vec{q}.\vec{r}} \left(\frac{\langle C_1(\vec{0},0)\ C_1(\vec{r},t)\rangle}{C_1^2} - 1 \right) \quad (4.24)$$

where $C(\vec{r},t)$ is the (conditional) monomer local concentration at time t, given a concentration $C(\vec{0},0)$ at the origin and at t=0. The function $\mathbb{H}_{11}(\vec{q},t)$ tends to zero at $t\to\infty$.

The counterions are considered to form clouds around polyions, which follow instantly the polyion motion. Consequently, we assume the time decay of (4.24) to be represented by a single exponential relaxation

$$\frac{\mathbb{H}_{11}(\vec{q},t)}{\mathbb{H}_{11}(\vec{q})} = e^{-t/\tau(q)} \tag{4.25}$$

where $\tau(q)$ is the wavenumber dependent relaxation time [Quasielastic light scattering experiments[20,21] have revealed the existence of a second relaxation time in the decay of the correlation function (4.24). The slow mode is identified with the mode to be discussed below. Interpretations are being proposed for the so called fast mode. (See also lecture notes by P. Madden, for a discussion on the electromagnetic interaction with an aqueous solution)]. We write

$$\frac{1}{\tau(q)} = D(q)\ q^2$$

where $D(q)$ is the polyion concentration diffusion coefficient. For charged systems, $D(q)$ is generally maximum at $q=0$, because overall electric neutrality constraints the concentration fluctuations to be rapidly compensated. As q increases, this constraint is less stringent and fluctuations tend to be controlled by chemical potentials. For both linear flexible and rigid spherical polyions, the decrease of $D(q)$ with q is very conspicuous. Data collected with light and neutron radiations give a consistent picture of the function $D(q)$ (see figure 11). This function ceases to decrease for $q \geqslant K$.

The coefficient $D(q)$ varies with concentration, and this fact is of interest because it reflects the successive rearrangements of the polyions in the solution. In the frame of the mode decoupling theory, $D(q)$ can be factorized into dynamic and static contributions[5]

$$D(q) = \frac{\mu(q)}{\beta\ c\mathbb{H}_{11}(q)}$$

This function is seen to be proportional to a mobility $\mu(q)$ and to an osmotic restoring force.

For neutral polymer solution, it is known that the restoring force in the limit $q \to 0$, increases with concentration, as an effect of short range repulsive interactions. For charged molecules, this force is independent of concentration, because the solution behaves as an ideal solution (see figure 2). The mobility of the polyion, in the limit $q \to 0$, is related to the solvent viscosity and to hydrodynamical backflow effects : the concentration diffusion coefficient $D(q \to 0)$ is in fact independent of concentration. At finite wave vector transfers, mobility reflects overall and internal motion of

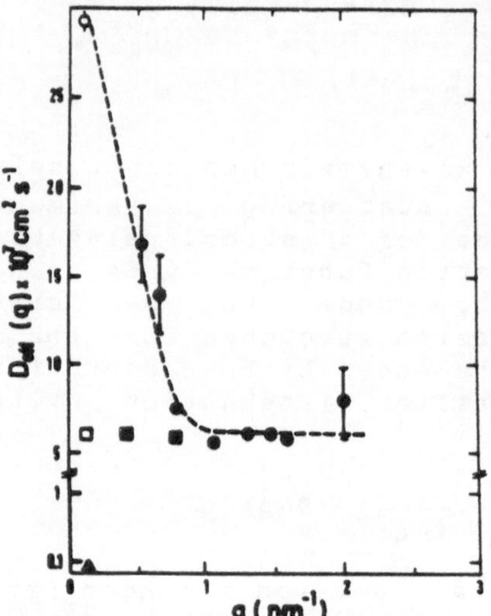

Figure 11. Variation of the solute concentration diffusion coefficient with wave vector q. ●, ■ neutron scattering : o, □ light scattering ; □ , ■ salt added to the solution ; o,● no salt added.(refs. 16 and 20).

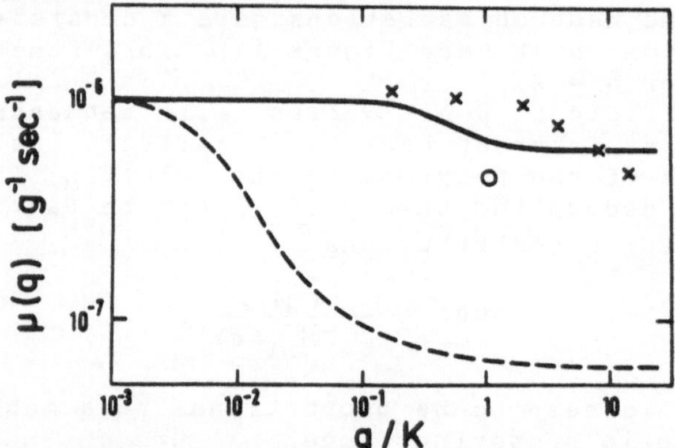

Figure 12. Polyion mobility as a function of wave vector
o light scattering : $M_w = 7.85 \times 10^5$
$\rho = 2.5 \times 10^{-2}$ g/ℓ (ref.20)
x Neutron scattering : $M_w = 2 \times 10^5$
$\rho = 70$ g/ℓ (ref.16)
Curves represent theoretical predictions
--- low polyelectrolyte concentration
_____ high polyelectrolyte concentration.

the polyion molecule. For rigid spherical polyions, only the overall motion is of importance ; the quasielastic scattering experimental data [19] are satisfactorily interpreted in terms of the ratio $\mu(q)/C\mathbb{H}_{11}(\vec{q})$, when such polyions are in the "liquid" state. A backflow contribution to the mobility is formed in the range $qR_G \ll 1$, and the center of mass pair correlation is derived directly from the data for $qR_G \gtrsim 1$.

In the case of linear flexible polyions, conformational diffusive motions generate a wave vector dependence on the mobility. The predictions are

$$\begin{aligned} \mu(q) &= \mu(0)/L_p\, q \quad , \quad L_p^{-1} < q < K \\ \mu(q) &= \mu(0)\; Ka \quad , \quad q > K \end{aligned} \tag{4.26}$$

At finite q, the polyion mobility (4.26) increases with concentration, as a result of the collapse of the rigid rod into smaller subunits. This increase should be important (see figure 12). The data derived from neutron[5,16] and light[20] scattering experiments show however small variations with concentration. We note that the theoretical expression of the mobility accounts only for intramolecular correlations. It may be necessary to evaluate intermolecular contributions to the quantity.

5. *CONCLUSION*

After the first significant observations of structures derived from conductivity, osmotic pressure and viscosity measurements in aqueous polyelectrolyte solutions, a major progress was obtained with neutron small angle scattering experiments. Partial structure functions and separate intra-, intermolecular correlations of polyions were ietermined for the first time. In the case of linear polyions, the overall scattering maximum or Coulomb peak, was reduced to the sum of a decreasing form function and a nearly monotonously increasing intermolecular structure 'unction. Screening length of the Coulomb intreraction and persistence length of polyions were evaluated. However, the ack of precision, both in the acquisition of data and in he derivation of the structure functions shades doubts on he results. Experimental progress is urged in this domain.

It is of interest to note that significant structural hanges in aqueous polyelectrolyte solutions are being bserved over more than four decades of concentration, when he polyion is made of linear flexible polymers. We have ointed out that the measured overlap ratio is a decreasing unction of concentration ! Screened Coulomb interaction etween elements of a self similar structure such as a

Brownian coil, is seen to produce many intermediate states between the extended rod configuration and the random walk. Somehow, the aqueous polyelectrolyte solution made of linear polyions is a system in a state of continuous melting over a large interval of concentrations. It is now necessary to formulate the corresponding intermolecular organisations.

REFERENCES

1. S. Hachisu, Y. Kobayashi and A. Kose ; J. of Colloïd and Interface Science, **42** (1973) 342.
2. J.W. Vanderhoff, H.J. Van Den Hul ; J. Macromol. Sci. Chem. **A7** (1973) 677.
3. D. Howe, S. Alexander, P.M. Chaikin and P. Pincus ; J. Chem. Phys. **79** (1983) 1474. and T. Ohtsuki, S. Mitaku and K. Okano ; Jap. J. of Applied Phys. **17** (1978) 627.
4. P.G. de Gennes, P. Pincus, R. Velasco, F. Brochard ; J. Physique **37** (1976) 1461.
5. J. Hayter, G. Jannink, F. Brochard, P.G. de Gennes J. Physique Lettres **41** (1980) L45.
6. T. Odijk ; Macromolecules **12** (1979) 688.
7. H. Eisenberg ; J. Polym. Sci. **30** (1958) 47.
8. A. Katchalsky ; Pure Appl. Chem. **26** (1971) 327.
9. A. Takahashi, N. Kato, N. Nagasawa ; J. Phys. Chem. **74** (1970) 944.
10. C. Wolff ; J. Physique **39** C2 (1978) 169.
11. M. Moan, Thèse université de Bretagne Occidentale, Brest 1976.
12. J. Plestil, J. Mikes, K. Dusek ; Acta Polymerica 30 (1979) 29.
13. J. Plestil, J. Mikes, K. Dusek, J.M. Ostanevitch, A.B. Kuchenko ; Polymer Bulletin **4** (1981) 225.
14. J. des Cloizeaux and G. Jannink ; "*Lrn polymèren en Solution ; leur Modelisation et leur Structure*"; Editions de Physique.
15. G. Jannink ; Makromol. Chem. Macromol. Symp. 1 (1986) 67.
16. F. Nallet, G. Jannink, J. Hayter, R. Oberthür, C. Picot ; J. Physique **44** (1983) 87.
17. S.S. Wijmenga ; Thesis, Leiden 1984.
18. M. Nierlich, F. Boué, A. Lapp, R. Oberthür ; J. Physique **46** (1985) 649. and J. Colloid Polymer Sci. **263** (1985) 955.
19. D.W. Schaeffer ; J. Chem. Phys. **66** (1977) 3980.
20. M. Drifford, J.P. Dalbiez, K. Tabtic, P. Tivaut ; J. Chimie Physique **82** (1985) 7.
21. R.S. Koene ; Thesis, Leiden 1983.

SUPERCRITICAL STUDIES

E.U. Franck
Institute of Physical Chemistry
University of Karlsruhe
12, Kaiserstraße
D7500 Karlsruhe
Federal Republic of Germany

ABSTRACT. Dense water at high subcritical and supercritical temperatures (T_c= 374 °C) is a unique electrolytic solvent. Certain thermophysical properties: PVT-relations, static dielectric constant and viscosity are discussed to 1000 °C. Above T_c and at supercritical densities the dielectric constant can still have values of 10, 20 or more. Self-ionization increases with P and T, so that water becomes an ionic fluid at 1000 °C and above 200 kbar. The molar electrolytic conductance of dissolved salts, acids and bases at supercritical conditions can be nearly ten times as high as at normal conditions. Examples together with the partial molar volume of sodium chloride at high pressures are shown. Supercritical water, even at high density is completely miscible with nonpolar partners like hydrogen, oxygen, carbon dioxide or methane. A discussion of phase diagrams, methods of equilibrium calculations and also dielectric constants of supercritical mixtures is given. Ternary systems like water - salt - carbon dioxide or methane are described. Combustion and "hydrothermal flames" can be produced to 1000 bar in the supercritical aqueous environment.

1. INTRODUCTION

Dense water at high temperature - subcritical and supercritical - is a unique electrolytic solvent. Not only has it obvious practical importance, for example in technology and geochemistry, it is also an outstanding medium to test and to extend electrochemical concepts and theories. This is particularly true at supercritical conditions, where thermophysical properties can be varied extensively and continuously with pressure and temperature from gas-like to liquid-like values. This could mean the transition from non-electrolyte to electrolyte behaviour as a solvent. In supercritical water there is a wide range of complete miscibility with otherwise unmiscible partners, although simultaneous ionic dissociation of other solutes may be possible. The critical temperature, pressure and density of water are 374 °C, 22.1 MPa and 0.32 g cm^{-3}. In order to generate substantial density variations at elevated temperatures, application of high pressure is

M.-C. Bellissent-Funel and G. W. Neilson (eds.), The Physics and Chemistry of Aqueous Ionic Solutions, 337–358.

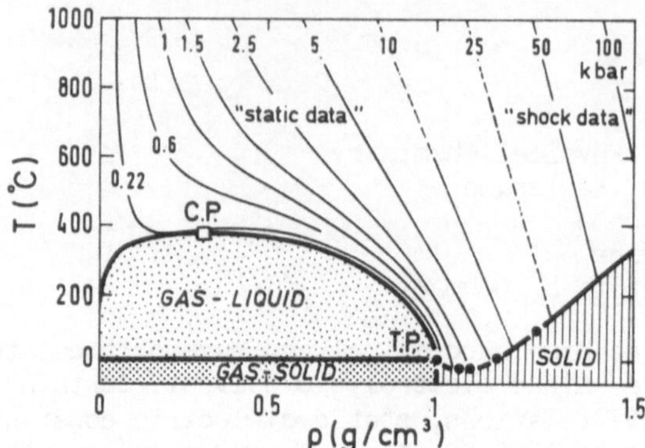

Figure 1. Temperature-density diagram of pure water (1)(13)(43). Full drawn isobars: Approximately the region covered by experiments. C.P., T.P.: Critical and triple points of water.

necessary, although pressure values up to several hundred MPa will be sufficient in most cases. It should be remembered, that pressures of this magnitude normally do not appreciably deform molecular and electronic structures. Experimental arrangements are often required, which permit precise measurements at conditions of elevated pressures, temperatures and high corrosion resistance.

With this presentation selected results of more recent, mainly experimental investigations shall be given and discussed. At first some thermophysical properties of pure water will be shown. Ionic fluids will follow and binary mixtures with nonpolar partners will be discussed. Ternary aqueous systems in the critical region shall be considered and finally the feasibility of "hydrothermal combustion" shall be demonstrated.

Pure water is probably the best investigated polar fluid as far as the pressure-volume-temperature relations are concerned. Modern editions of steam-tables for engineering purposes provide density data in dependence of temperature and pressure to nearly 1000 °C and 10 kbar = 1000 MPa. Above 3000 MPa shock wave data can be used (1)(13)(43). The situation is demonstrated with Fig. 1. The fully drawn isobars belong to the range of conditions covered by measurements either static or with shock waves. The intermediate part must for the present be covered by interpolations. It appears, however, that recent efforts to design equations of state for high temperature water, based on simple, rational molecular models can be successful to extreme conditions (2).

2. PURE WATER

Besides the PVT-relation, two other thermophysical properties are of particular importance for high temperature-high pressure water as an electrolytic solvent: The static dielectric constant and the viscosity. Both quantities are functions of temperature and density only.

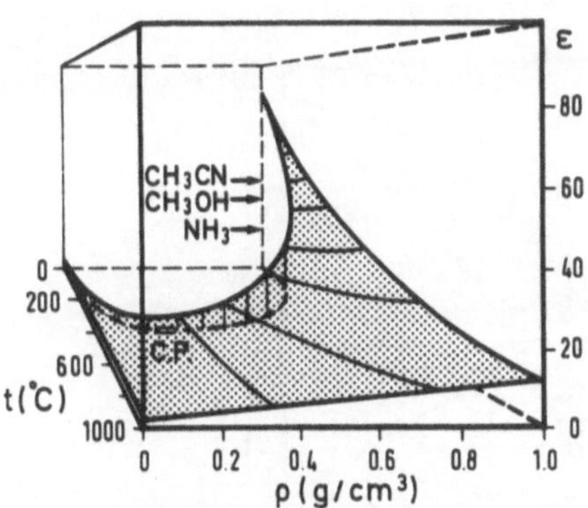

Figure 2. The static dielectric constant of water, ε , as a function of temperature t and density ρ in an approximate presentation (5)(7).

The dielectric constant of water at relatively normal conditions is widely discussed in the literature (see, for example (1)(3)(4)). Measurements have more recently been extended to 500 °C and 5 kbar (5)(6). A critical survey of all existing data and an equation of state for the dielectric constant has been presented (7). Fig. 2 gives a three-dimensional diagram of the constant to 1000 °C and to a water density of 1 g cm^{-3}. Above 550 °C the diagram is based on approximate calculations. It is evident, that the familiar high values occur only within a limited range of moderate temperature and high density. For comparison, values of liquid methyl cyanide, methanol and ammonia are indicated. It can also be seen, however, that an extended region of supercritical conditions exists, with dielectric constants between 10 and 25, where ionic dissociation of solutes can be expected. Efforts have been made, to predict the dielectric constant to higher temperatures and pressures (8)(9)(10)(11)(12). They are mainly based on a procedure suggested earlier (8), with an adjustment of the Kirkwood correlation factor as a simple temperature-density function to existing experimental data. A more recent, more rational computational method appears to be promising. High temperature dielectric constant values are particularly useful for geochemistry.

The viscosity of water and steam to several hundred degrees Celsius is also well investigated and discussed (1)(13). More sparse are the data at high pressures. Measurements have been reported to 500 °C and 5 kbar (14). In Fig. 3a three-dimensional diagram like the one in Fig. 2 is shown. It is of an approximate character above 500 °C. Again it is clear that the well-known high viscosity values are restricted to a relatively narrow range of low temperature and high density. At supercritical conditions one observes a range where the viscosity is only slightly temperature dependent and has values lower by a factor of ten than in the normal liquid. Thus, according to Walden's rule, one can expect ion mobilities ten-fold or more higher

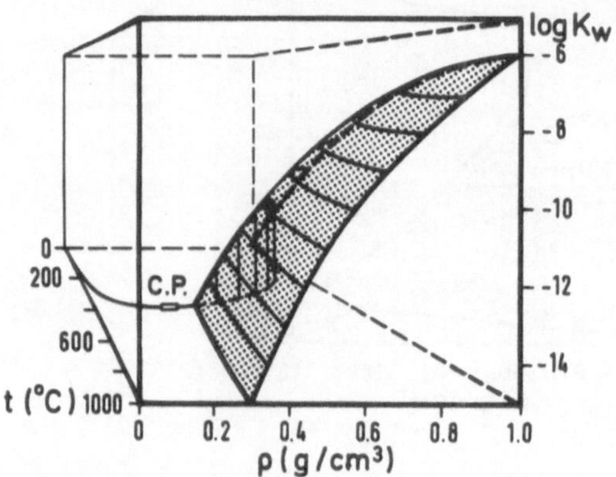

Figure 3. The viscosity of water, η, as a function of temperature t and density ρ in an approximate presentation (13)(14).

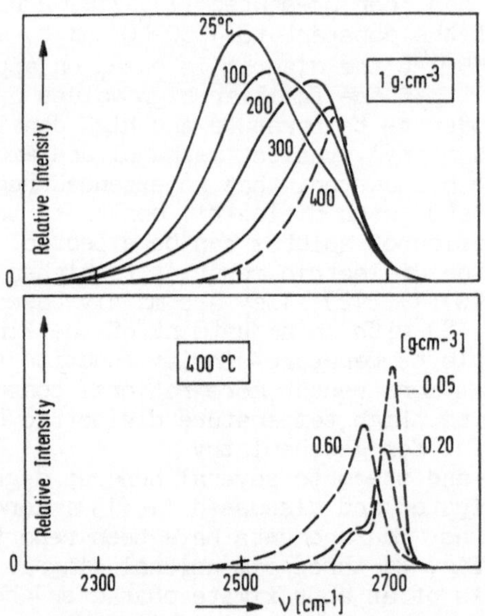

Figure 4. Raman spectra of the O-D-vibration of HDO, diluted in H_2O at a constant density of 1 $g{\cdot}cm^{-3}$ (above) and at a constant temperature of 400 °C (below) (17).

than at normal conditions at a "near-liquid" density. The calculation of dense-water viscosities is still unsatisfactory, but one can make reasonable estimates using Enskog's hard sphere treatment with effective, adjusted collision diameters.

The intermolecular structure of water is always of a certain interest in connection with the properties of aqueous ionic solutions. Numerous methods are being applied to obtain such structural information. Among these methods spectroscopic and scattering experiments are particularly important. Unfortunately, such experiments are difficult to perform at supercritical temperatures and high pressures. Efforts with neutron scattering are being made. Infrared measurements have been accomplished successfully to 400 °C and 4000 bar (15). They show a continuous change of absorption from high to low density. The bands at elevated pressures, however, are broad and strong and not very informative for structure details. Sample thicknesses must be very small. The Raman spectra have advantages. They can rather clearly show the two positions of hydrogen-bonded and non-bonded hydroxyl-group vibration bands. Thorough investigations of these bands have been made at moderate temperatures and high pressures (1)(16). Fig. 4 shows spectra obtained several years ago to the supercritical temperature of 400 °C (1)(17). For technical reasons the vibration of the oxygen-deuterium band of HDO, diluted in normal water, was measured. The upper part of Fig. 4 shows the change of the band with temperature at constant density. The lower part presents the change at 400 °C, caused by lowering the density to almost dilute steam values.

It is assumed that a low-temperature band around 2500 cm^{-1} indicates hydrogen-bonded hydroxyl groups. This band is reduced and blue-shifted with increasing temperature and decreasing density. Instead a second sharp band around 2700 cm^{-1} develops - that is at the position of the band of the stretching vibration of the hydroxyl group of water molecules in the dilute gas. No distinct change at the critical density of 0.32 g cm^{-3} is observed. Simplified quasi-chemical models with hydrogen-bonded and non-bonded water molecules in equilibrium have been suggested.

3. IONIC FLUIDS

The self-ionization of water leads to a highly ionic fluid, if density and temperature are sufficiently increased. The ion-dissociation of pure water is usually measured by the value of the "ion-product", the product of the activities of solvated hydrogen ions and hydroxyl ions in equilibrium. Fig. 5 gives a diagram of the logarithm of the ion product K_w as a function of density and temperature similar to those of Figs. 2 and 3 with the high temperature in the foreground. Near the rear right corner the familiar value of log K_w = -14 for room temperature and standard pressure can be seen. Electrochemical measurements of various kinds - at high temperature mainly the analysis of electrolytic conductance determinations - lead to a function which describes log K_w in dependence of temperature and density (18)(19). At 1000 °C and a density of 1 g·cm^{-3}, K_w is calculated to be

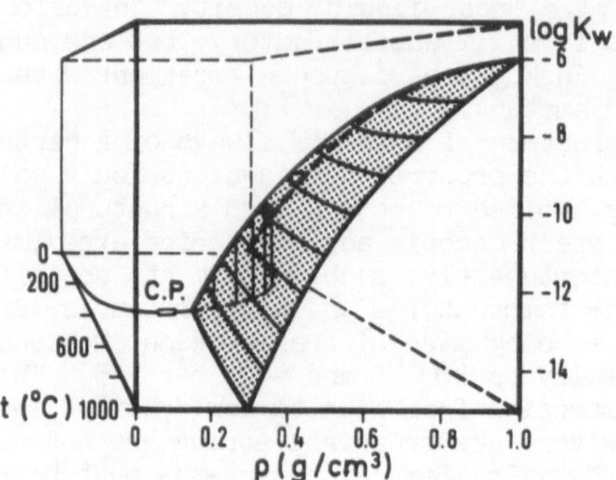

Fig. 5. The ion product of water, $\log K_W = \log [a(H^+) \times a(OH^-)]$ as a function of temperature t and density ρ in an approximate presentation (19).

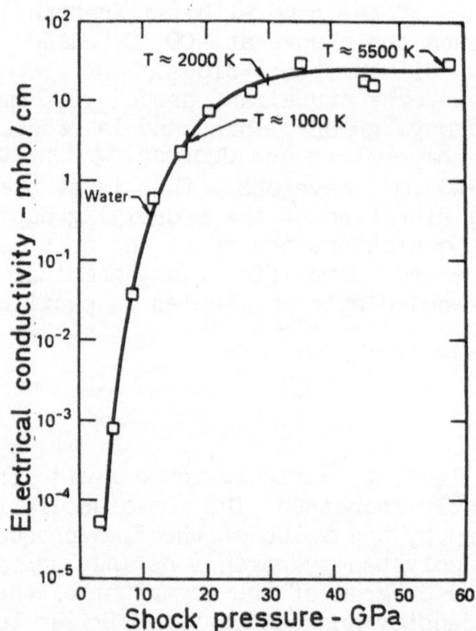

Figure 6. Specific electrical conductivity of water as a function of pressure and temperature measured in shock waves, determined by Mitchell and Nellis (21)(22). (10 GPa = 100 kbar).

10^{-6} $(mol \cdot dm^{-3})^2$. There may be an uncertainty of plus-minus one order of magnitude, but there is good reason to believe, that from room temperature to 1000 °C the ion product increases by a factor of 10^7 to 10^8. This conclusion is supported by static conductance measurements to 100 kbar (20) and more recently by conductance measurements in shock-wave compressed water to several thousand K and 500 kbar (21)(22). These shock-wave experiments appear to confirm an earlier suggestion, that water at 1000 K and above and at a density of 2 g cm^{-3} and more should become an ionic fluid like molten sodium hydroxide. It is possible, that compressed ammonia could also be brought to the state of an ionic fluid. Fig. 6 shows the specific conductance of water observed in shock-waves by Mitchell and Nellis (21). Above 30 GPa = 300 kbar the conductance remains fairly constant at a value which one would expect for liquid sodium hydroxide at such conditions.

It has been suggested, that fully ionized, high temperature, high density water may occur in the larger, outer planets (23). Uranus and Saturn, for example, may consist largely of ice and water with added helium and hydrogen. At a depth of about half the total radius, temperatures of several thousand K and pressures of 10 to 100 GPa are estimated to exist. Dense, ionic water at these conditions would have a high electric conductivity and would accordingly be of consequence for the electric and magnetic properties of these planets.

The specific electrolytic conductance of strong electrolytes in dilute aqueous solution increases with increasing temperature. The viscosity of the solvent decreases and this causes rising ion mobilities. With the application of high pressure the conductance of electrolyte solutions can be followed to much higher temperatures than those along the vapour-liquid saturation curve. Fig. 7 gives a number of isobars of

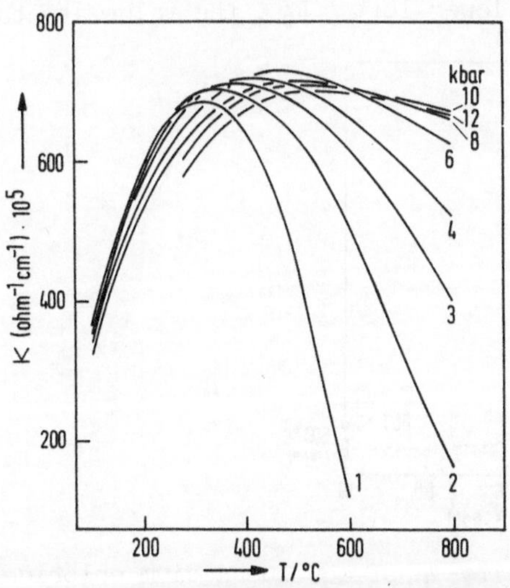

Figure 7. Specific electrolytic conductivity of 0.01 molal aqueous KCl-solutions as a function of temperature at different constant pressures (24).

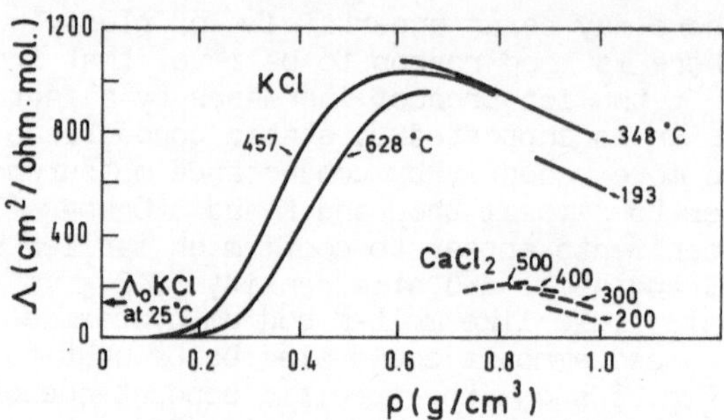

Figure 8. Molar electrolytic conductivity of 0.01 molal aqueous KCl-solutions and of 6 molal aqueous $CaCl_2$-solutions as a function of solution density ρ for constant temperatures (25).

the specific conductance of dilute KCl-solutions to 800 °C (24). After an initial deep rise all isobars pass through a maximum and decrease again: At higher temperature and, at constant pressure, at low density, the dielectric constant is lower and the concentration of effective charge carriers is reduced. Various kinds of ion association can be imagined. Making use of the equation of state data for water, the molar or equivalent conductance of the electrolyte can be plotted as isotherms in dependence of solution density. An example is given in Fig. 8, where a few selected, experimentally determined isotherms for dilute KCl-solutions are shown (25). At low water density only very little ionic dissociation of KCl is possible. With growing density the dielectric constant increases and the conductance reaches about eight times the "normal" value (see lower left , Λ_o , the value at standard

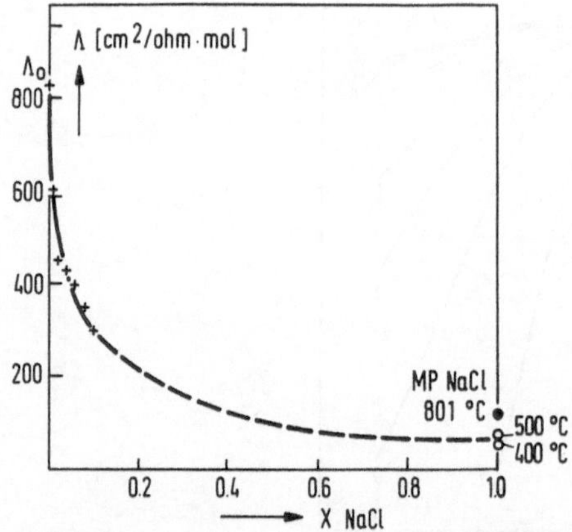

Figure 9. The molar conductivity Λ of aqueous sodium chloride solution as a function of the NaCl mole fraction x at 400 and 500 °C. —— and +++: experimental results. The difference of data at 400 and 500 °C is not relevant in this scale. •: experimental, O,O:calculated (28).

conditions). Viscosity is low and ion mobility high. Ionic dissociation may be not far from complete. Behind the maximum the conductance is lowered for kinetic reasons: The viscosity rises with further compression. Many more measurements of this kind have been made and with various other electrolytes (see, for example (26)(27)). The general behaviour is in general analogous.

Increased solute concentration can not increase the conductivity much further. Fig. 8 shows a few values for rather concentrated calcium chloride solutions. The low dielectric constant of the solvent reduces the ion formation considerably. This is more clearly demonstrated in Fig. 9, where conductance measurements with aqueous sodium chloride at concentrations up to a mole fraction x = 0.1 are shown (28) for 400 and 500 ^{o}C at a solution density near 0.8 g cm^{-3}. Beginning from the value of very dilute solutions (Λ_o,(27)) the molar conductance decreases very steeply. The conductance of molten NaCl at the melting point of 801 ^{o}C is known (see Fig. 9) and also its modest temperature dependence. A short extrapolation has been made, assuming that the liquid salt could be supercooled to 500 or 400 ^{o}C. A tentative interpolation has been made between the last solution value and the supercooled melt. Certainly a considerable uncertainty is involved. It is evident, however, that the 10 mole percent solution is already close to the fused salt in its conductivity. Very recently, conductance measurements could be made

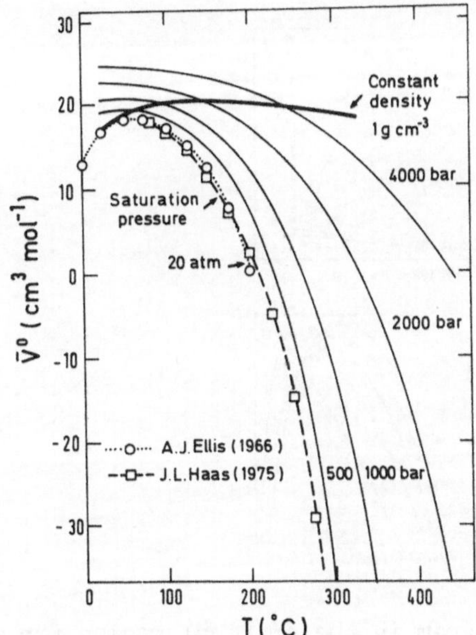

Figure 10. The partial molar volume of NaCl in aqueous solution at infinite dilution. Experimental isobars (29) compared with literature data at saturation conditions (Ellis (44), Haas (45)).

with the water - sodium hydroxide system over the complete range of concentrations from melt to dilute solutions at 400 °C. Data confirm the conclusion from Fig. 9. One may expect a comprehensive theory which covers, at least at high temperature, the complete range of ion-containing fluid from liquid salt to dilute solution.

An important source of information for water-ion interaction has always been the partial molar volume of electrolytes. Fig. 10 shows the variation of this quantity for sodium chloride within a wide range of supercritical conditions. The data have been derived from extended sets of molar volume measurements to 600 °C and 4000 bar (29(30)(43). The diagram in Fig. 10 gives values for sodium chloride extrapolated to infinite dilution. From a familiar value around 20 $cm^3 mol^{-1}$ the quantity decreases and reaches negative values if the temperature is elevated along the liquid-gas equilibrium curve for water. This reflects the growing difference between the density of the aqueous solvent and the high density of hydration shells. High pressure must reduce this difference. which is clearly demonstrated by the isobars of the diagram. If one can keep the total density constant at one $g{\cdot}cm^{-3}$, the partial molar volume remains positive and almost constant, as shown by the heavy curve in the upper part of the diagram. Average hydration numbers in the supercritical state can be derived (29)(30). An analogous investigation with aqueous sodium hydroxide solutions to very high concentrations is under way.

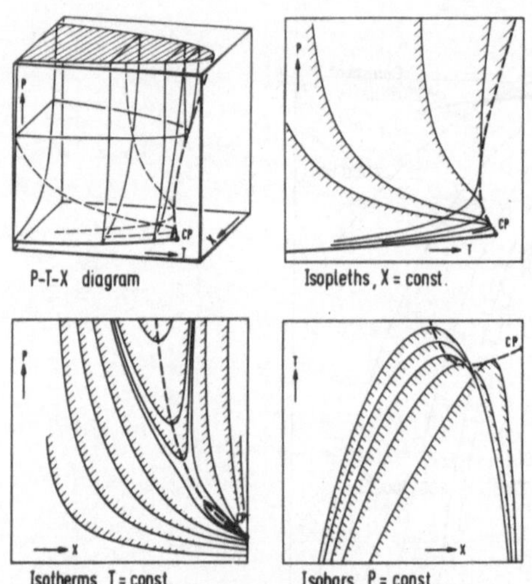

Figure 11. Schematic phase diagram in the critical region for a binary system water-nonpolar gas. Shaded area: Two-phase region. CP: Critical point of water. x: mole fraction of water. ---: Critical curve.

4. BINARY AQUEOUS MIXTURES - CRITICAL REGION

An interesting field to investigate is the behaviour of ion-forming electrolytes dissolved in water with varying contents of third, non-polar partners. Around normal conditions such mixtures can be prepared only in special cases. At sufficiently high temperature, however, partners of very different polarity can form homogeneous mixtures even of "liquid-like" densities, where particles are almost always within the range of their mutual interaction potentials. Thus even water can mix with entirely non-polar gases and liquid hydrocarbons.

Binary fluid systems have a critical curve, which extends in the three-dimensional pressure-temperature-composition diagram between the critical points of the two pure components. The curve is uninterrupted, if the components are rather similar. It is divided into two branches, if the partners are more different, for example in polarity. The upper branch of the critical curve begins at the critical temperature of the higher boiling component. The critical curve is an envelope of the two-phase, heterogeneous region. At temperatures above the critical curve the mixture is homogeneous. Systematic descriptions of such critical phenomena are available (see for example (31)(32)). Mixtures of water and non-polar components generally belong to the group with interrupted critical curves. Fig. 11 shows schematically a typical case which resembles water nitrogen or water-methane (see below). The upper left quarter gives a three-dimensional PTx-diagram. The composition is described by mole fractions x. The two-phase region is shaded. The critical curve has a shallow temperature minimum. The remaining three quarters of the diagram show the three two-dimensional projections: Isotherms, isobars and curves for constant mole fractions x: isopleths.

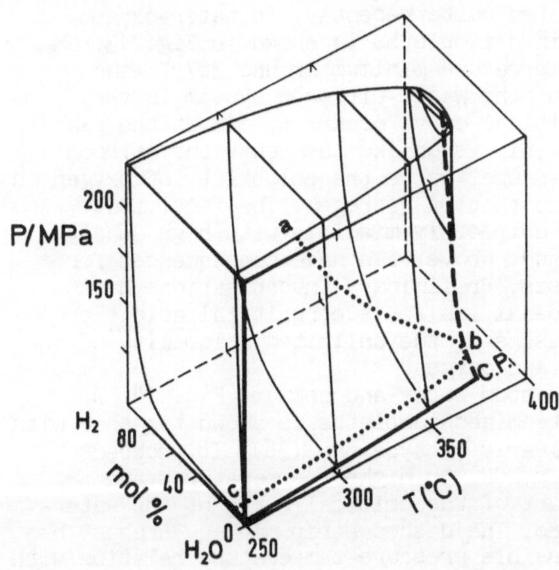

Figure 12. Experimentally determined PTx-diagram of the system water-hydrogen. The isotherms outline the two-phase equilibrium surface. abc: Isopleth for 90 mole percent water. ---: Critical curve (33).

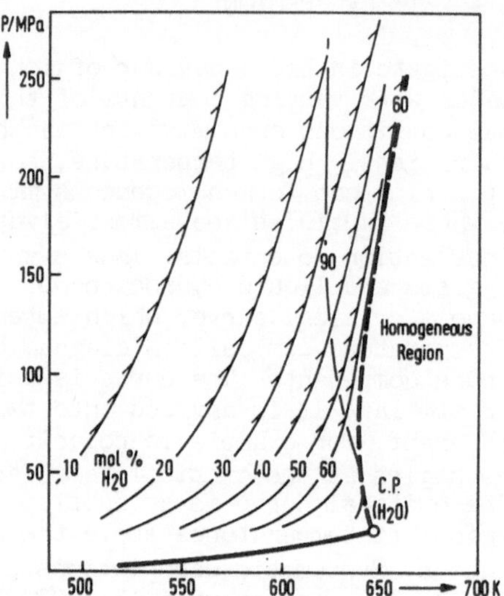

Figure 13. The phase diagram water-oxygen. Experimentally determined curves for constant water contents (isopleths) which outline the two-phase region in the PTx-space. ---: critical curve (34).

A real example of a phase diagram is given in Fig. 12 with the experimentally investigated system water-hydrogen (33). The curve abc is an actual isopleth for 90 mole percent of water. The projection of the critical curve in the pressure-temperature plane rises almost vertically. This critical curve would be of the so-called first type (31) (32). Another system, investigated quite recently, is water-oxygen. A set of experimentally determined isopleths is shown in Fig. 13. The critical curve has a slight temperature minimum around 367 °C and 750 bar (34). The PTx-diagram of the water-nitrogen system is very similar, the minimum of the critical curve occurs at almost the same pressure and temperature (35). This is unlike the behaviour at room temperature and atmospheric pressure, where the solubility of oxygen in water is nearly twice as high as that of nitrogen. The fact, that hydrogen as well as oxygen are completely miscible with high density supercritical water at 400 °C and above, can have consequences for future supercritical electrolysis. Preliminary investigations have resulted in very high current densities for supercritical acidic or basic solutions. One of the reasons is the unlimited solubility of the electrolysis products hydrogen and oxygen.

A further example is the system water and helium. Fig. 14. A selection of experimentally determined isopleths is shown together with the critical curve, which is clearly of type one (36). It proceeds from the critical point of water, CP, to higher temperatures and pressures. For comparison, a part of the critical curve of the water-hydrogen system (33) is included. The dashed strip marked "Uranus" has been calculated (23) as the possible pressure-temperature relation with

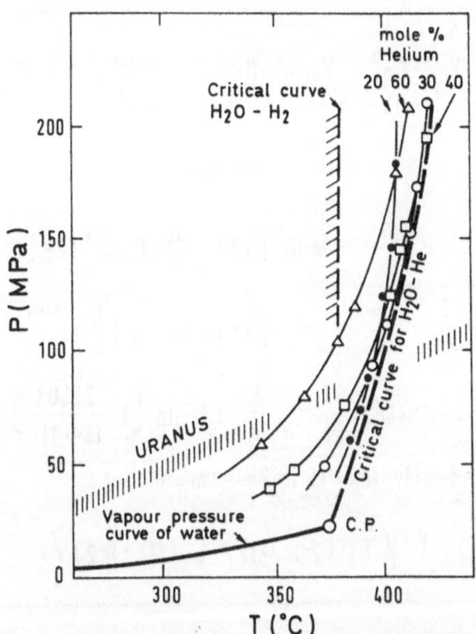

Figure 14. The phase diagram water-helium (36). Experimentally determined curves for constant helium contents (isopleths) which outline the two-phase region in the PTx-space. ---: The critical curves for water-helium and water-hydrogen. ////: Pressure-temperature relation inside Uranus, suggested by D.J. Stevenson (23).

increasing depth in the outer planet Uranus. Since the planet is believed to contain, besides highly compressed water, also helium and hydrogen, the knowledge of the intersection of the calculated pT-relation with the critical curve may permit conclusions concerning possible phase separation and stratification - provided the concentrations of helium or hydrogen can be estimated. The high degree of water self-ionization, discussed above in section 3, would occur at much higher pressures.

5. FLUID MIXTURES - CALCULATIONS

It is certainly highly desirable to have the possibility to calculate the important properties of the binary and perhaps ternary supercritical fluid systems. Tests of molecular interaction models can be made, predicitions given and the number of laborious experiments reduced. It is hoped also that it will become possible to calculate such thermophysical properties of supercritical water-gas mixtures which are relevant for electrochemical phenomena, for example the dielectric constant (see below). A considerable number of equations of state of various types for scientific and industrial use exist. Computer simulations and perturbation theory have been successfully applied. It is beyond the scope of this contribution to compare these efforts. Recent surveys are available (37). The equations are mostly not well

$$p(T,V_m) = \frac{RT}{V_m}\cdot\frac{V_m^3 + V_m^2\beta + V_m\beta^2 - \beta^3}{(V_m-\beta)^3} - \frac{4\beta RT}{V_m^2}(\lambda^3-1)\left[\exp\left(\frac{\varepsilon}{kT}\right)-1\right]$$

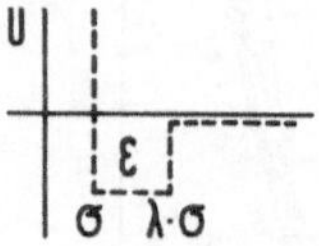

U, ε, σ, λ·σ

with $\beta(T) = \frac{\pi}{6} N_0 \sigma^3(T)$;

$$\beta(T) = \beta(T_c)\left(\frac{T_c}{T}\right)^{3/m}$$

$$\beta(T_c) = 0.04682\,\frac{RT_c}{p_c}\ ;\quad \frac{\varepsilon}{k} = T_c\cdot\ln\left[1+\frac{2.6503}{(\lambda^3-1)}\right]$$

λ: preferably from 1.5 to 2.5 ; m = 10

For mixtures : $\varepsilon_{12} = \xi\sqrt{\varepsilon_1\cdot\varepsilon_2}$; $\sigma_{12} = \zeta(\sigma_1+\sigma_2)/2$

Figure 15. Proposed new equation of state (CF-equation, (39)(40)) to calculate phase equilibria and thermodynamic functions of fluid mixtures with water or other polar components at high temperatures and pressures. V_m=Molar volume. N_o=Avogadro number.

suited for mixtures with highly polar molecules, like aqueous systems, however, although systematic investigations on the influence of polarity on critical curves exist (38).

For the special purpose, namely the prediction of phase equilibria and critical curves of mixtures which include highly polar partners to high temperatures and pressures, a new equation was designed (39) (40). This equation ("CF") contains only a limited number of parameters, which can all be interpreted on the basis of the applied "square well" model. The equation is of a modular type to permit later extensions and refinements. Fig. 15 shows the main characteristics of the equation with its repulsion term and added attraction term. The repulsion term is of the form derived by Carnahan and Starling from computer simulations (41). Only the particle parameter σ has been made slightly temperature dependent. The depth of the potential well, ε, and σ are derived from the critical data of the pure partners. ξ and ζ are adjustable parameters defined by combination rules, they can remain constant within certain groups of systems. Because of the relative simplicity of the CF-equation it is mainly applicable at elevated temperatures. It appears, however, that it can be extended to ternary systems, including such which contain limited concentrations of ionic solutes as third partners.

A test of the CF-equation with the system water-nitrogen is shown in Fig. 16, which was experimentally determined (35). The full black points were measured and show the course of the critical curve. The other curves are calculated spinodal isopleths for constant mole

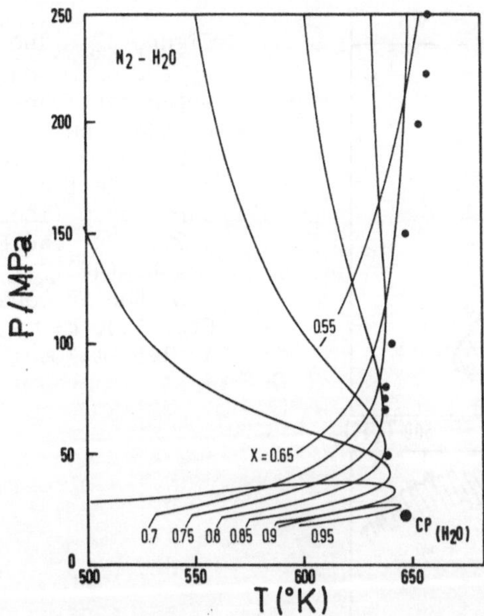

Figure 16. Pressure-temperature diagram for the binary system water-nitrogen. CP. Critical point of pure water. ●●●: Experimental points on the critical curve (35). Curves: Spinodal isopleths for several water mole fractions x, calculated with the CF-equation. (see Fig. 15). $\zeta=1.0$, $\xi=0.75$, $\lambda=2.5$, $m=10$.

fractions. The spinodals determine the boundary surface for diffusional or material instability. This surface coincides with the surface of mechanical stability, or binodal surface, along the critical curve. Thus the critical curve can be determined as an envelope of calculated spinodals in the PT-projection. λ , m and ζ (see Fig. 16) were set at fixed values. ξ was left as the only adjustable parameter. Best possible representation of the critical curve was obtained with ξ = 0.75. The agreement is good and probably within the range of experimental accuracy. It can be shown, that other, similar, binary aqueous systems can be described with the same unchanged set of parameters.

An important binary aqueous system is water-carbon dioxide. It is more difficult to describe, because of stronger interactions between water and carbon dioxide molecules. The latter have a high quadrupole moment. The two-phase equilibrium surface in the PTx-space and the critical curve have been determined experimentally to 2500 bar (42). The temperature along the critical curve changes by more than 100 K within this pressure region. To calculate the critical curve in this case the adjustment had to be made using a different set of parameters, as shown in the legend of Fig. 17. In the upper part of the Fig. 17 one calculated binodal isopleth (two-phase boundary curve) for 60 mole percent is shown with experimental points. The lower part of Fig. 17 gives one isobaric cross-section at 2000 bar, again with experimental points. The agreement is satisfactory.

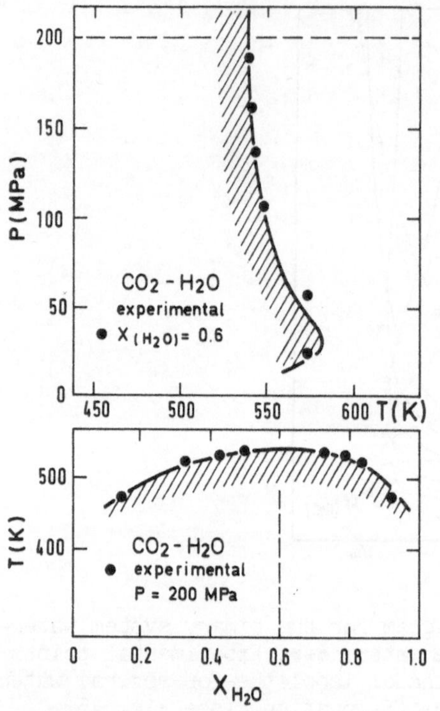

Figure 17. The binary system water-carbon dioxide. Upper part: The two-phase boundary curve (isopleth) for $x(H_2O) = 0.6$. Lower part: The two phase boundary curve (isobar) for p=200 MPa. ●●●:Experimental points (42). Curves from the CF-equation (see Fig. 15). ζ=0.99, ξ=0.94, λ= 1.5, m=10.

6. TERNARY FLUID MIXTURES

The vapour pressure of water is reduced by a dissolved solid electrolyte. It is to be expected, that the two-phase region and the critical curve of a binary water-gas system would also be shifted - presumably to higher temperatures - by the addition of a soluble salt, like sodium chloride, as a third component. This was first demonstrated by Takenouchi and Kennedy (46) for water - carbon dioxide - sodium chloride in the high pressure-high temperature range. A more extensive investigation of this system was made later (47). Six weight percent of sodium chloride, relative to water, shift the maximum temperature of the two-phase region to higher temperatures by about one hundred degrees at 1000 bar pressure. The effect is of considerable consequence in geochemistry.

Recently the system water - methane - sodium chloride was investigated to 500 °C and 2500 bar (48)(49). Fig. 18 shows a number of experimentally determined isopleths. The high temperature shift of the phase boundary curves for 17 and 51 mole percent methane, caused by an addition of only 0.53 and 2.61 mole percent of sodium chloride relative to water, is evident. The effect of such small amounts of electrolyte is very pronounced, particularly around 50 percent of methane. The figure shows also the critical curve for the binary water - methane system. Critical curves for the ternary systems cannot be given, since equilibrium tie lines across the two-phase region could not yet be

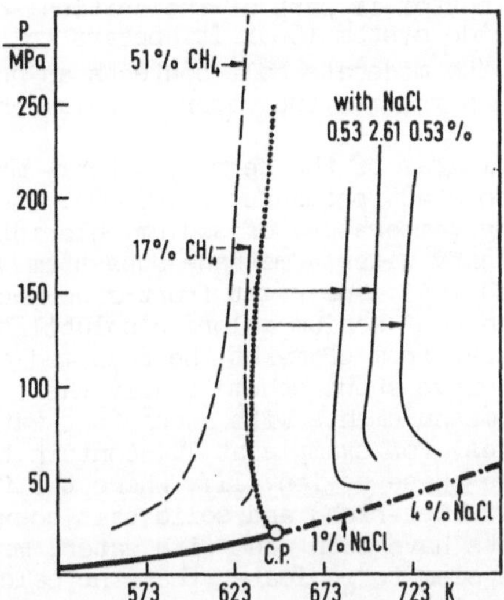

Figure 18. The ternary system water - methane - sodium chloride. Two isopleths (dashed curves) for the binary system water - methane with 17 and 51 mole percent methane. Two critical curves for water - methane (...) and water - sodium chloride (-.-.-). Full curves show the temperature shift caused by the addition of sodium chloride.

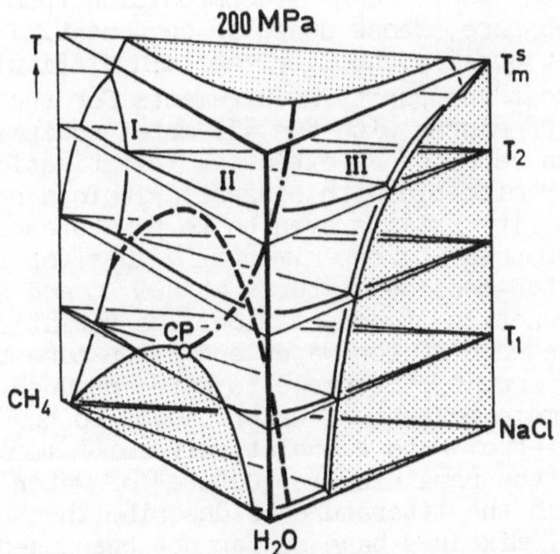

Figure 19. The ternary system water - methane - sodium chloride. Tentative composition-temperature diagram for a constant pressure of 200 MPa. Shaded areas on the two prism sides are two-phase regions. The course of measurements is indicated by the (---) curve. (48)(49).

determined. The curve at the bottom is part of the critical curve for the binary water-sodium chloride system (50). It appears to be possible, to treat the ternary system with moderate salt contents as quasi-binary, making use of the appropriate points on the water - sodium chloride curve.

Fig. 19 is a tentative diagram of the ternary water - methane - sodium chloride system at a constant pressure of 200 MPa. The prismatic diagram extends to the melting temperature of sodium chloride. The left, front side shows the binary water - methane subsystem which has been experimentally determined (51). The right front side shows the estimated temperature dependence of sodium chloride solubility at high pressure. The dash - point curve from CP marks the presumed course of a critical curve. The dashed curve shows schematically the trace of one set of phase equilibrium measurements with a constant water to salt ratio. The isothermal triangles, for example at T_2, contain two-phase regions (I and II) and a three-phase region, III, where two fluid phases (one methane-rich, one water-rich) and solid salt coexist. - A limited number of experiments have been made with water, methane and calcium chloride. The results seem to indicate, that the calcium chloride around 400 °C forms mainly mono-valued ions.

7. THE STATIC DIELECTRIC CONSTANT OF SUPERCRITICAL MIXTURES

For a discussion of ion formation and ionic behaviour in dense water-gas mixtures knowledge of the dielectric constant in such media is necessary. Supercritical aqueous mixtures permit to study the variation of the dielectric constant over the whole range of composition from pure supercritical water to the pure, dense nonpolar component. To perform measurements of sufficient quality, however, certain difficulties must be overcome in order to meet the purity requirements for the samples at high temperatures and pressures with the all-metal equipment. Water-benzene mixtures have been used for an extensive investigation (6). At 400 °C water is completely miscible with benzene. Within a certain pressure range the miscibility extends even to temperatures slightly below 300 °C. Fig. 20 presents experimental values for the dielectric constant of water-benzene mixtures at 400 °C and several constant pressures as a function of composition. The variation of the constant at 2000 bar is between 20 for water and 2 for pure benzene. The addition of 10 mole percent of benzene to water reduces the value to one half. It needs 50 mole percent of water. added to benzene, to increase the constant only from 2 to 4. Relatively small amounts of benzene apparently reduce the remaining "structure" of water considerably. Existing suggestions in the literature to describe theoretically the dielectric constant of mixtures have so far not been used to include wide variations of the total density. It appears, however, that an uncomplicated function can describe the mixture dielectric constant with pertinent values of the pure partners using suitably defined volume fractions (6)(4). It is interesting to make a comparison with water-dioxane mixtures at room temperature in Fig. 21. Dioxane (see formula) has no overall dipole moment, but two strong, opposed

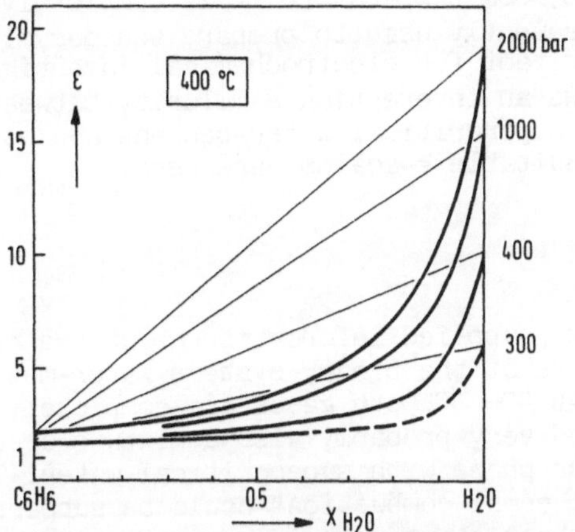

Figure 20. The static dielectric constant at supercritical mixtures of water and benzene at 400 °C and four different pressures as a function of the water mole fraction (6).

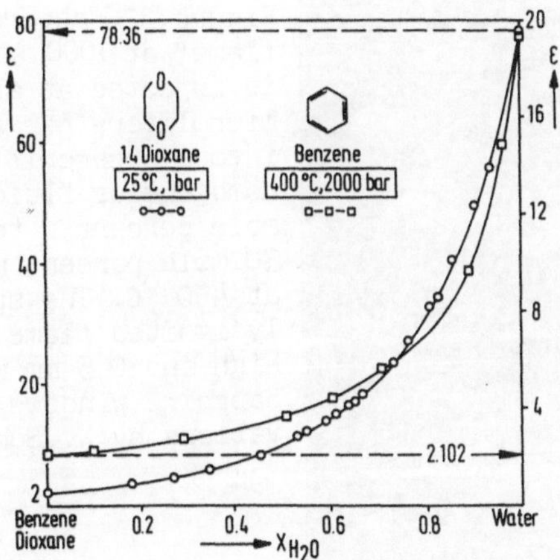

Figure 21. The static dielectric constant ε as a function of the water mole fraction for water-benzene mixtures at 400 °C and 2000 bar (right scale) and water-dioxane mixtures at 25 °C and 1 bar (left scale).

internal dipole moments. At room temperature it is completely miscible with water and has been frequently used to prepare aqueous mixtures with lowered dielectric constant for electrochemical investigations. The diagram of Fig. 21 shows an interesting similarity between the composition dependences of supercritical water-benzene and normal water-dioxane mixtures if suitable ε-scales are used.

8. HYDROTHERMAL COMBUSTION

Separate investigations have supplied information about the two-phase region and the critical curve of the binary systems water-methane and water-oxygen (see above). At 400 °C both gases are completely miscible with high pressure water and very probably also a mixture of both gases can form a homogeneous phase with supercritical water. Thus one should assume, that "hydrothermal combustion" could be supported in such phases. A project has been started to investigate such combustion. Through a narrow nozzle methane or oxygen are injected at rates of a few microliters per second in water or aqueous mixtures at temperatures between 400 and 500 °C and pressures up to 1000 bar. If oxygen is injected into a homogeneous mixture of water and 30 mole percent methane, spontaneous ignition occurs already at temperatures as low as 430 °C. Steadily burning hot flames are observed at 1000 bar. A photograph or such a flame is shown in Fig. 22.

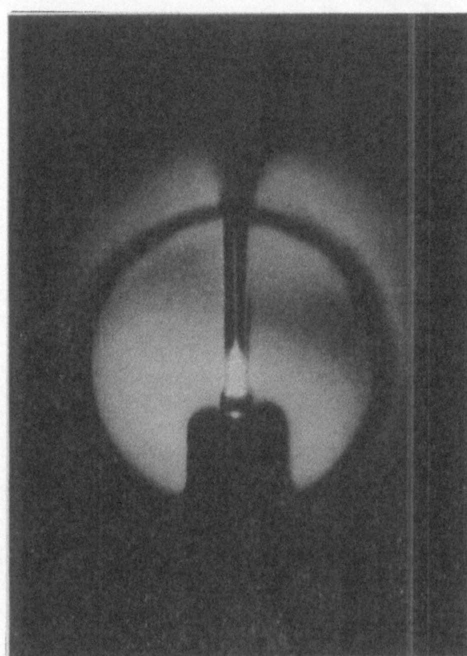

Figure 22. "Hydrothermal flame" at 1000 bar. Oxygen is injected at about 3 microliters per second into a supercritical homogeneous fluid of 70 mole percent water and 30 mole percent methane at 450 °C. The spontaneously ignited flame is 3 mm high and 0.5 mm wide. Sapphire windows (Observations by W. Schilling.)

Because the applied dense supercritical aqueous phases are still good electrolytic solvents, combinations of electrochemical phenomena with combustion or other chemical reaction processes can be envisaged.

9. CONCLUSION

It is expected that studies of supercritical aqueous ionic solutions can be expanded in the future to extended regions of conditions as well as to additional phenomena and methods. Theoretical approaches, based on computer simulations, and in close interaction with experiments will be of particular value in this field. Concerted application of experiment and theory could provide economic procedures in a difficult field. The results of investigations with the dense supercritical fluids, which make use of the unique variability at properties, may provide comprehensive descriptions of normally separated phenomena.

REFERENCES

1. F. Franks, Ed. "Water a Comprehensive Treatise", Vol. 1. Plenum Press, New York (1972).
2. F.H. Ree, J.Chem. Phys. 76, 6287 (1982).
3. D. Eisenberg, W. Kauzmann. "The Structure and Properties of Water". Clarendon Press, Oxford (1969).
4. C.J.F. Böttcher, "Theory of Electric Polarization". Vol. 1. Elsevier. Amsterdam (1973).
5. K. Heger, M. Uematsu, E.U. Franck, Ber. Bunsenges., Phys. Chem. 84, 758 (1980).
6. R. Deul, "Dielectric Constant and Density of Water - Benzene Mixtures to 400 °C and 3000 bar". Thesis. (1984). Institute of Physikcal Chemistry, University of Karlsruhe.
7. M. Uematsu, E.U. Franck. J. Phys. Chem. Ref. Data 9, 1291 (1980).
8. E.U. Franck. Z. Phys. Chem. N.F. 8, 107 (1956).
9. A.S. Quist, W.L. Marshall, J. Phys. Chem. 69, 3165 (1965).
10. D.J. Bradley, K.S. Pitzer, J. Phys. Chem. 83, 1599 (1979).
11. K.S. Pitzer, Proc. Natl. Acad. Sci., USA 80, 4575 (1983).
12. R.P. Beyer, B.P. Staples, in print in J. Solution Chemistry (1986).
13. U. Grigull, Ed. "Properties of Water and Steam in SI-Units", Springer-Verlag, Heidelberg. New York (1982).
14. K.H. Dudziak, E.U. Franck, Ber. Bunsenges. Phys. Chem. 70, 1120 (1966).
15. E.U. Franck, K. Roth, Disc. Far. Soc. No 43, 108 (1967).
16. G.E. Walrafen, M. Abebe, J. Chem. Phys. 68, 4694 (1978).
17. H.A. Lindner, "Raman Investigations with Water to 400 °C and 5000 bar", Thesis. Institute for Physical Chemistry. University of Karlsruhe (1970).
18. E.U. Franck, Z. Phys. Chem. N.F. 8, 192 (1956).
19. W.L. Marshall, E.U. Franck. J.Phys. Chem. Ref. Data 10, 295 (1981).
20. W.B. Holzapfel, E.U. Franck. Ber. Bunsenges. Phys. Chem. 70, 1105 (1966).

21. A.C. Mitchell, W.J. Nellis, J. Chem. Phys. 76, 6273 (1982).
22. M. Ross, Rep. Progr. Physics 48, 1 (1985).
23. D.J. Stevenson, Ann. Rev. Earth and Planetary Sci. 10, 257 (1982).
24. K. Mangold, E.U. Franck, Ber. Bunsenges. Phys. Chem. 73, 21 (1969).
25. E.U. Franck, Z. Phys. Chemie. N.F. 8, 92. 107 (1956).
26. E.U. Franck, Angewandte Chemie. 73, 309 (1961).
27. A.S. Quist, W.L. Marshall, J. Phys. Chem. 72, 684 (1968).
28. W. Klostermeier, "Electrical Conductivity of Concentrated Solutions to High Temperatures and Pressures". Thesis. Institute of Physical Chemistry. University of Karlsruhe (1973).
29. R. Hilbert, "PVT-Data of Water and Aqueous Sodium Chloride Solutions to 873K and 4000 bar", Thesis. Institute of Physical Chemistry University of Karlsruhe (1979).
30. M. Gehrig, H. Lentz, E.U. Franck. Ber. Bunsenges. Phys. Chem. 87, 597 (1983).
31. J.S. Rowhison, F.L. Swinton. "Liquids and Liquid Mixtures", III. Ed. Butterworth Scientific. London (1982).
32. G.M. Schneider, "High Pressure Phase Diagrams and Crital Properties of Fluid Mixtures", Spec. Per. Rep. Chem. Thermodynamics, Vol. 2, The Chemical Society. London (1978).
33. T.M. Seward, E.U. Franck. Ber. Bunsenges. Phys. Chem. 85. 2. (1981).
34. M.L. Japas, E.U. Franck, Ber. Bunsenges. Phys. Chemie 89, 1268 (1985).
35. M.L. Japas, E.U. Franck. Ber. Bunsenges. Phys. Chem. 89. 793 (1985).
36. N.G. Sretenskaja, M.L. Japas, E.U. Franck. to be published in Ber. Bunsenges. Phys. Chem. (1987).
37. S.M. Walas, "Phase Equilitria in Chemical Engineering". Butterworth Publishers. Boston (1985).
38. K.E. Gubbins, C.H. Twu. Chem. Eng. Sci. 33, 863. 879 (1978).
C.H. Twu, K.E. Gubbins. C.G. Gray, J. Chem. Phys. 64, 5186 (1976).
39. M. Christoforakos, E.U. Franck, Ber. Bunsenges. Phys. Chem. 90, 780 (1986).
40. M. Christoforakos, "Supercritical Aqueous Systems at High Pressures". Thesis. Institute for Physical Chemistry, Karlsruhe University (1985).
41. N.F. Carnahan, K.E. Starling, J. Chem. Phys. 51, 635 (1969).
42. K. Tödheide, E.U. Franck, Z. Phys. Chem. N.F. 37, 387 (1963).
43. R. Hilbert, K. Tödheide, E.U. Franck, Ber. Bunsenges. 85, 636 (1981).
44. A.J. Ellis, J. Chem. Soc. (London) A, 1579 (1966).
45. J.L. Haas, U.S. Geol. Survey. Reston, Virginia (1975). Rep. 75-675.
46. S. Takenouchi, G.C: Kennedy, Am. J. Sci. 263, 445 (1965).
47. M. Gehrig, H. Lentz, E.U. Franck, Ber. Bunsenges. Phys. Chem. 90, 525 (1986).
48. T. Krader, "The Ternary System Water, Methane and Sodium Chloride to 2.5 kbar and 800K", Thesis. Institute for Physical Chemistry, University of Karlsruhe (1985).
49. T. Krader, E.U. Franck, Physica, 139-140 B, 66 (1986).
50. S. Sourirajan, G.C. Kennedy, Amer. J. Sci. 260, 115 (1962).
51. H. Welsch, "The Systems H_2O-Xe and H_2O-CH_4 at High Temp. and Press." Thesis. Inst. Phys. Chem. Karlsruhe (1973).

THE STUDY OF AQUEOUS IONIC GLASSES

P. Chieux
Institut Laue-Langevin
156X Avenue des Martyrs
38042 Grenoble Cedex, France

The study of aqueous ionic glasses is by no means a recent one although for a long time these glasses were not really studied for themselves but for the supercooling properties of the corresponding liquids or as convenient materials to investigate the glass transition and glass forming ability. This is why a quick survey of these systems covers most of the questions related to the physics of glass formation but is also largely interdisciplinary (e.g. solution chemistry, cryobiology...) as is reflected by the variety of the literature sources. For didactic reasons we will, as much as possible, use the well-investigated LiCl.H_2O solutions as our reference system through this paper. As early as 1928, Tamman [1] chose the LiCl.H_2O solutions to follow the depression of the density maximum of water with addition of salt. The correspondence between the addition of a salt or a second component to water and the application of pressure had already been noticed and it was important to confirm the results of Brigdman [2] on the levelling off of the pressure dependence of the density maximum of water at about -5°C for pressures up to 3000 atmospheres. We see in figure 1 that the density maximum of

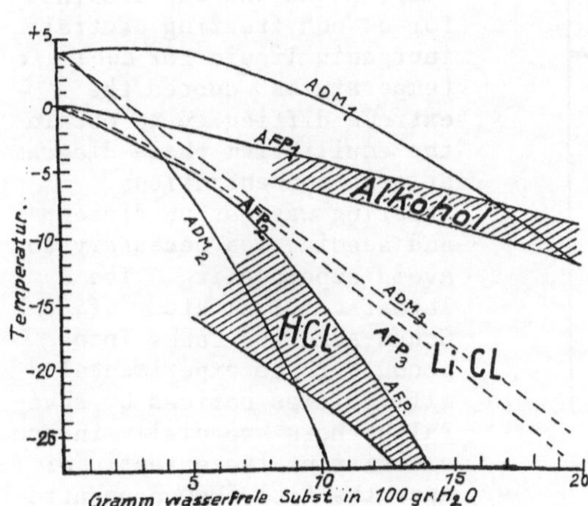

Figure 1. Concentration dependence of the density maximum (ADM) and of the melting point (AFP) for various aqueous ionic solutions [1].

M.-C. Bellissent-Funel and G. W. Neilson (eds.), The Physics and Chemistry of Aqueous Ionic Solutions, 359–377.

LiCl solutions remains above their known freezing point, which is not the case for alcohol and HCl to water mixtures. Moreover in the last two cases significant supercooling is observed (hatched area in the figure).

Tamman extended the measurements of the density maximum to concentrations of about 30 gr LiCl in 100 gr H_2O and observed it at temperatures below -50°C, however he did not report on supercooling properties of the LiCl solutions.

1. NOTES ON THE EQUILIBRIUM PHASE DIAGRAMS

A considerable effort has been made [3][4][5][6] to accurately determine the LiCl.H_2O equilibrium phase diagram. The figure 2, given by Moran in 1956 is nearly correct. These diagrams are normally obtained from warming curves in differential thermal analysis measurements (DTA) and more recently by differential scanning calorimetry (DSC). The eutectic point, the temperatures of transformation between the various hydrates and the melting points are usually well detectable. The composition of the hydrates is obtained from plots of the heats of transformation versus concentration (also called Tamman diagrams) which culminate at the compound composition (see also figure 15).

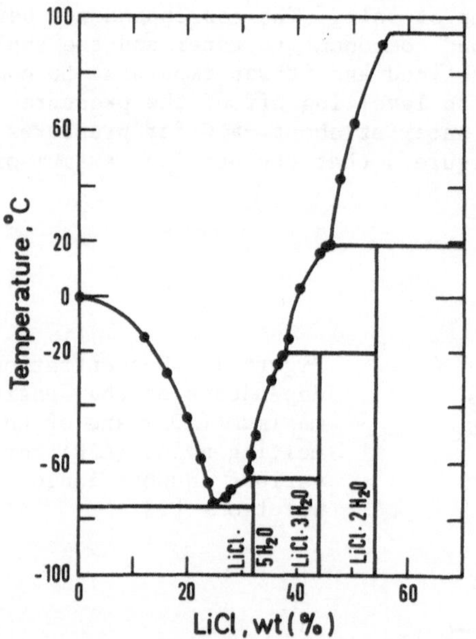

Figure 2. System LiCl.H_2O [4]

Interestingly, in the present case, the tetra and the hexahydrate do not exist. Moreover, Moran who was looking for a "non freezing neutral inorganic liquid for subartic temperatures" quoted the extreme difficulty to obtain the equilibrium phase diagram at some concentrations. Steering was not sufficient and seeding was necessary to avoid supercooling. The latest critical study of Cohen-Adad [6] takes into account these experimental difficulties noticed by several authors especially in the vicinity of the eutectic or near the LiCl.5H_2O concentration (see details of figure 4). However, it remains that equilibrium thermodynamics is somehow unsatisfactory when dealing with highly viscous media.

2. CONCENTRATION UNITS AND SAMPLE PREPARATION

Before proceeding further, let us consider some practical matters.

Different concentration scales were used in the above two figures and much more were developed through the years. Great care should therefore be taken when reading the papers. Table 1 summarizes the problem.

TABLE 1

Various concentration units	Conversions
a) wt (gr) salt/100 gr H_2O	$\text{moles }\% = \frac{\text{wt}\%}{2.33-0.013\text{wt}\%}$
b) wt (gr) salt/100 gr solution	
c) wt % or 100 (wt salt/wt solution)	wt % = 100(1/1+0.428R)
d) mole % or 100 (moles LiCl/(moles LiCl + moles H_2O))=100X (where X is the mole fraction)	moles % = 100(1/1+R)
e) mole ratio (R) or moles water/moles LiCl	R = (100/mole% - 1)
f) molarity (M) or moles LiCl/liter solution	R = 2.33 (100/wt%-1)
g) molality (m) or moles LiCl/moles H_2O in 1000gr solvent	m = 55.1/R

(N.B. the unit f) requires the density and is often avoided).

How easy is it to prepare aqueous ionic solutions or glasses ? One simply mixes weighted amounts of chemical grade salt and deionized water. However the "anhydrous" salts must be carefully checked for water content, especially when preparing deuterated solutions. Well-characterized hydrates obtained by recrystallization are also good starting materials. Depending on the experiment, some special sample handling might be necessary (e.g. filtering (light scattering), freeze-thaw-pump (O_2 removal for NMR)...). Quantitative analysis to an accuracy of 0.3% is routinely obtainable, and matches the accuracy achievable in sample preparation. A simple quench of gramme-sized sample to liquid N_2 provides a transparent glass if we are in the proper concentration range (see below). Cracks are normally observed in the glass for $T \lesssim T_g - 20K$ and might be healed by annealing (important for small angle scattering).

3. THE GLASS TRANSITION AND ITS DEPENDENCE ON ION CHARGE AND HYDRATION

In a search for optimum conditions for crystallization, Vuillard [5] made a systematic study of the glass transition temperature (T_g) and crystallization (T_c) versus concentration in $LiCl.H_2O$. The experiment involved a quench of the samples to liq.N_2 (50K/mn) followed by DTA on warming. A typical DTA curve is displayed in figure 3 [7].

T_g is not an equilibrium transition temperature, it might depend on experimental conditions; its definition is also conventional, the inflexion point of the DTA line being often used (arrow in figure 3) [24]. We see in figure 4 that T_g becomes concentration dependent above 14 mole %. Vuillard noticed that at concentrations around $LiCl.5H_2O$ the crystallization sometimes required a few days, i.e. with ordinary warming rates the system remained always in a disordered state e.g. respectively, glass, supercooled liquid and liquid.

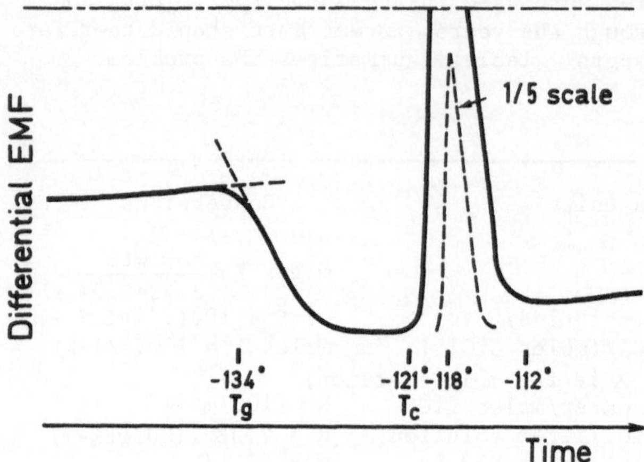

Figure 3. DTA trace for R=10, LiCl.H_2O solution, showing glass transition T_g and crystallization temperature T_c. Heating rate 8°C mn^{-1} [7].

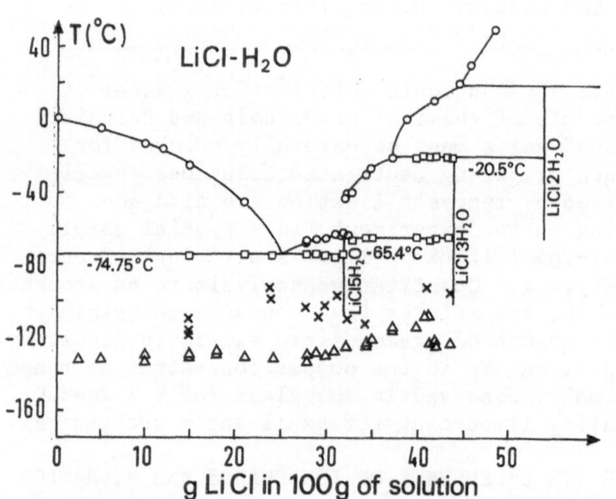

Figure 4. System LiCl.H_2O, T_g (Δ), T_c (X) [5].

However, in the most dilute end of the diagram, some ice crystallization did occur while quenching as was noticeable from the waxy colour of the samples. Near the LiCl.5H_2O composition, difficulties were encountered to reach equilibrium melting lines and it has since been shown that LiCl.5H_2O decomposes before melting [6]. The proof is therefore made, that, depending on concentration, rather stable glasses can be prepared from LiCl solutions. Of course the aqueous ionic solutions offer a large variety of possible glass formers where the effects of ion size or charge can be tested, and also where simple empirical relations such as the one found for high temperature glasses between the melting point temperature $T_{m.p}$ and T_g can be verified. Angell and Sare [8] undertook such a systematic study and figure 5 presents part of their findings.

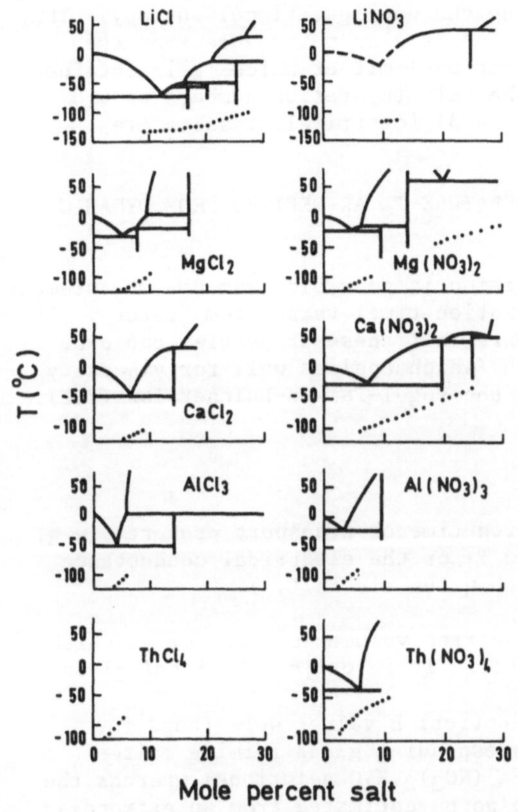

Figure 5. Phase diagrams for salt-water systems showing glass forming regions and glass transition temperatures (dots) [8].

Glasses are generally obtained in the eutectic regions and for a given anion, the ratio $T_{m.p}/T_g$ is nearly constant (e.g. 1.6 for NO_3^-). In figure 6 the T_g values are displayed on a concentration scale (Normality) which takes into account the formal electrostatic effects [8]. On that scale, although the concentration ranges of glass forming have become nearly identical, we see that the anions are playing a specific role. As a matter of fact T_g can be related to the viscosity coefficient of the anion, i.e. to non-electrostatic contributions to the friction, or to the pK_a, the strength of the acids of the anions.

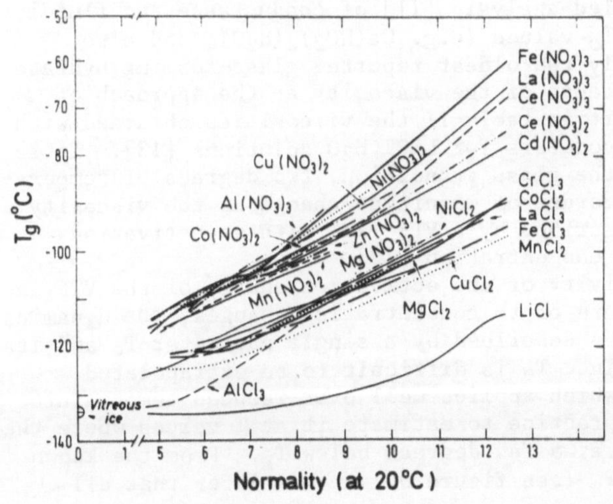

Figure 6. Glass transition for chlorides and nitrates versus normality [8].

In other words, T_g is related to the extent to which anions are ordering water around them and thus decreasing the configurational entropy. That hydration plays a role in glass forming is also noticeable from the relation between the lowest concentration limit at which T_g is obtained and the total hydration number of the salt (hydration numbers of 6-8 for monovalent, 17-20 for divalent, 24-31 for trivalent salts are determined in this manner).

4. THE IDEAL GLASS TRANSITION TEMPERATURE T_o AS DEFINED FROM DYNAMIC PROPERTIES

Another approach to the glass transition is possible from the measurement of some transport property (or relaxation time) versus temperature. It has been found that, for glass formers, all these properties could be described by an universal expression (which applies well for viscosity values $\eta < 10^8$ Poises) often called the Vogel-Tamman-Fulcher law (VTF), of the type [9][13] :

$$\kappa = \kappa_o \exp\left(\frac{B}{T-T_o}\right)$$

where κ is the measured relaxation time or transport property (e.g. the spin-lattice relaxation time T_1 or the electrical conductance Λ or the fluidity ϕ ($\phi = 1/\eta$ with η the viscosity))
κ_o and B are constants
T_o is the temperature at which no free volume is left, or at which the entropy of the system vanishes. It is called the ideal glass transition temperature.

It is important to note that identical B values were found for aqueous ionic glasses and for the corresponding glass forming molten salts [10] (e.g. nitrate melts and Ca $(NO_3)_2.H_2O$ solutions) whereas the T_o for the aqueous solutions could also be estimated from an extrapolation of the molten salts relation between T_o and charge to radius ratio. This to say that the details of the structure do not matter for these dynamic properties. A detailed analysis [11] of conductance and fluidity measurements gave identical T_o values (e.g. $Ca(NO_3)_2(H_2O)_4$ and also $Na_2S_2O_3.5H_2O$ which is probably the oldest reported glass-forming hydrate [12]). In view of the importance of the viscosity at the approach of the glass transition we report in table II the viscosities obtained with capillary and rotational viscometers for $LiCl.H_2O$ solutions [13]. It is important to note that near the glass transition, two degrees of temperature variation will bring an order of magnitude change in the viscosity. Again these data could be described by a VTF law, with an activation energy nearly independent of concentration [13].

In view of the insensitivity of the activation energy of the VTF law to the details of the structure or to concentration changes, the dynamics properties will essentially be described by a single parameter T_o and its concentration dependence. Since T_o is difficult to be extrapolated accurately from the VTF law which applies well over reduced temperature ranges only, it is a common practice to estimate it at T values where the viscosity $\eta \approx 10^{13}$ poise, i.e. a few degrees below T_g. From the known concentration dependence of T_g (see figure 6) we then infer that all

TABLE II. Smoothed viscosity values for aqueous lithium chloride solutions over the temperature interval 25–−125°C. [13]

t (°C)	$R=4.49$ η (P)	$R=5.03$ η (P)	$R=5.75$ η (P)
25	0.0490	0.0401	0.0326
10	0.0735	0.0592	0.0473
0	0.1009	0.0804	0.0636
−10	0.1458	0.1145	0.0898
−20	0.224	0.1740	0.1350
−30	0.375	0.286	0.219
−40	0.700	0.520	0.393
−50	1.498	1.079	0.803
−60	3.85	2.72	1.988
−70	12.8	8.68	6.24
−75	26.5	17.4	12.3
−80	61.5	38.7	26.6
−85	1.63×10^{2}	97.3	64.1
−90	4.9×10^{2}	2.83×10^{2}	1.78×10^{2}
−95	1.78×10^{3}	9.6×10^{2}	5.6×10^{2}
−100	7.9×10^{3}	3.93×10^{3}	2.11×10^{3}
−105	4.8×10^{4}	2.06×10^{4}	1.01×10^{4}
−110	4.0×10^{5}	1.51×10^{5}	6.8×10^{4}
−115	5.7×10^{6}	1.57×10^{6}	6.0×10^{5}
−120	1.29×10^{8}	2.52×10^{7}	7.7×10^{6}
−125	...	...	1.56×10^{8}

transport properties could be expressed as a function of some reduced temperature ($T-T_o$) and reduced normality ($N-N_o$) [14].

5. NUCLEATION AND GROWTH OF ICE IN THE DILUTE AQUEOUS IONIC GLASSES

It would be tempting to describe the concentration dependence of T_g for aqueous ionic glasses by some simple relation as the Jenkel's expression which applies in the case of water-alcohol mixtures [15][16]

$$T_g = T_{g_1} W_1 + T_{g_2} W_2 + K W_1 W_2$$

where W_i are the weight fractions, K is a constant, and T_{g_i} are the pure component values. (We note that for pure water T_g is estimated at - 137C. Amorphous ice can be prepared at very low temperature by techniques such as vapour deposition but is very unstable towards crystallization on warming [17]). A large variety of binary solutions with water seem to present T_g values which extrapolate well to the estimated pure water T_g [17]. On the other hand, for aqueous ionic glasses there are counter-examples where T_g becomes concentration independent on dilution and remains sometimes at values as high at -110°C [5]. This flattening of the T_g line is observed for $LiCl.H_2O$ or D_2O [5][21][24] (see figures 4,8,14). We have already quoted [5] that, in the most dilute cases, some ice crystallised while quenching. Is there a general tendency towards immiscibility in the supercooled domain, between a water rich and a salt rich phase, the T_g value being then fixed by the salt rich phase concentration ? This idea proposed by Angell and Sare [7] was shown to apply also to high temperature silicate glasses. Moreover light scattering experiments on $LiCl.H_2O$ at concentrations between 11% and 13% had revealed anomalies in the temperature dependence of the Landau Placzek ratio which were interpreted by concentration fluctuations or at least by temperature dependent equilibria between various hydrates [18]. Figure 7 summarizes this question on subliquidus fluctuations and segregation.

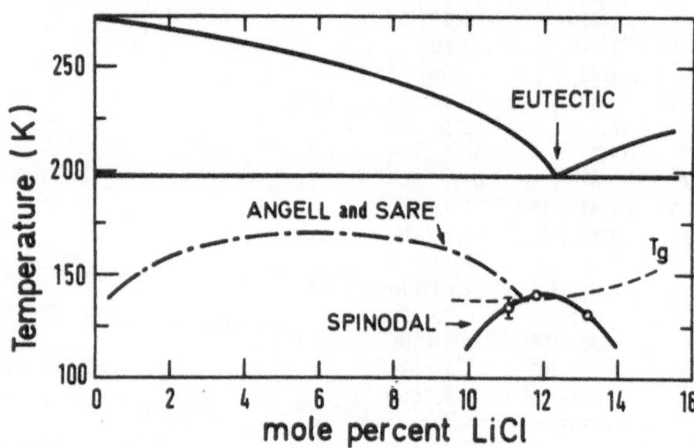

Figure 7. Phase diagram for $LiCl.H_2O$ with T_g line, subliquidus immiscibility line, and spinodal line from light scattering experiments [8]

However, in order to investigate concentration fluctuations in the supercooled regime, it is fundamental not to be perturbed by nucleation and growth of crystallites. An important step, in this respect, is to define the limits of the supercooling domain as a function of concentration. Here again, it is convenient to introduce a non-equilibrium temperature limit, the temperature of homogeneous nucleation (T_H) which is defined by a well-localized peak in steady state cooling thermograms [19], using microemulsions of salty water (with span 65 and saturated heptane) [20].

We should keep in mind that T_H is not an equilibrium property of the system and also that in most experiments we use rather large samples in which heterogeneous nucleation is dominant. Nevertheless, significant information are obtained from figure 8 in which the T_H and T_g lines are drawn for $LiCl.H_2O$ and D_2O [21]. The intersection of the T_H and T_g lines defines a concentration around which nucleation and growth will be considerably slowed down by the high viscosity of the medium and could then easily be controlled via small temperature changes (cf. section 4). At lower concentrations ice might precipitate on cooling, depending on the quench rate. At higher concentrations stable glasses are expected since the nucleation rate in the supercooled domain will be very small. The effect of pressure has also been investigated [22]. Pressure as we have seen, lowers $T_{m.p}$ as does addition of salt, it also and even more strongly lowers T_H, thus increasing the supercooling range. And since it slightly increases T_g, it altogether favours stable glass formation.

The preparation of $NaCl.H_2O$ glasses for example, which cannot be made by simple quench to liq N_2, becomes possible if pressure is applied [22].

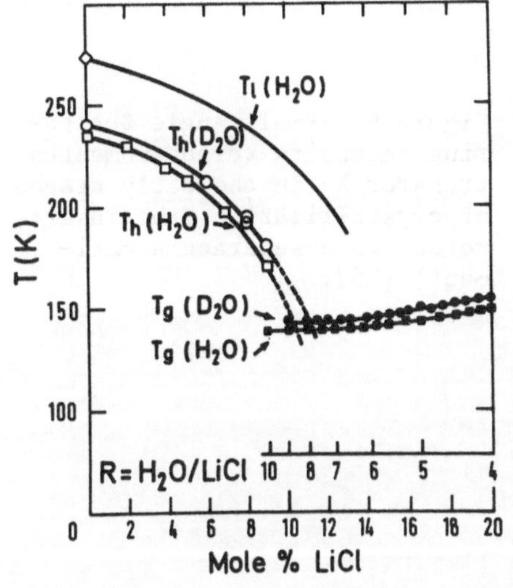

Figure 8. Liquidus line (T_L), T_H and T_g for $LiCl.H_2O$ and $LiCl.D_2O$.

Having defined the optimum conditions for preparing good aqueous ionic glasses, it is important to characterize them at the microscopic level. Considering that fluctuations or precipitates, in the vicinity of T_g, might be only of a few Angströms size, the most appropriate approach is the use of small angle scattering (SAS) techniques. Neutron SAS is especially adapted for light atom containing materials and easily allows for low temperature sample handling and in situ thermal treatment. As a matter of fact it will often be necessary to perform SAS, DSC and standard neutron diffraction experiments in order to achieve complete sample characterization. We briefly summarize the results [25] found on a 10.5% $LiCl.D_2O$ sample (optimum concentration for a good glass in which fluctuations or segregation might occur. D_2O is chosen to avoid the large incoherent neutron scattering of hydrogen). Whatever the thermal treatment no scattering intensity (I) could be described as a function of the momentum transfer (k) ($k = (4\pi\ \lambda) \sin \theta$, where λ is the neutron wavelength and 2θ is the scattering angle) by Ornstein-Zernike plots ($I^{-1} = f(k^2)$) which would have been the signature of concentration fluctuations [26]. The scattering never seemed to increase at the approach of some low lying critical point or immiscibility gap. On the contrary the signal always increased on warming and could well be described by the standard Guinier plots [$\ln I = f(k^2)$] characteristic of precipitating particles. This gave a direct measurement of the average particle radius of gyration (e.g. 100 Å at 140.4 K). Furthermore, annealing at temperatures slightly lower than T_g ($T_g \simeq 141$ K) allowed to control the distribution of nuclei which on subsequent warming produced a diffraction ring at very low k value (e.g. 0.9 10^{-2} Å^{-1}) characteristic of the average interparticular distance (e.g. 700 Å) (see figure 9) [25].

Figure 10 presents the intensity of the neutron SAS as a function of the thermal treatment. We see that, after annealing at 139 K, (about 2 degrees below T_g) the precipitation is detected around T_g. Then the signal goes down probably due to the relative density variation and its effect on the contrast matching between the scattering properties of the precipitates and that of the residual solution. Above 149 K, the signal raises again considerably and the particles grows into the micron size.

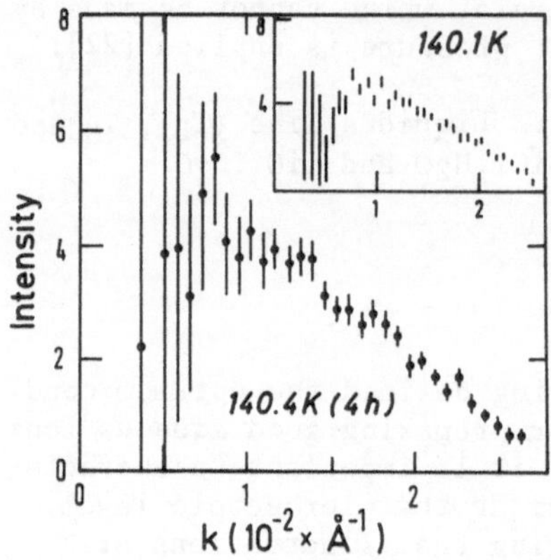

Figure 9. Small Angle Scattering intensity versus momentum transfer k, in the early stages of crystallization (the insert refers to a separate experiment) [25].

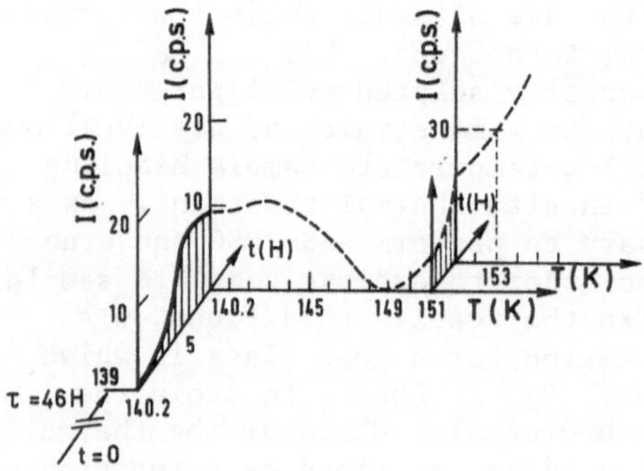

Figure 10. Time-temperature scattered neutron intensity [25].

The sample which was fully transparent up to 149K, changes colour going through light yellow, orange, reddish until it becomes white and opaque.

A complementary information is given by the large angle scattering experiment, allowing to identify the nature of the precipitates. We see in figure 11 that, at low temperature, the as-quenched sample is fully vitreous, then the peaks of the cubic ice (I_c) progressively appear, the stability of this phase being favoured by annealing below T_g. At higher temperature cubic ice continuously transforms into hexagonal ice (I_h) in a manner which can be fully controlled by small temperature changes [27].

Through all these steps the structure of the vitreous matrix (or of the residual solution) remains apparent and its main peak position progressively shifts as more ice precipitates. The concentration dependence of the main peak position of the aqueous ionic glasses has been calibrated on a series of as-quenched good glasses of various concentrations. We see in figure 12 that a very significant concentration dependence is observed and that on dilution, the temperature effect on the peak position during quenching becomes so large that it must imply considerable rearrangements and an opening of the structure. We understand that there is a point where the system will become unstable towards precipitation.

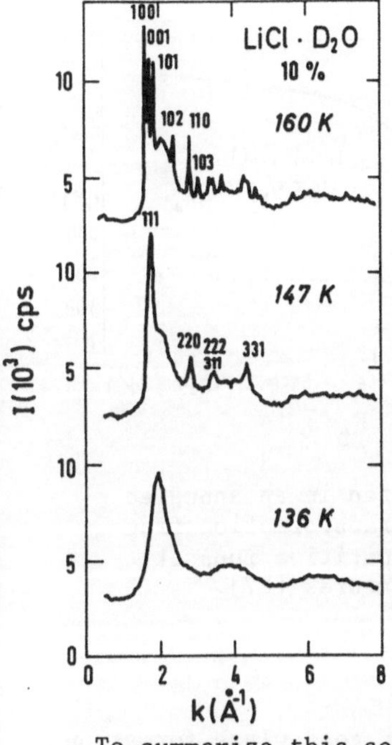

Figure 11. Typical diffraction patterns [27] of a $LiCl.D_2O$ 10 mole % sample.
a) as-quenched (136K)
b) after full precipitation of cubic ice (147K)
c) after $I_c \rightarrow I_h$ transformation (160K).

We therefore have several ways to estimate the fraction of precipitated ice, (i) the integrated intensity of the ice Bragg peaks (specific I_c or I_h ice peaks can be distinguished), (ii) the peak position of the remaining vitreous matrix (or residual solution), (iii) the average distance and size of the precipitated particles. If the sample thermal treatments are well under control and transferrable from one technique to the other, then it is possible to obtain at any stage of the treatment, the correlation between, the particle size, the fraction crystallized and even the percentage of cubic to hexagonal ice transformation. Examples are given in figure 13 [27].

To summarize this section, we have specified under what conditions of temperature and concentration, ice particles might nucleate and grow in glassy matrix or supercooled ionic solution. We have also seen how this nucleation could be fully controlled by proper annealing below T_g thus favouring cubic ice which is trongly correlated to crystal sizes less than about 250 Å. Transparency is therefore not a sufficient criterion for sample homogeneity and very small precipitates can indeed be detected in the glass. The idea of low lying immiscibility gap and corresponding concentration fluctuations is however not appropriate. This was further confirmed by recent light scattering experiments in

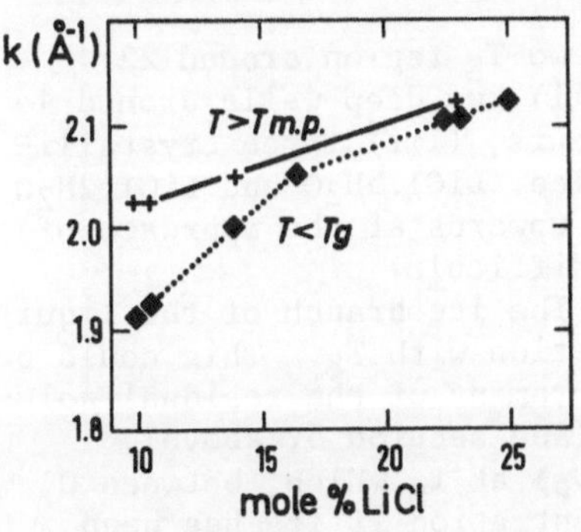

Figure 12. Concentration dependence of the main peak position of the structure factor, in the liquid and in the glassy state.

full agreement with some growth regime of particles [28]. The problem is then to distinguish more precisely the concentrations for good glass forming from the border line regions where the glasses are much more unstable towards nucleation and growth of metastable crystalline phases.

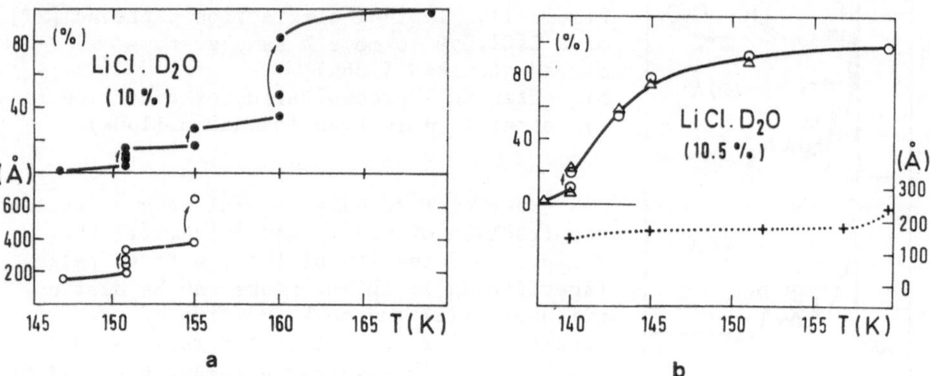

Figure 13. a) Percentage of cubic ice precipitated in an annealed sample and crystallite size b) percentage of transformation from cubic to hexagonal ice and crystallite size. Repetitive runs at intervals of 1 hour are reported for some temperatures [27].

6. NON EQUILIBRIUM PHASE DIAGRAMS

The most convenient way to present the domains of good glass formation and stability is to report them on the phase diagram. As a matter of fact, since we want here to describe non-equilibrium properties of the systems, it is more appropriate to construct directly some non-equilibrium phase diagram. This can be achieved by performing thermal analysis experiments (D.S.C.) with a conventionally fixed sequence of cooling and warming rates (e.g. quench to liq N_2 in 120 seconds followed by a 2K/mn warming) and reporting all thermal accidents such as Tg's and the crystallization of metastable phases (identified, if necessary, by Diffraction). Such a non-equilibrium phase diagram for $LiCl.H_2O$ is given in figure 14 [23].

We note (i) a S-shaped T_g line with a two-T_g region around 22 moles % (for further comments, see [23]), (ii) two deep wells around 14 and 20 moles %, where no crystallization occurs, (iii) three crystallization lines which have been identified as ice, $LiCl.5H_2O$ and $LiCl.2H_2O$ branches. These branches are strongly bent upwards at the approach of the wells, as crystallization is then more difficult.

Let us focus now on the ∿ 14 % well. The ice branch of the liquidus line has been drawn down to its intersection with Tg. This could be achieved in several ways, (i) from the measurement of the residual solution concentration after ice precipitation (see section 5, above), (ii) from the variation of heat capacity (ΔC_p) at T_g which, between 0 and 14 mole %, is a linear function of concentration if ice has been

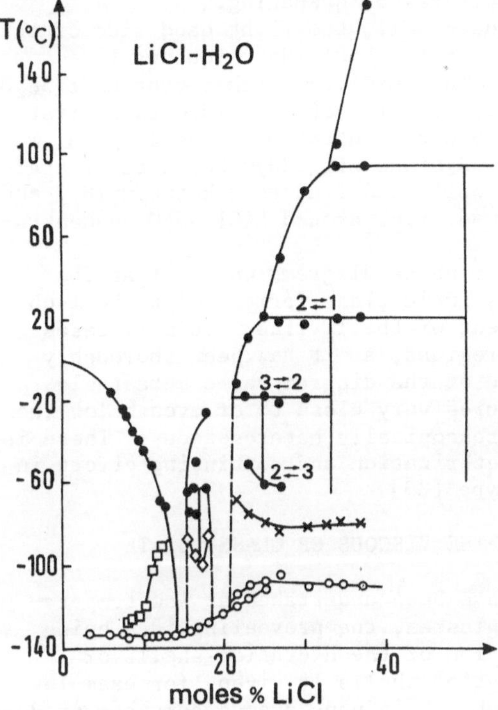

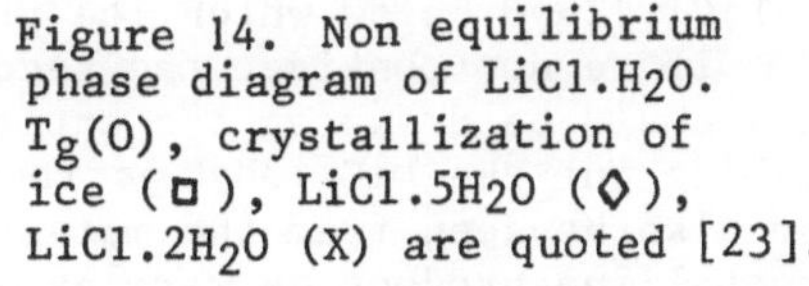
Figure 14. Non equilibrium phase diagram of $LiCl.H_2O$. Tg(O), crystallization of ice (□), $LiCl.5H_2O$ (◇), $LiCl.2H_2O$ (X) are quoted [23].

precipitated (this implies re-quenching the sample just after ice crystallization), (iii) from the extrapolation of an analytical expression of the liquidus line based on the equilibrium activity of the solute [24].
The right hand side of the well is much steeper and less easy to measure. There is however a simple way to delimit the bottom of the well from the concentration dependence of the heats of crystallization of ice and $LiCl.5H_2O$ (Tamman diagrams) [24]. See figure 15.

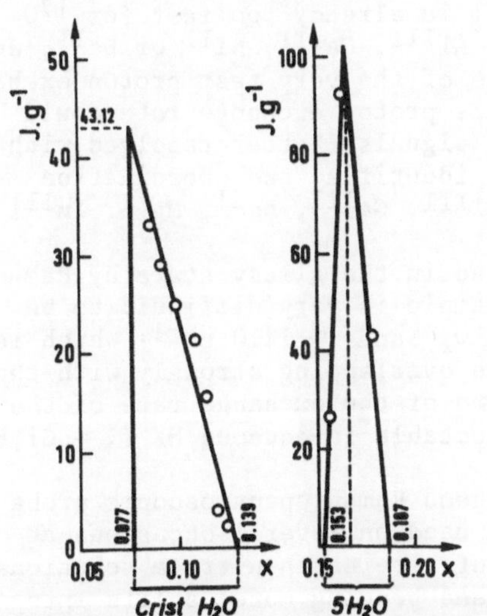

Figure 15. Heat of crystallization ΔH for ice and $LiCl.5H_2O$ as function of the $LiCl.H_2O$ concentration given in mole fraction [24].

From these data, good glasses are formed between 14% and 15.3 moles %, i.e. just around the $LiCl.6H_2O$ composition. We remind that all the above comments refer to a non-equilibrium diagram as specified by the quoted rates of temperature change. Interestingly also there is an abrupt limit at $LiCl.12H_2O$ below which ice precipitates on quenching.

If we now consider the second deep well, the right hand side of this well could be drawn as the extension of the $LiCl.2H_2O$ liquidus line washing out the $LiCl.3H_2O$ peritectic. That $LiCl.2H_2O$ takes over $LiCl.3H_2O$ can also be seen from the metastable formation of $LiCl.2H_2O$ as a first crystalline product on warming at those concentrations. Here again, we use the Tamman diagrams of the $LiCl.5H_2O$ and $LiCl.2H_2O$ heats of crystallization to define the width of the well, and find that between 18.7 and 21 moles % good glasses can be formed, i.e. around $LiCl.4H_2O$ concentration.

With the help of non-equilibrium phase diagrams we see that the concentration range for good aqueous ionic glass-formers might be much narrower than expected from the extent of the T_g line. In many cases we are likely to enter border line regions, as it has been thoroughly investigated on the ice forming side of the diagram where metastable crystals might nucleate at temperatures very close to or even below the quoted T_g and render the sample microscopically heterogeneous. There is place for more careful sample characterization and continuing effort in systematic investigations of this type [23].

7. SPECTROSCOPIC INVESTIGATIONS IN THE VISCOUS OR GLASSY STATE

Many spectroscopic investigations have been undertaken at low temperature on aqueous ionic solutions or glasses, the prevealing idea being the identification and characterization of the hydration shells or complexes. A direct access to hydration shells is given, for example, by proton magnetic resonance (PMR) which, in non-aqueous solvents produces separate signals for bulk and complexed solvent molecules. However, in the case of aqueous solutions, the exchange of the solvent molecule between the bulk and the hydration shell is already too fast for ^{17}O NMR studies, except for cations such as Al^{III}, Ga^{III}, Ni^{II} or Be^{II}, and the matter is even worse for PMR because of the very fast proton exchange. At low temperature ($-60 < T < -40°C$) this proton exchange rate could be reduced to 0.1 to 1 second and separate signals (better resolved with addition of acetone) could be obtained, identified and coordination numbers measured, for cations such as Al^{III}, Ga^{III}, Be^{II}, Mg^{II}, In^{III} [29][30].

Some nice results have been obtained in the glassy state by Raman Spectroscopy. The oxonium H_3O^+, for example is very difficult to be observed because of the weakness of its ν_2 band ($\sim$ 1210 cm^{-1}) which is the only observable, all the other bands overlapping strongly with those of water. Again through the slowing down of the exchange rate of the proton, the ν_2 band becomes clearly detectable in aqueous HX (X = Cl,Br) glasses [31].

The nitrate anion which is a very good Raman spectroscopic probe for studying complex formation has been used on several occasions at low temperature [32][33]. In the case of rare earth nitrate solutions, for example, the ν_4 band of NO_3^- (in plane bending mode) splits into two peaks, the $\nu_{4\ell}$ at $\sim$ 715 cm^{-1} belongs to free NO_3^-, while the ν_{4h} (740 - 755 cm^{-1}) is characteristic of the inner sphere complexes. The ν_4 peaks for light-rare-earth solutions do not change much with temperature from the liquid to the glassy state, while in the heavy-rare-earth glasses the ν_{4h} peak disappears. From this it is inferred that in heavy-rare-earth aqueous nitrato glasses, the ions are in a state of solvent separated ion pairs. It is very interesting also to note that the Ln^{3+} - OH_2 stretching Raman band ν_W (350-410 m^{-1}) characteristic of a water molecule in the inner rare earth shell, is only detectable when solutions are vitrified (because of strong Rayleigh wings at room temperature). This allowed to detect an anomaly in the value of ν_W versus ionic radius for the rare earth chlorides aqueous glasses. The $EuCl_3$ and $GdCl_3$ aqueous glasses present two ν_W peaks which have been attributed to two kinds of Eu^{3+} and Gd^{3+} ions with inner sphere hydration numbers of either 8 or 9 (this anomaly is not observed in the nitrato aqueous glasses series).

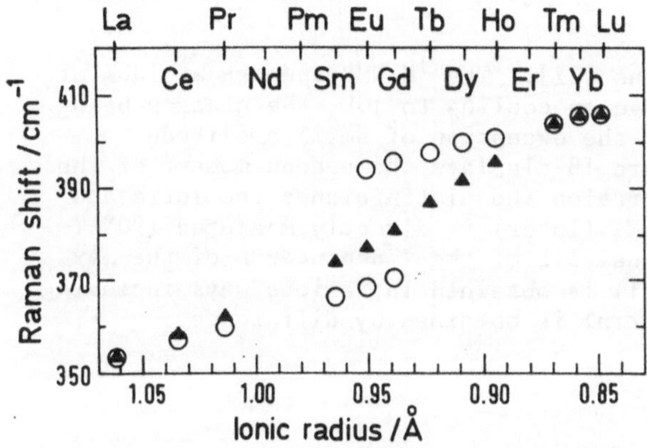

Figure 16. Frequencies of the ν_W bands for glassy rare-earth nitrate (Δ) and chloride (O) solutions [33]

Mössbauer absorption spectra [34] of $FeCl_2$ solutions and glasses have been performed in the border line region of metastability of the system towards ice nucleation. The interpretation of some spectra showing a transition in the quadrupole splitting of ^{57}Fe was given as a modification of the ice structure (cubic to hexagonal transformation) in which the $FeCl_2$ was trapped. However, a detailed sample characterization at the microscopic level as described in section 5 was not available at the time of the experiment.

A large amount of pulse radiolysis (γ-irradiation) experiments were performed on aqueous ionic glasses in order to investigate their transient products and, in particular, the trapped electrons. Indirect information could be gathered on the structure of the medium. For example, the decay of the I.R. absorption band has been used in order to characterize the depth of the electron traps. A shallow and a deep trap are often observed such as in irradiated 9.5M $LiCl.D_2O$ glass [35], the deep trap being related to the hydration shell. Again, detailed microscopic sample characterization, would be interesting when such bulk and hydration water are distinguished.

Neutron inelastic scattering and NMR of concentrated ionic solutions having been introduced by J. Dianoux [36] we shall just make a few remarks concerning the glassy state studies. Proton dipolar spin echo measurements have been performed on LiCl and LiBr aqueous glasses ($2 < R < 10$) to selectively measure the inter and intra molecular (H_2O)

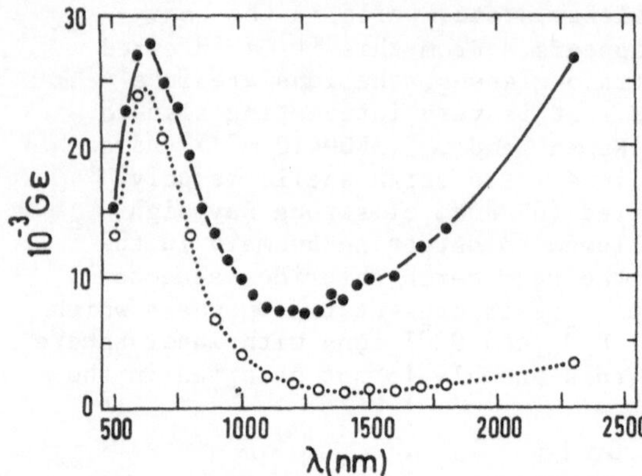

Figure 17. Transient absorption spectra in irradiated 9.5M LiCl.D_2O glass at 76K : ● at the end of the pulse, O, 150μs later (Pulse length 100ns, dose per pulse ∿ 7 krad). The band beyond 2400 nm is attributed to trapped electrons in D-defects [35].

proton-proton dipolar interactions [37]. The ^1H NMR spectra at 100K are roughly bell-shaped with no change on cooling to 10K, the glasses being rigid on the NMR time scale with the exception of small amplitude librations and vibrations. Figure 18 displays the second moment of the frequency spectra versus concentration and distinguishes the intra and inter molecular contributions. M_2 (inter) is directly measured (90°-τ-90° pulse experiment) from the analysis of the τ dependence of the max amplitude of the echo. M_2 (total) is obtained in various ways including the dipolar echo signal. M_2 (intra) is obtained by difference.

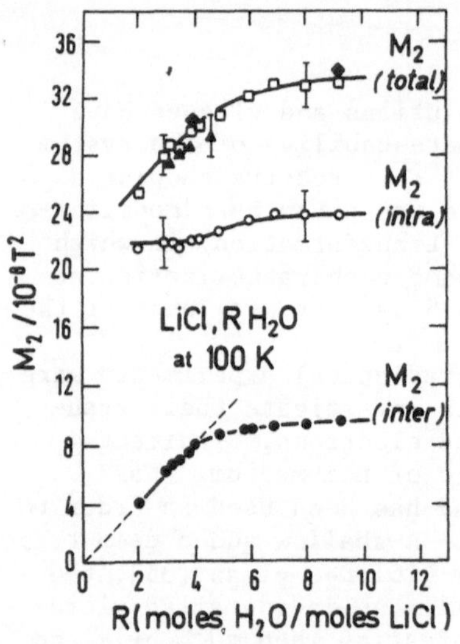

Figure 18. Second moments of the frequency proton NMR spectra versus concentration of LiCl aqueous glasses at 100K.

The concentration dependence of M_2 (inter) is very interesting, it can be written as follows :

for $R < 4$ M_2 (inter) = (R/4).A

for $4 < R < 10$ M_2 (inter) = (4/R).A + ((R-4)/R)B

where $A = M_2(\text{inter})_{R=4}$, and B are constants nearly independent of the anion.

For $R < 4$ it suggests that M_2 monitors the progressive hydration of Li^+ which is complete at $R = 4$. The interaction between H_2O's located on neighbouring Li^+ contributes significantly to M_2 and increases as $R/4$. This is confirmed by calculations of M_2 for an isolated complex $Li^+(H_2O)_4$ which gives a too small value for A. For $4 < R < 10$, a two-site model is satisfactory, in the first site, the water is in a $Li(H_2O)_4{}^+$ complex, the second site is consistent with water molecules in very small and dense water clusters (the value of B is slightly larger than what is measured for I_h ice) or "interstitial" water between the first type of complexes.

Of course, this description of the glassy state has been extensively extended to the study of the liquid relaxation mechanisms and their concentration dependence. Various NMR experiments (^7Li, ^1H, ^2H) [38][39] have shown that the Li hydration is basically the same at high and low temperature, with no anomaly at T_g and a lifetime much longer than the correlation times of the water molecule [40][41]. And the concentration dependence of the deuteron spin lattice relaxation time T_1, at room temperature, has been analyzed with the above two-site models [42]. Finally, in a recent study two relaxation times are invoked for hydrated water [43] corresponding to the reorientational motions of the water molecules by a diffusional reorientation about the Li-O axis (with dipole moment tilted away from the Li-O direction) and the isotropic overall reorientation diffusion of the hydration complex.

In any case, there is a point which needs some emphasis. The temperature dependence of all ^2H NMR relaxation times (for bulk as for hydrated water and for the hydration complex reorientation) as well as of various dynamic measurements such as the viscosity and the shear relaxation [44] is made via VTF laws, with T_o values well compatible with the observed T_g. Moreover the only concentration dependent parameter is T_o. This means that in supercooled aqueous ionic solutions at low temperature, the local structure has no much effect on the sample dynamics which becomes dominated by large scale configurational rearrangements leading to the freezing of all reorientation movements at the glass transition. This point can be illustrated by drawing in the non-equilibrium phase diagram of figure 14, the isoviscosity lines (extrapolated from table II) parallel to the T_g line and this in the supercooled regions defined between T_g and crystallization.

8. CONCLUSION

With the accent on down to earth questions such as sample characterization, range of supercooled regions and their stability towards nucleation, conditions for good glass preparation, we have reviewed some of the work and basic concepts on aqueous ionic glasses. And we have shown that in the case of the $LiCl.H_2O$ system, there are two narrow ranges for good glass forming around $LiCl.4H_2O$ and $LiCl.6H_2O$ concentrations. Why at these concentrations do we make good glasses although the hydrates $LiCl.5H_2O$ and $LiCl.3H_2O$ crystallize rather easily ? This is still not known at the microscopic level. Neither are known the structure and the microscopic mechanisms for the various hydrates and their transformations. There is obviously a need for more comparative structural and dynamic investigations between the hydrates and glass formers. Such a specific effort on aqueous ionic glasses would be welcomed in order to make some progress in the understanding of more general problems related to the physics of the glass transition [9][45] or to the tunneling centers observed in glasses at very low temperature [46] which remain largely unsolved.

ACKNOWLEDGEMENTS

I would like to thank A. Elarby-Aouizerat, J.-F. Jal and J. Dupuy (Département de Physique des Matériaux, Université Claude-Bernard, 69622 Lyon-Villeurbanne) J.M. Letoffe and P. Claudy (Laboratoire de Thermochimie, INSA, 69622 Lyon-Villeurbanne) with whom I have been involved in a friendly and fruitful collaboration.

REFERENCES

[1] G. TAMMANN, E. SCHWARZKOPF, Z. Anorg. Allgem. Chem. 174, 216 (1928)
[2] P.W. BRIDGMAN, Z. Anorg. Chem. 77, 387 (1912), Proc. Amer. Acad. 47, 544 (1912)
[3] G.F. HÜTTIG, W.D. STEUDEMANN, Z. Physik Chem. 126, 105 (1927)
[4] H.E. MORAN, J. Phys. Chem. 60, 1666 (1956)
[5] G. VUILLARD, Thèse, Paris (1957), Ann. Chim. 2, 233 (1957)
G. VUILLARD, J.J. KESSIS (1960), Soc. Chim. 5, 2063 (1960)
[6] R. COHEN-ADAD, J. LORIMER (1985), to be publ. in Pergamon Press
[7] C.A. ANGELL, E.J. SARE, J. Chem. Phys. 49, 4713 (1968)
[8] C.A. ANGELL, E.J. SARE, J. Chem. Phys. 52, 1058 (1970)
[9] M. CYROT, Phys. Letters, 83A, 275 (1981)
[10] C.A. ANGELL, J. Phys. Chem. 69, 2137 (1965)
[11] C.T. MOYNIHAN, J. Phys. Chem. 70, 3399 (1966)
[12] M. SAMSOEN, Ann. Phys. 9, 35 (1928)
[13] C.T. MOYNIHAN, N. BALITACTAC, L. BOONE, T.A. LITOVITZ, J. Chem. Phys. 55, 3013 (1971)
[14] C.A. ANGELL, J. Chem. Phys. 46, 4673 (1967)
[15] E. JENCKEL, R. HEUSCH, Kolloid-Z 130, 89 (1958)
[16] Don H. RASMUSSEN, A.P. MACKENZIE, J. Phys. Chem. 75 967 (1971)
[17] D.R. MacFARLANE, C.A. ANGELL, J. Phys. Chem. 88, 759 (1984)
[18] S.Y. HSICH, R.W. GAMMON, P.B. MACEDO, C.J. MONTROSE, J. Chem. Phys. 56, 1663 (1972)
[19] D. CLAUSSE, L. BABIN, F. BROTO, M. AGUERD, M. CLAUSSE, J. Phys. Chem. 87, 4030 (1983)
[20] C.A. ANGELL, J. DONNELLA, J. Chem. Phys. 67, 4560 (1977)
[21] C.A. ANGELL, E.J. SARE, J. DONNELLA, D.R. MACFARLANE, J. Phys. Chem. 85, 1461 (1981)
[22] H. KANNO and C.J. ANGELL, J. Phys. Chem. 81, 2639 (1977)
[23] J. DUPUY, A. ELARBY-AOUIZERAT, J.-F. JAL, P. CHIEUX, P. CLAUDY, J.M. LETOFFE, Rivista della stazione sperimentale del vetro 5 63 (1984)
A. ELARBY-AOUIZERAT, J.-M. LETOFFE, J.-F. JAL, P. CLAUDY, P. CHIEUX, J. DUPUY, to be published

[24] P. CLAUDY, J.M. LETOFFE, J.J. COUNIOUX, R. COHEN-ADAD, J. Thermal Anal. 29, 424 (1984)
[25] J. DUPUY, J-F. JAL, C. FERRADOU, P. CHIEUX, A-F. WRIGHT, R. CALEMCZUK, C.A. ANGELL, Nature, 296, 138-140 (1982)
[26] J-F. JAL, P. CHIEUX, J. DUPUY, J.P. DUPIN, J. de Physique 41, 657 (1980)
[27] A. ELARBY-AOUIZERAT, J-F. JAL, C. FERRADOU, J. DUPUY, P. CHIEUX, A. WRIGHT, J. Phys. Chem., 87, 4170 (1983)
[28] J. PELOUS, A. ESSABOURI, R. VACHER, J. de Physique, Colloque C8, 12, 46 (1985)
[29] N.A. MATWIYOFF, H. TAUBE, J. Amer. Chem. Soc., 90, 2796 (1968)
[30] A. FRATIELLO, R.E. LEE, V.M. NISHIDA, R.E. SCHUSTER, J. Chem. Phys., 48, 3705 (1968)
[31] H. KANNO, J. HIRAISHI, Chem. Phys. Letters, 107, 438 (1984)
[32] F. GUILLAUME, M. PERROT, G. ROTHSCHILD, J. Chem. Phys., 83, 4338 (1985)
[33] H. KANNO, J. HIRAISHI, J. Phys. Chem., 88, 2787 (1984)
[34] A.J. NOZIK, M. KAPLAN, J. Chem. Phys., 47, 2960 (1967)
[35] G.V. BUXTON, H.A. GILLIS, N.V. KLASSEN, Chem. Phys. Letters, 32, 533 (1975)
[36] J. DIANOUX (these proceedings)
[37] I.C. BAIANU, N. BODEN, D. LIGHTOWLERS, M. MORTIMER, Chem. Phys. Letters, 54, 169 (1978)
[38] E.J. SUTTER, J-F. HARMON, J. Phys. Chem., 79, 1958 (1975)
[39] J.F. HARMON, E.J. SUTTER, J. Phys. Chem., 82, 1938 (1978)
[40] H.G. HERTZ, M.D. ZEIDLER, Ber. Bunsenges Phys. Chem., 67, 774 (1963)
[41] S. ENGSTRÖM, B. JÖNSSON, B. JÖNSSON, J. Magn. Res., 50, 1 (1982)
[42] N. BODEN, M. MORTIMER, J. Chem. Soc. Faraday Trans. II, 74, 353 (1978)
[43] E.W. LANG, H.D. LUDEMANN, Ber. Bunsenges Phys. Chem., 89, 508 (1985)
[44] C.T. MOYNIHAN, R.D. BRESSEL, C.A. ANGELL, J. Chem. Phys., 55, 4414 (1971)
[45] M.H. COHEN, G.S. GREST, J. Noncryst. Solids, 61 & 62, 749 (1984)
[46] J. PELOUS, R. VACHER, A. ESSABOURI, U. REICHERT, M. SCHMIDT, in "Phonon Scattering in Condensed Matter", Ed. W. Eisenmenger, K. Lassmann, S. Döttinger, Springer-Verlag, p.398 (1984).

NATURAL AQUEOUS SOLUTIONS IN THE EARTH

Gil MICHARD
Laboratoire de Géochimie des Eaux
Université Paris 7
75251 Paris Cedex 05

1. INTRODUCTION

Geochemistry is a science which studies distributions of chemical elements in the different parts of the earth and exchanges of these elements between these different parts. Theoretical basis for geochemistry is mainly chemistry and the main difference between geochemistry and laboratory chemistry is that in geochemistry we only observe natural objects ; geochemist perform observations and chemist experiments.
Water is, by far, the most important and most widespread liquid in the superficial zone of the earth. The main part of waters is the oceans as very saline water. Amount of total dissolved salts in open ocean varies in a restricted range and the ratios between major constituants are strictly constant.
Physical chemistry of sea water has been developed using the ionic medium approach :
as the main active dissolved species are $H_2CO_3/HCO_3^-/CO_3^=$ sea water can be considered as dilute solution of sodium bicarbonate + sodium carbonate in a "solvant" consisting in sodium, potassium, magnesium and calcium chloride and sulfate plus water.
This approach and the related activity scale are well known by physical chemists and I will not deal a long time with this. On the contrary, I will focus on the problem of inputs and outputs of chemical elements in (or from) oceans.

M.-C. Bellissent-Funel and G. W. Neilson (eds.), The Physics and Chemistry of Aqueous Ionic Solutions, 379–397.

For longtime, inputs were restricted to input by streams and outputs by evaporation and deposition of solids. Since about twenty years, it has been suggested that the important chemical reactions occuring at the ocean-sediment interface result in important exchange of elements between waters and the bottom of the ocean. Still more recently (1979), discovery of hydrothermal submarine vents involved a complete reexamination of the budget of dissolved substances in the oceans.
We will present here two brief discussions : the first about chemical exchanges at the ocean-sediment boundary and the second on hydrothermal submarine vents.

2. CHEMICAL EXCHANGES AT THE WATER-SEDIMENT BOUNDARY

Fresh sediment deposited in a lake or in ocean contains more than ninety percent of water. Below a few centimeters, solid sediment is motionless and pore water is almost completely immobilized. Thus we go in a few centimeters from a highly stirred fluid to a motionless fluid.

Solid deposited derived from detrital particles brought by streams and from biological and chemical solids formed within the ocean. Organic debris represent about 1% of the solid phase ; important part of this organic matter is highly reactive and behaves as a strong reducer, which reacts on different oxidizing inorganic components (O_2,- MnO_2, ferric oxides, $SO_4^=$,NO_3^-). All these reactions are referred as diagenetic processes.
Sedimentation rates are highly variable and range from 1mm/year in coastal zones to about 1mm/10 years in central part of the ocean. It means that at 1m depth within a core the "age" of the solid deposit is in the range 10^3-10^6 years. Therefore, investigations in chemical change in pore water can afford an insight on geochemical kinetics, i.e. kinetics of reaction occuring in natural systems at geological time scales.

2.1. Pore water sampling

In the first studies, and still yet for deep ocean sites, pore water were extracted by squeezing or centrifugation of sediment cores. Owing to difficulties resulting from chemical changes related to temperature and pressure changes, on aeration of reduced waters, in situ sampling by devices called "harpoon" or "peeper" are today preferred. For measurements in shallow waters, peeper is generally used. Briefly, a peeper consists of a serie of compartments milled into a sheet of "Altuglass" deionized water was poured into each compartment. A membrane covering all the compartments is held in place by a Altuglass template with holes milled over each compartment. The three pieces are fastened together with nylon screws.
Peepers were flushed with N ; then they were inserted vertically into the sediment by scuba-divers. Equilibration time ranged from 15 to 20 days. Pore waters are sampled by directly piercing the membrane with pipette tips.

2.2. Silica

Silica profiles show sharp gradient at the sediment water boundary. Diatoms and radiolaria living in the surface waters of the oceans secrete skeletons consisting of amorphous silica. Because the energy provided by sunlight, they are able to do this even though seawater is undersaturated with respect to amorphous silica. Once they die, siliceous skeletons settle to the bottom and undergo dissolution.

From profiles studied mainly by Shink et al., (1975) and estimation of diffusion coefficient in pore waters,

$$D = 3.3 * 10^{-5} \text{ cm}^2 \text{ sec}^{-1}$$

it can be concluded that silica fluxes in both Atlantic and Pacific Ocean average to,

$$5 \text{ moles/year.cm}^2$$

This figure yield a total flux for silica from the sediment of about,

$$1.5 * 10^{13} \text{ moles/year}$$

which is greater than the estimated input by rivers ($6*10^{12}$ moles/year).

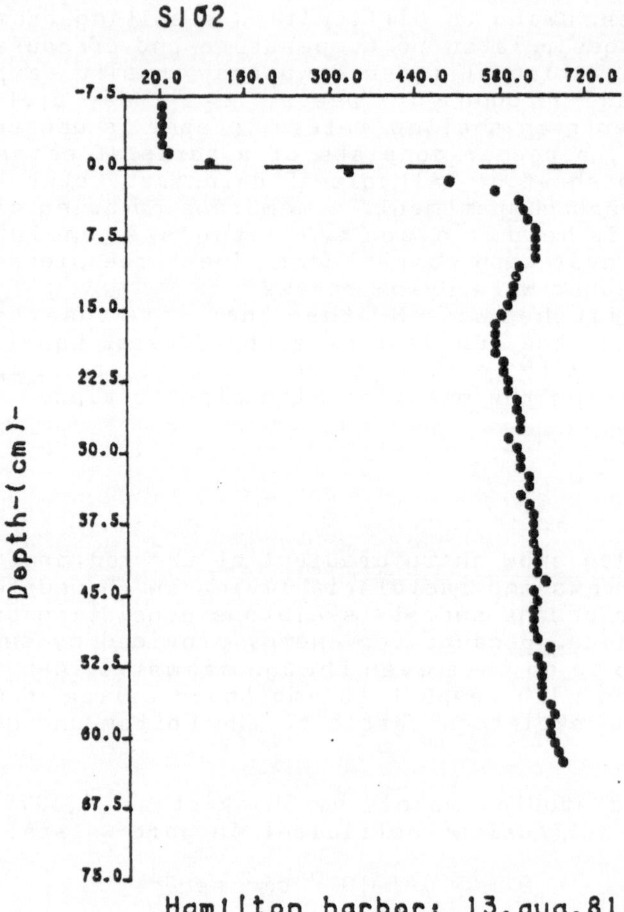

Figure 1. At the ocean-sediment interface, or at the lake-sediment interface, strong silica concentrations gradients are observed. (from Gaillard, 1982)

2.3. Reference frame :

In first approximation, diagenetic processes can be considered as a monodimensional (vertical) process. We are presented with a choice of two origins of depth coordinate : either a given layer in the sediment or the sediment waters interface. (Berner, 1980)

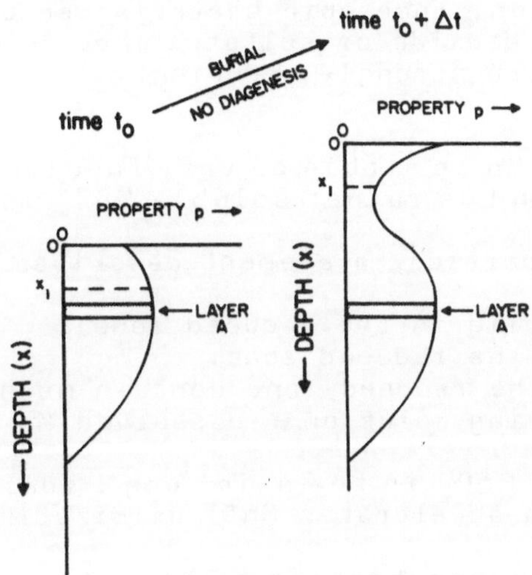

Fig 2a : If a sediment property p remains constant within a layer, it corresponds to a lack of diagenesis for this property.

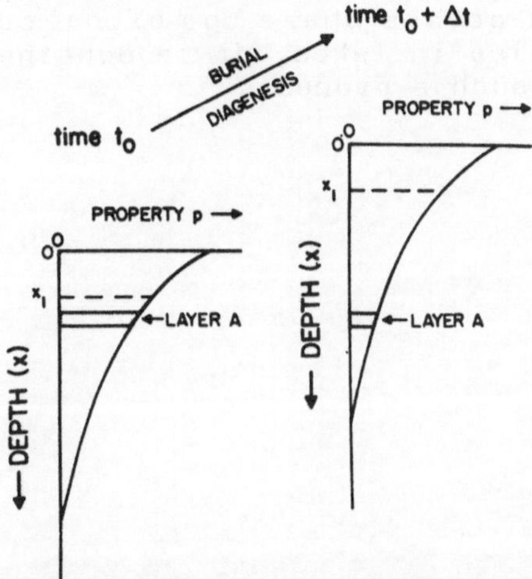

Fig 2b : The opposite situation is steady state diagenesis where p at a given depth remains constant.

2.4. Manganese diagenesis

The main process in early diagenesis is oxidisation of organic matter. In the first step, the oxidizing agent is oxygen. Removal of oxygen by diffusion in the sediment can cancel consumption by organic matter only at low depth within the sediment. In the steady state, oxygen disappears at a constant depth. Below, anaerobic bacteria use other oxidants such as iron hydroxide or sulfate. The environnement is therefore strongly reducing.

A trace element such as Mn is stable as very insoluble MnO_2 when oxygen is present and as rather soluble Mn^{++} ions in reducing environnement.
With these chemical properties, a element describes a cycle (Michard, 1971),
1) entering a Mn-rich solid in the reduced zone ;
2) dissolution of Mn in the reduced zone.
3) aqueous solution in the reduced zone contain much more dissolved Mn than overlying solution : dissolved Mn diffuses upwards.
4) Mn^{++} precipitates as MnO_2 in the oxic zone ; solid MnO_2 acts as a catalyst which accelerates Mn^{++} oxidization.

This cycle results in a concentration of Mn in the superficial zone of the sediment. In coastal zones, where the thickness of the oxidising layer is small (0-2dm), manganese is removed from the sediment to sea water ; exchange of dissolved manganese is from sea water to the sediment in the center of oceans where the oxidized layer is very thick. Some Mn nodules in lakes, or in continental margins are produced by such a process.

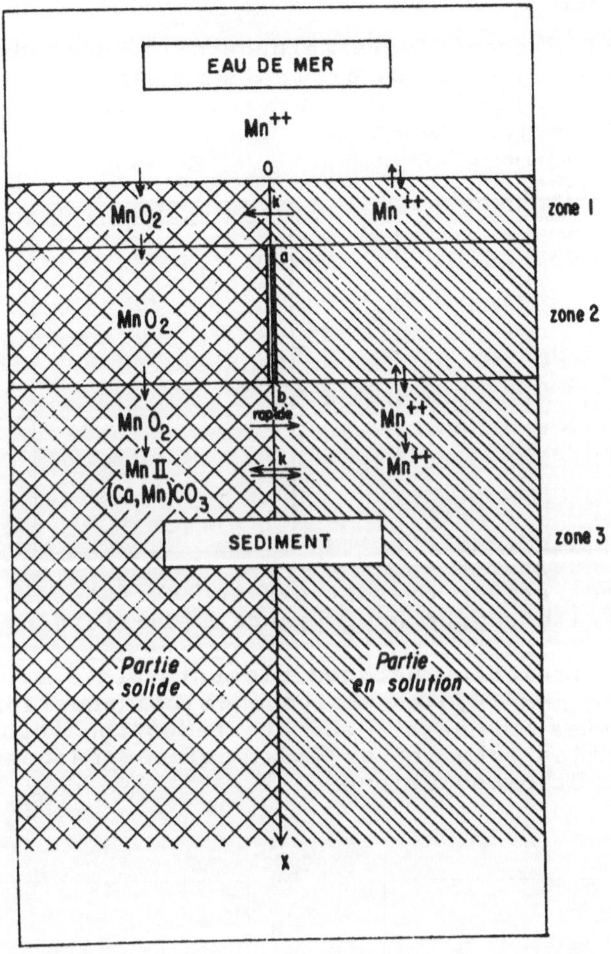

Fig 3. Schematic box model for manganese behavior in sediment. For clarity, solid phase and solution were separated.

Zone 1 : oxidizing layer, Mn^{++} is oxidized in MnO_2

Zone 2 : intermediate layer, no reaction

Zone 3 : reducing layer, solid MnO_2 is reduced in Mn^{++}. Concentrations of dissolved Mn are greater in zone 3 than in zone 1 and Mn diffuses from bottom to top.

Exchanges in the solid boxes are related to burial and constant depth of zones limits.

2.5. Global studies :

Development of microanalysis methods allow a study of a larger and larger number of elements in pore waters. Elderfield et al. (1981) and Gaillard et al. (1986) present studies of all major elements.

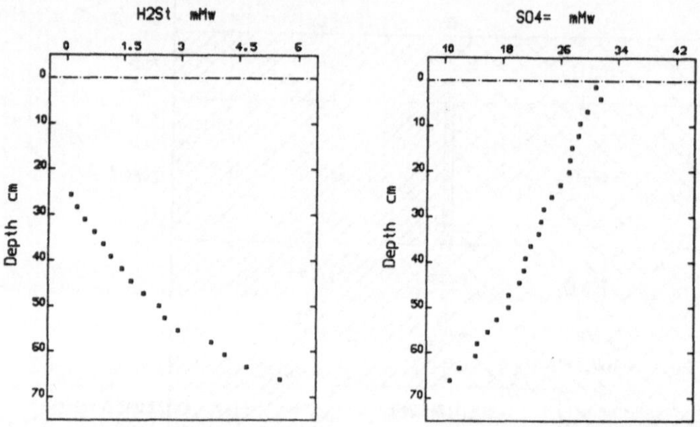

Fig 4a. Concentrations of sulfides and sulfates in pore waters from Villefranche bay, France. (Gaillard et al, 1986)

Rather good correlation between elements are used to derive stoichiometric models, i.e., write chemical equations relating chemical changes : sulfate reduction into H_2S, sulfate reduction into FeS or FeS_2, calcium carbonate precipitation or dissolution.

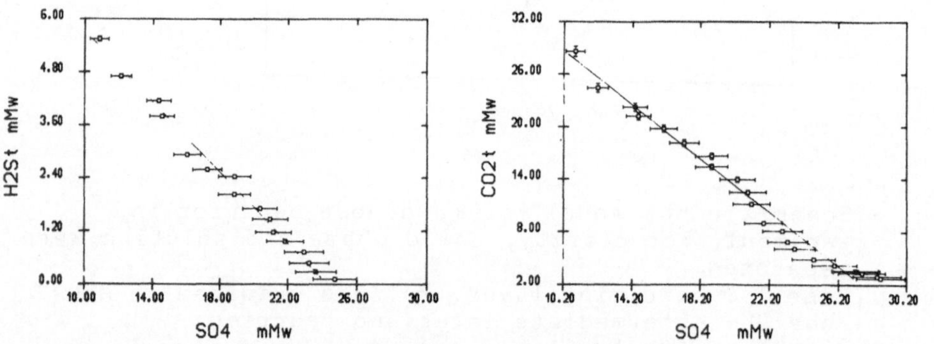

Fig 4b. Correlations between total CO_2, sulfide and sulfate concentrations in pore waters from Villefranche bay (Gaillard et al, 1986).

2.6. Flux gradients relationships :

Multicomponent diffusion is of considerable practical importance ; the diffusivity of a solute in a multicomponent mixture is significantly different from its binary diffusivity if solutes are electrolytes.
Relationships between gradients and fluxes are therefore a complex problem in a multicomponent electrolyte solution.
The diffusion coefficient of an ion is made of two terms one depending on the mobility of the ion and the second due to the other ions.
Gradient for some major elements are often difficult to measure.
Moreover, natural waters are often concentrated solutions and ions are associated in complexes or ion pairs ; theoretical approaches of diffusion in such complex mixtures become to be available, but numerical figures for the diffusion coefficients of these complex ions are still lacking.
In some natural systems, we can observe fluxes of an ion without any concentration gradient of this ion, due to cross coupling with gradients of other ions.
In some other cases, concentration gradients can be observed which result from coupling with diffusion of major ions : this has been observed above salt domes in Eastern Mediterranee (cf. fig. 5)

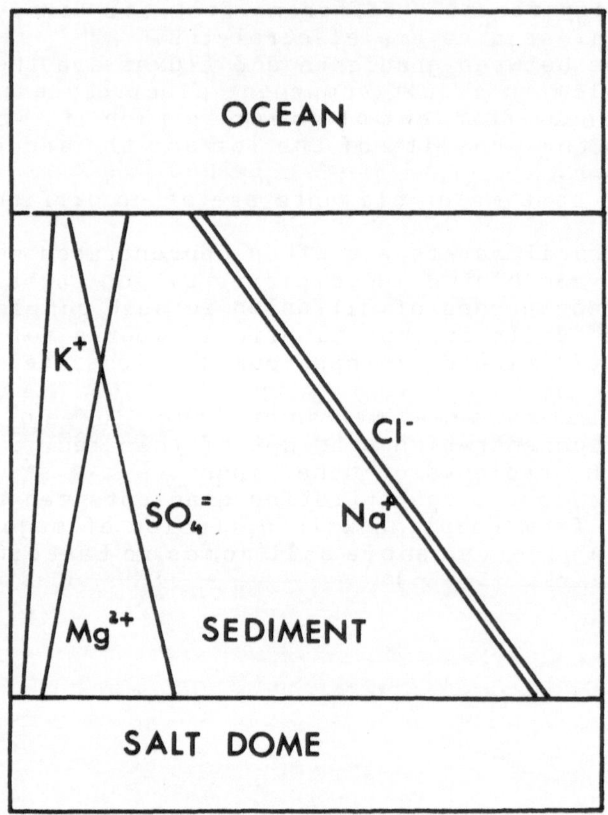

Fig 5. Schematic diagram of ion concentration profiles in pore waters above a salt (NaCl) dome in Eastern Mediterranee (adapted from Ten Haven et al, in prep). Junction potential created by Cl^- and Na^+ diffusion lead to gradients of anions (e.g. $SO4^=$) and cations (e.g. K^+ and Mg^{++}).

3. HYDROTHERMAL SUBMARINE VENTS

3.1. Some words about plate tectonics :

At the end of seventies, theory of plate tectonics was a revolution in earth sciences.
Briefly, the earth surface can be divided in 6 or 7 main plates.
According to the relative motion of adjacent plates, we can define three kinds of plate boundaries :
(i) zones of divergence, typically ocean ridges
(ii) zones of convergence or subduction zones
(iii) fracture zones or transform faults (Press and Siever, 1982).
Zones of divergence are boundaries along which plates separate. Partially molten material upwells along linear ocean ridges, and new lithosphere is created.
Zones of convergence are boundaries along which the leading edge of one plate overrides another, the overriden plate being subducted or thrust into the mantle where lithosphere is reabsorbed. Continental margin may or may not coincide with plate boundaries. If they do, the continental plate is the overriding plate because it is thicker and too buoyant to be subducted (e.g. West of South America along Peru-Chili trench). Where two plates with continents at the leading edges converge, great mountain ranges, such the Himalaya, are formed.
Transform faults are boundaries along which plates slidepast one another, with neither creation nor destruction of lithosphere.
As a result of this theory, heat flow can be calculated which indicates very high heat flow values along the ridges a theoretical result which was not confirmed by experiments :
The deficit in heat flow along the ridges was attributed by some geophysicists, to local large heat flow anomalies such as high temperature water springs.

Some evidence of hot water occurence in deep ocean along the ridges was observed in Galapagos where temperatures of water of about 15-20°C was recorded. (Edmond et al., 1979).

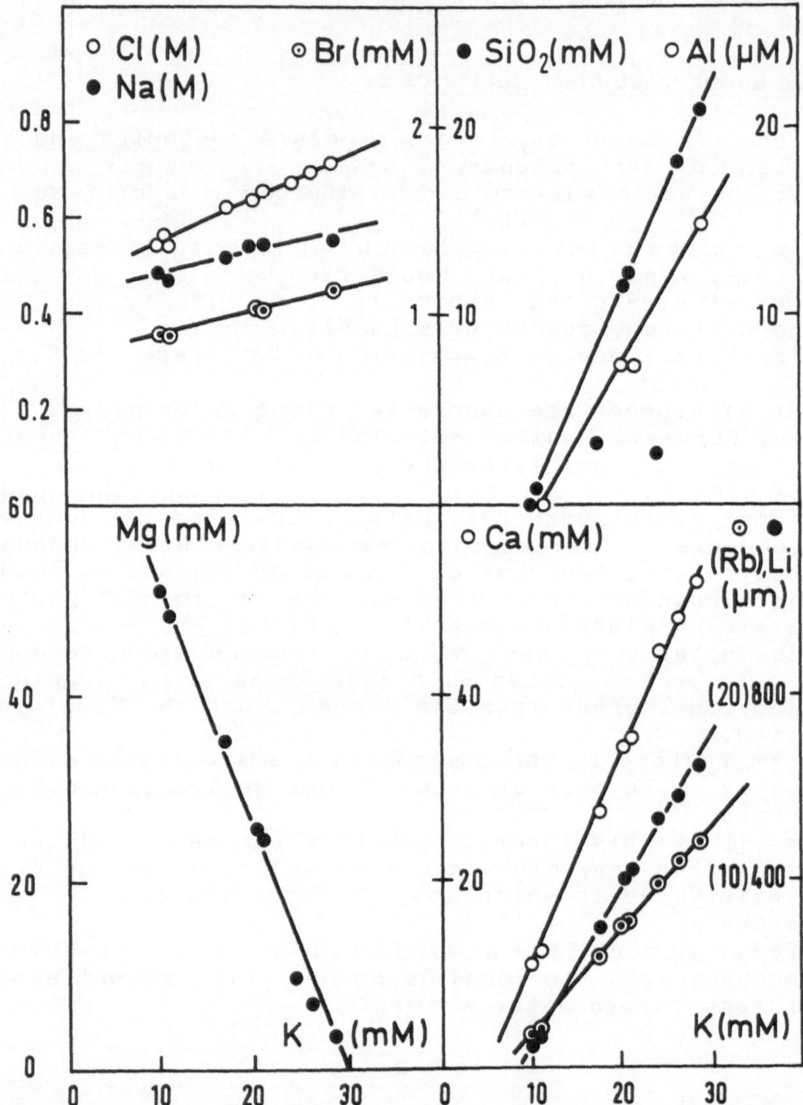

Fig 6. Waters sampled at the same vent are mixtures of hydrothermal water and sea water. "Good" samples have a very low Mg content. The composition of pure end member is rather well defined between the best sample and concentration corresponding to Mg = 0.

And finally springs spouting out at about 350°C were discovered along East Pacific Rise at 21°N in 1979, (Von Damm et al., 1985).
Hydrothermal waters were characterized by a low pH, very large amounts of dissolved silica, iron, manganese ... and by a large depletion in magnesium and sulfate.
The ratio of manganese in hydrothermal water/manganese in sea water was in the order of 10^6. Therefore, Mn is a very suitable tracer for a prospection of hydrothermal vents. It is true also for methane and ^3He
Prospections cruises over the East Pacific Rise led to the discovery of large anomalies of Mn,CH_4 and ^3He near 13°N.
A cruise with the diving saucer "Cyana" in 1982, allows the discovery of tens of vents with a temperature 300°C. This area was revisited in 1984 (Michard et al. 1984; Grimaud et al. 1984).
Water was sampled with a 1.5 litre graphite syringue with an inner coating of pyrolitic carbon. The syringue is filled through a graphite nozzle which can be directly inserted in the vent orifice.
All samples consist from a mixture of hydrothermal water and sea water, resulting in linear plots of one element versus another. In the less mixed samples, Mg concentration was less than 5% of the sea water value. By convention, the hydrothermal end member corresponds to (Mg)=0.

An important number of elements have been analysed in hydrothermal vents and the figure 7 groups the most interesting results.
Generally, hydrothermal waters contain more dissolved elements than seawater, but Mg, $SO_4^=$ and U are almost completely removed from hydrothermal waters.
The mechanisms of Mg removal are :
a) formation of magnesium hydroxy sulfate $Mg_3(SO_4)_2(OH)_2$
b) reaction with silica contained in the basalt and formation of a magnesium silicate.
These two reactions yield a large decrease of the pH of the water which is only partially cancelled by the reaction with the basalt.
From the deficit of heat flow on the ridges, a total amount of

$$1.5 * 10^{11}\ m^3/an$$

of water at 350° can be inferred.
This figure led to a sink of $8 * 10^{12}$ moles of Mg/year which can be compared with an input by river of 6.10^{12} moles/year.

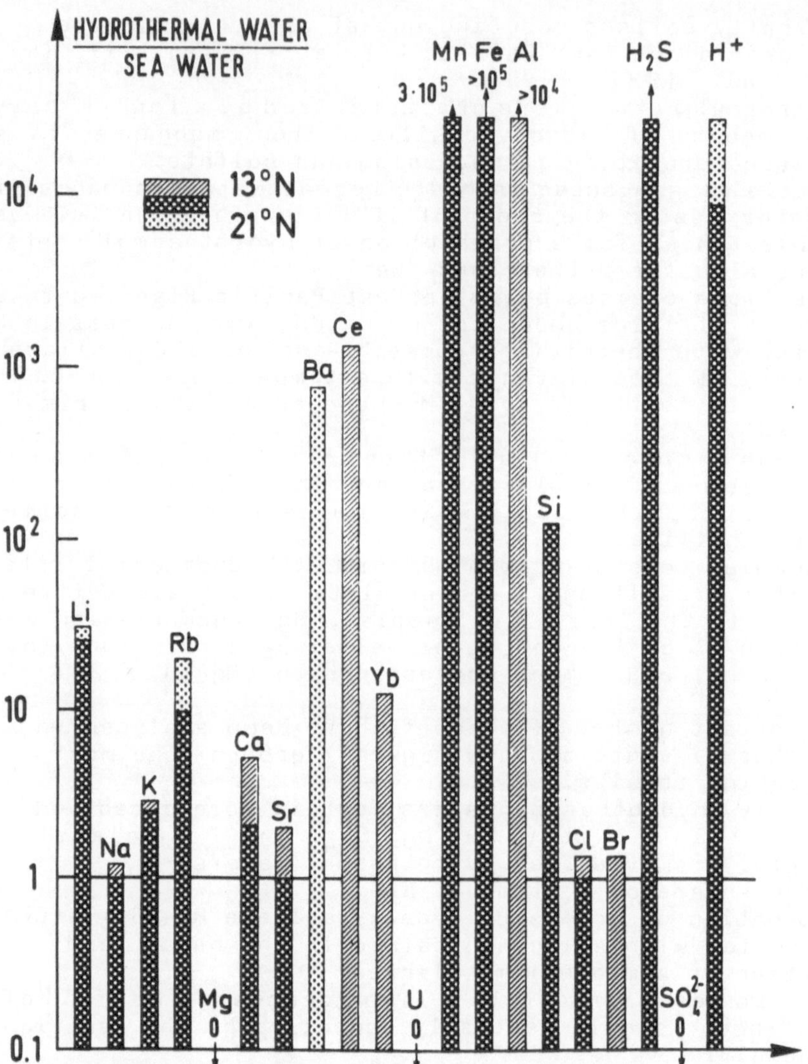

Fig 7. Comparison of chemical compositions of hydrothermal water and sea water. Some differences are observed between the two first sites which were studied : East Pacific Rise 21°N (Von Damm et al., 1985) and 13°N (Michard et al. 1983, 1984)

3.2. Temperature and pressure

High temperature water results from an interaction of sea water with basalt. Basalt was just solidified and its temperature was in the order of 1000 - 1100°C. Interaction with sea water at 0° is therefore performed at a temperature depending on the water/rock ratio.
We can use different chemical and isotopic indicators to try to determine the mean temperature of reaction.
Dissolved silica is known to equilibrate with quartz in the 150 - 400°C temperature range. Below about 280°C, solubility of quartz is only temperature dependent. As deposition of quartz during coaling is a slow process, silica content of the water can be used as a geothermometer. In the 300 - 400°C temperature range, quartz solubility is also pressure dependent. Fig. 8 gives the 21, 23 and 25 mmoles isopleths in a T,P diagram. The observed value at 13°N is 23 mmoles.
Isotopic fractionation between different soluble species (such as C isotopes between CO_2 and CH_4) can be used as indicator of temperature, as the ΔV_R for the isotopic exchange reaction is 0. In the case of 13°N the temperature calculated from C for CO_2/CH_4 is 375°C.
Thus, the silica content allows to determine pressure in the range 500 - 600 bars, corresponding to a depth of about 3 km below sea-basalt interface if pressure is assumed to the hydrostatic (open fractures).

3.3. Water/rock ratio

Temperature obtained in the previous section corresponds roughly to a 1/1 water/rock ratio. It means that 1 kg of water extracts heat from about 1 kg of rock.
This ratio which is important to predict behavior of elements in the reaction can be approached by other simplistic considerations.
Some elements such as alkali trace (Li Rb Cs) are quantitatively extracted from the rock ; the W/R ratio will be derived from the content of Li Rb Cs in both hydrothermal water and sea water if the amount of water loss by hydration of the rock can be estimated.
This amount can be derived from deuterium data.
Knowing isotopic fractionation between water and hydroxylated minerals (chlorite, epidote), D value measured in the waters led a value of water loss (of about 3 to 5 percent).

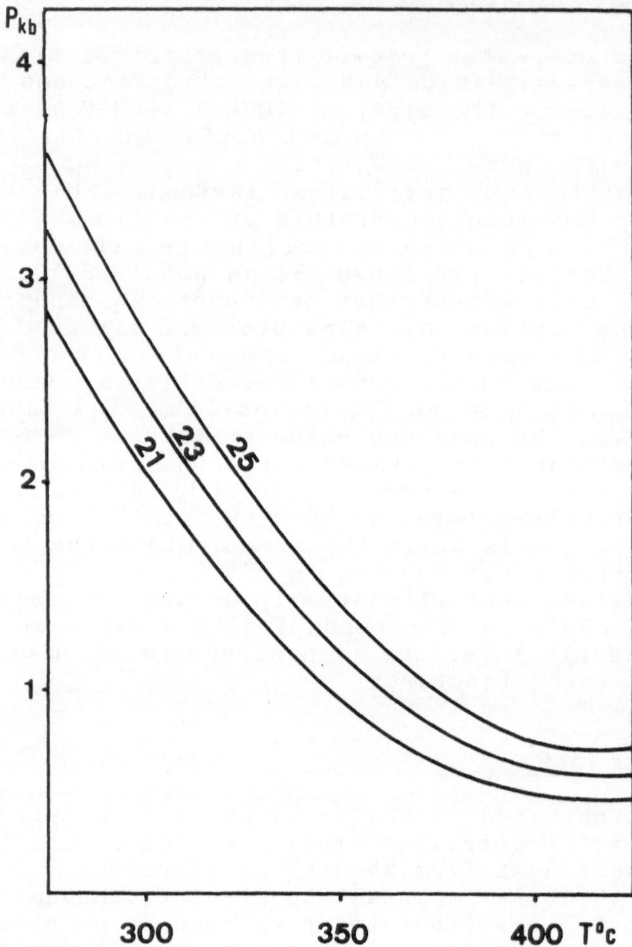

Fig 8. Isopleths (curves of constant concentration) of silica in a T,p diagram. 23 mmoles curve corresponds to waters at 13°N. From a temperature of 375°C calculated from C isotopic ratio in CO_2 and CH_4, a pressure of about 600 bars can be derived.

From these values, Li Rb K content of hydrothermal waters led to water/rock ratio in the range of 2 to 5.
Other estimation can be done with oxygen or strontium isotopes. For instance, ^{87} ^{86} Sr is about 0.709 in sea water and 0.703 in basalts. Concentration are about 8 and 300 ppm respectively. The basic assumption is that mixing of strontium from the two origins is complete ; after that, precipitation does not fractionate isotopes, from the observed Sr isotopes ratio, a W/R ratio of about 1.5 can be calculated.
Although all these estimates are simplistic, many of them are independent and yield the same order of magnitude. Furthermore, an additional limit can be provided by the $SO_4^=$ decrease ; this is related with $SO_4^=$ reduction by ferrous iron contained in the basalt, a maximum W/R ratio of 6 can be inferred.
With this low W/R ratio, is not possible to extract from the rock significant amount of elements sparingly soluble in this water (such as Rare Earth Elements).

3.4. The problem of chloride

Hydrothermal water, sampled at 21°N have a chloride concentration similar to sea water.
An unexpected result was obtained at 13°N where Cl- concentrations as high as 0.90 M/l (sea water concentration is 0.56 M.) were observed.
Such an increase cannot be explained neither by Cl extraction from the rock (basalts contain only small amounts of chlorine) nor by the water loss associated with rock hydration (which is less than 5-6%). More recently, strong depletions in the chloride content (0.35 M) was observed in other area.
Different hypotheses have been proposed to explain this Cl anomaly :
(a) in its deep course, water crosses the two phases boundary and a vapor-like phase separates from a brine. This implies a temperature greater than 405°C, the critical temperature of sea water. It means that water has an higher temperature at depth than this measured at the vent. This can be related with a cooling due to adiabatic expansion during arise of the water. The phase separation will led to a gas depletion in the brine ; this depletion was not observed in Cl rich waters of 13°N and hypothesis of unmixing is falsified.

(b) in sea water-basalt interaction experiments, Seyfried et al. (1985) observed strong reversible adsorption of chloride ion on the solid phase. Chloride rich solutions can result from the leaching of solids which have previously adsorbed chloride ions ;
(c) occurence of rock with an anomalous content of chloride or input of Cl-rich magmatic gases. This hypothesis implies a non steady state for Cl in the oceans and cannot explain Cl depletions.
Today, the problem is still open. The origin of chloride in many continental hot waters is an open problem too.

4. CONCLUSIONS

The two examples presented here can be completed by many other studies : surface and underground cold fresh waters, continental thermomineral waters, fluid inclusions, etc. All of these will lead to the same conclusion : our knowledge in physical chemistry of natural waters increased dramatically in the last ten years, but we need for an increasing knowledge especially on the nature of chemical species actually present in concentrated solutions typically at high temperature and pressure and on mechanisms (and kinetics) of dissolution and precipitation reactions.

REFERENCES

BERNER R.A. (1980). Early diagenesis. A theoretical approach. Princeton publ. 241pp.

EDMOND J., MEASURES C., MCDUFF R.E., CHAN L.H., COLLIER R., GRANT B., GORDON L.I. and CORLISS J.B. (1979). Earth Planet. Sci. Letters, **46**, 1.

ELDERFIELD H., MCCAFFREY R.J., LUEDTKE N., BENDER M. and TRUESDALE V.W. (1981). Am. J. Sci, **281**, 1021.

GAILLARD J.-F. (1982). Thesis, University of Paris (VI).

GAILLARD J.-F., PAUWELS H. and MICHARD G. (1986). Geochim. Cosmochim. Acta (submitted).

GRIMAUD D., MICHARD A. and MICHARD G. (1984). C. R. Acad. Sci. Paris, **299** serie II, 865.

MICHARD A., ALBAREDE F., MICHARD G., MINSTER J.-F, CHARLOU J.-L. (1983). Nature, **303**, 795.

MICHARD G. (1971). J. Geophys. Res., **76**, 2179.

MICHARD G., ALBAREDE F., MICHARD A., MINSTER J.-F., CHARLOU J.-L. and TAN N. (1984). Earth Planet. Sci. Letters, **67**, 297.

MOREL F.M.M. (1982). Principles of aquatic chemistry. 446pp. J. Wiley

PRESS F. and SIEVER R. (1982). Earth, 612pp. Freemann and Co.

SCHINK D.R., GUINASSO N.L. and FANNING K.A. (1975). J. Geophys. Res., **80**, 3013.

STUMM W. and MORGAN J.J. (1981). Aquatic chemistry. 582 pp. J. Wiley.

VON DAMM K.L., EDMOND J., GRANT B., MEASURES C.I., WALDEN B. and WEISS R.F. (1985). Geochim. Cosmochim. Acta, **49**, 2197.

THEORETICAL CALCULATIONS OF IONIC SOLUTIONS

D. LEVESQUE
Laboratoire de Physique Théorique et Hautes Energies
Université de Paris XI - Bâtiment 211
91405 ORSAY
France

ABSTRACT. In this lecture, the results recently obtained by two theoretical methods for the computation of the equilibrium and dynamical properties of ionic solutions are presented.

The first method allows an accurate calculation of the thermodynamic properties of an ionic solution with a solvent of polarizable molecules by a mean field treatment of the many body polarizability effects proposed by Carnie and Patey

The second method is the well-known molecular dynamics method applied to the calculation of the frequency dependent dielectric constant ($\varepsilon(\omega)$) and conductivity ($\sigma(\omega)$).

1 - MEAN FIELD TREATMENT OF MOLECULAR POLARIZABILITY.

In order to simplify the presentation of the theory, this will be done for the case of the pure solvent. The extension of the theory to ionic solution is very simple.

As a specific example, a model system of molecules with a dipole moment $\vec{\mu}$, a quadrupole $\overleftrightarrow{Q}$ and a polarizability $\overleftrightarrow{\alpha}$ is considered where the molecular interactions other than electrostatic are represented by a short-range potential $v_{SR}(r_{ij})$.

The instantaneous configurational energy of the system is :

$$U = \frac{1}{2}\sum_{i\neq j=1}^{N} v_{SR}(r_{ij}) - \frac{1}{2}\sum_{i=1}^{N} \vec{m}_i\cdot\vec{E}_i^{(m)} - \sum_{i=1}^{N} \vec{m}_i\cdot\vec{E}_i^{(Q)}$$

$$- \frac{1}{2}\sum_{i\neq j}^{N} \frac{1}{9}\, \overleftrightarrow{Q}_i : \overleftrightarrow{T}_{ij}^{(4)} : \overleftrightarrow{Q}_j + \frac{1}{2}\sum_{i=1}^{N} \vec{p}_i\cdot(\vec{E}_i^{(m)} + \vec{E}_i^{(Q)})$$

$\vec{E}_i^{(m)}$ and $\vec{E}_i^{(Q)}$ are respectively the local electric field due to dipole moments and quadrupole moments of the other molecules at molecule i :

M.-C. Bellissent-Funel and G. W. Neilson (eds.), The Physics and Chemistry of Aqueous Ionic Solutions, 399–408.

$$\vec{E}_i^{(m)} = \sum_{j\neq i=1}^{N} \nabla_i\nabla_i \frac{1}{|\vec{r}_i-\vec{r}_j|} \cdot \vec{m}_j ,$$

$$\vec{E}_i^{(Q)} = -\frac{1}{3}\sum_{j\neq i=1}^{N} \nabla_i\nabla_i\nabla_i \frac{1}{|\vec{r}_i-\vec{r}_j|} : \overleftrightarrow{Q}_j$$

and $\overleftrightarrow{T}_{ij}^{(4)}$ is equal to $\nabla_i\nabla_i\nabla_i\nabla_i(1/|\vec{r}_i-\vec{r}_j|)$.

The total dipole moment of a molecule i, $\vec{m}_i$, is the sum of the permanent dipole moment $\vec{\mu}_i$ and of the induced dipole moment $\vec{p}_i$:

$$\vec{m}_i = \vec{\mu}_i + \vec{p}_i = \vec{\mu}_i + \overleftrightarrow{\alpha}\cdot(\vec{E}_i^{(m)} + \vec{E}_i^{(Q)})$$

The average dipole moment $<\vec{m}_i>$ calculated in the molecular coordinate system, is equal to : $<\vec{m}_i> = \vec{\mu}_i + \overleftrightarrow{\alpha}\cdot <\vec{E}_i>$. For a linear molecule or molecules with C_{2v} symmetry, $<\vec{E}_i>$ is non-zero only in the direction of $\vec{\mu}_i$; with the notation $<\vec{m}_i> \equiv \vec{m}'_i = m'\frac{\vec{\mu}_i}{|\mu|}$ (m' is identical for all molecules), $<\vec{E}_i>$ can be written : $<\vec{E}_i> = C(m')\vec{m}'_i$. C(m') is a scalar function dependent upon m' and other parameters (Q, density, ...). The expression of m'_i in term of C(m') is :

$$\vec{m}'_i = \vec{\mu}_i + C(m')\overleftrightarrow{\alpha}\cdot(1-\overleftrightarrow{\alpha}C(m'))^{-1}\cdot\vec{\mu}_i$$

The last equation is equivalent to the pair of equations :

$$\vec{m}'_i = \vec{\mu}_i + C(m')\vec{\mu}_i\cdot\vec{\alpha}' \qquad (1)$$

$$\overleftrightarrow{\alpha}' = \overleftrightarrow{\alpha} + C(m')\overleftrightarrow{\alpha}\cdot\overleftrightarrow{\alpha}' \qquad (2)$$

Using the definition of m' and defining $C_m(m')$ and $C_Q(m')$ from $<\vec{E}_i^{(m')}> = C_{m'}(m')\vec{m}'_i$ and $<\vec{E}_i^{(Q)}> = C_Q(m')\vec{m}'_i$, the average value of U is given by :

$$<U> = U_{SR} + U_{QQ} - \frac{1}{2}N\mu m'C_{m'}(m') - N\mu m'C_Q(m') - \frac{1}{2}\sum_i <\vec{p}_i\cdot\vec{E}_i^{(Q)}>$$

where U_{SR} and U_{QQ} are respectively the average values of $\sum_{i<j=1}^{N} v_{SR}(r_{ij})$ and $\sum_{i<j=1}^{N} \overleftrightarrow{Q}_i:\overleftrightarrow{T}_{ij}^{(4)}:\overleftrightarrow{Q}_j$. This expression of $<U>$ is exact and the first approximation of the mean field treatment of polarizability effects by Carnie and Patey is to replace $<\vec{p}_i\cdot\vec{E}_i^{(Q)}>$ by

$$<\vec{p}_i\cdot\vec{E}_i^{(Q)}> \simeq <\vec{p}_i>\cdot<\vec{E}_i^{(Q)}> = (m'_i-\mu_i)\cdot C_Q(m')m_i = m'(m'-\mu)C_Q(m')$$

and consequently < U > becomes equal to

$$< U > = U_{SR} + U_{QQ} - \frac{1}{2} N\mu m' C_{m'}(m') - \frac{1}{2} N\mu m'(m'+\mu)C_Q(m')$$

The important remark is that this expression of < U > depends on density ρ, temperature T, $\overleftrightarrow{Q}$, $\vec{\mu}$ and upon $\overleftrightarrow{\alpha}$ through the average value of $|<\vec{m}_i>| = m'$, (notice that if $\overleftrightarrow{Q}$ is zero, the expression of < U > above is exact). This dependency of < U > upon $\overleftrightarrow{\alpha}$ suggested to Carnie and Patey[1] to substitute to U an effective energy U_{eff} depending upon $\overleftrightarrow{Q}$, $\vec{\mu}$ and a permanent effective dipole $\vec{m}_e$. The value of $\vec{m}_e$ should be chosen in order to be consistent with the average values of the electric field $\vec{E}_i^{(m)}$ and $\vec{E}_i^{(Q)}$ associated to U_{eff}.

In U, the fluctuating dipoles occur in three ways $\vec{m}_i$, $\vec{m}_i \cdot \vec{m}_j$ and $\vec{m}_i \cdot \overleftrightarrow{Q}_j$. Carnie and Patey propose to replace $\vec{m}_i \sim m_o \vec{\mu}_i/\mu$ (m_o identical for all molecules), $\vec{m}_i \cdot \vec{m}_j \sim < m_i^2 > \vec{\mu}_i \cdot \vec{\mu}_j$ and $\vec{m}_i \cdot \overleftrightarrow{Q}_j \sim |< m_i^2 >| \vec{\mu}_i \cdot \overleftrightarrow{Q}_j$. The expression of $< \vec{m}_i^2 >$ is

$$< \vec{m}_i^2 > = \vec{\mu}_i^2 + 2 < \vec{p}_i > \cdot \vec{\mu}_i + < \vec{p}_i^2 > = \vec{m}'^2 + < \vec{p}_i^2 > - < \vec{p}_i >^2$$

Following Pratt[2], Carnie and Patey write $<\vec{p}_i^2> - <\vec{p}_i>^2 = 3\alpha' k_B T$. With the hypothesis that $\overleftrightarrow{\alpha}$ and $\overleftrightarrow{\alpha}'$ have the form $\alpha\overleftrightarrow{U}, \alpha'\overleftrightarrow{U}$ ($\overleftrightarrow{U}$ unit tensor) and k_B Boltzmann constant), then $< \vec{p}_i^2 > = 3\alpha' kT + (m_o - \mu)^2$. Making the approximation $m_e = |< \vec{m}_i^2 >| \sim m_o$, the expression of U_{eff} is :

$$U_{eff} = \frac{1}{2} \sum_{i \neq j=1}^{N} v_{SR}(r_{ij}) - \frac{1}{2} \frac{m_e^2}{\mu^2} \sum_{i \neq j=1}^{N} \vec{\mu}_i \cdot \overleftrightarrow{T}_{ij}^{(2)} \cdot \vec{\mu}_j$$
$$- \frac{m_e}{\mu} \sum_{i=1}^{N} \vec{\mu}_i \cdot \overleftrightarrow{T}^{(3)} : \overleftrightarrow{Q}_j - \frac{1}{2} \sum_{i \neq j=1}^{N} \frac{1}{9} \overleftrightarrow{Q}_i : \overleftrightarrow{T}_{ij}^{(4)} : \overleftrightarrow{Q}_j + U_{pol}$$

where $U_{pol} = \frac{1}{2} N < \vec{p}^2 > /\alpha$ is a constant. Using U_{eff}, the values of $C_{m_e}(m_e)$ and $C_Q(m_e)$ are calculated, one finds :

$$C_{m_e}(m_e) = - \frac{2 < U_{DD} >}{m_e^2 N}, \quad C_Q(m_e) = - \frac{< U_{DQ} >}{m_e^2 N}$$

where U_{DD} are the energy between the dipoles and U_{DQ} the energy between the dipoles and the quadrupoles. The value of $\vec{m}_e$ which is consistent with the given values of $\vec{\mu}$, $\overleftrightarrow{Q}_T$, $\overleftrightarrow{\alpha}$ at fixed values of T and ρ, is obtained by the following process. A first estimate of m_e (m_e^o) is used for computing $C_m(m_e^o)$ and $C_Q(m_e^o)$ from $U_{eff}(m_e^o)$, which leads to a new estimate of m_e, and α' by the relations (1) and (2). The process is iterated and it is stopped when self-consistent values of m_e and α' are found.

The self-consistent mean field (SCMF) theory of Carnie and Patey is similar in spirit to the theories proposed by Wertheim[3], and Hoye and Stell[4]. Recently it has been tested using Molecular Dynamics (MD) simulations for the calculation of the self-consistent value of m_e[5]. The system considered was composed by polarizable molecules interacting by a v_{SR} potential equal to a Lennard Jones potential of parameters ε_{LJ} and σ_{LJ}. The system was at density $\rho\sigma_{LJ}^3 = 0.8$, temperature $k_BT/\varepsilon_{LJ} = 1.35$, with $(\mu^2/\varepsilon_{LJ}\sigma_{LJ}^3)^{1/2} = 1.5$, $(Q^2/\varepsilon_{LJ}\sigma^5)^{1/2} = 0.72$ and $\alpha\sigma_{LJ}^3 = 0.05$. The agreement for different thermodynamic quantities between the MD results for the polarizable system and SCMF theory results is rather satisfactory (see table 1).

Another test of the SCMF theory has been done for an ionic solution composed of a solvent identical to the pure solvent system described above[6]. The ions interacted with the solvent molecules by a L.J. potential and had a charge q such that $(q^2/\varepsilon_\mu\sigma_\mu) = 8$ (ε_μ and σ_μ : L.J. parameters of the solvent). The ratios σ_i/σ_μ and $\varepsilon_i/\varepsilon_\mu$ (σ_i and ε_i : L.J. parameters of the ions) were 1.47 and 1. Due to the limitation on computer time presently only the first step of the iterative determination of m_e was done in using $m_e^o = 1.92$. The agreement for the ion energies obtained by MD simulation and SCMF theory is rather good and should be improved by the convergence of the iterative process, (see table II).

In conclusion, the proposal of taking into account the many body polarizability effects by an effective dipole is certainly not a new idea, but the SCMF theory provides a well defined procedure to calculate it and it gives a good estimate of the thermodynamic properties of a polarizable solvent.

2 - MICROSCOPIC CALCULATION OF $\varepsilon(\omega)$ AND $\sigma(\omega)$

The calculation of $\varepsilon(\omega)$ and $\sigma(\omega)$ by MD simulation supposes that the problem of relating these experimental quantities to correlation functions which are the quantities computable by a numerical simulation is solved. This relation is shown to depend upon the boundary conditions of the system due to the fact that, for instance, the time dependent dipole-dipole correlation function has a long range spatial extent.

Fulton[7] provides a general formalism to establish the expression of $\varepsilon(\omega)$ and $\sigma(\omega)$ in term of correlation functions. The formalism is applied here in a classical form to the derivation of an expression valid for systems of non-polarizable molecules. The outline is as follows : consider an ionic solution in volume Λ (maybe infinite) which, under the influence of an external field $\vec{E}_o(\vec{r},t)$, acquires a macroscopic polarization $\vec{P}$ and a macroscopic current density $\vec{j}$; $\vec{P}$ and $\vec{j}$ are related to the electric Maxwell field $\vec{E}$ inside Λ by the constitutive relations :

$$4\pi \vec{P}(\vec{r},\omega) = [\varepsilon(\omega)-1] \vec{E}(\vec{r},\omega) \qquad (3)$$

$$\vec{j}(\vec{r},\omega) = \sigma(\omega) \vec{E}(\vec{r},\omega) \qquad (4)$$

where $\vec{P}(\vec{r},\omega)$ and $\vec{j}(\vec{r},\omega)$ are the time Fourier Laplace transforms of $\vec{P}(\vec{r},t)$ and $\vec{j}(\vec{r},t)$. In the relations (3) and (4) the dielectric tensor $\overleftrightarrow{\varepsilon}(\vec{r},\vec{r}',t)$ and the conductivity tensor $\overleftrightarrow{\sigma}(\vec{r},\vec{r}',t)$ are assumed to be local and to have the form

$$\overleftrightarrow{\varepsilon}(\vec{r},\vec{r}',t) = \varepsilon(t)\delta(\vec{r}-\vec{r}')\overleftrightarrow{U} \qquad r \in \Lambda$$
$$\overleftrightarrow{\sigma}(\vec{r},\vec{r}',t) = \sigma(t)\delta(\vec{r}-\vec{r}')\overleftrightarrow{U} \qquad r \in \Lambda$$

and
$$\varepsilon = 1 \qquad r \notin \Lambda$$
$$\sigma = 0 \qquad r \notin \Lambda$$

These last relations are valid for a system surrounded by vacuum, but ε and σ could be taken equal to the dielectric constant and conductivity of any type of continuum.

The polarization $\vec{P}$ and the current density $\vec{j}$ are the average values in presence of $\vec{E}_o$ of the microscopic polarization and current $\vec{P}$ and $\vec{j}$, respectively equal to

$$\vec{P}(\vec{r},t) = \sum_{i=1}^{N_s} \vec{\mu}_i(t)\delta(\vec{r}-\vec{r}_i(t)) \text{ and } \vec{j} = \sum_{i=1}^{N_i} q_i \vec{v}_i\delta(\vec{r}-\vec{r}_i(t))$$

(N_s number of solvent molecules and N_i number of ions).

The averages $< \vec{P}(\vec{r},t) >_{E_o}$ and $< \vec{j}(\vec{r},t) >_{E_o}$ are calculated by the linear response theory, which provides an expression of $\vec{P}(\vec{r},t)$ and $\vec{j}(\vec{r},t)$ in term of $\vec{E}_o(\vec{r},t)$ and correlation functions of $\vec{P}$ and $\vec{j}$ (susceptibilities). Then the relations between the susceptibilities and $\varepsilon(\omega)$ and $\sigma(\omega)$ are obtained by eliminating $\vec{E}_o$ in favor of $\vec{E}$.

The interaction Hamiltonian H_I between the ionic solution and $\vec{E}_o(\vec{r},t)$ is :

$$H_I = -\frac{1}{c}\int_\Lambda d\vec{r}\; \vec{j}(\vec{r},t)\cdot\vec{A}(\vec{r},t) + \int_\Lambda d\vec{r}\; \rho_c(\vec{r},t)\phi(\vec{r},t)$$
$$- \int_\Lambda d\vec{r}\; \vec{P}(\vec{r},t)\cdot\vec{E}_o(\vec{r},t)$$

where $\vec{A}(\vec{r},t)$ and $\phi(\vec{r},t)$ are the vector and scalar potentials associated with $\vec{E}_o(\vec{r},t)$ and $\rho_c(\vec{r},t) = \sum_{i=1}^{N_i} q_i\, \delta(\vec{r}-\vec{r}_i(t))$.

Martin[8] has shown that the linear response theory is gauge invariant, this allows to choose $\phi \equiv 0$. Then a standard application of linear response theory leads to :

$$\vec{P}(\vec{r},\omega) = \int d\vec{r}'\, \overleftrightarrow{\chi}_{Pj}(\vec{r},\vec{r}',\omega)\cdot\frac{\vec{A}}{c}(\vec{r}',\omega) + \int d\vec{r}'\, \overleftrightarrow{\chi}_{PP}(\vec{r},\vec{r}',\omega)\cdot\vec{E}_o(\vec{r},\omega)$$

$\overleftrightarrow{\chi}_{Pj}$ and $\overleftrightarrow{\chi}_{PP}$ are the average values of the Poisson brackets of $\vec{P}$ and $\vec{j}$:

$$\chi^{\alpha\beta}_{Pj} = -\int_0^\infty dt\; e^{i\omega t} < \{ P^\alpha(\vec{r},t), j^\beta(\vec{r}',0)\}> ,$$

$$\chi^{\alpha\beta}_{PP} = - \int_0^\infty dt\; e^{i\omega t} <\{P^\alpha(\vec{r},t),\; P^\beta(\vec{r}',0)\}>$$

in the absence of $\vec{E}_o(\vec{r},t)$.

The expression of $\vec{j}$ (cf. Martin[8]) is :

$$\vec{j}(\vec{r},\omega) = \int d\vec{r}'\; \overleftrightarrow{\chi}_{jj} \cdot \frac{\vec{A}(\vec{r}',\omega)}{c} + \int d\vec{r}'\; \overleftrightarrow{\chi}_{jP} \cdot E_o(\vec{r}',\omega) - \omega_P^2 \frac{\vec{A}(\vec{r},\omega)}{c}$$

where $\omega_P^2 = < \sum_{i=1}^{N_i} \frac{q_i^2}{m_i} \delta(\vec{r}-\vec{r}_i) >$ (plasma frequency).

Now a condensed notation is introduced ; an expression similar to

$\int d\vec{r}'\overleftrightarrow{\chi}_{PP}(\vec{r},\vec{r}',\omega)\cdot\vec{E}_o(\vec{r}',\omega)$ will be written $\overleftrightarrow{\chi}_{PP}\bullet\vec{E}_{o,\omega}$

Using $\vec{E}_o = -\frac{1}{c}\frac{\partial}{\partial t}\vec{A}(\vec{r},t)$ and also the definitions $\vec{j}_T = \vec{j} + \frac{\partial\vec{P}}{\partial t}$, $\overleftrightarrow{C}_{Pj_T} = \frac{1}{c}\overleftrightarrow{\chi}_{Pj_T}$ and $\overleftrightarrow{C}_{jj_T} = \frac{1}{c}\overleftrightarrow{\chi}_{jj_T} - \frac{\omega_P^2}{c}\,\mathbb{1}$

($\mathbb{1}$ is the tensor $\delta(\vec{r}-\vec{r}')\overleftrightarrow{U}$), the expressions of $\vec{P}_\omega$ and $\vec{j}_\omega$ become

$$\vec{P}_\omega = - i\,\frac{c}{\omega}\,\overleftrightarrow{C}_{Pj_T}\bullet\vec{E}_{o,\omega} \quad\text{and}\quad \vec{j}_\omega = -i\,\frac{c}{\omega}\,\overleftrightarrow{C}_{jj_T}\bullet\vec{E}_{o,\omega} \qquad (5)$$

The constitutive relations rewritten with the notation $\overleftrightarrow{D}_P = \frac{i\omega}{c}\frac{\varepsilon(\omega)-1}{4\pi}\,\mathbb{1}$ $\overleftrightarrow{D}_j = \frac{i\omega}{c}\sigma(\omega)\,\mathbb{1}$ and $\overleftrightarrow{D}_{j_T} = \overleftrightarrow{D}_j - i\omega\overleftrightarrow{D}_P$, take a form similar to (5)

$$\vec{P}_\omega = -\frac{ic}{\omega}\overleftrightarrow{D}_P\bullet\vec{E}_\omega,\; \vec{j}_\omega = -\frac{ic}{\omega}\overleftrightarrow{D}_j\bullet\vec{E}_\omega \quad\text{and}\quad \vec{j}_{T,\omega} = -\frac{ic}{\omega}\overleftrightarrow{D}_{j_T}\bullet\vec{E}_\omega \qquad (6)$$

The elimination of E_ω in favor of $E_{o,\omega}$ is made by solving the Maxwell equation :

$$(-\frac{\omega^2}{c^2} - \nabla^2 + \nabla\,\nabla)\vec{E}_\omega = -\frac{4\pi\,\omega_i}{c^2}(\vec{j}_{T,\omega} + \vec{j}_{o,\omega}) \qquad (7)$$

where $\vec{j}_{o,\omega}$ is the external current associated with $\vec{E}_{o,\omega}$.

A formal solution of (7) is :

$$\vec{E} = \frac{i\omega}{c}\overleftrightarrow{G}_o\bullet(\vec{j}_{T,\omega} + \vec{j}_{o,\omega}) = \vec{E}_{o,\omega} + \frac{i\omega}{c}\overleftrightarrow{G}_o\bullet\vec{j}_{T,\omega}$$

where $\overleftrightarrow{G}_o$ is equal to :

$$\overleftrightarrow{G}_o = [\overleftrightarrow{U} + (\frac{c}{\omega})^2 \nabla \nabla] \frac{e^{i\omega|\vec{r}-\vec{r}'|}}{c|\vec{r}-\vec{r}'|}$$

and is the Green's function of equation (7). From (6), $\vec{E}_\omega$ is easily obtained :

$$\vec{E}_\omega = [\mathbb{1} + \overleftrightarrow{G}\bullet\overleftrightarrow{D}_{j_T}]\bullet\vec{E}_{o,\omega} \qquad (8)$$

where $\overleftrightarrow{G} = \overleftrightarrow{G}_o\bullet(\mathbb{1} - \overleftrightarrow{D}_{j_T}\bullet\overleftrightarrow{G}_o)^{-1}$, $\overleftrightarrow{G}$ depends on boundary conditions through $\overleftrightarrow{D}_{j_T}$. Neglecting the propagating modes associated to $\overleftrightarrow{G}_o$ the expression of $\overleftrightarrow{G}_o$ is :

$$\overleftrightarrow{G}_o \simeq -\frac{4\pi}{c}(\frac{c}{\omega})^2 \overleftrightarrow{\delta}_L(\vec{r}-\vec{r}')$$

with $\overleftrightarrow{\delta}_L = -\frac{1}{4\pi}\nabla_r \nabla_{r'} \frac{1}{|\vec{r}-\vec{r}'|}$ and $\overleftrightarrow{G}$ takes the form

$$\overleftrightarrow{G} \simeq \overleftrightarrow{G}_s = -\frac{4\pi}{c}(\frac{c}{\omega})^2 \overleftrightarrow{\delta}_L\bullet(1 + \Sigma\, \overleftrightarrow{\delta}_L)^{-1}$$

where $\Sigma(\omega) \equiv \frac{4\pi i}{\omega}\sigma_T(\omega) = \varepsilon(\omega) - 1 + \frac{4\pi i\sigma(\omega)}{\omega}$

From (7), the relations between susceptibilities and $\varepsilon(\omega)$ and $\sigma(\omega)$ are easily obtained :

$$\overleftrightarrow{C}_{Pj_T} = i\,\omega\,\frac{\varepsilon(\omega)-1}{4\pi}[\delta(\vec{r}-\vec{r}')\overleftrightarrow{U} + \overleftrightarrow{G}(\vec{r},\vec{r}',\omega)\frac{i\omega}{c}\sigma_T(\omega)] \qquad (9)$$

$$\overleftrightarrow{C}_{jj_T} = \frac{i\omega}{c}\sigma(\omega)[\delta(\vec{r}-\vec{r}')\overleftrightarrow{U} + \overleftrightarrow{G}(\vec{r},\vec{r}',\omega)\frac{i\omega}{c}\sigma_T(\omega)] \qquad (10)$$

Because the susceptibilities are not known as functions of r and r' but rather as an average value on a domain of $D \subset \Lambda$, the equation (9) and (10) are integrated on Ω volume of D. For spherical or cubic shapes of D, and using the general relation between average values of Poisson brackets and correlation functions in the canonical ensemble $<\{A,B\}> = \beta<\dot{A}\,B>$ ($\beta = (k_BT)^{-1}$, $\dot{A}$ time derivative of A), and the definitions of $\vec{M}_D$ and $\vec{J}_D$,

$$\vec{M}_D = \sum_{i\in D}\vec{\mu}_i(t) \quad\text{and}\quad \vec{J}_D = \sum_{i\in D} q_i\,\vec{v}_i(t)$$

the formulas (9) and (10) lead to :

$$\frac{\beta}{3\Omega}[<\vec{M}_D^2> + i\omega<\vec{M}_D\cdot\vec{M}_D>_\omega + <\vec{M}_D\cdot\vec{J}_D>_\omega]\overleftrightarrow{U}$$

$$= \frac{\varepsilon(\omega)-1}{4\pi}[\overleftrightarrow{U} + \frac{1}{\Omega}\int_D d\vec{r}d\vec{r}'G(\vec{r},\vec{r}',\omega)\frac{\omega^2}{4\pi c}\Sigma(\omega)]$$

$$\frac{\beta}{3\Omega}[<\vec{J}_D\cdot\vec{J}_D>_\omega + i\omega <\vec{J}_D\cdot\vec{M}_D>_\omega]\overleftrightarrow{U}$$

$$= \sigma(\omega)[\overleftrightarrow{U} + \frac{1}{\Omega}\int_D d\vec{r}\, d\vec{r}'\, \overleftrightarrow{G}(\vec{r},\vec{r}',\omega)\, \frac{\omega^2}{4\pi c}\, \Sigma\ (\omega)]$$

An explicit expression of the integral $\int_D d\vec{r}d\vec{r}'\ \overleftrightarrow{G}(\vec{r},\vec{r}',\omega)$ can be found in the approximation $\overleftrightarrow{G} = \overleftrightarrow{G}_s$, for three boundary conditions : infinite system, spherical systems embedded in a continuous medium of dielectric constant $\varepsilon^*(\omega)$ and conductivity $\sigma^*(\omega)$, and periodic boundary conditionswith an Ewald summation of Coulomb interactions.

For an infinite system $\overleftrightarrow{G}_s = -\frac{4\pi}{c}(\frac{c}{\omega})^2\frac{1}{1+\Sigma}\overleftrightarrow{\delta}_L$ as was shown by Fulton[7], then (11) and (12) become :

$$\frac{\varepsilon(\omega)-1}{4\pi}[1-\frac{1}{3}\frac{\Sigma(\omega)}{1+\Sigma(\omega)}] = \frac{\beta}{3\Omega}[<\vec{M}_D^2> + i\omega <\vec{M}_D\cdot\vec{M}_D> + <\vec{M}_D\cdot\vec{J}_D>]_\omega$$

$$\sigma(\omega)\ [1-\frac{1}{3}\frac{\Sigma(\omega)}{1+\Sigma(\omega)}] = \frac{\beta}{3\Omega}[<\vec{J}_D\cdot\vec{J}_D>_\omega + i\omega <\vec{J}_D\cdot\vec{M}_D>_\omega]$$

The expressions of (11) and (12) for a periodic boundary condition with an Ewald evaluation of coulombic interactions are :

$$\frac{\varepsilon(\omega)-1}{4\pi} = \frac{\beta}{3\Omega}[<\vec{M}_D^2> + i\omega <\vec{M}_D\cdot\vec{M}_D>_\omega + <\vec{M}_D\cdot\vec{J}_D>_\omega]$$

$$\sigma(\omega) = \frac{\beta}{3\Omega}[<\vec{J}_D\cdot\vec{J}_D>_\omega + i\omega <\vec{J}_D\cdot\vec{M}_D>_\omega]$$

because Neumann[9] has proven in this case

$$\int_D d\vec{r}'\ \overleftrightarrow{G}_s(\vec{r},\vec{r}',\omega) = 0 \quad \text{and then } \vec{E}_\omega = \vec{E}_{o,\omega}$$

in the limit ω equal to 0, $\varepsilon(0)$ and $\sigma(0)$ are respectively, for an infinite system and a system with periodic boundary condition with Ewald interactions :

$$\sigma(0) = \frac{3}{2}\frac{\beta}{3\Omega} <\vec{J}_D\cdot\vec{J}_D>_{\omega=0},\quad \frac{\varepsilon(0)-1}{4\pi} = \frac{3}{2}\frac{\beta}{3\Omega}[<\vec{M}_D^2>+<\vec{M}_D\cdot\vec{J}_D>_{\omega=0}]$$

and $$\sigma(0) = \frac{\beta}{3\Omega}<\vec{J}_D\cdot\vec{J}_D>_{\omega=0},\quad \frac{\varepsilon(0)-1}{4\pi} = \frac{\beta}{3\Omega}[<\vec{M}_D^2> + <\vec{M}_D\cdot\vec{J}_D>_{\omega=0}]$$

This last expression used for calculating $\varepsilon(0)$ in a MD simulation of an ionic solution of 256 molecules (10 ions and 246 solvent molecules) identical to the system described in the section 1 leads to the result[6] :

$$\frac{\varepsilon(0)-1}{4\pi} \simeq \frac{\beta}{3\Omega} < \vec{M}_D^2 >$$

because the correlation function $< \vec{J}_D(t)\cdot\vec{M}_D(0) >$ is zero in the statistical uncertainties of the simulations.

It is worthwhile to remark that this value of $\varepsilon(0)$ equals the "static dielectric constant" α_s of an ionic solution as defined by Adelman[10] ($(\varepsilon_s - 1) = 4\pi\beta < \vec{M}_D^2 >/3\Omega$).

The conclusions of the second part of this lecture are : (i) is the result $\varepsilon(0) \simeq \varepsilon_s$ a general result for ionic solution ? (ii) in the interpretation of experimental data for $\sigma_T(\omega)$ in terms of microscopic correlation functions the use of formulae taking into account boundary condition is essential.

TABLE I

Comparison between internal energies of a system of polarizable particle calculated by MD simulation and SCMF theory (ε dielectric constant)

	MD	SCMF
$< U_{DD} + U_{DQ} + U_{Pol} >/(\varepsilon_{LJ}N)$	- 7.46	- 7.28
$< U_{QQ} >/\varepsilon_{LJ}N)$	- 1.20	- 1.16
$U_{SR}/(\varepsilon_{LJ}N)$	- 4.09	- 4.15
m'_e	1.87	1.86
α'	0.062	0.062
ε	26 ± 3	27 ± 2

TABLE II.

Same as in table I but for an ionic solution. U_{qq} is the ion-ion energy, U_{qD} is the ion-dipole energy and U_{qQ} is the ion-quadrupole energy (ε dielectric constant in the Adelman sense, see text).

	MD	SCMF
$U_{qq}/(N\varepsilon_{LJ})$	- 0.67	- 0.75
$U_{qD}/(N\varepsilon_{LJ})$	- 1,28	- 1.12
$U_{qq}/(N\varepsilon_{LJ})$	- 0.26	- 0.29
ε(Adelman)	20 ± 2	24 ± 2

ACKNOWLEDGMENTS

I thank J.M. Caillol and J.J. Weis for fruitful collaboration.

REFERENCES

1. S.L. Carnie and G.N. Patey, Mol. Phys. 47, 1129 (1982)
2. L.R. Pratt, Mol. Phys. 40, 347 (1980)
3. M.S. Wertheim, Mol. Phys. 25, 211 (1973)
4. J.S. Hoye and G. Stell, J. Chem. Phys. 73, 461 (1980)
5. G.N. Patey, D. Levesque and J.J. Weis, Mol. Phys. 57, 337 (1986)
6. J.M. Caillol, D. Levesque and J.J. Weis, J. Chem. Phys. 85, 6645 (1986)
7. R.L. Fulton, J. Chem. Phys. 68, 3089, 3095 (1978)
8. P.C. Martin in Many-Body Physics, edited by C. DeWitt and R. Balian (Gordon and Breach, New York 1968) p. 37
9. M. Neumann, Mol. Phys. 50, 841 (1983)
10. S.A. Adelman, J. Chem. Phys. 64, 724 (1976).

COMPUTER SIMULATION OF ELECTROLYTE SOLUTIONS

P. Turq
Laboratoire d'Electrochimie UA 430. Tour 74.
Université Pierre et Marie CURIE
8 Rue Cuvier. 75005 Paris.
France

ABSTRACT. The developments of computer simulation in physics and chemistry allows for the treatment of some practical problems of electrolyte solutions, such as that of structural and dynamic properties. Ionic association processes, such as those appearing in transport and thermodynamic properties, can even be formulated in terms of rate constants of the corresponding chemical reactions. This analysis holds for multiply charged ions, as well as polyions.

1. INTRODUCTION

The methods of computer simulation, as applied to physical and chemical problems have been extensively developed in the last years, and allow now for the treatment of relatively complex multicomponent systems such as electrolyte solutions.

As regards the microscopic simulation of statistical mechanical problems, the most commonly used methods are the following:

- The molecular dynamics (MD)
- The Monte-Carlo method (MC)
- The Brownian dynamics (BD)
- The alternative methods

The molecular dynamics method start from a classical Hamiltonian and gives the time and space dependent properties for an assembly of interacting particles. The Monte-Carlo method is restricted to space dependent properties and does not use explicitly the time variable. The specific field of chemistry covers especially the chemical reactions and the behaviour of multicomponent systems. Since most of such difficult problems are out of the range of possibilities for present computers, some simplified treatments have been proposed. As an example, if we consider aqueous solutions, most of the water models do not take account explicitly of the self-ionization of the water molecule. If we go a step further, solvent averaged models can be used, in which

M.-C. Bellissent-Funel and G. W. Neilson (eds.), The Physics and Chemistry of Aqueous Ionic Solutions, 409–415.

only the solute species are considered explicitly. That is the basis of the so-called Brownian dynamics [1][2] . Alternative methods were introduced, in which part of the solvent molecule is considered explicitly, as in molecular dynamics, the bulk being treated as a continuous medium, as in Brownian dynamics [3][4]. In the case of electrolyte solutions, as in other ionic systems, the occurence of long range coulombic interactions leads to a supplementary difficulty. Since the practical limitation in system size has introduced some peculiar techniques to extend the apparent number of molecules under consideration, such as the use of periodic boundary conditions, a natural way to treat properly coulombic interactions is to use the methods established by crystallographers for ionic solids, as the Ewald method. A presentation of this method for the computation of electrostatic interactions in computer simulation is given in [5].

The main differences between the computer simulations of electrolyte solutions come from the level of description for the solvent:

- The discrete solvent models wich take into account the full molecular nature of the solvent can be treated either by molecular dynamics or by Monte-Carlo techniques. At normal concentra-concentration of electrolyte solutions (≤ 1 mole/liter) there is a large difference between the number of solute molecules and that of solvent: e.g. a 1M aqueous solution contains 55 solvent molecules per molecule of solute. The explicit consideration of the discrete molecular nature of the solvent has therefore been restricted to systems in which it plays a direct role for the properties under consideration. For chemical reactions in ionic solutions we have to mention the work of Jorgensen [6][7], who treated by Monte-Carlo methods the problem of nucleophilic substitutions (SN2 reactions).

- The solvent averaged models in which only the solute particles are described explicitly. These particles interact by the means of a solvent averaged Hamiltonian, the MacMillan-Mayer Hamiltonian [8][9]. For structural and thermodynamic properties, the Monte-Carlo method can be used in a natural way and gives a reference for integral equations and other approximation methods [10][11].

If one wishes to consider explicitly time dependent properties such as transport coefficients or rates of chemical reactions between the solute particles, e.g. pair formation reactions, the MacMillan-Mayer Hamiltonian is insufficient, since it does not involve solute-solvent interactions and in particular solute solvent collisions, which fix the dynamics of solute particles, even at infinite dilution. A simple way to introduce these solute-solvent collisions is to consider the limiting value of the friction coefficient of the solute particles at infinite dilution. Most transport coefficients are related to this friction coefficient. The self-diffusion coefficient, for example, is given by:

$$D/\Omega = kT$$

where D is the diffusion coefficient of the solute species, whose mobility Ω, the reciprocal of the friction coefficient and kT is the Boltzmann factor.

If one introduces a mechanism in which friction forces are balanced by random forces which thermalize the solute particles, one has then a "Brownian machinery" in which the interactions between the solute particles can be treated by using the solvent averaged MacMillan-Mayer hamiltonian. This is the starting point of all Brownian dynamics methods [1][2][12].

In the next paragraphs we will present the main features of solvent averaged models of electrolyte solutions and of Brownian dynamics methods. We will then give some examples of application of these methods to ionic association problems, for multiply charged electrolytes as well as polyelectrolyte solutions. The principle of evaluation of the rate constant for these reactions will then be given.

2. CONTINUOUS SOLVENT MODELS AND BROWNIAN DYNAMICS

In order to emphasize the features of continuous solvent models we will refer to solute-solute interactions in discrete solvent models. For simplicity we will consider central forces potentials.

In such a system, the potential of interaction between two ions i and j is given by:

$$V_{ij}=V^{SR}_{ij}+V^{CB}_{ij}$$

$$\text{with } V^{CB}_{ij}=Z_iZ_je^2/r_{ij}$$

The total potential is the sum over a short range repulsive and a coulombic repulsive or attractive contribution. It should be noticed that this coulombic part does not involve any contribution of the solvent.

If we newt consider a solvent averaged system, not only the short range repulsive part is modified in order to take into account the solvation of the ions, but also the coulomb part is replaced by:

$$V^{CB}_{ij}=Z_iZ_je^2/\epsilon r_{ij}$$

We see that the coulomb interaction is divided by a factor ϵ which is the dielectric constant of the solvent at zero frequency.

The use of this form for the coulomb part of the solvent averaged solute-solute interaction was first introduced in an heuristic way in the Debye Huckel theory and has been justified at least for long distances (10 Å for 1-1 elecrolytes in aqueous solutions). It is clear that for intermediate distances the coulomb interactions are not completely screened and have no reason to be divided by the macroscopic dielectric constant. The usual way to treat the physico-chemical properties of electrolytes is to adjust the parameters of the short range interactions in order to fit some thermodynamic or nonequilibrium properties such as activity or osmotic coefficients or sometimes electrical conductance.

The simplest way to treat the short range interactions is the so-called primitive model of electrolyte solutions, in wich the short range repulsion is treated as a hard-sphere interaction.

If the ionic radii are taken as equal, we have the so-called restricted primitive model. If one wishes to treat the dynamics of such systems, the mixture of long range coulombic and short range hard sphere interactions is not very convenient and usually the hard sphere repulsion is replaced by a continuous potential in $1/r^n$. It is clear that the main progress which can occur from discrete solvent models will be a realistic, even if not real, potential of average force between the solute ions. Most of the works in this field are devoted to this problem [13].

The basic tool in the Brownian dynamics is the Langevin equation, which is presented here in its simplest form:

$$p'_i(t)=-p_i(t)/\Omega_i m_i+R_i(t)+F_i(t)$$

The left hand term is an acceleration term. The first term of the right hand side represents the friction force, which is balanced by the random force $R_i(t)$. The last term represents the contribution of the interionic forces. As indicated above these interionic forces involve both short and long range interactions.

There are several kinds of Brownian dynamics:

a) If the inertial term is neglected [12][14], one has the so-called Smoluchowski dynamics, which allows the use of large time steps if the short range interations are neglected.

This method leads to tractable results if one has to consider repulsive solute particles in a background, as for polyelectrolytic or micellar objects in an excess of supporting electrolyte. It is insufficient, in its simplified form, to study ion association between two ions of opposite charge, as well as counterion condensation on polyions.

The algorithm of integration becomes in this case:

$$\underline{r}(t+\delta t)-\underline{r}(t)=\Omega[\underline{R}(t)+\underline{F}(t)]\delta t$$

δt is the time step and the acceleration term can be neglected only if:

$$\delta t \gg m\Omega$$

This condition implies that only long range, slowly varying forces (coulomb) are under consideration.

The above equation can then be integrated step by step and one obtains the finite differences counterpart of a Smoluchowski equation for single particle motion.

b) If the inertial terms are not neglected, the Brownian dynamics appears as a variety of molecular dynamics in which the Newton equation is replaced by a Langevin equation (supplementary friction and random forces). Since molecular dynamics simulation consists of replacing the Newton equation by a difference equation and integrating by stepwise integration, one has to do the same for Brownian dynamics. Unfortunately the type of Langevin equation presented above describes instantaneous friction and random forces and is not suitable directly to define a finite differences algorithm.

One introduces then a generalized Langevin equation which is suitable for finite time steps integration:

$$p'(t) = - \int_0^t f(t-s)p(s)ds + R(t) + X(t)$$

In this equation the memory kernel f(t) is related to the time correlation function of the random forces:

$$\langle R(t)R(0)\rangle = \langle p^2\rangle f(t)$$

If one chooses a particular form for the correlation function of the random forces, the same form holds for the memory kernel. In the strict Langevin equation this memory kernel is simply a δ function.

With this Brownian dynamics approach, we shall see in the next paragraph that it is possible to describe the dynamics of ionic association in electrolyte and polyelectrolyte solutions. However, as indicated previously, the alternative between discrete and continuous solvent models is not so drastic. One can use an intermediate approach, i.e. molecular dynamics with stochastic boundary conditions [3][4]. In this model, one distinguishes the internal or reaction region which is treated by full molecular dynamics, from the intermediate or buffer or stochastic region which is treated by stochastic dynamics. The external or secondary region is characterized by repulsion forces. This model has been proved as efficient in the treatment of hydrophobic interactions, but has not yet given a description of the reaction region which could be considered as representative from the bulk and is for the moment strongly dependent on the choice of the boundary conditions.

3. IONIC ASSOCIATION

The Brownian dynamics study of 2-2 electrolytes in water like $ZnSO_4$ exhibits strong evidence for ionic association [2].

The simplest quantity to analyse is then the osmotic coefficient:

$$\Phi = P/P_{ID} \approx 0.5$$

Φ shows than we have about two times less particles than it would be expected in the case of complete dissociation.

The equivalent electrical conductance, defined by the current autocorrelation function, can be analyzed in analogous terms.

These results lead to a simple chemical model:

$A^- + B^+ = AB$

This chemical model which discriminates "free" and "paired" ions can be related to the coordination numbers coming from the pair distribution function:

$$N_{+-}(R)=4\pi\int_0^R g_{+-}(r)r^2dr$$

At short distance this function is very small because of the excluded volumes, and at long distance varies as r^3. In the intermediate region it presents often an inflexion point in the case of ionic association. This inflexion point defines the separation between associated ions at short distances and unpaired ions at long distances. This approach of ionic association is identical to the Bjerrum association concept. Moreover BD gives not only the proportion of associated ions (ionic clusters), but also their life-time. It is then possible to analyze the life-time of clusters corresponding to the simplest model of 2-2 electrolytes in water (repulsive potential in $1/r^n$ plus coulomb interaction). For this particular model [2], noticeable triplets proportion can occur in 1M solutions. We shall see that it is possible to express these results in terms of the rate constants of the corresponding chemical model and to separate explicitly the diffusive contributions.

The Brownian dynamics simulation has been applied to a linear polyelectrolyte model in which two consecutive sites on the polyelectrolyte chain are related by an appropriate constraint. The analysis of the counterion motion shows a strong anisotropy of the counterion displacements with respect to the chain orientation, as well as a strong dependence of the chain conformation with its charge density. All these results constitute a generalization of existing polyelectrolyte theories and in particular for counterion condensation of Manning's theory.

4. CLUSTER MODEL FOR RATE CONSTANTS

The relation of computer simulation to the rate constant of chemical reaction is a difficult problem which is under development.

However in the particular case mentioned above, the definition of cluster densities allows for a tractable procedure for the evaluation of these rate constants.

If we define a cluster density (in contrast to a species density which is a given quantity defined by the number of particles introduced in the computer simulation box), it will correspond to the instantaneous number of clusters of a given type (free ions, pairs, triplets....), defined by any arbitrary geometric or energetic criterion as the Bjerrum one mentioned above.

In the computer simulation box with periodic boundary conditions, we have neither density nor charge fluctuations for the ionic species and the only possible fluctuations are the cluster fluctuations. The time correlation function [15] for these fluctuations can be related to the rate constants of the corresponding cluster formation reactions.

REFERENCES

[1] P. TURQ, F. LANTELME & H.L. FRIEDMAN, J. Chem. Phys., **66**,7,3039 (1977)

[2] P. TURQ, F. LANTELME & D. LEVESQUE, Mol. Phys., **37**,1,223 (1979)

[3] M. BERKOWITZ & J.A. Mc CAMMON, Chem. Phys. Letters, **90**,215 (1982)

[4] A.C. BLECH & M. BERKOWITZ, Chem. Phys. Letters, **113**,278 (1985)

[5] S.W. DE LEEUW, J.W. PERRAM & E.R. SMITH, Proc. Roy. Soc. London, **A373**,27 (1980)

[6] J. CHANDRASEKHAR, S.F. SMITH & W.L. JORGENSEN, J.A.C.S., **106**,3049 (1984)

[7] ibid. **107**,154 (1985)

[8] J.C. RASAIAH & H.L. FRIEDMAN, J. Chem. Phys., **48**,2742 (1968)

[9] P.S. RAMANATHAN & H.L. FRIEDMAN, J. Chem. Phys., **54**,1086 (1971)

[10] D.N. CARD & J.P. VALLEAU, J. Chem. Phys., **52**,6232 (1970)

[11] J.C. RASAIAH, D.N. CARD & J.P. VALLEAU, J. Chem. Phys., **56**,248 (1972)

[12] D.L. ERMAK, J. Chem. Phys., **62**,4189,4197 (1975)

[13] CECAM WORKSHOP, 'Computer simulation of Chemical Reactions', Orsay, FRANCE (1985)

[14] T. AKESSON & B. JONSSON, J. Phys. Chem., **89**,2401 (1985)

[15] G. CICCOTTI, P. TURQ & F. LANTELME, Chem. Phys., **88**,333 (1984)

STRUCTURED MONOLAYERS OF CHARGED AND POLAR MOLECULES AT THE LIQUID/AIR INTERFACE

David Andelman[†*], Francoise Brochard[*] and Jean-Francois Joanny[*]
[†]Exxon Research and Engineering Company, Route 22 East, Annandale, NJ 08801, USA
[*]Physique de la matière Condensée, Collège de France, 75231 Paris Cedex 05, France

ABSTRACT. Effective dipoles in charged monolayers and permanent dipoles in neutral ones are shown to have a drastic effect on the structure and phase transitions of insoluble Langmuir monolayers. These long-range and repulsive dipolar interactions stabilize undulating phases in thermodynamic equilibrium. Results are presented for two cases: (i) Close to a liquid-gas critical point. (ii) At low temperatures. Possible implications of the former on the liquid-gas transition and of the latter to the liquid-solid and liquid expanded-liquid condensed transitions are discussed. In an ionic solution, the undulation periodicity can be controlled by the strength of the ionic solution.

1. INTRODUCTION

Monolayers of insoluble amphiphilic molecules, such as surfactants, fatty acids and phospholipids at the liquid/air interface (Langmuir monolayers) have been studied quite extensively over the last sixty years. They are of fundamental interest because they exhibit a rich variety of (quasi-) two dimensional phase transitions. In addition, they are also of applied interest since they serve as a simple model for biological membranes.

Until recently, most of the experiments done on monolayers were isothermal measurements of the surface pressure as a function of the area per molecule [1]. For very low surface pressure (< 0.1 dynes/cm^2) the monolayer behaves like a two-dimensional gas. As the surface pressure increases, the area per molecule decreases monotonicaly and in some cases, e.g. pentadecanoic acid [2], a further increase in the pressure induces a first-order transition to a liquid state. In the coexistence region, the isotherm has a plateau and then, in the liquid, it starts to increase again. This two-dimensional liquid-gas coexistence region which ends with a critical point similar to a bulk liquid-gas transition, was observed for pontadecanoic acid around room temperature [2].

M.-C. Bellissent-Funel and G. W. Neilson (eds.), The Physics and Chemistry of Aqueous Ionic Solutions, 417–427.

In the liquid state, many experiments [1] reported a peculiar "kink" in the isotherms. However, the origin of this singularity is not clear. Furthermore, it is not known whether it separates two phases (so called "liquid expanded" and "liquid condensed") which are in thermodynamic equilibrium. Such a kink in the isotherms can be interpreted as a signature of a second-order transition and was thus explained [3-5] as an orientational ordering of the molecular tails. On the other hand, quite recently Middleton et al [6] found a plateau instead of the kink in some amphiphic systems. Their claim is that a kink indicates presence of impurities, poor control of water vapor pressure, retention of the spreading solvent and nonequilibrium determination of the isotherms.

The controversy of the liquid expanded-liquid condensed transition could not be resolved without having some other structural information on the monolayer. Hence other experimental techniques such as: surface potential measurements [7-9], viscoelastic measurements [10], non linear optics [11], epifluorescence microscopy [12-13] and recently X-ray diffraction from a synchroton source [14-15] were used. The epifluorescence microscopy, for example, allows a direct visualization of the monolayer on length scales of microns. In lipid monolayers, an organization of liquid-like and solid-like regions that repeats periodically is seen for high concentrations. The same epifluorescence technique was also used to study the liquid-gas transition of fatty acid monolayers [16].

In this paper, theoretical calculations that take into account dipole interactions [17-18] are presented. The dipoles can be permanent in neutral monolayers or induced in charged monolayers where an electrical double layer is formed. The intensity of the dipoles can be varied in charged systems by changing the ionic strength of the aqueous solution or by changing the Debye-Huckel screening length. Experimentally, these electrostatic effects are measured by the so-called surface potential (or "zeta potential") [1,7].

2. ELECTROSTATIC INTERACTIONS IN LANGMUIR MONOLAYERS

2.1 Neutral Monolayers

Most neutral surfactant molecules carry a permanent dipole moment. Here we will assume that the dipoles $\vec{\mu}$ are oriented perpendicular to the liquid/air interface and that they are immersed in the liquid close to its surface where locally the dielectric constant is ε (a quantity that is not well known close to the liquid/air interface). Such a dipole density will cause a jump, ΔV, in the electrostatic potential

$$\Delta V = \frac{\mu\phi}{\varepsilon}\,, \qquad (1)$$

where ϕ is the inplane monolayer concentration. By using the appropriate boundary conditions, the electrostatic field $E(r)$ can be calculated (details are found in Ref. [18]) for a monolayer concentration that oscillates with wavevector q, $\phi(x) = \phi_o + \phi_q \exp(iqx)$. The electrostatic free energy of the dipoles subject to this electrostatic field E(r) is

$$F_{el} = -\frac{1}{2}\int \phi\, \vec{\mu}.\vec{E}\, d^2\vec{r} = -\frac{1}{2}\frac{\varepsilon_o}{\varepsilon(\varepsilon+\varepsilon_o)}|q|\mu^2\phi_q^2 . \qquad (2)$$

This result, Eq. (2), can also be obtained in terms of image dipoles and is proportional to the Fourier transform in two dimensions of the dipole-dipole interaction which varies as r^{-3}. A strong dielectric constant, $\varepsilon \gg \varepsilon_o$, reduces the dipole interactions, Eq. (2). These interactions are stronger if the dipoles are associated with the molecular tails (above the liquid interface where the dielectric constant is ε_o) rather than with the polar heads that are immersed in the liquid.

2.2 Charged Monolayers

For charged monolayers, we calculate the electrostatic free energy using a linearized Poisson-Boltzmann approximation. The surface charge density is $\sigma = e\phi$, where e is the charge per molecule and the ionic solution is characterized by the Debye-Huckel screening length κ^{-1}. If the charge concentration of the monolayer oscillates with wave vector q, $\delta\phi = \phi_q \exp(iqx)$, the effective screening length in the solution is

$$\kappa'^2 = \kappa^2 + q^2, \qquad (3)$$

and the electrostatic energy of the amphiphilic molecules is [18]

$$F_{el} = \frac{e^2\phi_q^2}{2(\varepsilon\kappa' + \varepsilon_o|q|)} . \qquad (4)$$

Two limits of Eq. (4) can be distinguished at this point:
(i) the limit of small wave vector, $|q| \ll \kappa$, where the charged monolayer is described in terms of an effective dipole moment. This is the limit applicable to strong ionic solutions where the electrostatic free energy is very similar to Eq. (2)

$$F_{el} \simeq \frac{e^2\phi_q^2}{2\,\varepsilon\,\kappa}\left(1-\frac{\varepsilon_o}{\varepsilon}\frac{|q|}{\kappa}\right) . \qquad (5)$$

This case will be employed throughout the remainder of this article.
(ii) In the limit of large wave vector, $|q| \gg \kappa$, the interactions

are Coulomb-like rather than dipole-like. This limit is relevant to low ionic strength solution and one can expect the formation of Wigner crystals similar to the two-dimensional colloidal crystals studied by Pieranski [19].

3. SUPERCRYSTAL STRUCTURE AND PHASE TRANSITIONS

3.1 Ginzburg-Landau Expansion close to a Critical Point

The dipole interactions discussed in the previous section have a drastic effect on the phase transition between two homogeneous phases (e.g. liquid and gas) that can be described close to the critical point by a Ginzburg-Landau expansion. We write an expansion in powers of the order parameter $\psi = \phi - \phi_c$, where ϕ_c is the critical value of the inplane concentration. Including the dipolar contribution, Eq (2), the total free energy is

$$\frac{F}{T} = \int \left\{ \frac{1}{2}\alpha\, \psi^2(\vec{r}) + \frac{1}{4} u\psi^4(\vec{r}) + \Sigma_o^2\, (\nabla\psi)^2 \right\} d^2\vec{r} + \frac{1}{2T} \int \psi(\vec{r})\, g(|\vec{r}-\vec{r}\,'|)\, \psi(\vec{r}\,')\, d^2\vec{r}\, d^2\vec{r}\,' . \tag{6}$$

The first two terms are the usual Landau expansion with $\alpha \sim T-T_c$, and u being a constant. The third term is the lowest order gradient term and expresses the additional cost of inhomogeneity in ψ, where Σ_o is the area per polar head. The last term is exactly the dipolar contribution calculated above with

$$g(r) = \frac{T}{2\pi}\left(\frac{b}{r}\right)^3 = \frac{\varepsilon_o}{\varepsilon(\varepsilon+\varepsilon_o)}\,\frac{\mu^2}{2\pi r^3} . \tag{7}$$

In Fourier space, $\psi(r) = \Sigma_q \phi_q \exp(i\vec{q}.\vec{r})$, and Eq. (6) can be written as

$$\frac{F}{T} = \int \left\{ \frac{1}{2}\alpha\, \psi^2(\vec{r}) + \frac{1}{4} u\psi^4(\vec{r}) \right\} d^2\vec{r} + \frac{1}{2}\Sigma_q \left(\Sigma_o^2\, q^2 - b^3|q|\right)\phi_q^2 , \tag{8}$$

where b is defined in Eq. (7). Due to the competition between the last two terms in Eq. (8), there is an optimum q-vector that minimizes the free energy. Its magnitude is

$$|q^*| = \frac{b^3}{2\,\Sigma_o^2} . \tag{9}$$

Thus, in addition to the homogeneous phases (liquid and gas) we must consider these phases with undulations, Eq. (9). An approximation that can be justified close to a critical point [20-21], considers only the optimal q-vector, $|q| = |q^*|$, in Eq. (8).

Following Garel and Doniach [20] and Brozovskii [21], the two simplest solutions with undulations are:

(i) the stripe (smectic) phase

$$\psi_s = \psi_o + \phi_q \cos qx \tag{10a}$$

(ii) the hexagonal phase

$$\psi_H = \psi_o + \sum_{i=1}^{3} \phi_q \cos(\vec{k}_i \cdot \vec{r}_i) \ , \qquad \text{with } |\vec{k}_i| = q \ , \ \sum_{i=1}^{3} \vec{k}_i = 0 \ . \tag{10b}$$

Substituting the two proposed solutions, Eqs. (10), into the free energy expansion (8), we get the following two rescaled free-energy densities

$$f_s = \frac{\delta}{2} M_o^2 + \frac{1}{4} M_o^4 + M_q^2 (\delta - 1 + 3M_o^2) + \frac{3}{2} M_q^4 \tag{11a}$$

$$f_H = \frac{\delta}{2} M_o^2 + \frac{1}{4} M_o^4 + M_q^2 (3\delta - 3 + 9M_o^2 + 12M_o M_q) + \frac{45}{2} M_q^4 \tag{11b}$$

for the stripe and hexagonal phases respectively, where

$$\delta \equiv \frac{4\alpha}{n^2} \ , \ M_o^2 \equiv \frac{4u}{n^2} \psi_o^2 \ , \ M_q^2 \equiv \frac{u}{n^2} \phi_q^2 \ \text{and} \ n^2 \equiv \frac{b^6}{\Sigma_o^3} \ .$$

The free energy of these undulating phases is compared with the free energy of the homogeneous solution $f_I = \frac{\delta}{2} M_o^2 + \frac{1}{4} M_o^4$.
In Fig. 1, the phase diagram in the reduced temperature δ - reduced average concentration ($M_o \sim \langle\phi\rangle - \phi_c$) plane is shown. The usual coexistence region between liquid and gas ($M_o^2 = \delta$) in the absence of dipolar interactions is largely modified. A novel critical point at $M_o^* = 0$, $\delta^* = 1$, is the termination point of five distinct phases: gas (G), dilute hexagonal (H) that consists of droplets of liquid in gas, stripe (S), inverted dense hexagonal (IH) and liquid (L). All the transition lines below the critical point (M_o^*, δ^*) are first-order, hence four regions of two-phase coexistence exist between the phases. At low enough temperatures, ($\delta < \delta^*$) the stripe (S) and the hexagonal (H, IH) phases disappear, as seen in Fig. 1. We believe that this is a defect of the single mode Ginzburg-Landau expansion that we used here. In the next section, a direct calculation of the free energy at low temperatures will show that the undulating phases

are expected to remain stable even at low temperatures over some range of concentrations.

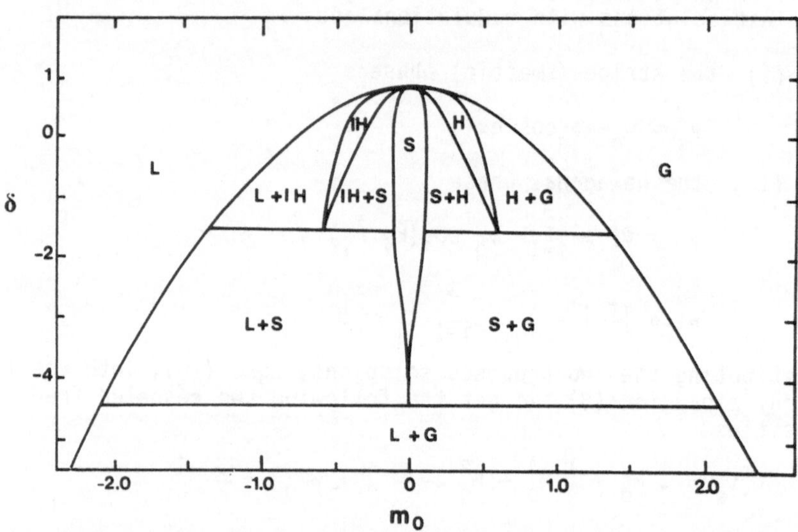

Figure 1. Phase diagram in the (M_o, δ) plane where $\delta \sim T-T_c$ is the reduced temperature and $M_o \sim \langle\phi\rangle-\phi_c$ is the reduced concentration. The two homogeneous phases: liquid (L) and gas (G) are separated by the hexagonal (H), stripe (S) and inverted hexagonal (IH) phases. Two-phase coexistence regions are also indicated. This phase diagram was obtained from Eqs. (11) and is valid only close to $(\delta^*=1,\ M_o{}^*=0)$.

3.2 Undulating Phases at Low Temperatures

It is of interest to compare the relative stability of the undulating and the homogeneous phases at low temperatures; our treatment of the previous section was valid only close to a critical point. In what follows, we make few assumptions that simplifies the formulation; a more detailed calculation can be found in Ref. [18].

The undulating phase is assumed to be a stripe phase with sharp domain boundaries where more condensed regions ("liquid") of size D_ℓ and more dilute regions ("gas") of size D_g alternate, as is shown in Fig. 2. Thus the periodicity of the pattern is $D = D_\ell + D_g$. In principal, one should take into account both the concentration of the liquid-like ϕ_ℓ, and the gas-like regions ϕ_g ; here we will approximate $\phi_g = 0$. (This assumption was not, however, made in Ref. [18]). The dipolar contribution to the total internal energy for this structure is easily derived from Eq. (6)

$$\frac{f_{el}}{T} = \frac{b^3}{\pi a} x\phi_\ell^2 - \frac{b^3}{\pi D} \phi_\ell^2 \left\{ \ln \frac{D}{a} x - \sum_{p=1}^{\infty} \ln \frac{p^2}{p^2 - x^2} \right\}. \quad (12)$$

where $x \equiv D_\ell/(D_\ell + D_g) = D_\ell/D$ and $a \simeq \sqrt{\Sigma_o}$ is a molecular cutoff. The first term in Eq. (12) represents the average contribution of the dipole interactions and is independent of the periodicity D. The second and third terms represent the intra- and inter-stripes electrostatic interactions respectively. However, the leading contribution comes from the interactions within the same stripe; the effect of the inter-stripe interactions is to renormalize the line tension at the liquid/gas interface and following Ref. [22], this term can be summed analytically.

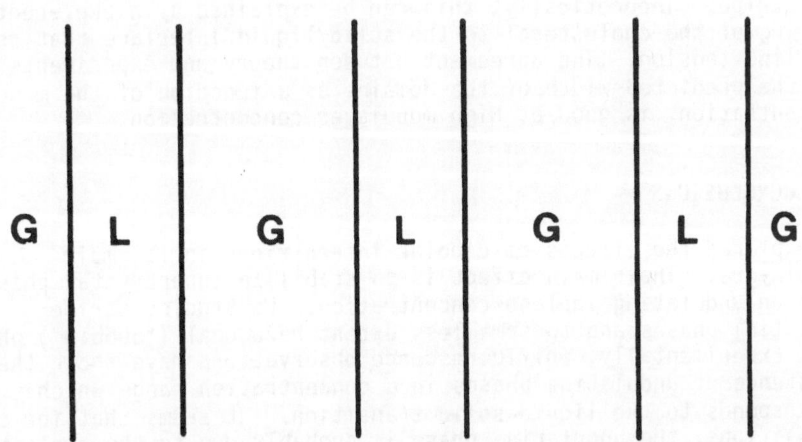

Figure 2. The stripe phase is shown schematically. Domain walls (which are sharp only at low temperatures) separate denser liquid (L) from dilute gas (G).

A further contribution to the free energy difference Δf between the stripe phase and the homogeneous phase comes from the line tension γ that accounts for the concentration variation at each liquid/gas interface. The total free energy difference between the stripe and homogeneous phase is thus

$$\Delta f = -T \frac{b^3}{\pi D} \phi_\ell^2 \left\{ \ln \frac{D}{a} x + \ln \frac{\sin \pi x}{\pi x} \right\} + \frac{2\gamma}{D} . \quad (13)$$

The equilibrium periodicity of the stripe structure is given by the minimum of Δf

$$D^* = \frac{a}{x}\,\frac{\pi x}{\sin \pi x}\,\exp\left(\frac{2\pi\gamma}{Tb^3\,\phi_\ell^2} + 1\right) . \tag{14}$$

The exponential dependence of the periodicity D^* on the ratio between line tension γ and the dipolar interaction coefficient $b^3\phi_\ell^2$ makes it difficult to give accurate estimates of it since neither γ nor b (which depend on the local dielectric constant ε) are accurately known.

In a recent paper, Keller et al [22] presented calculations for the electrostatic free energy that are similar to ours [18] at low temperatures. In addition, they compared their predictions with epifluorescence experiments [23] done on phospholipid monolayers where stripe-like solid and liquid domains coexist. It was also observed that trace amounts of cholesterol can reduce the width of the stripe. Theoretically, this can be explained by a preferential binding of the cholesterol to the solid/liquid interface that reduces the line tension. The agreement between theory and experiments [23], for the predicted width of the domains as a function of the monolayer concentration, is good at high monolayer concentration.

4. CONCLUSIONS

We explored the effects of dipolar interactions in Langmuir monolayers. Their main effect is to stabilize supercrystal phases with an undulating inplane concentration. We studied stripe (smectic) phases and to some less extent hexagonal ("bubble") phases.

Experimentally, epifluorescence observations have shown the existence of undulating phases in a concentration range which corresponds to the liquid-solid transition. It seems that for such transitions, the undulating phase is probably due to the nucleation of two-dimensional solid regions in a liquid background and the observed hexagonal phases are, most likely, a non-equilibrium phenomenon. However, since the sizes of the solid domains depend strongly on the ionic strength of the solution, we believe that the electrostatic interactions play a major role in determining the structure of these undulating phases [17,22].

The theories proposed here both close to a critical point and for low temperatures are more applicable to fluid phases rather that to the solid-liquid region. This can be both in the liquid-gas or the liquid expanded-liquid condensed regions. Due to the low surface pressure and concentration in the former case, fewer effects can be measured (for neutral monolayers). The period of the undulations can be estimated from Eqs. (9) and (14); it varies from 1000Å to 1μm according to the precise magnitude of the dipoles on the surface (or equivalently the ζ potential). It also depends strongly on the local dielectric constant at the interface and the line tension γ which is not very well known.

For charged monolayers, the effective dipole can be tuned by varying the ionic strength in the solution. Our theory gives an explicit dependence of the undulation period and the critical temperature as function of the screening length κ^{-1}. However, this does not represent the entire dependence on the ionic strength, since the charge per molecule e also depends strongly on the ionic strength. This must be taken into account when a comparison with experiments is done.

At the liquid expanded - liquid condensed transition, the model does not give a definite explanation for the kink observed in many experiments. It does, nevertheless, demonstrate the role played by an additional order parameter - in our case, the undulation amplitude. As was explained in the introduction, it is not clear if the experimental isotherms show a kink or a more complicated structure as proposed here. Moreover, the jump in the pressure between the two extreme transitions in Fig. 1 is of order of $\varepsilon\Delta V^2/D$, where ΔV is the potential jump across the monolayer and D is the period of the undulations. Estimating its magnitude we get a very small value of the order of 0.01 dynes/cm.

We also would like to point the connection between the dipolar monolayer and two other magnetic systems: a thin uniaxial magnetic film [24] and a thin layer of ferrofluid [25] both subject to a normal magnetic field. In these magnetic analogs, a competition between dipolar forces and domain wall energies or line tension also destabilizes the homogeneous state of the system. The ferrofluid is more like our system at zero temperature since thermal fluctuations do not play an important role, whereas for the thin magnetic film thermal fluctuations are important. These three systems are remarkably similar although the length scale of the undulations is very different; it can reach a few millimeters in the ferrofluid (subject to magnetic fields of only several hundred Gauss), whereas it can be as small as few hundred Å for the monolayer and in the intermediate micron range for the thin solid magnetic films.

In conclusion, it should be noted that the effect of the long-range dipolar forces is important for most of the monolayer properties and not only for the equilibrium phase diagram as was studied in this work. From an experimental point of view, non-equilibrium phenomena seem to be of particular importance in Langmuir monolayers. Thus, the inclusion of dipolar forces in the kinetics of domain growth is of relevance and will be addressed in a separate study.

REFERENCES

1. For a general review see: A. W. Adamson, Physical Chemistry of Surfaces (Wiley, New York, 1982); G. L. Gaines, Insoluble monolayers at liquid gas interfaces (Wiley, New York, 1966).

2. G. A. Hawkins and G. B. Benedek, Phys. Rev. Lett. **32**, 524 (1974); M. W. Kim and D. S. Cannell, Phys. Rev. Lett. **33**, 889 (1975); M. W. Kim and D. S. Cannell, Phys. Rev. A **13**, 411 (1976).

3. See e.g., G. M. Bell, L. L. Combs, and L. J. Dunne, Chem. Rev. **81**, 15 (1981); J. P. Legre, G. Albinet, J. L. Firpo, and A. M. S. Tremblay, Phys. Rev. A **30**, 2720 (1984) and references therein.

4. I. Langmuir, J. Chem. Phys. **1**, 756 (1933).

5. J. G. Kirkwood, Publ. Am. Assoc. Advmt. Sci. **21**, 157 (1943).

6. S. R. Middleton, M. Iwasaki, N. R. Pallas, and B. A. Pethica, Proc. Royal Soc. (London) **A396**, 143 (1984); N. R. Pallas and B. A. Pethica, Langmuir **1**, 509 (1985).

7. M. Plaisance, Ph.D. Thesis, Univ. of Paris (unpublished); M. Plaisance and L. Terminassian-Saraga, C. R. Acd. Sci. (Paris) **270**, 1269 (1970).

8. M. W. Kim and D. S. Cannell, Phys. Rev. A **14**, 1299 (1976).

9. S. R. Middleton, B. A. Pethica, J. Chem. Soc. Faraday Symp. **16**, 109 (1981).

10. B. M. Abraham, K. Miyano, S. Q. Xu and J. B. Ketterson, Phys. Rev. Lett. **49**, 1643 (1985).

11. Th. Rasing, Y. N. Sen, M. W. Kim, and S. Grub, Phys. Rev. Lett. **55**, 2903 (1985).

12. H. M. McConnell, L. K. Tamm, and R. M. Weis, Proc. Natl. Acad. Sci. **81**, 3249 (1984).

13. M. Losche and M. Mohwald , European Biophysics **11**, 35 (1985); M. Losche , E. Sackmann, and M. Mohwald , Ber. Bunsenges. Phys. Chem. <u>87</u>, 848 (1983); M. Losche and M. Mohwald , J. Phys. Lett. (Paris) **45**, L785 (1984).

14. K. Kjaer, J. Als-Nielsen, C. A. Helm, L. A. Laxhuber, and M. Mohwald , preprint 1986.

15. P. Dutta, J. B. Peng, B. Lin, J. B. Ketterson, and M. Prakash, preprint 1986.

16. B. Moore, C. Knobler, D. Broseta, and F. Rondelez, Faraday Symp. Chem. Soc. **20**, xxx (1985).

17. C. A. Helm, L. Laxhuber, M. Losche, and M. Mohwald , J. Coll. and Pol. Sci., in press.

18. D. Andelman, F. Brochard, P. G. DeGennes, and J. F. Joanny, C. R. Acd. Sci. (Paris) **301**, 675 (1985); D. Andelman, F. Brochard and J. F. Joanny, preprint 1986.

19. P. Pieranski, Phy. Rev. Lett. **45**, 569 (1980).

20. T. Garel and S. Doniach, Phys. Rev. B **26**, 325 (1982).

21. S. A. Brazovskii, Zh. Eksp. Teor. Fiz. **68**, 175 (1975) [Sov. Phy. JETP **41**, 85 (1975)].

22. D. J. Keller, H. M. McConnell, and V. T. Moy, J. Phys. Chem. **90**, 2311 (1986).

23. R. M. Weis and H. M. McConnell, J. Chem. Phys. **89**, 4453 (1985).

24. The equivalent magnetic problem for thin magnetic films has been considered by C. Kooy and U. Enz, Phillips Res. Rep. **15**, 7 (1960).

25. R. E. Rosensweig, M. Zahn, and R. Shumovich, J. Mag. and Mag. Mat. **39**, 127 (1983); For a review on ferrofluids see R. E. Rosensweig, Ferrohydrodynamics, (Cambridge Univ. Press, New-York, 1985).

CHEMICAL EQUILIBRIUM BETWEEN MINERALS AND NATURAL WATERS

Christophe Monnin and Jacques Schott
Laboratoire de Minéralogie et Cristallographie
Université Paul-Sabatier
31062 Toulouse Cédex
France

ABSTRACT. The purpose of this paper is to review some of the models of electrolyte solutions which are currently used to calculate chemical equilibrium between minerals and natural waters.

After a brief description of the ion association approach to the calculation of activity coefficients of aqueous species, we introduce Pitzer's ion interaction model and emphasize its links with the ion pairing phenomenology. The capacity of this model to calculate the thermodynamic properties of complex electrolyte solutions at high concentration is illustrated by two examples of mineral formation in salted lakes.

The last part of the paper addresses the chemistry of metamorphic fluids. It is shown that because of the decrease of the dielectric constant of water, the aqueous ionic species tend to associate into neutral complexes. The description of such fluids may thus be simpler than that for the earth surface conditions.

1. INTRODUCTION

Recently there has been a growing awareness of the role of aqueous phases as agents of the major mass transfers observed in the earth. Movement of fluids are accompanied by continuous chemical reactions as different physico-chemical environments are successively penetrated. Depending on the rate of chemical reactions versus the rate of transport, equilibrium may or may not be achieved in each environment and this is responsible to a large extend for the infinite chemical variety encountered on the earth. Thus the accurate characterization of chemical equilibrium between rocks and natural waters is a requirement for a better understanding of the mechanisms of mineral formation (and dissolution) and for the quantitative modeling of heat and mass transfers in the earth crust.

The aqueous phases involved in geochemical processes are solutions of mainly alkali and alkaline earth chlorides, sulfates and carbonates (Table). Besides the widely different temperature and pressure conditions of their geological environments, these solutions show a broad

M.-C. Bellissent-Funel and G. W. Neilson (eds.), The Physics and Chemistry of Aqueous Ionic Solutions, 429–440.

Table.- Composition of some concentrated natural waters (molalities).

	Chott el Jerid (Tunisia) T = 30°C	Lake Magadi (Kenya) T = 40°C	Geothermal brine Salton-Sea (California) T = 320°C P ≃ 400 bars	Oil-field brine Ekofisk (Norway) T = 140°C P ≃ 500 bars
Na	5.2	7.1	3.1	1.5
K	0.22	$6.0\ 10^{-2}$	0.57	$1.0\ 10^{-2}$
Ca	2.10^{-2}	∿ 0	1.0	0.50
Mg	0.33	∿ 0	$5.5\ 10^{-4}$	$5\ 10^{-2}$
Cl	5.87	2.53	5.90	2.63
SO_4	0.12	$2.5\ 10^{-2}$	∿ 0	$5\ 10^{-4}$
HCO_3	$1.5\ 10^{-3}$	$1.7\ 10^{-2}$	$1.5\ 10^{-2}$	$3.2\ 10^{-5}$
CO_3	∿ 0	2.28	∿ 0	∿ 0
SiO_2	6.10^{-5}	$2.1\ 10^{-2}$	9.10^{-3}	$1.5\ 10^{-3}$
Ionic strength	6.6	9.4	6.7	3.1

range of concentration going from that of river waters to the high density brines of the fluid inclusions trapped in minerals or to the NaCl saturated waters of the salted lakes. Thus any attempt to model chemical reactions and mass transfers in such geochemical environments will have to take into account first the complexity of multicomponent electrolyte solutions in a large spectrum of concentration.

Essential to such modeling calculations are a) general equations which accurately describe the excess free energy of the aqueous solution, and b) an algorithm for solving this set of equations to find the equilibrium composition of the solution in question. Two different approaches meet in the same need of calculating the excess quantities of the aqueous solution. First, given a water sample the composition of which is known, one would like to know the solid phases which are likely to control the concentration of the various solutes. By the calculation of the saturation indices of minerals (i.e. the ratio of the product of the activities of the aqueous species and the solubility product of the mineral formed by these species) we can know the state of saturation of the solution with respect to them (saturated, super- or undersaturated).

On the other hand, in some cases like the study of fossil geothermal systems or metasomatic processes, aqueous solutions have obviously circulated through the host rock and have left behind a number of secondary minerals as a signature of their passage. In this case when the solution cannot be sampled anymore, the understanding of the evolution of the water-rock system up to its present and observable stage goes through the calculation of the solubility of minerals and their stability with respect to the fossil solutions. Besides the calculation of activity coefficients, this also requires algorithms to find the equilibrium configuration of the system examined.

The purpose of the present paper is to review some of the models currently used in geochemistry for the calculation of chemical equilibrium in natural waters, with a particular emphasis on the effect of increasing ionic strength and temperature.

2. THE ION PAIRING APPROACH

The very first attempt to use this approach in geochemistry was made by Garrels and Thomson in the early sixties (5). They applied the concept of ion association to estimate the ionic forms of the major solutes present in sea water as well as their activity coefficients. This was the starting point of a whole generation of the ion pair models still of common use in geochemistry. The main assumption of this approach is that the long range electrostatic forces between two ions of opposite charges can overcome the thermal energy and attract them together to form an individual species. This phenomenon can be accounted for by a chemical equilibrium between the free ions i and j and the ion pair ij.

$$K_{ij} = \frac{a_{i^+} \cdot a_{j^-}}{a_{ij}} \qquad (1)$$

where a stands for activity.

The activity product on the molality scale is

$$\begin{aligned} a_{i^+} \cdot a_{j^-} &= m_{i^+,T} \cdot m_{j^-,T} \gamma^2_{ij,T} \qquad (2) \\ &= m_{i^+,F} \cdot m_{j^-,F} \gamma^2_{ij,F} \end{aligned}$$

The subscript T denotes the total amount of the considered species while F refers to the free ions. Also the total amount of the ion i in the system is the sum of the molalities of the free ion and the molalities of the ion-pairs containing i

$$m_{i,T} = m_{i,F} + \sum_j m_{ij} \qquad (3)$$

Then one needs an expression for the activity coefficients of the free ions as well as the ion-pairs neutral or charged. The most common expressions used in geochemical models are more or less modified Debye-Hückel functions for charged ions. A detailed review of this topic can be found in Whitfield (22).

The solution of equations 1, 2 and 3 gives the sought activity coefficients and the distribution of the species between free ions and ion pairs.

This approach has been and is still widely used in geochemistry Many routines based on this phenomenology like the WATEQ code (21) have been built. One can find in (17) a comparative description of many of these computer programs.

The ability of some of these codes to predict the solubility of minerals in aqueous solutions has been tested (10) against experimental data and it has been shown that they succeed in giving most of the time only the right order of magnitude of the solubilities, which may be sufficient in many cases.

Three main reasons can be found to explain this lack of accuracy. First, most of these programs use more or less modified Debye-Hückel expressions for the activity coefficients of charged species, incorporating the ion size parameter å, which has been shown (7) inconsistent with the Gibbs-Duhem relationship. Thus it is not possible to accurately calculate the activity of water, a quantity of primary importance when dealing with hydrated minerals. Secondly, there is often some inconsistency between the data base of the model (free energy of formation of minerals and association constants for ion pairs) and the selected expression for the activity coefficients of the aqueous species. For example, the activity coefficients at saturation may be different when calculated from the free energy data or by the chosen analytical expression. Third, the most common equations for the activity coefficients, like the Debye-Hückel function, become less accurate when the concentration increases above 0.1M.

All this results in a fair agreement between calculated and experimental results, but nevertheless these models provide useful guidelines for the geochemistry of natural waters.

3. THE ION INTERACTION APPROACH

An alternate approach to the ion pairing phenomenology has been introduced by Guggenheim (7) and extensively developped by Pitzer (see (19) for a review). The excess free energy of the solution is written as the sum of two terms. The first one arises from the contribution of the electrostatic forces while the second one represents the contribution of short range forces which are likely to become important at high concentration

$$\Delta G^{ex} = \Delta G^{ex}_{coul.} + \Delta G^{ex}_{s.r.}$$

Pitzer represented the first term by a Debye-Hückel function, while the other is a series of the molalities by analogy to the virial expansion used for non ideal gases. Then the classical thermodynamic relationships allow to derive expressions of the activity coefficients of the solutes and the activity of water through the osmotic coefficient of the solution. The expression of the activity coefficient of a salt MX in a binary solution is given below.

$$\ln \gamma_{MX} = \left| z_M z_X \right| f_{DH} + m \left(2 \frac{\nu_M \nu_X}{\nu_M + \nu_X} \right) B^{\gamma}_{MX} + m^2 \left(2 \frac{(\nu_M \nu_X)^{3/2}}{\nu_M + \nu_X} \right) C^{\gamma}_{MX} \qquad (4)$$

z et ν are respectively the charge and the stoichiometric coefficient of the considered species ; m is the molality ;

$$f_{DH} = -A_{\phi} \left[\frac{\sqrt{I}}{1 + 1.2\sqrt{I}} + \frac{2}{1.2} \ln(1 + 1.2\sqrt{I}) \right] \qquad (5)$$

with A_{ϕ} the Debye Hückel slope and I the ionic strength $= \frac{1}{2} \sum_i m_i z_i^2$;

$$B^{\gamma}_{MX} = 2\beta^{(0)}_{MX} + \beta^{(1)} g(\alpha_1\sqrt{I}) + \beta^{(2)} g(\alpha_2\sqrt{I}) \tag{6}$$

$$g(x) = \frac{2}{x^2}\left[1 - (1 + x - \frac{x^2}{2})\exp(-x)\right]$$

α_1 and α_2 depend only on the charge of the ions ($\alpha_1 = 2$ and $\alpha_2 = 0$ for 1-1, 2-1 and 1-2 salts ; $\alpha_1 = 1.4$, $\alpha_2 = 12$ for 2-2 salts).

The parameters $\beta^{(0)}$, $\beta^{(1)}$, $\beta^{(2)}$ and C^{γ} (in Eqs. 4 and 6) are specific to each solute and are estimated from experimental data. They are tabulated by Pitzer (19) and Harvie et al. (11) for a large number of salts. Eqs. 4 to 6 allow the accurate representation of experimental values of activity coefficients up to typically 6M.

It has been known for long from statistical mechanics that the quantity B_{MX} is related to the interaction between two dissolved species and hence to their ability to form an ion pair. This has been emphasized for example by Harvie (9) who gave the following simple illustration of the link between the ion association and the ion interaction approaches.

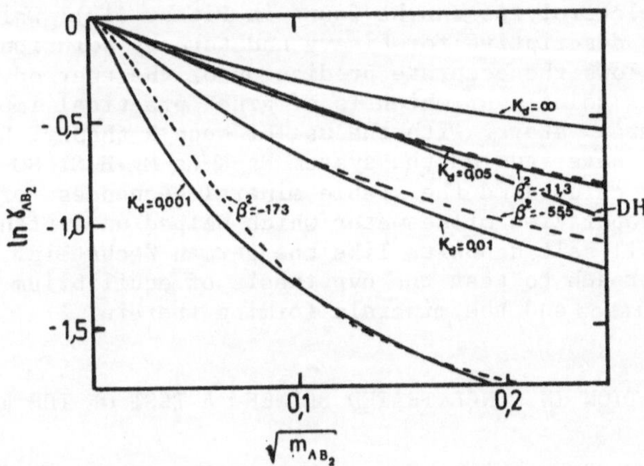

Figure 1. Plot of the calculated dilute solution behavior of the thermodynamic mean activity coefficient for a hypothetical electrolyte AB_2 versus the square root of the total AB_2 concentration (see text ; DH = Debye-Hückel limiting law) [after Harvie (9)].

He considers a fictitious electrolyte AB_2 that forms in the solution the ion pair $AB^+_{(aq)}$, the dissolution constant of which is

$$K_d = \frac{a_{A^{++}} \cdot a_{B^-}}{a_{AB^+}} \tag{7}$$

The activity coefficients of the aqueous species are calculated by

$$\ln\gamma_i = -\frac{Az_i^2\sqrt{I}}{1 + 1.85\sqrt{I}} \tag{8}$$

Then the variation of the mean stoichiometric activity coefficient of the salt γ_{AB_2} can be calculated for various values of K_d, which generates the set of solid curves plotted in Fig. 1. (Note that the more important the association (the lower the K_d value) the larger the deviation from the Debye-Hückel limiting law). Then with a simplified form of Eqs. 4 and 6 (with $\beta^{(0)} = \beta^{(1)} = C^{\gamma} = 0$), one can calculate $\beta^{(2)}$ in order to reproduce the shapes of the solid curves of Fig. 1. The adjusted values of $\beta^{(2)}$ are negative and the larger their absolute values the smaller the dissociation constant K_d. The ion interaction model takes into account the ion pairing phenomenon by the inclusion of interaction parameters in the activity coefficient expression instead of including an explicit chemical equilibrium. It thus extends the notion of strong electrolytes. The main chlorides, sulfates and carbonates can be treated this way. For weaker electrolytes and strongly associating systems (like H_3PO_4, Pitzer and Silvester (20)), ion association has to be considered.

For mixed electrolytes, the effect of mixing is taken into account by the additional parameters θ and ψ which are calculated from experimental data on ternary solutions with a common ion. The general equations for mixed electrolytes can be found in Pitzer (19). While Pitzer's formalism is only descriptive for binary and ternary solutions with a common ion, it allows the accurate prediction of the thermodynamic properties of complex solutions, which is of great practical importance for the study of natural waters. With the use of such a theory, Harvie and others (10), (11) have studied the system Na-K-Ca-Mg-H-Cl-SO_4-HCO_3-CO_3-OH-H_2O at 25° and calculated the stable mineral sequences forming during the isothermal evaporation of seawater which helped understanding the formation of fossil salt deposits like the german Zechstein. We have used the same approach to test the hypothesis of equilibrium between the waters of salted lakes and the minerals forming therein.

4. MINERAL FORMATION IN CONCENTRATED BRINES : A TEST OF THE EQUILIBRIUM HYPOTHESIS

The waters of salted lakes can be classified into two groups : sulfate-chloride brines like the Chott El Jerid (Tunisia) and sodium carbonate waters like Lake Magadi (Kenya). We have used the ion interaction approach to compute saturation indices of various minerals in such waters.

4.1. Gypsum formation in Chott El Jerid (Tunisia)

Gypsum ($CaSO_4$, $2H_2O$) is one of the main constituents of the salt deposit of Chott El Jerid. Its saturation index as well as that of anhydrite and the activity of water, is plotted in figure 2 versus the ionic strength of the waters which concentrate by evaporation from the dilute inflow to highly concentrated ($I \simeq 7.5$) brines. We have used the Harvie-Weare (10) model for the system Na-K-Ca-Mg-Cl-SO_4-H_2O at 25° to calculate these quantities (15). In the calculations we have included a refinement of Pitzer's equations : the variation of the mixing parameters θ_{ij} with

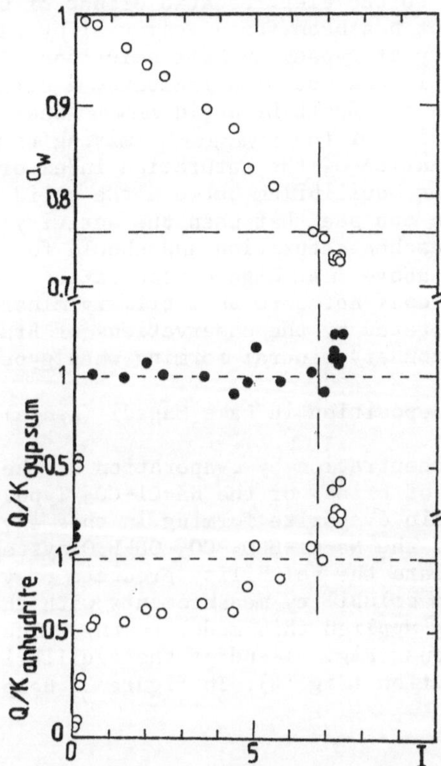

Figure 2. Activity of water and gypsum and anhydrite saturation index in Chott El Jerid [after Monnin (15)].

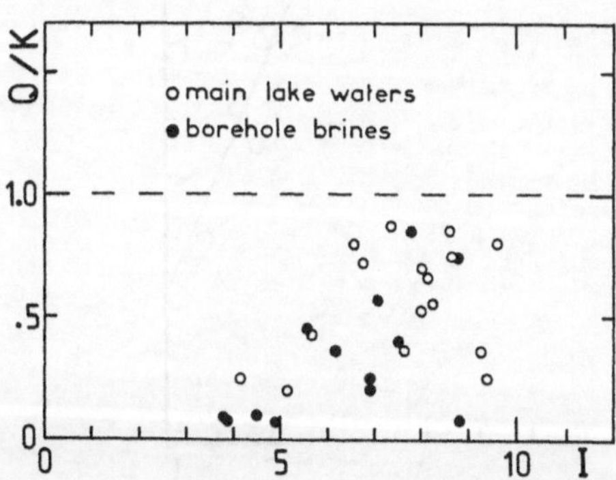

Figure 3. Trona saturation index for Lake Magadi brines [after Monnin and Schott (16)].

the ionic strength due to the electrostatic effect of unsymetrical mixing (18). This effect has been found essential by Harvie and Weare in modelling the solubility of gypsum in NaCl solutions. In previous calculations (6) where it has not been included, a slight super-saturation of gypsum in the Chott El Jerid waters was found, but this is corrected by the inclusion of the asymmetric mixing terms, as shown by figure 2 where the variation of the saturation index of gypsum with the ionic strength indicates equilibrium between the solid and the waters. Also in this figure one can see that when the activity of water is lower than 0.76, anhydrite reaches saturation and should form. But its saturation index increases above 1 at high concentration. This leads to believe that anhydrite does not form as a primary mineral at this temperature. This is corroborated by the observations of Braitsch (2) showing that anhydrite is a secondary mineral forming when gypsum dehydrates.

4.2. Sodium carbonate deposition in Lake Magadi (Kenya)

In Lake Magadi, the concentration by evaporation of the inflow waters leads to the formation of brines of the Na-Cl-CO_3 Type. Trona (Na_2CO_3, $NaHCO_3$, $2H_2O$) is the main evaporite forming in this environment. We have in a first step studied the Na-Cl-HCO_3-CO_3-OH-H_2O system between 0 and 50°C in order to calculate the solubility products of various sodium carbonate minerals from solubility measurements with the use of Pitzer's equations (16). We then applied this model to the calculation of the saturation index of trona (Fig. 3) and of the equilibrium partial pressure of CO_2 of the solution (Fig. 4). In figure 3, despite some scatter

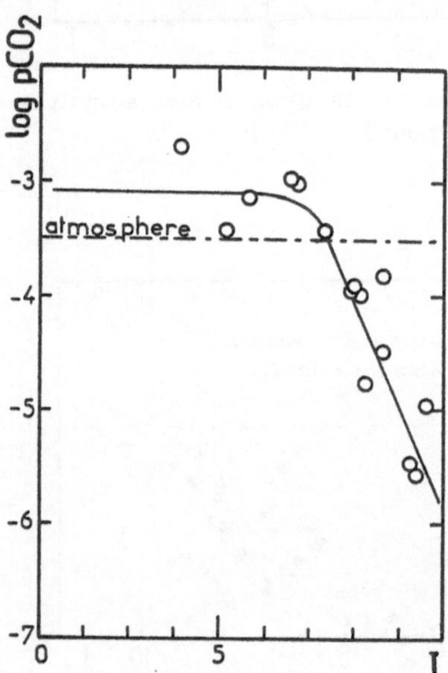

Figure 4. Equilibrium CO_2 pressure for Lake Magadi surface waters [after Monnin and Schott (16)].

of the points due to a lack of accuracy of the bicarbonate determination (13), one can see that Trona reaches equilibrium for an ionic strength of about 7. Then (Fig. 4), the equilibrium pCO_2 starts decreasing rapidly because of the consumption of the carbonated aqueous species by trona formation. Also it appears that diffusion of CO_2 from the atmosphere into the waters is too slow to compensate the bicarbonate decrease. This shows a case of equilibrium between the solid (Trona) and the solution but not between the atmosphere and the lake waters.

These two examples of mineral formation illustrate the need of accurate modelling of the thermodynamic properties of natural waters in low temperature geochemistry and the conclusions that can be drawn from such a modelling. The extension of the ion interaction model to high temperatures and pressures is still a subject of current research.

5. THE MODELLING OF AQUEOUS FLUIDS UNDER METAMORPHIC CONDITIONS

Under metamorphic conditions, as a consequence of the dramatic decrease in the dielectric constant of H_2O with increasing temperature at constant pressure and/or decreasing pressure at constant temperature, solutes ionization constants become much smaller (12) and ion association equilibria have to be substituted to Pitzer's approach. In the last part of this paper, we will present results on recent experiments on the solubility of minerals at high temperatures which bring new insights for the speciation of solutes of important rock - forming components such as silica, alumina and alkalis.

A spectacular illustration of changes in the speciation of solutes under metamorphic conditions can be seen on Fig. 5 where the increase

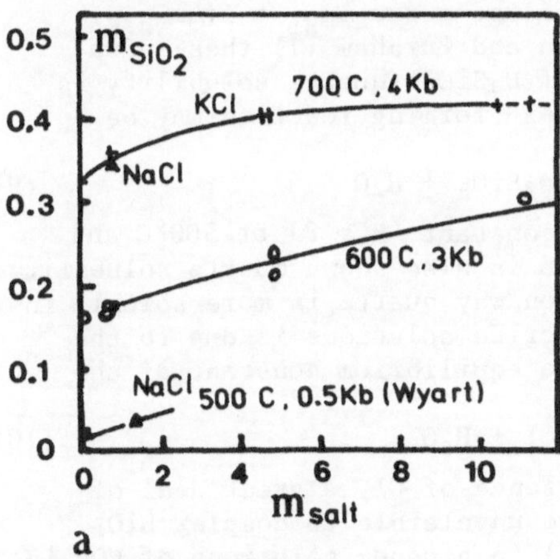

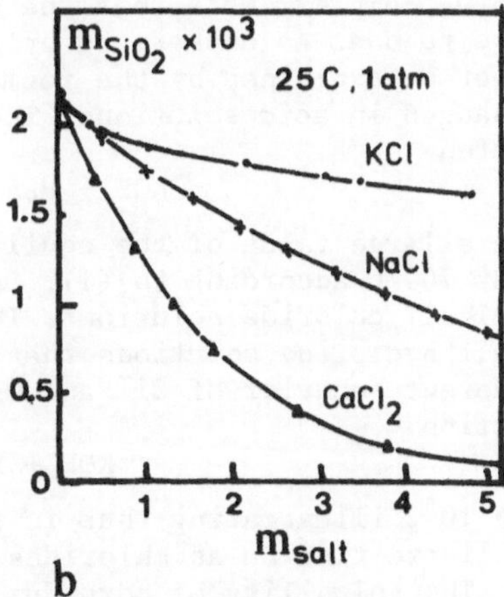

Figure 5. a) Solubility of quartz (molality) in KCl and NaCl solutions (data from Anderson and Burnham (1) and Wyart and Sabatier (23)).
b) Solubility of amorphous silica (molality) in KCl, NaCl and $CaCl_2$ solutions at 25°C and 1 atm (data from Marshall and Warakomski (14)).

in solubility of SiO_2 in aqueous solutions of alkali at high temperature and pressure is contrasted with the decrease at 25°C, 1 atm. There is no KH_3SiO_4 (or NaH_3SiO_4) complexes at 25°C but there evidently is at high T, P. For example the data of Fig. 6 show that addition of KOH or NaOH

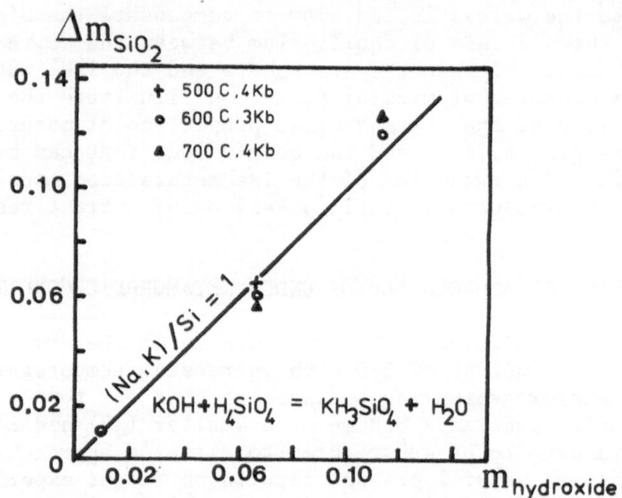

Figure 6. Change in solubility of quartz (molality) [Δm_{SiO_2} = solubility in KOH solution - solubility in pure water] on adding KOH at several P, T conditions [after Anderson and Burham (1)].

increases the solubility of quartz due to the formation of a K-SiO_2 or Na-SiO_2 complex. Note that the slope of the curve $\Delta m_{Si} = f(\Delta m_{Na,K})$ is close to one. As emphasized by Anderson and Burnham (1) these data cannot be explained by the formation of $H_3SiO_4^-$ (quartz solubility is unchanged in acid solutions). The complex - forming reaction may be written

$$KOH + H_4SiO_4 = KH_3SiO_4 + H_2O \quad (9)$$

with a large value of the equilibrium constant : K = 81 at 500°C and 27 at 700°C according to (1). On Fig. 6 is also shown quartz solubility in alkali chloride solutions. The reason why quartz is more soluble in alkali hydroxide solutions than in chloride solutions is due to the different behavior of Cl^- and OH^- ; the equilibrium constant of the reaction

$$KOH + HCl = KCl + H_2O \quad (10)$$

is > 10^3, illustrating that in the presence of Cl^- a great deal of alkali are tied up as chlorides and are unvalaible to complex SiO_2.

The solubility of corundum (Al_2O_3) in aqueous solutions of KOH is shown in Fig. 7. It can be seen that K (and Na) similarly complex with Al at high T and P (formation of $KH_2Al_2O_3$). Similarly recent data (1), (3) on albite and on the system K-feldspar + quartz indicated the evidence for the existence of a dissolved complex which has the feldspar

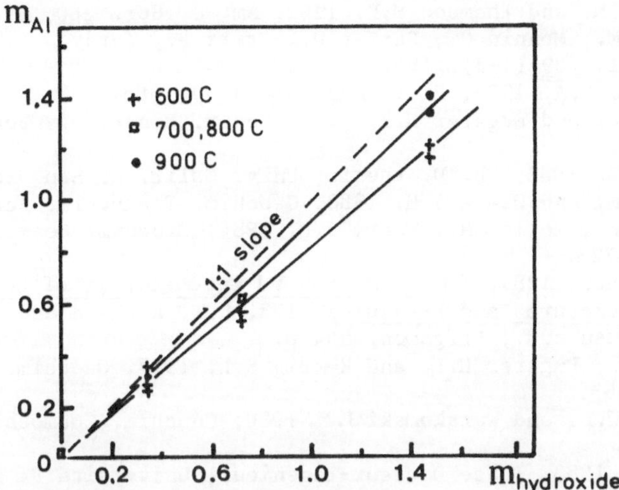

Figure 7. Solubility of corundum (molality) in KOH solutions at several temperatures [after Anderson and Burham (1)].

stoichiometry (SiO_2 is bound to the alkali - Al_2O_3 group forming a complex such as $(Na,K)AlSi_3O_8$).

If such uncharged simple groupings of atoms dominate in these P-T conditions, then it is impossible to predict metamorphic solution compositions based on presently known hydrothermal equilibria while pH measurements become much less useful than they are at lower P, T. The problem is now to determine the free energy of formation of the important solute species. This can be achieved by simply solubility measurements, a method well within present experimental capabilities (4), (8). Thus, it might be very encouraging for future works that the thermodynamic modelling of fluids in metamorphic conditions appears to be simpler than that for the earth surface environments.

ACKNOWLEDGEMENTS

We acknowledge the financial support of C.N.R.S. via A.T.P. Géochimie. G.M. Anderson and J.H. Weare are thanked for helpful discussions.

REFERENCES

1. Anderson G.M. and Burnham C.W. 1983, Am. J. Sci. 283-A, 283.
2. Braitsch O. 1971, Salt deposits : their origin and composition, Springer-Verlag, 279 p.
3. Burnham C.W. 1967, in Geochemistry of Hydrothermal Ore Deposits, 34, H.L. Barnes, ed., Holt, Rinehart and Winston.
4. Frantz J.D. and Popp R.K. 1979, Geochim. Cosmochim. Acta 43, 1223.

5. Garrels R.M. and Thomson M.E. 1962, Am. J. Sci. 260, 57.
6. Gueddari M., Monnin C., Perret D., Fritz B., Tardy Y. 1983, Chem. Geol. 39 (1-2), 165.
7. Guggenheim E.A. 1935, Phil. Mag. 19 (127), 588.
8. Gunter W.D. and Eugster H.P. 1978, Contr. Mineralogy Petrology 66, 271.
9. Harvie C.E. 1982, Ph. D. thesis, Univ. Calif. at San Diego.
10. Harvie C.E. and Weare J.H. 1980, Geochim. Cosmochim. Acta 44, 981.
11. Harvie C.E., Moller N., Weare J.H. 1984, Geochim. Cosmochim. Acta 48, 723.
12. Helgeson H.C. 1981, in Chemistry and Geochemistry of Solutions at high Temperatures and Pressures, 133, D.T. Rickard and F.F. Wickman eds., Pergamon, 564 p.
13. Jones B.F., Eugster H.P. and Rettig S.L. 1977, Geochim. Cosmochim. Acta 41, 53.
14. Marshall W.L. and Warakomski J.M. 1980, Geochim. Cosmochim. Acta 44, 915.
15. Monnin C. 1983, Thèse Docteur-Ingénieur, Université P. Sabatier, Toulouse (France).
16. Monnin C. and Schott J. 1984, Geochim. Cosmochim. Acta 48, 571.
17. Nordstrom D.K., Plummer L.N., Wigley T.M.L., Wolery T.J., Ball J.W., Jenne E.A., Basset R.L., Crerar D.A., Florence T.M., Fritz B., Hoffman M., Holdren G.R. Jr., Lafon G.M., Mattigod S.V., McDuff R.E., Morel F., Reddy M.M., Sposito G. and Thraikill J. 1979, A.C.S. Symp. Series 93, 857.
18. Pitzer K.S. 1975, J. Sol. Chem. 5 (4), 269.
19. Pitzer K.S. 1979, in Activity Coefficients in Electrolyte Solutions, I (7), 157, R.M. Pytckowicz ed., CRC Press, 265 p.
20. Pitzer K.S. and Silvester L.F. 1977, J. Sol. Chem. 5 (4), 269.
21. Truesdell A.H. and Jones B.F. 1974, J. Res. U.S. Geol. Surv. 2, 233.
22. Whitfield M. 1979, in Activity Coefficients in Electrolyte Solutions, II(3), 153, R.M. Pytckowicz ed., CRC Press, 300 p.
23. Wyart J. and Sabatier G. 1955, CR. Acad. Sc., Paris 240, 2157.

SOLID AQUEOUS SOLUTIONS

J. Klinger
Laboratoire de Glaciologie et de Géophysique de l'Environnement
B.P. 96
F-38402 St. Martin d'Hères CEDEX
France

ABSTRACT. The physical properties of ice grown from aqueous ionic solutions are briefly reviewed. Special attention is given to HF, NH_3 and NH_4F that are likely to enter the lattice of ice in substitutional positions at least at very low concentrations. Incorporation in interstitial positions must be considered in order to explain the high value of the diffusion coefficient of HF in ice. In some cases formation of impurity clusters occurs.

1. INTRODUCTION

The purpose of this paper is to discuss what happens when ice crystals are formed from an aqueous ionic solution. This problem has been intensively studied for HF, NH_4F and NH_3 solutions during the sixties. The reason is that these compounds are considered to be key species for the investigation of the electrical properties of ice.

Ice is a hydrogen bond solid and thus a protonic semiconductor in which the dielectric relaxation and the conductivity is explained by the interplay of two types of specific defects (1,2,3,4). The protons that are responsible for the bonding are placed following the Bernal-Fowler rules (5) :

- The H_2O molecules are conserved in the ice structure.
- The H_2O molecules are oriented so that two O-H bonds are directed approximately toward two of the four neighbouring oxygen atoms.
- Only one proton lies approximately along the axis between adjacent oxygen atoms.

According to Pauling (6) all orientations of the water molecules have approximately the same probability.

The defects responsible for the electrical properties of ice are violations of the Bernal-Fowler rules. These defects are of two types:

- Orientatational defects (Bjerrum defects)

M.-C. Bellissent-Funel and G. W. Neilson (eds.), The Physics and Chemistry of Aqueous Ionic Solutions, 441–446.

- Ionic defects

The Bjerrum defects occur as bonds occupied by two protons (D-defect) or as bonds without protons (L-defect).

The ionic defects as usual are water molecules with a deficit or an excess of one proton.

In order to increase the concentration of Bjerrum defects Graenicher proposed to dope ice crystals with HF, NH_3 and NH_4F. The first of these compounds should provide an excess L-defect, the second an excess D-defect, the third leaves the total amount of protons unchanged. This evidently will be the case only if the impurities are placed in a substitutional position in the ice latice. In the following we try to have a closer look on the problem of substitution of water molecules by ionic impurities.

2. PHYSICAL PROPERTIES OF ICE CONTAINING IONIC IMPURITIES

2.1 Making ice from an aqueous ionic solution.

In naturally forming polycrystalline ice a very high concentration of impurities is found in the grain boundaries. An important example of such a system is sea ice.

When a single crystal is built up from an aqueous ionic solution an important amount of the ionic impurities is rejected to the mother solution. The amount of the solute that is rejected depends on its chemical nature, the initial concentration and the method of crystal growth.

If we use the classical Bridgeman type crystal growth method (7) an important rejection of the impurites takes place. As an exemple we can say that when we make ice crystals from a solution of HF the crystal only absorbs about 10^{-3} of the impurity content of the mother solution. The upper limit of the mother solution that gives clear crystals even at growth rates as low as one centimeter a day is of the order of $5 \cdot 10^{-4}$ molar ratios. The HF concentration in the crystal thus will be about $5 \cdot 10^{-7}$ molar ratios. A much more efficient method of producing highly doped ice crystals is the Czochralski method (8). By this method clear crystals containing up to $4 \cdot 10^{-5}$ molar ratios of HF and $2 \cdot 10^{-6}$ molar ratios of NH_3 have been made (8,9).

It has been found that chloride and fluoride of ammonium is more easily incorporated in ice crystals than the corresponding alkali salts (10).

Now let us have a closer look on the way that ice crystals absorb impurities. An analysis of local concentrations of HF in crystals grown by the Bridgeman method showed that a great deal of the impurities are concentrated in clusters containing between 10^{15} and 10^{16} molecules

(11). Unfortunately the precise nature of these clusters has not been investigated up to now.

In crystals obtained by the Czochralski method the distribution of the impurities has not been studied but the fact that clear crystals with higher impurity concentrations can be obtained indicate that these crystals have a higher degree of homogeneity.

2.2 Diffusion of ionic impurities through ice single crystals.

Haltenorth and Klinger (12) measured the diffusion coefficient of HF in ice between 183 and 268 K. At 263 K they obtained $D=(1.08\pm0.01)\cdot10^{-7}$ cm^2/s. The activation Energy was found to be 0.2 eV. Comparatively the Activation energy of the selfdiffusion in ice measured with tritium as a tracer has an activation energy of 0.63±0.03 eV with $D=2.5\cdot10^{-12}$ cm^2/s at 263 K (13). The selfdiffusion measured with oxygen-18 as a tracer showed an activation energy of .68±0.13 eV (14). It turned out that the self diffusion of ice was independant of the concentration of HF or NH_4F in the crystal. Obviously the diffusion mechanism of HF is very different from that of the self diffusion thus indicating that the main part of the impurities is not in substitutional positions (15).

Barnaal and Slotfeldt-Ellingsen (16) studied the diffusion of several ionic impurities in ice. They found for NaCl, HCl and HNO_3 an approximate diffusion rate of $4\cdot10^{-9}$ cm^2/s at 258 K. The plots of the spin-lattice relaxation time versus temperature for HCl, KCl and NaCl exhibited activation energies comparable to that of HF.

The solubility of HF in ice crystals doped by diffusion is of $2.75\cdot10^{-5}$ molar ratios at 265 K. This solubility increases with falling temperature down to 230 K where it reaches a maximum value of $5.45\cdot10^{-5}$ molar ratios and than falls with falling temperatures reaching $1.8\cdot10^{-5}$ molar ratios at 180 K (17). The maximum concentration of HF absorbed by an ice crystal fits a phenomenological law as :

$$C_{max}=5.93\cdot10^{-5}-1.95\cdot10^{-8}(T-231)^2 \qquad /1/$$

Where T is the temperature in Kelvin and C the HF concentration in molar ratios. It is remarkable that these concentrations are in the same order of magnitude than that obtained in crystals grown by the Czochralski method.

A striking fact is the outdiffusion that has been observed in highly HF doped crystals as well as in crystals doped with NH_3. It has been reported that NH_3 as well as HF doped ice crystals obtained by the Czochralski method loses about 50% of their impurity content overnight when they are stored in free air (9). A similar outdiffusion has been found in crystals doped with helium and neon (18).

Spin-Lattice Relaxation studies performed by Bilgram et al. (19) revealed that the diffusion coefficient of HF in ice is strongly concentration dependent. At low concentrations a value of the diffusion coefficient comparable to that of the selfdiffusion has been found. It must be said that in an earlier work a hint was given that goes in the same sense.It has in fact been shown by Klinger (11) that the points of the diffusion profiles showed an important scattering when the HF concentration of the diffusion "cap" brought in contact with the ice sample was lower than the concentration given by the phenomenological law /1/.

Bilgram et al.(19) explain the fast diffusion by an interstitial mechanism as it has been suggested for the diffusion of helium and neon in ice (20).

2.3 Electrical properties of ice containing ionic impurities.

Electrical measurements of pure ice as well as of ice containing impurities often suffer from electrode effects. This is related to the very nature of ice as a protonic semiconductor in which the charge transport is associated with a mass transport. This problem is partially overcome by using electrode systems with two opposite properties:

- Injecting electrodes for exemple hydrogen charged palladium electrodes eventually combined with hydrogen absorbing electrodes.
- Blocking electrodes, for example gold electrodes.

An other difficulty in electrical measurements is the discrimination between surface and volume conductivity. This problem has been solved using electrodes with guard rings.

A rather complete review of electrical properties of pure and doped ice has been given by Hobbs(21). More recently several studies on electrical properties of ice containing ionic impurities have been published (see for exemple 10, 22).

Gross et al. (10) studied electrical conductivity and relaxation in ice crystals with known content of different impuririties in the frequency range of 20 Hz to 100 kHz. They distinguish between conductivity enhancing impurities as HCl, HF, NaCl, KF, NH_4F and conductivity depressing solutes (NH_4OH, NH_4Cl, NH_3CO_3, $NaHCO_3$). In the first group are found those impurities that according to the theory by Graenicher and Jaccard should create excess Bjerrum defects. In the case of bases the static conductivity depends only weakly on temperature. The dielectric conductivity contribution strongly depends on the impurity concentration but only weakly on the chemical nature of the solute. The principal relaxation time is found to be reduced by most solutes with respect to that of pure ice.

2.4 Heat conduction of ice crystals containing ionic impurities.

Low temperature heat conduction is a classical technique for studying defects in crystals. It can be applied for the investigation of the incorporation mechanism of ionic impurities in ice.

Up to now only one of such kind of studies is reported in the literature. This work has been done on a single crystal of ice doped with HF by diffusion. In a first series of measurements performed one month after doping, only a slight decrease of the low temperature heat conduction with respect to the virgin sample has been revealed. The same sample stored for 32 months at 250 K had a heat conduction coefficient roughly speaking 90% less in the temperature range between 2 and 10 K. In other words : despite of the fact that the major part of HF should have been lost by outdiffusion an important perturbation of crystal structure occurs leading to similar characteristics as that of polycrystalline ice (23). It should be mentionned in this context that the spin-lattice relaxation time of aged HF doped ice increases with aging but does not reach the value of pure ice even after four years (16).

3. DISCUSSION AND CONCLUSION

When aqueous ionic solutions freeze, several processes of impurity incorporation are observed. These processes depend on the chemical nature of the solute and on the way the solution freezes. When polycrystalline ice is formed the main part of the solute is rejected to the grain boundaries. The grain boundary structure in the presence of impurities is still a subject of speculation.

HF, NH_4F and HCl are readily incorporated in single crystals. For HF, molar ratios of approximately $4 \cdot 10^{-5}$ can be obtained. The absorption of NH_3 and of alkali halides is possible but less easy.

Some impurities as HF, HCl, NH_4F and NH_3 enter the lattice in substitutional positions as far as the concentrations are low, most likely in the order of magnitude of the concentration of the intrinsic lattice defects. In higher concentrations these impurities form interstitials that diffuse rapidly through the lattice. Sometimes the formation of impurity clusters is obsrved. The structure of such kind of clusters has never been investigated. they may be amorphous. In some cases hydrates may be formed locally.

REFERENCES

1. Bjerrum, N. 1951. Koniglige Danske Videnskabernes Selskab. Matematisk-fysiske Meddelelser 27, 1.
2. Graenicher, H. 1958. Z. Kristallographie 110, 432.
3. Jaccard C. 1959. Helvetica Physica Acta 32, 89.
4. Jaccard C. 1964. Physik der kondensierten Materie 3, 863.
5. Bernal J.D. and Fowler R.H. 1933. J.Chem. Phys. 1, 515.
6. Pauling L. 1935. J. Am. Chem. Soc. 57, 2680.

7. Schulz H. 1961. Diplomarbeit Technische Hochschule Stuttgart.
8. Bilgram J. 1973. Physics and Chemistry of Ice Whalley et al. eds. p. 275.
9. Hubmann M. 1978. J. of Glaciology 21, 161.
10. Gross G. W., Hayslip I. C., and Hoy R. N. 1978. J. of Glaciology 21, 143.
11. Klinger J. 1966. Diplomarbeit Technische Hochschule Muenchen.
12. Haltenorth H. and Klinger J. 1969. Physics of Ice Riehl N. et al. ed. 579.
13. Blicks H.,Dengel O., and Riehl N. 1966. Physik der kondensierten Materie 4, 375.
14. Delibaltas P., Dengel O., Helmreich D., Riehl N., and Simon H. 1966. Physik der kondensierten Materie 5, 166.
15. Dengel O., Jacobs E., and Riehl N. 1966. Physik der kondensierten Materie 5, 58.
16. Barnaal D., and Slotfeldt-Ellingsen D. 1983. J. Phys. Chem. 87, 4321.
17. Haltenorth H., and Klinger J. 1977. Solid State Communications 21, 533.
18. Haas J., Bullemer B., and Kahane A. 1971. Solid State Communications 9, 2033.
19. Bilgram J., Roos J., and Graenicher H. 1976. Z. Physik B23, 1.
20. Kahane A., Klinger J., and Philippe M. 1969. Solid State Communications 7, 1055.
21. Hobbs P. V. 1974. Ice Physics Clarenton Press Oxford.
22. Camplin G. C., Glen J. W., and Paren J.G. 1978. Journal of Glaciology 21, 123.
23. Klinger J. 1973. Physics and Chemistry of Ice Whalley et al. eds. p. 114.

VITRIFICATION AND CRYSTALLIZATION OF WATER

J. Dupuy*, A. Elarby-Aouizerat*, P. Claudy, J.-F. Jal,**
J.M. Letoffe, P. Chieux*****
* Département de Physique des Matériaux 69622 Villeurbanne
** INSA de Lyon
*** I.L.L. Grenoble

Abstract. The ability of electrolytic glasses to control the ice crystallization is presented with an analysis of ice transformation in terms of polymorphism. During the devitrification of such glasses, an intermediate disordered water phase is always present until the stable ice I_h is formed.

Electrolytic solutions form glasses easily(1). The ionic influence on the water is such that, in some systems, the crystallization of water can be obtained without interference with the crystallization of another defined compound. In that case ice phases are formed in a glass and can be easily identified. The **density differences** between the various forms of H_2O phases and the concentrated glasses, with **their structural diffraction patterns,** permit easy and complete analysis of the out of equilibrium properties by diffraction at small and large angles scattering(2).

Two cations Be^{2+} and Li^+ with the same anion Cl^- are particularly interesting. In equivalent out-of-equilibrium conditions (degree of supercooling, rate of cooling, warming) and at two equivalent concentrations, the behavior of the ice nucleation and ice crystallization is different.

The aim of this lecture will be **to raise the question of the correspondence between the existence of polymorph phases in crystallized ice and the existence of several different amorphous phases of this same compound(3).**

1. GROWTH OF CRYSTALLIZED ICE AND VITRIFIED WATER

1.1. Crystallization in aqueous solutions

Figure 1 shows the equivalence of the out-of-equilibrium phase diagram of $BeCl_2$, RH_2O (5) and LiCl, RH_2O systems(4) in the range of water crystallization. (The ratio R = mole H_2O/mole salt, x is the mole fraction of salt). Be^{2+} and Li^+ have equivalent electron configuration and an equivalent charge to radius ratio. The example given in this

M.-C. Bellissent-Funel and G. W. Neilson (eds.), The Physics and Chemistry of Aqueous Ionic Solutions, 447–452.

lecture will be at almost equivalent concentration x = 0.06 for $xBeCl_2 + (1-x)D_2O$ and x = 0.10 for $xLiCl + (1-x)D_2O$ (figure 1).

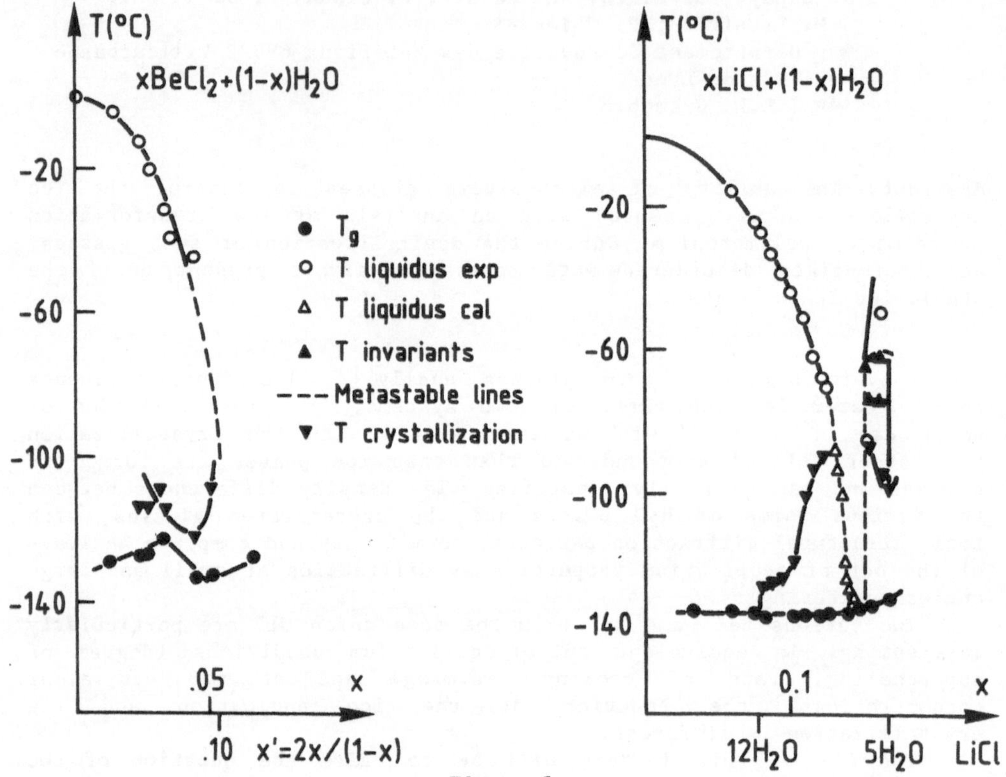

Figure 1

The first characteristics of water crystallization (measured by neutron diffraction, figure 2) are as follows :

a) - The pattern of the stable ice I_h is not observed. There is no evidence of the (102) Bragg peak nor of any change in relative intensity of the triplet (100), (101), (002) peaks.

b) - Typical spectra for Bragg peaks (10ℓ) are analyzed[6] **as non correlated planes of I_h,** (in analogy with bidimensional crystals[7]). The shape of peaks can be analyzed on the basis of a Warren formula[7].

c) - With Li^+, **the cubic form I_c** (metastable form of I_h) appears **quite pure** and **the transformation $I_c \rightarrow I_h$** is continuous by the mechanism of correlation between the planes (10ℓ) of the hexagonal structure. This means that there are **stacking faults.**

d) With Be^{2+} and the same thermal history as for LiCl, RD_2O, the pure cubic structure is never observed.

Work is now in progress to measure size of the ice crystallites. There is certainly a **correspondence between the correlation of planes (10ℓ) and the size of the ice crystal.**

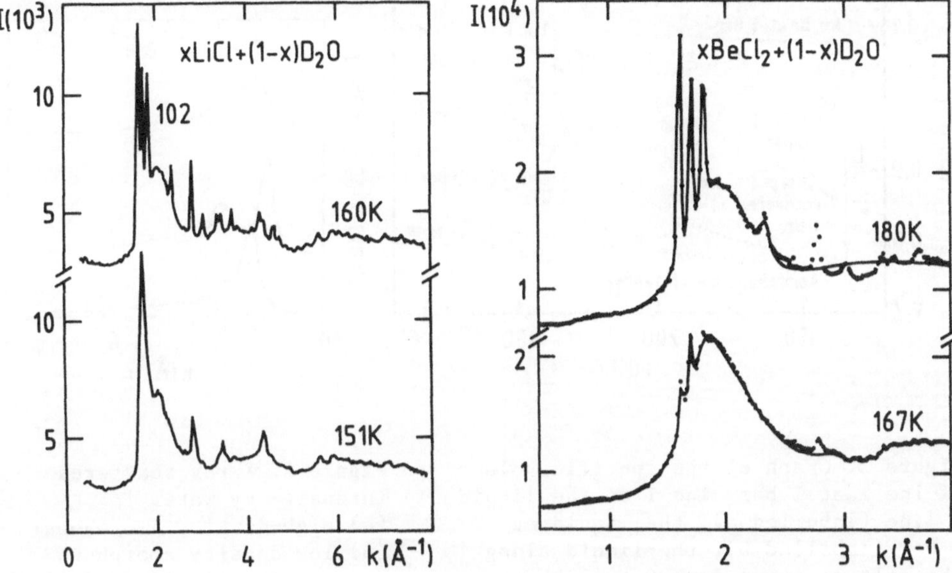

Figure 2. Neutron diffraction spectra recorded at the beginning of ice crystallization (151 K and 167 K) and before complete crystallization of I_h form. ($xBeCl_2 + (1-x)D_2O$ was measured with a neutron wavelength λ=1.108 Å).

This typical behavior was already observed in the neutron diffraction spectra recorded at 77 K on ice I_c, prepared by the transformation of ices II, V and IX[8].

We establish the **existence of a polymorphism between the two forms of pure ice I_c and I_h**. This polymorphism is influenced by the ions of the glass matrix. These ions may also have an effect on the size of the crystals formed. In fact it is simply the result of the interaction between ion and coordination shells.

1.2. (H_2O) vitrified or supercooled water

A recent publication[9] discusses the very exciting possibility **of melting ice I_h at 77 K and at 10 kbar.** This procedure according to graph 3 gives a **high density form of disordered ice,** and a continuous range of vitrified water, whose density lies between 0.91g/cm^3 to 1.31 g/cm^3. The structure can be accommodated, according X-ray experiment[10] by the bending of hydrogen bonds. (figure 4).

These phases may be seen as polymorphs.

This conclusion raises the question of the correspondence between crystals and amorphous phases of ice.

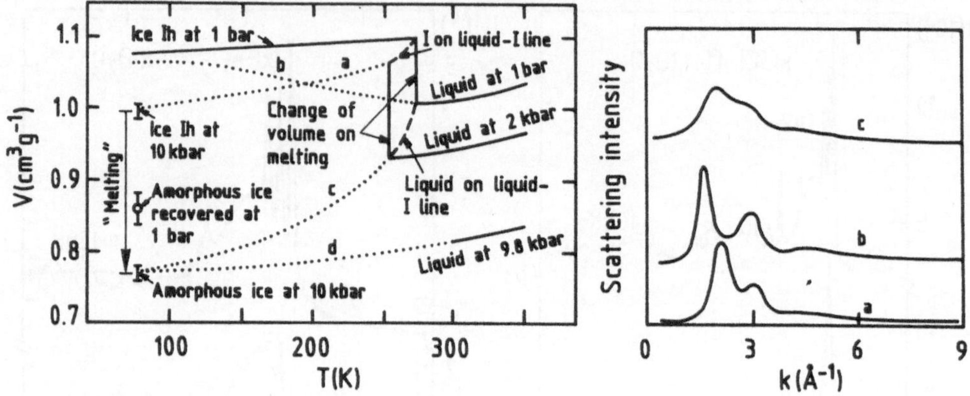

Figure 3. Graph of the specific volume of ice I at 1 bar, ice I on the liquid-I line (labelled a), the liquid at 1 bar (labelled b), the liquid along the melting curve from 0 to 2 kbar (labelled c), the liquid at 9.8 kbar (labelled d) and reasonable extrapolations to 77 K. Measured values are represented by full and dashed lines and extrapolated values by dotted lines.

Figure 4. X-ray scattered intensity by water
(a) high-density amorphous
(b) low density amorphous and (c) liquid.

2. ICE NUCLEATION

Spectra recorded on crystallization of ice (shown for example in figure 5) cannot be fitted using a glassy matrix and crystallized ice. A gaussian component **with a momentum transfer at 1.73 $Å^{-1}$** has to be summed with Bragg peaks. **This gaussian component disappears when Ih is formed.** This can be related to **a disordered metastable structure seen as an intermediate phase which is always present until the stable crystalline form Ih is obtained.** From a beautiful analysis of calorimetric data, D. Clausse and coworker[11] have concluded to the evidence of such an intermediate. This phase coexists also with Ic in a free jet expansion of water[12]. Analysis (by differential scanning calorimetry and neutron diffraction) of the behavior of the glass at the glass transition allows us to study the beginning of ice nucleation and the possible role of the ion on this mechanism[5].

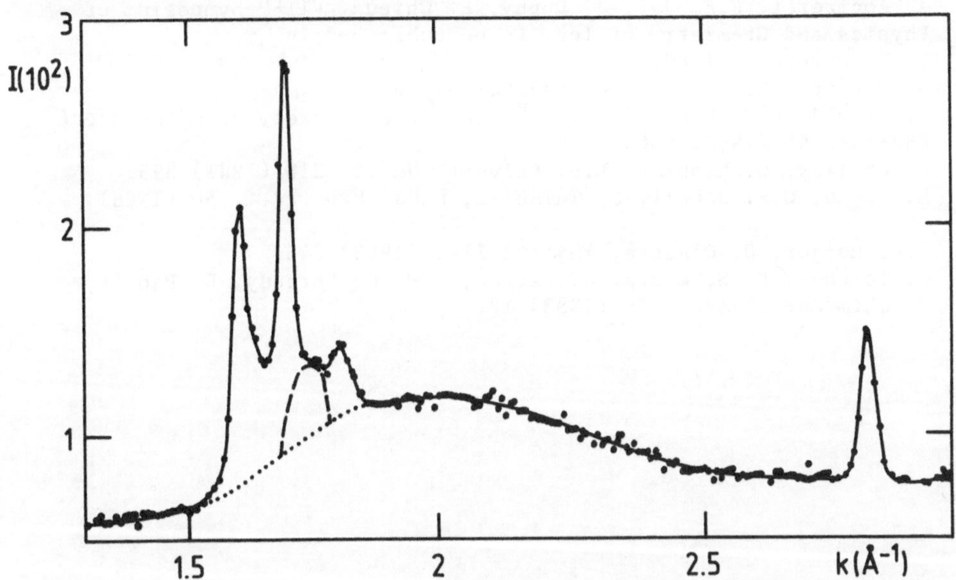

Figure 5. Neutron diffraction spectra of $xLiCl + (1-x)D_2O$ at molar fraction $x = 0.06$ recorded during ice crystallization. The fit of the data is given with Warren and gaussian components for Bragg peaks, gaussian components for the residual glassy matrix. In addition a gaussian component (----) at $q = 1.73$ Å^{-1} ($\theta = 20.5°$ with neutron wavelength $\lambda = 2.52$ Å) is necessary.

We wish to aknowledge the quite perfect experimental neutron diffraction facilities at Institut Laue Langevin in Grenoble and Laboratoire Léon Brillouin in Saclay.

References

1. G. Vuillard, *Ann. Chim.* (Paris) **2** (1957) 233.
   C.A. Angell, E.J. Sare, *J. of Chemical Physics* **52** (1970) 233
2. A. Elarby-Aouizerat, J.F. Jal, C. Ferradou, J. Dupuy, P. Chieux A. Wright, *J. of Physical Chemistry* **87** (1983) 4170.
3. B. Luyet, *Annals N.Y. Academy of Science* (1958) 502.
4. A. Elarby-Aouizerat, P. Chieux, P. Claudy, J. Dupuy, J.F. Jal, J.M. Letoffe, *J. Physique Coll. C8, Suppl. n°12,* **46** (1985) C8-629
5. P. Claudy, J.M. Letoffe, A. Elarby-Aouizerat, J.F. Jal, J. Dupuy, M.-C. Bellissent, to be published.

6. A. Aouizerat, J.F. Jal, J. Dupuy, P. Chieux, VIIth Symposium of Physics and Chemistry of Ice, to be published in J. de Physique (1987).
7. B. Croset, Thèse d'Etat (1985) Marseille
8. G. Arnold, E.D. Finch, S.W. Rabideau, R.G. Wenzel, *J. of Chemical Physics*, **45** (1968) 4365
9. E. Whalley, C. Mishima, L.D. Calvert, *Nature* **310** (1981) 393.
10. L. Bosio, G.P. Johari, J. Teixeira, *Phys. Rev. Lett.* **56** (1986) 460.
11. P.H. Meijer, D. Clausse, *Physica* **119B** (1983) 243.
12. G. Torchet, P. Schwartz, J. Farges, M.F. de Feraudy, B. Raoult, *J. Chemical Physics.* **79** (1983) 12.

Posters presented at the NATO Summer School
"The Physics and Chemistry of Aqueous Ionic Solutions"

Polyelectrolytes , Primitive Model and HNC Equation
Luc Belloni
L. Belloni, Chem. Phys. 99,43 (1985) ; J. Chem. Phys. 85, 519(1986); Phys.Rev. Lett., 57, 2026 (1986).

On the Dynamics of Polyelectrolyte Solutions
Moustapha Benmouna

Study of Aqueous KH_2PO_4-Solutions by Raman Spectroscopy
Alejandro del Valle Gonzalez
F. Rull, A. del Valle Gonzalez, S. Veintemillas and F. Sobron
Bol.Soc.Esp.Min, in press
F. Rull, J.A. de Saja, J. Raman Spectrosc. 17,167(1986).

Collective Excitations in Electrolyte Glasses
Anna Aouizerat-Elarby, Jean-François Jal, Josette Dupuy, Albert José Dianoux and Pierre Chieux.

Equilibrium and Non-Equilibrium Diagrams of Salt Water Systems: Ion Influence
Anna Aouizerat-Elarby, Jean-François Jal, Josette Dupuy, Pierre Claudy, Jean-Marie Letoffe and Pierre Chieux
A. Aouizerat-Elarby, J.F. Jal, C. Ferradou, J. Dupuy, P. Chieux and A. Wright, J.Phys.Chem. 87, 4170 (1983);
A. Aouizerat-Elarby, P. Chieux, P. Claudy, J. Dupuy, J.F. Jal and J.M. Letoffe, Third International Conference on the Structure of Non Crystalline Materials, July 1985, Grenoble, France; J. de Physique (Paris) C8, 629 (1985).

Hydration of Ions in Aqueous Solutions Studied by Infrared Spectroscopy
Anders Eriksson, Olof Kristiansson, Jan Lindgren and Czeslawa Paluszkiewicz
O. Kristiansson, A. Eriksson and J. Lindgren, Acta Chem. Scand. A38, 609 (1984).

Cross Coupling Effects During Diffusion Processes
Jean-François Gaillard, Jean-Pierre Simonin and Pierre Turq
R. Mills, A. Perera, J.-P. Simonin, L. Orcil and P. Turq
J. Phys. Chem. 89, 2722 (1985)
P. Turq, L. Orcil, M. Chemla and R. Mills, J. Phys. Chem. 86, 4062 (1982)

On Water's Ability to Dissociate Ion Pairs
Saul Goldman and Peter Backx
S. Goldmann and P. Backx, J. Chem. Phys. 84, 2761 (1986).

Remarks on the McMillan–Mayer Theory of Solutions
Juan L. Gomez Estévez
H.L. Friedman, A Course in Statistical Mechanics, Prentice Hall 1985, Chapter 10

Electron Spectroscopy for Liquids and Solutions
Sven Holmberg and Robert Moberg
H. Siegbahn, J.Phys.Chem. 89,897 (1985)
H. Siegbahn, M. Lundholm, M. Arbmann and S. Holmberg, S.Phys.Scr. 27,241 (1983)

Ion–Ion Correlation Effects in Double Layers between Two Walls
Roland Kjellander and Stjepan Marcelja
R. Kjellander and S. Marcelja, J.Phys.Chem. 90, 1230 (1986); J. Chem. Phys. 82, 2122 (1985); Chem. Phys. Lett. 127, 402 (1986)

Light Scattering from Concentrated Dispersions of Charged Silica Particles in Salt Free, Weak Polar Solvent Mixtures Determination of Static and Hydrodynamic Interactions
Albert Philipse
H.M. Fijnaut, C. Pathamanoharan, E.A. Neiuwenhuis and A. Vrij, Chem. Phys. Lett. 59, 352 (1978)

Nonhomogeneous Coulomb Liquids: A Functional Integral Approach
Rudi Podgornik and B. Zeks
R. Kjellander and S. Marcelja, J.Chem.Phys. 82, 2122 (1985)
S.F. Edwards and A. Lenard, J. Math. Phys. 3, 778 (1962)

Water Activity Measurement of Ion Hydration in Supersaturated Lithium Halide Solutions, or
The Stokes Bithermal Isopiestic Method Applied to a Levitated Microscopic Droplet
Charles B. Richardson
J.F. Spann and C.B. Richardson, Atmospheric Environment 19, 819 (1984)
C.B. Richardson and C.A. Kurz, J. Am. Chem. Soc. 106, 6615 (1984)

Reorientational Dynamics and Local Order in Aqueous Nitrate Solutions
Barbara Rosi and M.P. Fontana

Structure of an Electrolyte Solution near a Hard Wall: Solute Adsorption and Ionic Solvation
Vincent Russier, M.L. Rosinberg and J.P. Badiali
J.P. Badiali, Mol. Phys. 55, 939 (1985)
J.P. Badiali, M.L. Rosinberg and V. Russier, Mol. Phys. 56, 105 (1985)

Ionic Structure in Aqueous Electrolyte Solutions by the Difference Method of X-ray Diffraction
Neal Skipper, John Herdman, Philip Gullidge, George Neilson and Pamela A. Walker
N.T. Skipper, S. Cummings, G.W. Neilson and J.E. Enderby, Nature 321, 6065, 52 (1986)

Molecular Dynamics Simulation of the Water Structure near an Electrolyte Solution/Solid Interface
Eckhard Spohr
E. Spohr and K. Heinzinger, Chem. Phys. Lett. 123, 218 (1986)
E. Spohr, Ph.D. Thesis, Johannes Gutenberg Universität Mainz 1986

A Comparison of the Bending and Stretching Modes of Water Adsorbed on Silver Electrodes at Different Potentials
Zhong-Qun Tian
M. Fleischmann and I. Hill, J. Electroanal. Chem. 146, 367 (1983)
M. Fleischmann, G. Sundholm and Z.Q. Tian, Electrochem. Acta, in press

Application of Polyelectrolytes in Industrial Wastewater Treatment
Aysen Türkmann and Füsun Sengül
S.D. Faust and O.M. Aly, Chemistry of Water Treatment, Butterworth 1983
F. Sengül, Thesis, Dokuz Eylül University, Izmir 1983

Phase Equilibria in a Microemulsion System
George A. van Aken
J.Th.G. Overbeek, P.L. de Bruyn and G.J. Verhoecks in "Surfactants", edited by R.F. Tadros, Acad. Press 111 (1984)
J.Th.G. Overbeek, Proc. Kon. Ned. Akad. Wetenschap., B89, 61 (1986)

On the Structure and Dynamics of Water in $AlCl_3$-Solutions, a NMR Contribution
Johan R.C. van der Maarel, H. de Boer, J. de Bleijser and J.C. Leyte,
J.R.C. van der Maarel, D. Lankhorst, J. de Bleijser and J.C. Leyte
Chem. Phys. Lett. 122, 541 (1985); J. Phys. Chem. 90, 1470 (1986)

The studies of the Simultaneous Very Weak and Very Strong Complexation Equilibria in Aqueous Solutions
Michal Wilgocki
M. Wilgocki and J. Bjerrum, Acta, Chem. Scand. A37, 307 (1983)
M. Wilgocki, J. Coord. Chem. 14, 39 (1985); ibid 14, 151 (1985)

Pair Potentials for Strongly Interacting Colloidal Particles
Clifford Woodward and Bo Jönsson

Proton Transfer in the Model System $(HONH_3HOH)^-$
Tai-he Xia, Xiu-fan Shi and Xian-shan Ni
H. Umeyama et al, J. Mol. Biol. 150, 409 (1981)
S. Scheiner, Int. J. Quant. Chem. XXIII, 739 (1983)

POSTSCRIPT AND SUMMARY

J.E. Enderby

At the last NATO-ASI on liquids held in Corsica, (August 1977), Professor **de Gennes**, in his summary of that meeting, suggested that the next ASI should concentrate on some specific aspect of the subject and mentioned explicitly ionic solutions as one possibility. The challenge was taken up by **Marie-Claire Bellissent-Funel** and **George Neilson** ; I am sure that all the participants would wish to congratulate our two colleagues for putting together an outstanding programme of lectures, round tables and poster session.

The theory which underlies the subject was covered by four leading authorities : **J.-P. Hansen** (Paris) set out the general framework in terms of the statistical mechanics of bulk and surface properties ; **H.L. Friedman** (Stony Brook) focused attention on ionic liquids at equilibrium, and **J.B. Hubbard** considered non-equilibrium properties such as the electrical conductivity and ionic friction coefficients. Finally, the basic theory of polyelectrolytes treated as charged linear polymers in aqueous solution was presented by **J.M. Victor** (Paris).

One very important development in recent years has been the development of clear links between theoretical objects like friction coefficients and correlation functions and experimental objects like the d.c. conductivity, diffraction intensities and thermodynamic coefficients. This has been greatly assisted by numerical statistical mechanics in which microscopic simulations of ionic solutions are carved out principally by the methods of molecular dynamics or Monte-Carlo techniques. **P. Turq** (Paris) gave a survey of the general approach and a detailed study of several ionic solutions was presented by **P. Bopp** (Darmstadt). The latter carefully reviewed the limitations of both the simulations and the experiments and drew interesting conclusions about areas where more effort was required. Other important theoretical contributions were made by **D. Levesque** (Paris) and **P. Madden** (Oxford), the latter again drawing attention to the interplay of theory and experiment.

Turning now to the experimental side, emphasis was placed on the importance of neutron scattering. **A.J. Dianoux** (Grenoble) showed how quasi-elastic and inelastic neutron spectroscopy was a complementary technique to N.M.R. **J.E. Enderby** (Grenoble) explained the use of "the method of differences" whereby

M.-C. Bellissent-Funel and G. W. Neilson (eds.), The Physics and Chemistry of Aqueous Ionic Solutions, 457–459.

individual correlation functions can be determined, and **P. Chieux** (Grenoble) and **J. Dupuy** (Lyon) described the structural significance of glasses formed from the aqueous phase. The application of small angle scattering (both X-ray and neutrons) and of quasi-elastic scattering to the polyelectrolyte problems was described by **G. Jannink** (Saclay) and here again important links between theory and experiment were established.

The study of liquid surfaces (liquid gas, liquid/liquid, liquid/solid) is a field of considerable interest and difficulty. The presence of ionic charges complicates the problem but microscopic theories for both the electrolyte/electrode and the electrolyte/air surfaces have been developed by several groups. The papers presented by **M.P. Tosi, P. Ballone** and **G. Pastore** (Trieste) and by **D. Andelman** (Exxon, New Jersey), **F. Brochard and J.F. Joanny** (Paris) described theoretical approaches which were microscopic in their philosophy. Contact with experiment is a feature of this work and the series of lectures by **R. Parsons** (Southampton) on metal/electrolyte interfaces emphasised both the practical and theoretical problems which must be solved if a fuller understanding of this class of problem is to be achieved. New techniques such as Surface Enhanced Raman Spectroscopy will become of increasing significance, but sight should not be lost of the traditional methods of electrochemistry. Indeed, many speakers drew attention to the fact that, as our understanding at the microscopic level increased, a re-examination of some of the older work would be rewarding. After, a roundtable animated by **M. Rosinberg** (Paris) was agreed that mixed projects involving, say, neutron scattering and classical measurements could form an excellent training programme for graduate students.

The physics and chemistry of aqueous solutions under extreme conditions were described by **E.U. Franck** (Karlsruhe), the lecture series ending with a spectacular film of "hydrothermal flames" in a supercritical aqueous environment. Much of the data presented will provide a challenge for theorists and experimentalists for several years. The solidification of aqueous solutions into ice crystals was discussed by **J. Klinger** (St Martin d'Hères) and the significance of a theoretical problem posed by natural aqueous solutions in the geophysical context was described by **G. Michard** (Paris) and by **C. Monnin** and **J. Schott** (Toulouse). Finally, **J. Finney** (London), **D. Middendorf** (Oxford) and **R. Oberthur** (Grenoble) presented us with vivid accounts of the variety of roles which ions can play in biochemical systems.

Aqueous solutions **are** complicated liquids but our progress in understanding them has, over the past decade, been quite outstanding. Scientists from many disciplines have contributed to this progress and the willingness of one group to learn from other groups has been vital. We should, however, always remember Friedman's remark (Milton not Harold) that "there is no such thing as a free lunch". If we need a really deep understanding, we have to work terribly hard; the trick is to decide when we can safely decide that enough is enough.

SUBJECT INDEX

A

Ab initio potential 221
Activation - controlled reactions 61,89
Activation energy 176,364,443
Activity 341
Activity coefficients 63,429-434
Adelman's Solution Theory 78
Adsorption - desorption process 271
 - specific adsorption 272
Algorithm 430
Amphiphilic molecules 417
Anisotropic systems 218,232
Anomalous X-ray scattering 142
Anti-Stokes scattering 185
Aqueous Ionic Glasses 359,365
 - pulse radiolysis 373
 - dipolar spin echo 373
Aqueous Ionic Solutions
 - NMR 168-176,373
 - Simulation 217-240
 - Neutron spectroscopy 147-168,373
 - X-ray scattering 142-144
 - On surfaces 245-254,255-290
Aqueous Solutions
 - Non polar 232
Arrhénius plot 159
Auger spectroscopy 268
Autocorrelation function (ACF) 17-19,24, 112,209,211,212,218,413

B

Badger-Bauer Rule 230
Bardeen model 246
Bending motions 236,238
Bernal - Fowler rules 441
Binary collision approximation 117
Binary mixture 338,347
Binodal surface 351
Biological membranes 40
Biological systems 221,224
Bjerrum association 414,415
Bjerrum defects 441
Bjerrum length 292,301
Blob 303,306
Bogolubov-Born-Green-Yvon hierarchy (BBGY) 249
Boltzman factor 63
Born Oppenheimer level 105,106,113,124
Branched polymer 313
Bridge diagram 252
Bridgeman crystal growth method 442
Brillouin lines 193,194,213
Brownian Dynamics (BD) 33,102,409
Brownian motion 21,24,95,100
Brownian particle 114

C

Capacitance 246
 inner layer- 247
Capacitance curve 248
Capacity 271
 - of the electrode 261
 - of the diffuse layer 262
Catalytic activity 276
Cavity - dipole 124
 - field polarization dynamics 125
 - ion 124
Characteristic length 167
Charge density fluctuations 100
Charge dynamical structure factor 34
Charged monolayer 417
Charge response function 33,34
Charge structure factor 35,36,38
Chemical equilibrium 429,430,434
Chemical potential 259
Chemisorbed monolayer 283
Chemisorption with charge transfer 273
Classical plasma 247
Cluster - density 414
 - expansion 66,86,249
 - model for rate constants 414
 - theory 61,77
Coherent scattering lengths (table) 130,133
Colloids 40
Combination rules 222,223
Common ion mixtures 83
Compensation potential 256
Compressibility equation 9,36,45
Computer simulation 95,181,217,357,409
Concentrated brine 434
Concentration diffusion coefficient 333
Concentration fluctuations 366
Condensation 291,293,294,301
Conductance 95
Conductivity 27,315,399
Constitutive relations 402,404

Contact theorem 248,249
Continental thermomineral water 396
Continuity – condition 97
– equation 101
Continuum theory 120
Convective fluctuations 108
Coordination number 130,414
– running 129
Correlation function 95,99,109,148,152,237
– direct 67,82
– partial 91
– technique 236
Correlation – length 302,303,305,308,309
– time 169,171,172
Coulomb – interaction 85,105
– potential 230,233
Counterion 291,292,294,295,301,304,306,314
Critical – curve 347,348,350–352
– density 341
Critical exponent 291
Critical point 338,347,367,420
Cross-section 151
Cryobiology 359
Crystallization 361,447
– temperature 362
Current autocorrelation function 18,20,22,27,29
Cyclic voltammetry 268,273,286
Czochralski method 442

D

D-defect 442
Debye dispersion 126
Debye equation for dielectric relaxation 123
Debye-Falkenhagen-Onsager theory (DFO) 95,96,101,120
Debye-Huckel theory 38,39,41,69,431–434
Debye model 164,198
– relaxation time 123
– screening 126
– screening length 36,70,100,102,262
– shielding length 87,246,419,425
Debye-Waller factor 149,153
Density-density correlation function 4,7
Density fluctuation 184
– correlation function 98,99
Density functional theory (DFT) 245
Density profile 2,13
Density response function 14,18
Depolarized – Raman 164–166
– Rayleigh 184,199,207
– scattering 183,187,188,190,196,203
Deuteron 204
– spin lattice relaxation time 375
Dielectric – coefficient 61,75,81,82
– constant 32,48–50,53,124,200,232,337–339,344,345,349,399,407,437
– friction 117,127
– function 33
– properties 221
– spectroscopy 181,190,197,200,211
– susceptibility 123
Differential scanning calorimetry (DSC) 360
Differential thermal analysis (DTA) 360
Diffuse layer 248
Diffusion – coefficient 25,27,97,99,100,107,124,234,443
– equation 21,95,98
– rotational 107
– tensor 110
Dipole-dipole interaction 47,168
Dipole-induced-dipole (DID)
– mechanism 188–190,209,211
– polarizability 189,207,208
– scattering 208
Dipoles orientation 260
Direct correlation function 5,7,8,10,13,35,37–39,41,43,44,52,53
Direct coupling 163
Discrete solvent model 51,52,54
Dissipative noise 108
Dissociation constant 434
Dissolution constant 434
Distribution function 95,206
Domain walls 423
Double layer 245,259

Dynamical – properties 217,218,364,399
- structure factor 19,28,98

E

Earth surface 439
Effective – charge 314
- polyion charge 315
Effective – potential 52,78
- potential model 221,238
- BJH 221,222,227,229,230,233, 238
- Central Force (CF) 221,222,229
- Dang and Pettitt 221
- Lie and Clementi 221
- MYC 221,230,233
- RWK2M 221
- ST2 221–223,225,227,229,230, 232, 236
- TIP4P 221,223
- Toukan and Rahman 221
Elastic incoherent structure factor (EISF) 154
Electrical – conductance 95,364
- double layer 40,54
Electrocapillary curve 262
Electrocatalysis 279
Electrochemical – biography 269
- potential 256
Electrode 267,268,270,271,273
- charge 261
/electrolyte interface 245
- reactions 280
Electrohydrodynamics 127
Electrolyte dynamics 96,100
Electrolyte–mixture/Mixed electrolytes 61,83
Electrolyte solutions (see also Aqueous Ionic Solutions) 409
- complex 429
- multicomponent 430
- non equilibrium theories 95
Electrolytic conductance 337,341–344
Electrolytic glasses 166,447
Electron exchange 90
- overshoot 259
- vibration coupling function 164
- transfer 281
Electroneutrality 73,96,98,102
Electronic work function 256
Electro–oxidation 288
Electrophoresis 104
Electrophoretic – correction 104,106
- force 105
- light scattering 107,111
Electrostatic – capacity 260
- forces 432
Energy loss 126
Ensembles – canonical 63
- grand canonical 63
Enskog's hard sphere 341
Enstein relation 22,23,107,121
Epifluorescence microscopy 418
Equation of state 339,349,350
Equilibrium – constant 231
- pCO2 437
- properties 399
Eutectic – point 360
- region 363
Ewald simulation 219,406
EXAFS 142
Excess free energy 430,432
Exchange current density 281
Excluded volume 291,298,299,301,304,309

F

Far–infrared absorption 197,202,203
Fast exchange 155
Fast ion regime 126
Fermi gas 246
Fermi pseudo potential 151
Ferrofluid 425
Field emission microscope 259
First order difference method 134
Flexible molecules 219
- model 206,237
Flory radius 295,296,299,303–306,309
Fluctuating – friction coefficient 115
- resistance 114
Fluctuation – dissipation theorem 14,25,26,28,34,95,114
- spectrum 110
- theory 65
Fluid – inclusion 396,430
- mixture 349
Fluidity 364
Force autocorrelation function 118
Form function 322
Fractal network 167
Fractons 167
Free energy of formation 439

Free - ions 431
- surface 260
Frequency shift 218
Fresh water 396
Friction coefficient 97,107,114,115, 117,119-121
Frictional torque 122
Fuell cell 285
Functional differentiation 4-6,10

G

Gaussian overlap potential 221
Generalized frequency distribution 163,164,166,167
Geochemical kinetics 380
Geochemistry 337,339,352,379
Geothermal systems 430
Gibbs Duhem relation 432
Ginzburg-Landau expansion 420,421
Glass - formation 359,363,366,370
- formers - ionic 372
- HX 372
- preparation 361
- spectroscopic investigations 372
Glass transition 359,362,364,376
- temperature 361,363
- dependence on ion charge 361
- dependence on hydration 361
Glassy ionic solutions (see also electrolytic glasses) 166,447
Gouy-Chapman theory 41-43,246,261,272
Graph - Mayer f bonds 66
- h-allowed 67
- theory 61
Green-Kubo relation 234
Guinier plots 367
Gurney - parameter 88
- term 77

H

Hard force 114
Heat capacity 370
Henry's constant 231,232
Heterogeneous electrocatalytic reactions 282
- surface 285
High pressure 221,337
Hindered translation 233
Hydration 224,239,314,320,364
- anionic 136
- cationic 136
- number 156,224,346
- shell 155,172,225-227,346,372
- layer 234,237
Hydrodynamic momentum transfer 105
- limit 21,22
- pressure 105
Hydrogen - bond 181,182,222,228-230,236,240,241
- bonded systems 199,206,221
- evolution 282
Hydrothermal - combustion 337,338
- flame 337
- submarine vents 380,389
Hypernetted chain - approximation (HNC) 69,249
- closure 11-13,37,38,43,48, 49,52

I

Ice 441-451
- amorphous 365,449
- Bragg peak 448
- crystallization 447
- crystallized 369
- cubic 368-370,448
- cubic - hexagonal transformation 369
- diffusion of ions 443
- electrical properties 444
- heat conduction 444
- He,Ne diffusion 444
- hexagonal 368-370
- nucleation 365,447
- physical properties 442
Image potential 247
Immiscibility gap 367
Incoherent neutron experiment 107,197,367
Incompressible flow 105
Inelastic neutron scattering 20,148, 162-164,177,181,200,204,210
Inertial model 149
Infrared absorption 186
- coefficient 184,185
Infrared spectroscopy 181,190,199,204,238, 240,288,341,373
Interaction - energy 218
- parameter 434

- potential 218,219
- Born–Oppenheimer 221
Interface 221,260
- lake sediment 382
- liquid/air 417
- liquid/vapour 247
- metal/electrolyte 225,267,269
- ocean–sediment 382
Intermediate scattering – function 18
- law 152,163
Intermolecular structure function 330
Internal noise 108
Interparticle interaction 217
Interstitial water 375
Ion – association 227,344,409,413,429, 431,433,437
- dissociation 345
- interaction 432–434
Ion–ion distribution 140
Ion – pair 429,431,433,434
- product 341,342
- size parameter 432
Ion – solvent – interaction 95
- dipole interaction 114
- dynamics 116
Ionic atmosphere–Debye Huckel 70
Ionic – fluid 338,341
- friction coefficient 95,127
- friction continuum theory 121
- membrane 158
- medium 379
Ionic solutions – theory 61–91,399–408
- supercooled 369
Ionic strength 431,434
Isomorphic substitution 143
Isotope 133
- effect 231,232
Isotopic – fractionation 393
- substitution 227

J

Jellium model 246,247,251
Jump length 157

K

Kinetic theory 29
Kirkwood – correlation factor 339
- g factor 48
- hierarchy 249
Kuhn–chain 296,297,299,309
- segment 304,306,309

L

L–defect 442
Landau-Placzek ratio 194,213,366
Langevin – equation 22–24,96,97,113,412
- model 149
Langmuir – electrostatic interactions 418
- isotherm 283
- monolayers 417
Larmor angular frequency 168
Laser light scattering 95
Lennard – Jones – potential 233
- spheres 222
Lewis – Randall system 87
Librational motions 199,200,218,237
Lifetime 239
Light scattering 96,99,181–216
- experiments 107
Limiting law 85
- electrical conductance 106
Lindemann rule 314
Line tension 423
Linear response theory 14,24–28,34,403
Liouville – equation 25
- operator 112
Lippmann's equation 262
Liquid crystals 44,192
Liquid – metal surface 247
- surfaces 255
Liquidus line 371
Local density 62
Lone pair orbital 229
Long range – interactions 70
- order 275
Low energy electron diffraction 268,275

M

Mc Millan Mayer – approximation 76,82,85
- hamiltonian 410
- pair potential 74
- system 87
- theory 61,73,82
Macroion 100

Macromolecular solution 104
Macromolecules 95,313
Magnetic field 425
Manning's theory 414
Mass spectrometer 288
Mass transfer 429
Material frame indifference 127
Maxwell time 99
Maxwellian distribution 150
Mean – field treatment 399
– spherical approximation (MSA) 13,38,39,43,48,52,54
Melts 127
Memory – function 24,28-30,111,113
– kernel 111-119,413
Metamorphic – conditions 437,439
– fluid 429
Metasomatic processes 430
Metastable formation 372
Method of differences 134
Microemulsions 366
Mineral formation 434
Minerals 429,431,432,434
Minimum distance convention 218
Mirror particles 218
Mixed – electrolytes 434
– solvent effects 128
Mixing – coefficient 84
– parameter 434
Mobility 25,158,333
Mobility fluctuations 108,111
– electrophoretic 111
– correlation function 110
Mobility – ion 101,104,345
– metal surfaces 273
– polyion 314,332,335
Mode coupling 28,29
Models 217,218,229,430
– flexible 218
– polarization 222
Molality (m) 361
Molarity (M) 361
Molecular dynamics simulations (MD) 113,148,155,158,166,177,217,219,220,399,402,409
Molecular – motions 148
– orientations 82
– pair distribution function 43,44,46
– reorientation 107
Molten salts 227,364
Monoatomic ions 224
Monte-carlo simulations (MC) 219,220,409
Multi component diffusion 387

N

n-particle density 5,6,9
Natural waters 387,396,429,431,432,434,436,437
Navier-Stokes – equation 29,95
– hydrodynamics 127
Nematic 308
Nernst-Einstein – approximation 315
– relation 28
Neutral complexes 429
Neutron – diffraction 90,206,367
– scattering 150,177,190,193
– scattering experiments 95,218,227,229
– spectroscopy 148
Newton's equation 220
Noise-multiplicitive 114
Non correlated planes 448
Non electrocatalytic reactions 281
Nuclear magnetic resonance (NMR) 168-176,181,197,206
– relaxation 190
– relaxation time 200,375
Nucleation 370
– heterogeneous 366,372
– homegeneous 366
– rate 366

O

Ocean – ridges 389
– sediment boundary 380
One-component-plasme (OCP) 60,307
One-particle density 3,4,6,7,9,11,14,44
Onsager reciprocal relations 95
Open shell ions 91
Orientation polarization 122
Orientations 227,229,230
Ornstein – Zernike relation 7,12,35,38,44,46,48,52,61,68,367
Osmotic – coefficient 77,87,432
– pressure 38,39,42,74,316
– equilibrium 73
Out of equilibrium 447
Outdiffusion 443
Outer Helmholtz plane (oHP) 261

Overlap ratio 330
Oxygen adsorption 275,276

P

Pair correlation function 12,129,224
- time dependent 101
Pair distribution function (p.d.f.) 5,11,15,45,220
Pair potential - solvent averaged 61,76,77,88
Pairwise - additivity 64,71,74,82,219
- non additivity 75
Partial - molar volume 337,346
- pair ditribution function 36,37
- pressure 436
- structure factor (pair correlation function) 35,65,132
Percus - Yevick - closure 12,13
- equation 40
Periodic boundary conditions 218,219
Persistence length 291,295,296,299,300,301, 304-306,309,328,330
pH measurements 439
Phase diagram 306,310,314,346,360,376,447
- non equilibrium 370-372
Phase transitions
- expanded liquid-condensed liquid 417
- liquid-gas 417,418
- liquid-solid 417
Photon experiments 95
Pitzer's - approach 437
- equations 434,436
- ion interaction 429
Planets 343,349
Plate tectonics 389
Poisson-Boltzmann - approach (MPB) 249
- approximation (PB) 37,41,42,419
—equation 102,261,291, 293,294,300
Poisson's equation 97,103
Polar liquid 13,43,47
- solvents 47
Polarizability 50,181,182,184,186,187-190,195,196,203,212,213
- density 206
Polarizable molecules 399,402
Polarization - charge atmosphere 126
- current 124
- density 185,186
- fluctuations 96,106,107
- flux 125
- relaxation 120,122
Polarized scattering 182,183,187,188,191
Polycrystalline ice 442
Polyelectrolyte solutions 291,293,302,306, 308,309,312
Polyion 291,294,302,304-309
Polyion - deuterated 326
- hydration 314
- mobility 314,332
- osmotic compressibility 325
- structure function 321,323,325,328
Polymer 291,295,299,301-304,309,313
Polymer solutions 127
Polymorph phases 447
Polymorphism 449
Pore water 381
Potential of - mean (average) force 64
- zero charge (pzc) 261,270,271
Power spectra 218,235,237,238
Primitive model (PM) 32,38-40,43,48,51,53,411
- restricted 245
Probability flux 97,101
Protein 107,291
- solvent interface 107,111
Proton magnetic resonance (PMR) 196,372
Proton relaxation measurements 237
Protonic semiconductor 441
Pulsed gradient method 174

Q

Quadrupolar - interaction 168
- relaxation 172
Quantum - correction 66
- effects 218,231
Quasi chemical model 264
Quasi elastic scattering 332
- neutron 148,149,152,156,158, 161,162,172,177,181
Quench rate 366

R

Radial distribution functions 224,227
- ion-ion 226
- partial 227

Radius of gyration 367
Raman - coupling function 165,166
- spectra 340,345
- spectroscopy 164,183,199,372
Random - force 24
- jump diffusion model 157
Random phase approximation (RPA) 11,12,39
Rare earth - aqueous nitrato glasses 373
- solutions 372,373
Rate constants 89,90,415
Rayleigh-Brillouin-scattering 20,194,213
Rayleigh - scattering 183,193
- wings 373
Reaction field method 219
Real potential 256
Reflection spectroscopy 265
Reflectivity 269
Relaxation 101,199
- correction 104
- effect 104,117
- structural 205,209,212,213
Relaxation time 99,157,159,200,364
- spin-lattice 168-173,176,364,445
- spin-spin 173,174
Renormalisation 296
Reorientation 187,189,211,212,237
- time 220
Residence time 157,159,236,237
Rigid molecules model 206,221
Rotational - diffusion coefficient 156
- invariants 45,47,49-51
- motion 121,149,153,154,156,187, 191,192,201,202,222,237,238
- time 149,227,241
Running average 224

S

Saturation index 430,434-436
Scaling 304,307
Scattering - experiments 227,341
- - functions 220
Scattering law 154
- fast exchange 155
- slow exchange 156
Scattering length 65,130,132,133
- coherent 151
— incoherent 151
Scattering vector 130
Screening 35,38,41,42,48,53
- length 314
Second order difference method 136,227
Selective deuteration 163
Self-consistent mean field theory (SCMF) 402
Self-diffusion 174
- coefficient 149,157-159,172,176, 218,220,234,235
Self-ionization 337,341
Setchenow coefficients 80
Shear - relaxation 375
- waves 31
Shifted force potential method 219
Short range forces 432
Single - crystal electrodes 287
- electrode potential 257
Single ion - properties 255
- values 218,220,238
- solvation energy 257
Site-site - central force potential 221
- pair distribution function 46
Slow exchange 155,172
Small angle scattering (SAS) 320,367,368
Smoluchowski equation 107,108,115
Soft force 114
- correlation time 118
Solubility 430,437-439
- of minerals 431
- product 430
Solvated ion complex 96
Solvent - averaged models 409
- isotope 128
- orientation 263
Speciation of solutes 437
Spectroscopic experiments 341
Spin-echo measurement 174
Spin-relaxation 89,90
Spin-rotation interaction 168
Spinodal line 366
Stacking faults 448
Standard - rate constant 280
- value of the real potential 257
Static - correlation function 2,17,29
- field gradient method 174
Steam - tables 338

Stillinger – Lovett – moments 69,70,75
– sum rules 36
Stochastic model 149
– dynamics 413
Stokes scattering 185
Stokes – Einstein law 23
Stretching motions 236,238,239
– symmetric 240
– assymmetric 240
Strong electrolytes 343,434
Structural – properties 217,218,221
– relaxation 166
Structure factor 14,47,99
Structure function 227,321,323
– counterion 314
– polyion 314
– X-ray 227
Subcritical temperature 337
Supercooling 360
Supercooled – domain 184,365,366
– electrolyte solution 172
Supercritical – mixtures 354
– temperature 337,341
Supercrystal structure 420
Surface 276
– blocking 288
– composition 269
– dipole 259
– enhanced Raman effec 264
– potential 259
– reconstruction 275
Susceptibility 25,26,95,405

T

Tamman diagrams 360
Ternary – fluid mixtures 352
– solutions 434
– system 337,338,352,353
θ-temperature 299
Thermodynamic – limit 63
– properties 220,409,429
Thermoionic work 255
Thermophysical properties 338,349
Thomas-Fermi screening length 269
Three-body forces 219,222
Time-correlation function (TCF)
15-17,28,181,182,185,197,203,209
– dipole density 186
– orientational correlation g_2 195
– translational 197
– velocity 209-210
Torque fluctuations 107
Total dipole potential 263
Tracer diffusion 156,162
Transfer coefficient 281
Translational – diffusion 233,234
– dipole motion 121
– motion 149,153,189,191, 192,201,202,208,209,218,238
Transmembrane 233
Transport – coefficients 91,96,120
– properties 409
Trapped electrons 373
Triple ions 87
Triple point 338
Two-particle density 4,7,10,11,44
Two-state model 155

U

UHV system 268
Undulating phases 421,422
Unsymmetrical mixing 436
UV/visible spectroscopy 269

V

Van der Waals – forces 116
– interactions 230
– liquids 117
Van Hove function 19,21,107
Vapour deposition 365
Velocity autocorrelation function (VACF) 22,23,28,29,96,113-116, 233,234
Vibrational – density of states 164
– motion 149,153,218,222
– properties 221
– spectroscopy 222
Vibrations 188,189,191,206,237
Virial – coefficients 79,80,85
– expansion 432
Viscoelastic measurements 418
Viscosity 30,95,105,176,194,337-340,345,364-366
– coefficient 363
– reduced 318
– shear 105
– solvent 120
Vitrification 447
Vogel-Tamman-Fulcher law (VTF) 364,375
Volcano curve 284

W

WATEQ code 431
Water – activity 432,434–436
– clusters 375
– density maximum 359,360
– dielectric constant 339
– dimers 265
– electrical conductivity 342
– hydrothermal 390–392,395
– ionic product 342
– monolayers 264
– natural 429
– Raman spectra 340
/rock ratio 393
– supercooled 157
Wolynes ansatz 114
Work function 270
Wormlike chain model 297–299,304

X

X-ray – atomic form factor 130,133
– diffraction 206,418
– reflectivity 247
– scattering experiments 218

Z

Zeeman Hamiltonian 168
Zeroth moments 72
Zeta potential 96,107,418,424

Formula and Compound Index

A

Al 390,392

Al^{3+} 90, 372

$AlCl_3$ 363

Al $(NO_3)_3$ 363

Al_2O_3 (Corundum) 438,439

AN (Acetonitrile) 282

Albite 438

Alkali - carbonates 429
- chlorides 429
- halides 445
- salts 442
- sulfates 429
- trace 393

Alkaline earth - carbonates 429
- chlorides 429
- sulfates 429

Aluminia 437

Ammonia (NH_3) 339,343,441-443,445

Ammonium - chloride 442
- fluoride 442

Anhydrite 434-436

Argon (Ar) 81,116,208,209,232

B

Ba 392

$BaCl_2$ 84

BF_4^- 270

Be^{2+} 372,447,448

$BeCl_2.RH_2O$ 166,167,447,448

Br^- 142,237,390,392

Bu_4NBr 77

Benzene (C_6H_6) 81,354-356

C

Ca 133,390,392,430

^{40}Ca 133

^{44}Ca 133

Ca^{2+} 138,225,226,238-240

$CaCl_2$ 138,139,223,226,238-240,344, 345,354,363,437

$CaCO_3$ 385

$Ca(NO_3)_2$ 170,176,177,363

$Ca(NO_3)_2.H_2O$ 364

$Ca(N0_3)_2(H_20)_4$ 364

CCl_4 194

$CdPO_4$ 39

$CdSO_4$ 39

CF_3SO_3H 98

$(CH_3)_4N^+$ 90

C_6H_6-$(CH_3)_4N^+$ 81

Cl 133,430

^{35}Cl 133

^{37}Cl 133

Cl^- 64,88,139-141,223,225,227229,234, 237,238, 272,273,388, 395,438,447

ClO_4^- 224,237

CO 46,280,289

CO_2 288,293,294,337,351,352,436,437

CO_3 430,434

CO_3^- 379

Cs 247,393

Cs^+ 237

Cs_2 182

CsCl 81

CsF 223,224,226

Cu 133

^{63}Cu 133

^{65}Cu 133

Cu^{2+} 138

$CuCl_2$ 138-140,363

$Cu(ClO_4)_2$ 138

$Cu(NO_3)_2$ 363

$CuSO_4$ 87

Carbonates 429,434

Chlorides 363,434

D

D 133

D_2O 135-137,141,165-167,172-174,196, 204,205,264,448,449,451

Dy^{3+} 138
$DyCl_3$ 138
Deuterium 135,136,138,139,151,163
Deuteron 172,175
Diethylether 271
Dimethyl phosphate anion 224
DMF 282
DNA 291,292,294,300,301,305
Dioxane 354-356

E

Et_4NBr 77
Et_4NClO_4 272
Eu^{3+} 373
$EuCl_3$ 373
Ethanol 260

F

F^- 223,225,237,273
Fe 133,391
^{54}Fe 133
^{56}Fe 133
^{57}Fe 133,373
Fe^{2+} 90
Fe^{3+} 90,138
$FeCl_2$ 373
$Fe(CN)_6^{3-}$ 282
$Fe(CN)_6^{4-}$ 282
$Fe(NO_3)_3$ 138
FeS 386
FeS_2 386
Fatty acids 417
Feldspar 438
Ferric oxides 380
Formic acid 285,286,288

G

Ga^{3+} 372
Gd^{3+} 373
$GdCL_3$ 373
Glycine 263
Gramicidin A 233
Gypsum 434-436

H

H 133,375
^2H 375
H^+ 162,313
HCO_3 430
HCO_3^- 379
H_2CO_3 379
HCl 162,360,443-445
$HClO_4$ 88,275
HDO 204,340,341
HF 441-445
HNO_3 443
H_2O 44,50,140,158,160,165,166,170
176,184,196,203,204,210,264,282,283,
340,352,437,441,447,449
H_3O^+ 372
H_3PO_4 434
H_2S 210,386
HSO_4^- 275
$H_3SiO_4^-$ 438
H_4SiO_4 438
HX 372
Helium 231,232,343,348,349,443,444
^3He 391
Heptane 366
Hydrates 361,366,445
Hydrogen 151,163,177,232,337,343,
347-349
Hydrothermal water 390-392,395

I

I^- 234-238
In^{3+} 372
Iron 391,392
Iron chloride 140

K

K 133,430
^{41}K 133

K^+ 237,388
KCl 81,343,344,437,443
KF 444
$KH_2Al_2O_3$ 438
KH_3SiO_4 438
KNO_3 282
KOH 282,438,439
KPF_6 270
K-SiO_2 438
$K_2S_2O_8$ 313
K-feldspar 438

L

^6Li 133
^7Li 133,172,175,375
Li^+ 64,88-90,96,120,128,138,172, 223,225,229,234-238,375,447,448
$Li^+(H_2O)_4$ 375
LiBr.H_2O 373
LiCl 81,88,138,139,141,166,167, 171-173
LiCl.D_2O 223,226-228,230,231, 447,448,451
LiCl.H_2O 359-375
LiI 223,224,226,232,234-237
$LiNO_3$ 363
Ln^{3+}-OH_2 373

M

Mg 390,391,430
Mg^{2+} 144,238,372,388
$MgCl_2$ 144,172,174,223,228,363
$Mg(NO_3)_2$ 363
$Mg_3(SO_4)_2(OH)_2$ 391
$M(H_2O)_6$ 90
Mn^{2+} 88,284,285
$MnCl_2$ 88
$Mn(NO_3)_2$ 363
MnO_2 380,384,385
MnO_4^- 282
MnO_4^{2-} 282
Manganese 384,385,391
Mannitol 264
Methane (CH_4) 116,336,347,352-354, 356,391,393,394
Methanol 96,120,285,287,288,339
Methyl cyanide 339
Mercury 251
Minerals 429,431,432,434

N

N 133
^{14}N 133
^{15}N 133
Na 247,430
Na^+ 120,223,225,229,233,237,312 313,388
NaCl 81,139,172,174,223,228,272,337 344,345,352-354,367,388,430,437,443, 444
$NaClO_4$ 223,224
NaCl-CO_3 436
Na-Cl-HCO_3-CO_3-OH-H_2O 436
Na-K-Ca-Mg-H-Cl-SO_4-HCO_3-CO_3-OH-H_2O 434
Na-K-Ca-Mg-Cl-SO_4-H_2O 434
$NaHCO_3$ 444
NaH_3SiO_4 438
$(Na,K)AlSi_3O_8$ 439
NaOH 313,343,346,438
Na_2SO_4 84
$Na_2S_2O_3.5H_2O$ 364
Na-SiO_2 438
$Na_{40}P$ 77
Nd^{3+} 138
$NdCl_3$ 138,139
ND_4^+ 138
NDS^{2-} 90
ND_4Cl 138
NH_3CO_3 444
NH_4^+ 224,237
NH_4Cl 223,224,444
NH_4F 441-445
NH_4OH 444

Ni 133
^{58}Ni 133
^{60}Ni 133
^{62}Ni 133
^{64}Ni 133
Ni^{2+} 89–91,136,138,140,141,144,372
$NiCl_2$ 135–141,144,155,156,363
$Ni(ClO_4)_2$ 138
$Ni(D_2O)_6^{2+}$ 141
$Ni(H_2O)_6^{2+}$ 140
$Ni(H_2O)_5Cl^+$ 140
$Ni(NO_3)_2$ 363
NO_3^- 363,364,372,380
n–butanol 271
Nafion membrane 158,160,161
Neon 443,444
Nickel phosphate 39
Nitrogen (N_2) 46,182,232,347,348,350, 351,361,367,370

O

^{17}O 173
^{18}O 361,443
OH^- 438
Oxygen (O_2) 135,136,138,139,144,232,268, 337,348,356,361,380

P

PF_6^- 270,273
PMA 317,324
PSSNa 291,312,317,324
PSSTMA 291,326
Perchloric acid 274,288
Phospholipid 417
Platinum (Pt) 232,233,282
Pore water 380,381,386,388
Proton 172,175,201,202,204
Pyridine 264,272

Q

Quartz 437,438

R

Rb^+ 237
RbCl 81,139
$Ru(NH_3)_4bpy^{2+/3+}$ 88,91

S

SO_4^{2-} 380,386,388,391,395,430
Sea water 379,384,390–393,396
Silica (SiO_2) 166,381,382,391,393,430, 437–439
Silicates 366
Silicon (Si) 392
Sodium – carbonate 434,436
– dodecylsulfate 313
– hydroxide (NaOH) 343,346
– sulfonated polystyrene 317,324
– sorbitol 264
Span 65,366
Strontium (Sr) 395
Styrene–polymer 313
Sulfates 429,434
Sulfides 436
Sulphuric acid 274,275,288
Surfactants 417

T

$ThCl_4$ 363
$Th(NO_3)_4$ 363
Tobacco Mosaic Virus (TMV) 307
Tritium 443
Trona (Na_2CO_3,$NHCO_3$,$2H_2O$) 435–457

U

Uranium (U) 391

V

Vanadium 157,160

W

Water (H_2O,D_2O,HDO) (see also subject index for more details) 43,46,47,50,51,

54,90,91,96,120,134,148,155–158,160–162, 164,166, 172,174,176,181–183,189,192–195, 198–200,203–210,213,218,219,221–224,230 232,234,235,237–239,251,259,260,263–265, 271,273–276,337–344,346–356,379,380,409, 413,430–432,435,441,442,447,448,450

Y

Yb 392

Z

Zn 133
^{64}Zn 133
^{68}Zn 133
$ZnBr_2$ 142
$ZnCl_2$ 139–141,156–158,164–166,170,176
$Zn(NO_3)_2$ 363
$ZnSO_4$ 413
Zechstein 434